IEEE COMMUNICATIONS SOCIETY'S

TUTORIALS IN
MODERN
COMMUNICATIONS

IEEE COMMUNICATIONS SOCIETY'S

TUTORIALS IN MODERN COMMUNICATIONS

Edited by

Victor B. Lawrence,

Bell Telephone Laboratories,
Holmdel, New Jersey

Joseph L. LoCicero,

Department of Electrical Engineering,
Illinois Institute of Technology,
Chicago, Illinois

and

Laurence B. Milstein,

Department of Electrical Engineering
and Computer Sciences,
University of California at
San Diego, La Jolla, California

 PITMAN

PITMAN PUBLISHING LIMITED
128 Long Acre, London WC2E 9AN

Associated Companies
Pitman Publishing New Zealand Ltd, Wellington
Pitman Publishing Pty Ltd, Melbourne

First published in Great Britain 1983
First published in USA 1983

© 1983 *Computer Science Press, Inc.*
11 Taft Ct.
Rockville, Maryland 20850

This book was first published in 1983 by
Computer Science Press, Inc.
11 Taft Ct.
Rockville, Maryland 20850

Printing   1   2   3   4   5   87   86   85   84   83   Year

Library of Congress Cataloging in Publication Data
Main entry under title:

Tutorials in modern communications.

   1.   Digital communications—Addresses, essays,
lectures.   2.   Data transmission systems—Addresses,
essays, lectures.   3.   Image processing—Digital
techniques—Addresses, essays, lectures.   4.   Spread
spectrum communications—Addresses, essays, lectures.
5.   Cryptography—Addresses, essays, lectures.
I.   Lawrence, Victor B.   II.   LoCicero, Joseph L.
III.   Milstein, Laurence B.
TK5103.7.T87   1982   621.38   82-10599
ISBN 0-914894-48-X
UK ISBN 0-273-08599-9

PREFACE

During the past decade, the growth in communications and the changes that many aspects of communication systems have undergone have been tremendous. From practical data compression techniques, to schemes providing secure communication such as spread spectrum and cryptography, to the impact of VLSI on signal processing, communications today is noticeably different from what it was when analog systems dominated the scene.

For the communication engineer, keeping up with such a rapidly changing area is oftentimes difficult, if not virtually impossible. For this reason, the IEEE Communications Society has decided to put together a series of books, each containing reprints of tutorial articles originally published in the *IEEE Communications Society Magazine*. These tutorial articles span a broad range of topics related to communications and signal processing. A new book in the series is planned at periodic intervals.

In this first volume, a total of thirty-seven tutorials are presented, broken down into the following six broad categories: (1) Quantization and Switching, (2) Data and Modulation Techniques, (3) Computer Communications, (4) Transmission Systems, (5) Signal Processing, and (6) Secure Communications. Each of the six sections is preceded by an introduction highlighting the particular category and the tutorial articles devoted to that category.

ACKNOWLEDGEMENT

The editors would like to acknowledge two groups of people who are in some way connected with this book. The first group of people are intimately connected with the material content of the book, and the second group are associated with the birth and production of the book.

The first group contains, of course, all the authors of the articles assembled here, and the various editors of the *IEEE Communications Magazine* who were responsible for the original publication of these articles. These editors include Drs. S. B. Weinstein, F. S. Hill, Jr., A. Gersho, F. J. Ricci, F. W. Ellersick, and the many technical editors who served the *Magazine* over the years. We gratefully acknowledge their contributions.

The second group starts with Dr. D. L. Schilling, COMSOC President when this book was conceived, who strongly encouraged its undertaking. Dr. J. Garodnick, COMSOC Director of Publications, was instrumental in the continual negotiations needed to make this book a reality. The IEEE *Magazine* staff, particularly Ms. C. M. Lof and Ms. J. A. Raposa, were very helpful in the compilation of the book material. Finally, Ms. B. B. Friedman and her publication staff at Computer Science Press, Inc. were very cooperative in the final production of this book. Our sincere appreciation goes to all these people.

Victor B. Lawrence
Joseph L. LoCicero
Laurence B. Milstein

CONTENTS

Section 1

Quantization and Switching

Editors' Comments
V. B. LAWRENCE, J. L. LoCICERO, and L. B. MILSTEIN

Quantization
A. GERSHO

Digital Television Transmission Using
Bandwidth Compression Techniques
H. KANEKO AND T. ISHIGURO

Techniques For Digital Switching
J. C. McDONALD

Integrated Circuits for Local Digital
Switching Line Interfaces
P. R. GRAY AND D. G. MESSERSCHMITT

EDITORS' COMMENTS

We are living in what appears to be a digital world. This trend toward digitization is reinforced as we are awakened in the morning at the precise time set on our digital clock radios. It is seen as calculators, some small enough to carry around in a billfold or wear as a wristwatch, permeate our entire lives. And we shall feel it more and more as the personal digital computer enters our home. However, much of the information that we are interested in still comes from analog sources. Some examples of analog information are audio and video signals; still others are bioelectric signals (such as an electroencephalogram) and geophysical signals (such as a seismogram).

In the realm of digital communications we must first convert the analog information into digital signals before we begin to transmit, store, or process these signals. Here we are constrained by the sampling theorem which says that the sampling rate must be at least twice the bandwidth of the information signal. However, we are also limited by the accuracy achieved in converting the analog signal into a digital format, where typically we might use eight, ten, or twelve bits per sample. The accuracy is in turn dependent on the bit rate that we are permitted in encoding the analog source. Generally, in a communications environment, the accuracy is reflected in the signal-to-quantization noise ratio, or alternately, the distortion is given in terms of the mean-square quantization error.

In this first section of tutorials we alternate between the digitization of information sources and the digital switching of communication channels. Although we cannot completely cover these two important areas, you'll find that each of the four papers has a good set of references that can be used to further expand each field. The first two papers are devoted to the quantization process and the digital encoding of television signals. The last two papers cover principles of digital switching as found in telephony and the realization of some of the elements of a digital switch using the ever growing integrated circuit technology.

Notably omitted under the heading of quantization is a thorough article on the basics of differential source encoding. Here the quantizing system, i.e., the encoder, operates at a rate greater than the minimum sampling frequency, but uses fewer bits per sample. A bit rate reduction is achieved without loss of fidelity because the encoder utilizes the high correlation between closely spaced samples of the information signal. Another technique that we would expect to find under quantization is parametric encoding, such as linear predictive coding (LPC). In an LPC system, the source is modeled as a time varying digital filter and an excitation. Periodically we need transmit only the parameters which specify the filter and the type of excitation needed, in order to reconstruct a quantized version of the information signal. Future tutorials on quantization and source encoding will undoubtedly include these topics.

The lead article in this section is an expository paper authored by A. Gersho and simply entitled "Quantization." The foundation for the digitization process is firmly laid as the paper covers uniform, companded, robust, and optimum quantization. Quantizer performance is related to both the mean-square distortion measure and the signal-to-noise ratio, where the dependence on the amplitude probability density function of the analog signal is clearly seen. Utilizing straightforward mathematics, Gersho shows why robust quantizers employ a logarithmic companding law, and the effect on the quantizer performance. He concludes the paper with a comparison of quantizer performance, where the signal-to-noise ratio is given as a function of the average number of bits per sample.

The second paper, "Digital Television Transmission Using Bandwidth Compression Techniques," was jointly written by H. Kaneko and T. Ishiguro. It can be considered a detailed example of digitization, with an emphasis on bit rate reduction, where the analog source is a video signal. The basic principle of interframe and intraframe coding of a television signal is presented. Here we take advantage of the redundancy within one frame or between two successive frames to reduce bandwidth by encoding a difference signal. Kaneko and Ishiguro also introduce adaptive bit sharing, which is conceptually very similar to time assignment speech interpolation. With this technique the average bit rate for a video signal can be reduced by utilizing the statistical difference among several television channels transmitted simultaneously. Overall, the paper is systems oriented, replete with encoder and decoder block diagrams, but it also cites the bit rates needed for video signals ranging from very high quality NTSC color television transmission through less stringent teleconferencing.

The next two papers turn our attention to digital switching of communications signals. The basic premise in digital switching is that we are presented with signals representing messages that have already been converted to digital format. We must transmit these messages to one of a number of remote locations through a switch. The challenge is to apply a combination of the many switching techniques, such as time and space division multiplexing, so as to create an efficient, reliable digital switching system. In addition, we must not only effectively manage or control the switching operation, but we must also provide the hardware for its realization. The operation of the hardware should be easily monitored and tested; and it should lend itself to replacement with improved components as the state-of-the-art advances.

In "Techniques for Digital Switching," J. C. McDonald presents the two major components of any switching system as the switching network and the control. The most expository part of the paper is the review of the elements of a digital switch as needed in a telephone network. McDonald covers digital voice encoding using companded pulse code modulation, time division multiplexing including control signaling, and the terminations needed in telephony for both analog and digital signals. The operation of a time slot interchanger and a time-space-time network example clearly show the importance of time divided networks. Finally, the elements of control and digital signal processing complete the description of a digital switch. In the last section,

McDonald discusses the performance standards for a digital network. He includes information impairments such as bit errors, misframing of a trunk of communication signals, and synchronization slip as a result of clock drift causing a frame to be repeated or deleted. He concludes with the effects of loss of transmission, noise, and echo in a telephone channel, and how these impairments can be controlled.

The last paper in this section is an excellent follow-up to McDonald's article. "Integrated Circuits for Local Digital Switching Line Interfaces," authored by P. R. Gray and D. G. Messerschmitt, takes some of the switching elements described by McDonald and presents the details of their realization. A discussion of the advances in integrated circuit technology enabling manufacturers to fully integrate a voice encoder/decoder (codec) on one monolithic chip acts as a good introduction to the important considerations governing codec architectures. This is followed by a presentation of the three primary approaches to the implementation of digital-to-analog converters, i.e., current switching, resistor string, and charge redistribution. The last implementation very naturally leads to charge coupled device and switched capacitor monolithic filter realizations that can be used in transmit and receive low-pass filters. Gray and Messerschmitt conclude the article with the primary functions of a line interface and the current efforts being made to realize an effective, economical monolithic line interface circuit.

QUANTIZATION *

Allen Gersho, Member, IEEE

Abstract   Quantization is the process of replacing analog samples with approximate values taken from a finite set of allowed values. The approximate values corresponding to a sequence of analog samples can then be specified by a digital signal for transmission, storage, or other digital processing. In this expository paper, the basic ideas of uniform quantization, companding, robustness to input power level, and optimal quantization are reviewed and explained. The performance of various schemes is compared using the ratio of signal power to mean-square quantizing noise as a criterion. Entropy coding and the ultimate theoretical bound on block quantizer performance are also compared with the simpler zero-memory quantizer.

I. INTRODUCTION

The processing and transmission of digital signals is rapidly approaching a dominant role in communication systems. Nevertheless, the physical origin of many information-bearing signals (speech, image, telemetry, seismic, etc.) is intrinsically analog and continuous-time in nature. Therefore, an effective interface between the analog and digital worlds is of crucial importance in modern signal processing. Very often the quality of analog-to-digital (A/D) conversion is the critical limiting factor in overall system performance. A clear understanding of quantization, the essential mechanism of A/D conversion, is needed to answer such questions as how many bits per second (or bits per sample) are really needed, or how much distortion (or quantizing noise) is inevitable for a given bit rate.

Analog-to-digital conversion may be viewed as being made up of four operations: prefiltering, sampling, quantizing, and coding. In this paper we focus on quantization, and specifically on "zero-memory" quantization.

Quantization begins with the availability of analog samples. Each sample may in general take on any of a continuum of amplitude values ranging from $-\infty$ to $+\infty$. The quantizer replaces each of these sample values with an output value which is an approximation to the original amplitude. The key feature is that each output value is one of a *finite* set of real numbers. Hence a symbol from a finite alphabet can be used to represent and identify the particular output value that occurs. A distinct n-bit binary word can be associated with each output value if the set of output values contains no more than 2^n members. With this procedure a sequence of analog samples can be transformed into a sequence of binary words suitable for storage, transmission, or some other form of *digital* signal processing. A receiver having the table of output values (sometimes called "quanta" or "quantum levels") associated with the set of binary words can then reconstruct an approximation to the original sequence of samples. Hence with some appropriate form of interpolation, a continuous waveform can be created which approximates the waveform originally applied to the A/D system. The reconstruction process is called digital-to-analog (D/A) conversion.

The simplest and most common form of quantizer is the *zero-memory* quantizer. In this case the output value is determined by the quantizer only from one corresponding input sample, independent of the values taken on by earlier (or later) analog samples applied to the quantizer input. More sophisticated (but less well understood theoretically) is the *block* quantizer which looks at a group or "block" of input samples simultaneously and produces a block of output values, chosen from a finite set of possible output blocks, approximating the corresponding input samples. In general, for a given number of bits per sample representing the output values, a better quality approximation can be achieved by block quantization. Of theoretical interest is the limiting case where the block length approaches infinity. Studying this limiting situation provides information about the ultimate quality of approximation achievable for a given bit rate. Another class of quantizers which could be described as *sequential quantizers* includes such well-known digitization schemes as delta modulation, dit-

This work was performed while the author was visiting the Department of System Science, University of California, Los Angeles, CA. He is with Bell Laboratories, Murray Hill, NJ 07974.

*Reprinted from *IEEE Communications Society Magazine,* September 1977, Vol. 15, No. 5, pp. 20-29.

ferential PCM, and other adaptive versions. A sequential quantizer stores some information about the previous samples and generates the present quantized output using both the current input *and* the stored information. In this paper we shall focus primarily on zero-memory quantization. Quantization with memory will be discussed only for the purpose of examining how much can be gained through the use of memory.

II. ZERO-MEMORY QUANTIZATION

A zero-memory N-point quantizer Q may be defined by specifying a set of $N+1$ *decision levels* $x_0, x_1, \ldots, x_N$ and a set of N *output points* $y_1, y_2, \cdots, y_N$. When the value X of an input sample lies in the ith quantizing interval, namely,

$$R_i = \left\{ x_{i-1} < X < x_i \right\},$$

the quantizer produces the output value y_i. Since y_i is used to approximate samples contained in the interval R_i, y_i is itself chosen to be some value in the interval R_i. The end levels x_0 and x_N are chosen equal to the smallest and largest values, respectively, that the input samples may have. Usually, the sample values are unbounded, which we henceforth assume, so that $x_0 = -\infty$ and $x_N = +\infty$. The N output points always have finite values. If $N = 2^n$, a unique n-bit binary word can be associated with each output point, yielding an "n-bit quantizer."

The input-output characteristic $Q(x)$ of a quantizer has a staircase form. The midtread characteristic shown in Fig. 1 produces zero output for input samples that are in the neighborhood of zero; the midriser characteristic shown in Fig. 2 has a decision level located at zero. A quantizer is simply a memoryless nonlinearity whose characteristic may be viewed as a staircase approximation to the "identity" operation $y = x$.

When the input sample is located in the end regions R_1 or R_N the quantizer is said to be *overloaded*. All other quantizing intervals R_i are finite in size.

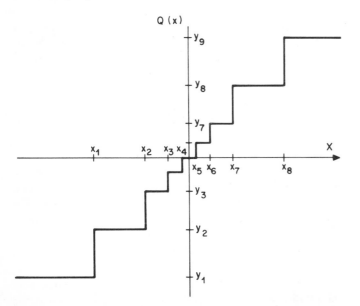

Fig. 1.   Input-output characteristic of a midtread quantizer with $N=9$.

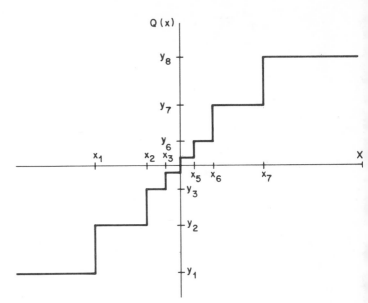

Fig. 2.   Input-output characteristic of a midriser quantizer with $N=8$.

Fundamental to an analytical study of quantization is the recognition that the input samples must be regarded as random in character. The input samples are not known in advance and thus can be regarded as information-bearing. Quantization is actually a mechanism whereby information is thrown away, keeping only as much as is really needed to allow reconstruction of the original signal to within a desired accuracy as measured by some *fidelity criterion*. We define $p(x)$ as the first-order probability density function (hereafter pdf) of each input sample to the quantizer. Assume for convenience that the mean value of the input samples is zero and that $p(x)$ has even symmetry about zero. The zero mean assumption implies that any dc bias has been removed. The symmetry assumption is satisfied by most common density functions including the Gaussian (normal) density. With the symmetry assumption, the quantizer characteristic $Q(x)$ is normally chosen to have odd symmetry.

The quantization process can be modeled as the addition of a random noise component $e = Q(x) - x$ to the input sample, as indicated in Fig. 3. Unlike the usual signal-plus-noise models in communication theory, here the noise is actually dependent on the signal amplitude. The quantization noise may be re-

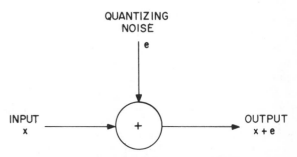

Fig. 3.   Additive noise model of quantization. The quantizing noise e is often approximated as being independent of the input samples when the number of levels is large.

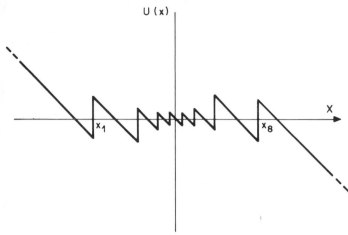

Fig. 4. Quantizing error as a function of input sample value for the quantizer of Fig. 1.

garded as the response when the input sample is applied to the nonlinear characteristic

$$U(x) = Q(x) - x$$

shown in Fig. 4. When the input sample lies within the interval $x_1 < x < x_{N-1}$, the output noise is described as *granular* noise and is bounded in magnitude. When the input lies outside this interval, the output is described as *overload* noise and the amplitude is unbounded. It is often convenient to artificially model quantization noise as the sum of granularity and overload noise as if they were two separate noise sources.

An effectively designed quantizer should be "matched" to the particular input probability density function, to the extent that this density is known to the designer. In particular, for a fixed number N of levels, the choice of overload levels x_1 and x_{N-1} controls a tradeoff between the relative amounts of granularity and overload noise.

In modeling quantization error as an additive noise source as in Fig. 3, it is often convenient to treat the noise as having a flat spectral density and as being uncorrelated with the input samples. This idea was used by Widrow [1] for uniformly spaced quantization levels. More generally, it may be shown that the quantizing noise is approximately white (i.e., successive noise samples are uncorrelated) and uncorrelated with the input process if: 1) successive input samples are only moderately correlated, 2) the number of output points N is large, and 3) the output points are very close to the midpoints of the corresponding quantization intervals. For a more precise treatment of the spectrum of quantizing noise, see Bennett [2].

III. PERFORMANCE MEASURES

Since the quantization error is modeled as a random variable, a measure of the performance of a quantizer must be based on a statistical average of some function of the error. Most common is the mean-square distortion measure D, defined by the usual expectation of the square of $U(x)$ above:

$$D = \int_{-\infty}^{\infty} [Q(x) - x]^2 \, p(x) dx. \tag{1}$$

This quantity can be used to measure the degradation introduced by the quantizer for a fixed input pdf $p(x)$. Frequently, it is more useful to describe the quantizer's performance by the "signal-to-noise ratio," often defined as

$$SNR = 10 \log_{10} (\sigma^2 / D) \tag{2}$$

where σ^2 is the variance of the input samples. Other error criteria have also been considered in the study of quantization, such as the expectation of the kth power of the error magnitude. Frequently, the performance measure adopted is a subjective evaluation, and psychological studies are used to determine preferred quantization schemes among a set of schemes considered. Another approach is to consider the quality of approximation of a segment of the reconstructed waveform to the original waveform. Mean-square distortion may be viewed as a special case of this approach where the performance measure is the expectation of the sum of the squared errors for all sampling instants of the waveform segment. However, this measure does not distinguish between different approximations having the same total squared error. For example, it might be subjectively preferable to have a very high squared error at one isolated sampling instant than to have moderately high squared errors at several adjacent sampling instants. Hence, a more sophisticated distortion measure might be more meaningful than the usual mean-square distortion criterion.

In most applications of quantization, the number of levels N is very large so that a sufficiently high SNR is obtained. A useful formula for mean-squared error can then be used. Equation (1) can be written in the form

$$D = \sum_{i=1}^{N} \int_{x_{i-1}}^{x_i} (y_i - x)^2 \, p(x) dx \tag{3}$$

by breaking up the region of integration into the separate intervals R_i and noting that $Q(x) = y_i$ when x is in R_i. For large N, each interval R_i can be made quite small (with the exception of the overload intervals R_1 and R_N which are unbounded). Then it is reasonable to approximate the probability density $p(x)$ as being constant within the interval R_i. On setting $p(x) \cong p(y_i)$ when x is in R_i and approximating $p(x) \cong 0$ for x in the overload regions, the integral for each term of the sum (3) is readily found, and we get

$$D = \frac{1}{12} \sum_{i=2}^{N-1} p(y_i) \, \Delta_i^3 \tag{4}$$

where $\Delta_i = x_i - x_{i-1}$, the length of interval R_i. This approximate formula is based on the assumption that, for N large, a sufficient number of quantizing levels are available for both the granularity and overload noise to be very small. Equation (4) implies that the overload points x_0 and x_N are chosen so that overload noise is negligible compared to granular noise. Equation (4) will be used later to derive an integral formula for distortion.

Of frequent interest is the special case of *uniform* quantization where the decision levels are equally spaced so that the intervals R_i are of constant length, i.e., $\Delta_i = \Delta$, sometimes called the *step size* of the quantizer. In this case, the staircase quantizer characteristic of Fig. 1 has equal width and equal height steps. The expression for mean-square error simplifies to

$$D = \frac{\Delta^2}{12} \sum_{i=2}^{N-1} p(y_i)\Delta.$$

But

$$\sum_i p(y_i)\Delta \approx \int p(s)ds = 1,$$

so that

$$D \approx \frac{\Delta^2}{12}. \tag{5}$$

Thus the mean-square distortion of a uniform quantizer grows as the square of the step size. This is perhaps the most often used result concerning quantization. This expression may be obtained directly by regarding the granularity noise as a uniformly distributed random variable over the interval $-\Delta/2$ to $+\Delta/2$ and neglecting overload noise.

A symmetric uniform quantizer is fully described by specifying the number of levels and either the step size Δ or the overload level V where $V = x_N = -x_0$. To avoid significant overload distortion, the overload level is chosen to be a suitable multiple, $y = V/\sigma$, called the *loading factor*, of the rms signal level σ. A common choice is the so-called *four-sigma loading* where $y = 4$. Then the step size is $\Delta = 8\sigma/(N-2)$ since the total amplitude range of the quantizing intervals is 8σ and there are $N-2$ levels in that range. Then, for an n-bit quantizer with $N = 2^n$ and $N \gg 2$, we find using (2) and (5) that

$$SNR = 6n - 7.3. \tag{6}$$

This linear increase of SNR with the number of bits of quantization was noted by Oliver, Pierce, and Shannon [3] in 1948. Note that changing the loading factor modifies the constant term 7.3, but does not alter the rate of increase of SNR with n. (The rate is actually $20 \log_{10} 2 \approx 6.0$.)

Varying the loading factor for a particular input power level σ^2 is equivalent to varying the input power level for a fixed loading factor. In Fig. 5, the dependence of signal-to-noise ratio on input power level is sketched for a uniform quantizer with $N = 128$. The curve takes into account the effect of overload noise which rapidly becomes dominant as the signal level reaches a critical value. The curve is based on the assumption of a Laplacian pdf,

$$p(x) = \frac{1}{\lambda} e^{-2|x|/\lambda}$$

which is occasionally used to approximate the pdf of speech. In this case it may be seen that the best performance is achieved when the loading factor of 6.1 (15.7 dB) is used. If the input power level deviates a few decibels from the anticipated value (used in designing the quantizer), a substantial drop in SNR will result.

IV. COMPANDING

Uniform quantization is not in general the most effective way to achieve good performance. For a given number of quantizing intervals, taking into account the input probability

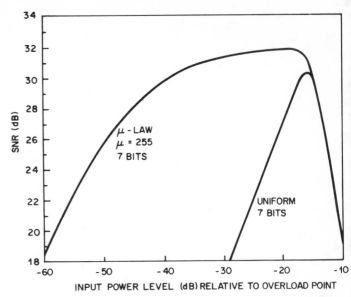

Fig. 5.   Dependence of signal-to-noise ratio on the input power level for uniform and μ-law quantizers both having 7 bits of quantization (128 levels). For a minimum acceptable quality of 25 dB, it can be seen that the μ-law quantizer has a dynamic range of about 40 dB, while the uniform quantizer has a range of about 10 dB. The curves may also be used to show how SNR depends on the choice of overload point when the input power level is fixed. Curves are based on a Laplacian input pdf.

density, nonuniform spacing of the decision levels can yield lower quantizing noise and less sensitivity to variations in input signal statistics. An effective technique for studying nonuniform quantization, used by Bennett [2], is to model the quantizer as a memoryless nonlinearity $F(x)$, the "compressor," followed by a uniform quantizer as shown in Fig. 6. The nonlinearity spreads out low-amplitude sample values over a larger range of amplitudes while shrinking the higher amplitude values into a smaller range. This compressed signal is then uniformly quantized. The effect is to allocate more quantizer levels to the lower amplitudes, which generally have higher probability, and fewer levels to the less frequently occurring higher amplitudes. The output values are then applied to the inverse nonlinearity $F^{-1}(x)$, producing an approximation to the signal originally applied to the compressor. The overall scheme in Fig. 6 is called *companding*, a term combining the words "compressing" and "expanding."

The characteristic $F(x)$ is a monotonically increasing function having odd symmetry, ranging from values $-V$ to $+V$, and with $F(V)=V$ and $F(0)=0$. This nonlinear operation, being monotonic, is completely invertible. That is, an input sample x applied to the compressor produces the response value $F(x)$;

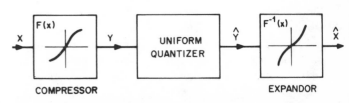

Fig. 6.   Companding model of nonuniform quantization.

the original value x could be recovered by applying the value $y = F(x)$ to the inverse nonlinearity, the "expandor" $F^{-1}(y)$, and obtaining x again. Thus, there is no loss of information due to the nonlinear operation itself. The uniform quantizer is chosen to have $N-2$ ($\approx N$) intervals, not including overload regions, so that $\Delta = 2V/N$. The combined effect of the compressor and the uniform quantizer is equivalent to the operation of a particular nonuniform quantizer whose decision levels and output points are determined by the shape of the compressor. Every possible nonuniform quantizer can be modeled in this way by a suitable choice of the function $F(x)$. Fig. 7 shows how the nonuniform quantizer decision levels are related to the uniform quantizer levels.

An important approximate formula for mean-square error in nonuniform quantizers can be derived based on the preceding model of nonuniform quantization. For large N, we approximate the curve of $F(y)$ in the ith quantizing interval by a straight-line segment with slope $F'(y_i)$, the derivative of $F(y)$ evaluated at y_i, where y_i is the output point of the equivalent nonuniform quantizer.

Then

$$F'(y_i)\Delta_i \approx F(x_i) - F(x_{i-1}) \approx 2V/N$$

so that, defining the slope of the compressor curve,

$$g(y) \stackrel{\Delta}{=} F'(y),$$

we have

$$\Delta_i \approx \frac{2V}{Ng(y_i)}. \qquad (7)$$

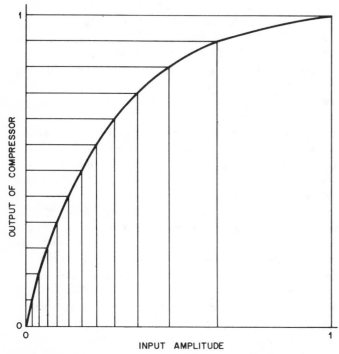

Fig. 7. Compressor mapping of decision levels. (Shown for positive amplitudes only.)

Now applying (4) yields

$$D = \frac{V^2}{3N^2} \int_{-V}^{V} \frac{p(s)}{[g(s)]^2}\, ds. \qquad (8)$$

This formula, due to Bennett [2], is based on the assumption that N is large and that the overload distortion is negligible. Given a proposed compressor characteristic $F(x)$ and choice of overload point V, the formula (8) can be used to evaluate the resulting quantizer distortion. The formula is also of analytic value for optimizing the compressor characteristic. (See Section VI.)

For speech signals as well as many other analog sources, lower amplitude values occur with higher probability than the higher amplitude values so that it would be reasonable to have quantizer levels more densely packed in the low signal region. For very low signal levels, the relevant step sizes will be approximately uniform with size

$$\Delta_0 = \frac{2V}{Ng(0)}.$$

The improvement in performance of the nonuniform quantizer for low signal level inputs over the uniform quantizer is then determined by the ratio

$$c_A = \frac{\Delta}{\Delta_0} = g(0)$$

which is called the *companding advantage*. This quantity is frequently used in comparing different compressor characteristics. Increasing the companding advantage concentrates more levels in the low amplitude region and improves the SNR for weak signal inputs. At the same time, a higher companding advantage means fewer levels in the high amplitude region, tending to reduce the SNR for strong signal inputs.

V. ROBUST QUANTIZATION

In certain applications, notably in speech transmission, the same quantizer must accommodate signals with widely varying power levels. The use of "robust" quantizers, which are relatively insensitive to changes in the probability density of the input samples, has become of great practical importance.

To obtain robust performance, the signal-to-noise ratio of the quantizer should ideally be independent of the particular pdf of the input signal. If the slope of the compressor curve were chosen to be

$$g(x) = \frac{V}{b|x|} \qquad (9)$$

then (8) reduces to

$$D = \frac{b^2}{3N^2}\sigma^2$$

so that the signal-to-noise ratio σ^2/D reduces to the constant $3N^2/b^2$, which is in fact independent of $p(x)$. Integrating (9) gives

$$F(x) = V + c \log(x/V) \qquad (10)$$

for $x > 0$ where c is a constant. This result shows that such a logarithmic compressor curve would give the desired robust performance. Of course, the formula (8) neglects overload noise so that the SNR will not remain constant but will begin to drop when the input power level becomes large enough. Also, the compressor curve (10) is not in fact realizable since $F(0)$ is not finite. To circumvent the latter difficulty, a modified compressor curve is used which behaves well for small values of x and retains the logarithmic behavior elsewhere.

A compressor curve widely used for speech digitization is the *μ-law* curve (see Fig. 8) given by

$$F(x) = V \frac{\log (1 + \mu x/V)}{\log (1 + \mu)} \tag{11}$$

for $x > 0$. As always, $F(x)$ is an odd function so that $F(x) = -F(-x)$ for negative x. This characteristic was first described in the literature by Holzwarth [5], studied extensively by Smith [12], and reportedly was used by Bennett as early as 1944 in unpublished work. For $\mu \gg 1$ and $\mu x \gg V$, $F(x)$ approximates the form (10). From (8), the mean-square granular quantizing noise can be calculated, leading to the result

$$D/\sigma^2 = \frac{[\log (1+\mu)]^2}{3N^2} \left\{ 1 + 2\alpha y/\mu + (y/\mu)^2 \right\} \tag{12}$$

where α is the ratio of mean absolute value to rms value of the input samples and y is the loading factor defined earlier. The effects of different choices of μ (corresponding to different companding advantages) has been examined by Smith [12]. Typical values of μ are 100 for 7-bit and 225 for 8-bit speech quantizers. PCM systems in the United States, Canada, and Japan use μ-law companding.

Another robust logarithmic characteristic due to Cattermole [6] is *A-law* companding where

$$F(x) = \begin{cases} \dfrac{Ax}{1 + \log A}, & 0 \le x \le V/A \\[2ex] \dfrac{V + V \log (Ax/V)}{1 + \log A}, & V/A \le x \le V. \end{cases} \tag{13}$$

A typical value for A is 87.6 for a 7-bit speech quantizer. The A-law characteristic is used in European PCM telephone systems. Both A-law and μ-law have the desired robust quality and can achieve more or less the same performance.

To illustrate the advantage of a robust quantizer, Fig. 5 shows curves of SNR versus input signal power level for both uniform and μ-law quantizers when the number of levels is 128. For a wide range of power levels, a high SNR of the μ-law quantizer is maintained, while the SNR of the uniform quantizer drops rapidly with diminishing power levels. In order to achieve the same quality over a significant dynamic range, an 11-bit uniform quantizer must be used. Thus a saving of 4 bits per sample is achieved by using nonuniform quantization.

In practice, companders are now designed as piecewise linear approximations to a desired characteristic. These "segmented" companding laws are conveniently implemented with digital circuitry. The coded binary word has certain bits that identify to which segment the analog sample belongs and the remaining bits identify which level within the segment represents the analog sample.

VI. OPTIMUM QUANTIZATION

For applications where one particular probability density function is known to describe adequately the distribution of input samples to be quantized, it is natural to seek the best possible quantizer characteristic for that density. Two approaches have been taken to this problem: one uses the assumption that N is large and leads to explicit solutions; the other is valid for any N, and leads to algorithmic procedures for finding the optimum decision levels and output points. We begin with the latter approach.

In a little known Polish article, Lukaszewicz and Steinhaus [7] in 1955 found necessary conditions for optimality of a set of decision levels and output points for both the mean-square and the mean-absolute error criterion. [In the latter case, $[Q(x)-x]^2$ is replaced by $|Q(x)-x|$ in (1).] Independently, in 1957 Lloyd [4], using the mean-square error criterion, found necessary conditions for optimality and an effective algorithm for computing the optimal solution. In 1960, Max [8] independently formulated the necessary conditions for optimality for a kth absolute mean error criterion (including $k=2$), and rediscovered the same algorithm used by Lloyd. In addition, Max examined the optimization of the step size for uniform quantization. Max also tabulated the optimum quantizer levels for the Gaussian distribution for various values of N.

For the mean-square error criterion, with some fixed value

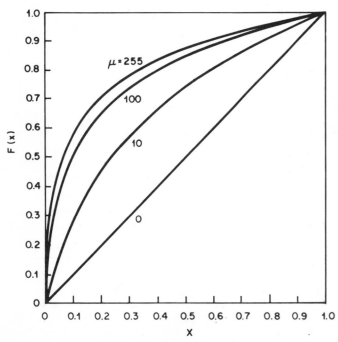

Fig. 8. μ-law compressor curve.

of N, the necessary conditions for optimality on the values of $x_1, x_2, \cdots, x_{N-1}$ and $y_1, y_2, \cdots, y_N$ are found simply by setting derivatives of D as given in (3) with respect to each of these parameters to zero. The resulting conditions are as follows.

1) Each output level of y_j must be the *centroid* or center or mass of the interval R_j with respect to the input density $p(x)$. In other words, y_j is the conditional mean value of the input random variable x given that x is in the region R_j.

2) Each decision level x_j must be halfway between the two adjacent output points.

These conditions do not give the optimum values explicitly, since the value of the output point y_j for an interval R_j depends on the value of the decision levels x_{j-1} and x_j defining R_j, and the decision levels x_j depend on the output levels y_j and y_{j+1}. However, these conditions are used in the Lloyd-Max algorithm (see Max [8]) for computing iteratively a set of parameters that simultaneously satisfy both conditions. Using the Lloyd-Max algorithm, Paez and Glisson [9] tabulated the optimum quantizer parameters for the Laplacian and a particular form of the gamma density.

Lloyd also observed that the conditions, while necessary, are not sufficient conditions for a minimum. In fact, he gave a counterexample of a probability density function and an associated quantizer that satisfies the conditions and is not optimal. Fleischer [10] obtained sufficient conditions which, if satisfied, will guarantee that the quantizer is in fact optimal. In particular, he showed that, if the input density $p(x)$ satisfies the property that

$$\frac{d^2}{dx^2} [\log p(x)] < 0 \qquad (14)$$

for all x, in other words, if $\log p(x)$ is concave, then only one quantizer exists which satisfies the Lloyd-Max conditions 1) and 2) and that quantizer is indeed optimal. It should be noted that the converse is not true, so that it is possible to have a density $p(x)$ not satisfying (14) and yet a unique optimal quantizer may exist. Nonetheless, condition (14) holds for the Gaussian density as well as for many other common densities. Hence, the tabulated quantizer parameters given by Max for the Gaussian density are in fact unique and optimal.

An alternate approach to the search for optimal quantizers begins with the use of Bennett's formula (8), which is based on the assumption that N is large. Minimization of (8) over the class of all curves of compressor slope $g(x)$ that satisfies a suitable constraint yields the result that the optimum compressor slope $g^*(x)$ is proportional to the cube root of the pdf:

$$g^*(s) = c_1 [p(s)]^{1/3}.$$

By integrating $g^*(s)$, one obtains the compressor characteristic

$$F^*(s) = c_1 \int_0^s [p(\alpha)]^{1/3} \, d\alpha, \qquad \text{for } s > 0 \qquad (15)$$

where c_1 is the constant chosen so that $F(V) = V$. Equation (15) was first obtained by Panter and Dite [11] in a classic and often overlooked paper. Their approach started with (4) and did not make use of Bennett's formula. Direct minimiza-

tion of (8) was first examined by Smith [12]. Roe [13], while unaware of the works of Panter and Dite and Smith, derived a formula for the optimal decision levels that is equivalent to the result (15), but does not use the companding model. Algazi [14] used the companding model to obtain results on optimal quantizers for a general class of error criteria. His results include (15).

Finally, we note that (15) determines the optimum quantizer for a given choice of overload point V. A separate one-dimensional minimization of D can be used to obtain the best overload point. (See [14].) From Fig. 7, it is evident that once the compressor curve is known, the decision levels and output points are readily obtained by a mapping of the uniform quantizer parameters. Computation of the minimum mean-square error obtained with this approach leads to values in good agreement with Max's tabulations (for the Gaussian pdf), even for values of N as small as 6. For $N = 6$, the individual decision levels are within 3 percent of the correct values (see Roe [12]). Naturally, as N increases, the discrepancy approaches zero, since (15) is based on the assumption that N is large.

An example of optimum quantization studied by Smith [12] is based on the Laplacian pdf. The optimum compressor according to (15) has the form

$$F(x) = \frac{V(1 - e^{-mx})}{1 - e^{-mV}}, \qquad \text{for } x > 0$$

giving rise to the "m-law" quantizer. Fig. 9 shows the dependence of SNR on input power level for the robust μ-law quantizer with $\mu = 255$ and for the optimum m-law quantizer with $m = 10$ when the input density is Laplacian. Comparison of SNR performance of μ-law and m-law quantizers shows that, for $\mu = 255$ and $m = 10$, the m-law curve has a 5 dB advantage at the power level for which it is designed. The μ-law quantizer maintains its reasonably high SNR over a broad range of power levels, while the m-law quantizer becomes inferior for input power levels about 10 dB below or 5 dB above the designed value. Comparing the m-law SNR curve in Fig. 9 with the uniform quantizer SNR curve in Fig. 5 shows that there is less than a 6 dB improvement in using the optimum quantizer rather than the uniform quantizer. In some applications this gain might not justify the extra cost of implementing a specially designed nonuniform quantizer as opposed to the simpler uniform quantizer.

A convenient and general way to describe the performance of optimal quantizers is based on the application of the optimal compressor slope $g^*(y)$ to the Bennett formula (8) for mean-square distortion. The result is that the minimum granular distortion for optimal N-point quantization and for large N is given by

$$D = \left\{ \frac{1}{12N^2} \left[\int_{-Y}^{Y} \left[p_0(x) \right]^{1/3} dx \right]^3 \right\} \sigma^2 \qquad (16)$$

where $p_0(x)$ denotes the input density normalized to have unit variance, and y is the loading factor discussed above. This formula, first derived by Panter and Dite [11], is useful for

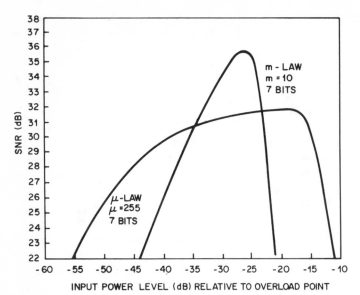

Fig. 9.   Dependence of SNR on input power level for μ-law and m-law quantizers when the input pdf is Laplacian and the number of levels is 128 (7 bits). The m-law quantizer is optimal for the Laplacian pdf with input power level 26.5 dB. At this power level there is a 4 dB improvement in SNR over the suboptimum but more robust μ-law quantizer.

estimating the number of quantization levels needed for a desired performance (i.e., SNR specification). The integral in square brackets, L, depends only on the shape of the input density function and not on the actual power level. It can be seen from (16) that the SNR for an optimal quantizer has the form $10 \log_{10} N^2 - C$ where C is a constant determined by $p_o(x)$ using (16). Letting $n = \log_2 N$ gives the result

$$\text{SNR} = 6n - C$$

where $C = 10 \log_{10} (L^3/12)$. The 6 dB per bit improvement in SNR is the same as for the nonoptimum uniform quantizer where the SNR is given by (6). However, the value of C obtained from (16) is as small as possible since (16) gives the minimum granular distortion attainable for any zero-memory quantizer. For the Gaussian density, C is found to be 4.35 dB by approximating (16) using $V = \infty$.

VII. QUANTIZATION WITH MEMORY

In block quantization, more commonly considered for image digitization rather than speech, a block of k input samples $(x_1, x_2, \cdots, x_k) = x$ (which may be regarded as a vector in k dimensions) is simultaneously quantized, producing an output "point" or vector $(y_{i1}, y_{i2}, \cdots, y_{ik}) = y_i$ approximating x. Thus, the output y_{ij} is an approximation to x_j for each $j = 1, 2, \cdots, k$. An N-point quantizer selects one of N output "points" $y_1, y_2, \cdots, y_N$ to approximate x. Unlike zero-memory quantization, the value y_{ij} depends not only on the corresponding input sample x_j, but also on the values of all other samples x_i in the block. Even if the input samples are statistically independent, an advantage can be gained by quantizing a block at a time rather than one sample at a time. A convenient measure of the distortion of the block quantizer is

$$D = \frac{1}{k} \sum_{i=1}^{k} \overline{e_i^2}$$

where $\overline{e_i^2}$ is the mean-square error in the ith sample. The performance of block quantization could be compared with zero-memory quantization by examining how the *bit rate* or *average number of bits per sample*, $B = (\log_2 N)/k$, depends on D, the distortion per sample. Clearly, as the block length k increases, the minimum bit rate needed for a given distortion will decrease. In the limit as $k \to \infty$, the minimum bit rate B approaches a limiting value R depending on D. The function $R(D)$ is the *rate-distortion* function due to Shannon (who defined it in a different way). For certain classes of input process x_i, explicit solutions for $R(D)$ have been found and for many other cases upper and lower bounds are available. For a treatment of rate-distortion theory, see Berger [15].

One simple technique for reducing the bit rate without the full complexity of block quantization is by *entropy-coding* the successive output symbols of a zero-memory quantizer. The output of an N-point zero-memory quantizer is one of N different symbols $y_1, y_2, \cdots, y_N$, each having a corresponding probability $p_1, p_2, \cdots, p_N$ of occurring. Instead of transmitting $\log_2 N$ bits per sample (or the next largest integer if $\log_2 N$ is not an integer) to identify each output sample, variable-length codes such as the Huffman code can be used. Such a code assigns a word with more bits to a low probability symbol and fewer bits to a high probability symbol. The resulting average number of bits per sample attainable approaches or equals the entropy of the quantizer output,

$$H = - \sum_{1}^{N} p_i \log_2 p_i.$$

This scheme requires buffering in order to produce a steady output bit stream.

In general, optimal quantizers do not result in equal probabilities for the output symbols, in which case H is always smaller than $\log_2 N$. For example, a 16-point optimal quantizer for Gaussian samples produces output symbols with entropy 4.73 bits (from Max [8]) compared to the 5 bits per symbol needed for equal-length coding. Once entropy coding is to be used, the preceding optimization theory is no longer relevant. It is more appropriate to find a compressor curve which leads to minimum mean-square error for a constraint on output entropy rather than on number of output points. This leads to the surprising result that the uniform quantizer is nearly optimal! See Gish and Pierce [16].

Finally, another class of quantizers with memory are the sequential quantizers such as delta modulation, differential PCM, and the various adaptive versions of these schemes. In essence, all of these schemes take advantage of correlation in the successive input samples by using a feedback loop around the quantizer. However, this is a subject for a separate paper.

VIII. QUANTIZER PERFORMANCE

From a user's viewpoint, the performance of a quantizer is determined by the number of bits per sample needed to digitize a given analog source so that it can be reproduced with a prescribed maximum amount of distortion (or minimum SNR). Alternatively, the performance is determined by how high an SNR can be achieved for a prescribed average bit rate B measured in bits per sample. We take the latter approach here and survey some key results on achievable quantizer performance. For convenience, we focus only on the case of input samples with a Gaussian pdf and the mean-square distortion measure. The issue of robustness is not considered in this discussion.

From rate-distortion theory it is known that, in the limit as the block length approaches infinity, block quantization of a Gaussian source with statistically independent samples can achieve the bit rate

$$B = \frac{1}{2} \log_2 (\sigma^2/D)$$

where D is the average mean-square distortion per sample (and $D < \sigma^2$). Converting to SNR then gives the result

$$SNR_1 = 6B, \qquad (17)$$

which is also a lower bound on attainable SNR for any realizable quantization scheme regardless of the input pdf as long as the samples are independent. If the source samples are correlated, a higher SNR can always be achieved. See Berger [15].

If a zero-memory uniform quantizer is used with entropy coding of the output symbols, an efficient quantization scheme is achieved. By optimizing the overload point, Goblick and Holsinger [17] found that H^*, the highest output entropy attainable with uniform quantizing, satisfies the equation

$$H^* = \frac{1}{4} + \frac{1}{2} \log_2 (\sigma^2/D).$$

With entropy coding, H^* may be taken as the attainable bit rate B so that solving for SNR gives

$$SNR_2 = 6B - 1.50. \qquad (18)$$

Hence the uniform quantizer with entropy coding achieves an SNR only 1.5 dB below the very best attainable performance with block quantization.

If the best nonuniform quantizer for minimizing distortion is combined with entropy coding, taking the bit rate as the output entropy gives the result

$$SNR_3 = 6B - 2.45 \qquad (19)$$

where (19) was empirically found to fit the data tabulated by Max for quantizers with more than eight levels. Clearly, the nonuniform quantizer is inferior when entropy coding is being used.

Of course, entropy coding adds a significant amount of complexity to the implementation of a quantizer. Without entropy coding, we have seen that the highest SNR achievable with nonuniform quantization is given using (16) by

$$SNR_4 = 6B - 4.35. \qquad (20)$$

Recall that (16) is based on the assumption of large N and it neglects overload noise. The exact SNR values for N between 2

and 36 can be obtained from Max's tables. It turns out that (20) is about 3.6 percent too small for $N = 12$ and becomes progressively more accurate as N increases. See Fig. 10.

Finally, the simplest quantization scheme, using a uniform quantizer without entropy coding, gives the least favorable performance. Using a loading factor of 4 and neglecting overload noise led to the SNR formula (6):

$$SNR_5 = 6B - 7.3. \qquad (21)$$

However, the optimum loading factor depends on the number of levels used and the effect of overload distortion. Goblick and Holsinger [17] fitted the curve

$$B = 0.125 + 0.6 \log_2(\sigma^2/D)$$

to Max's tabulated data for uniform quantizers with optimized loading factor. Converting this expression to a SNR formula gives

$$SNR_6 = 5B - 0.63. \qquad (22)$$

Since Max's tables go up to $N = 36$, it is not known how accurate (22) is for $B > 5.2$. For $B < 6.7$, (22) gives higher SNR values than (21), which shows that four-sigma loading is not an optimal choice for a Gaussian pdf.

Summarizing, we have seen that uniform quantizing followed by entropy coding can achieve SNR values within 1.5 dB of the best performance theoretically attainable with any quantization scheme whatever. For an additional 3 dB penalty

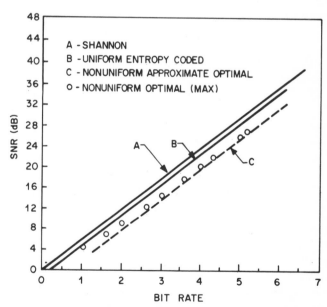

Fig. 10. Quantizer performance in SNR as a function of the average number of bits per sample needed to encode an analog source. Curve A is the best theoretically attainable performance for a Gaussian source with independent samples. It is also a lower bound for any source with independent samples. Curve B is the performance achieved for Gaussian samples with a uniform quantizer followed by entropy encoding. Curve C, based on (16), is asymptotically for a large number of levels the optimal performance obtainable with nonuniform quantization of Gaussian samples without entropy coding. Circled points are based on Max's tabulated values for optimal nonuniform quantization of Gaussian samples without entropy encoding.

in SNR, an optimum nonuniform quantizer without the complexity of entropy coding can be used. Simplest of all, the uniform quantizer can achieve an SNR within 7 dB or so of the best performance theoretically attainable.

It should be emphasized that this modest difference in performance between the simplest and most complex quantization schemes is based on the assumption that the input samples are statistically independent. Zero-memory quantization can be grossly inadequate when there is substantial correlation between successive input samples. However, the utility of zero-memory quantizers does not end when the input is correlated. In such situations, the zero-memory quantizer is still used as a component part of more sophisticated quantization schemes. Sequential quantization schemes all use a zero-memory quantizer of one form or another imbedded in a feedback loop. Also, block quantization schemes generally attempt to transform the vector of input samples into a new vector with independent components. These components are then individually quantized with a zero-memory quantizer. Indeed, the basic zero-memory quantizer plays a ubiquitous role in the digital coding of analog sources.

REFERENCES

[1] B. Widrow, "A study of rough amplitude quantization by means of Nyquist sampling theory," *IRE Trans. Circuit Theory,* vol. CT-3, pp. 266-276, 1956.

[2] W. R. Bennett, "Spectrum of quantized signals," *Bell Syst. Tech. J.,* vol. 27, pp. 446-472, July 1948.

[3] B. M. Oliver, J. R. Pierce, and C. E. Shannon, "The philosophy of PCM," *Proc. IRE,* vol. 36, pp. 1324-1331, 1948.

[4] S. P. Lloyd, "Least-squares quantization in PCM," unpublished memorandum, Bell Laboratories, 1957 (copies available from the author).

[5] H. Holzwarth, "PCM and its distortions by logarithmic quantization" (in German), *Arch. Elekt. Ubertragung,* vol. 3, pp. 277-285, 1949.

[6] K. W. Cattermole, *Principles of Pulse Code Modulation.* Elsevier, London: Iliffe, 1969.

[7] J. Lukaszewicz and H. Steinhaus, "On measuring by comparison" (in Polish), *Zastos. Mat.,* vol. 2, pp. 225-231, 1955.

[8] J. Max, "Quantizing for minimum distortion," *IRE Trans. Inform. Theory,* vol. IT-6, pp. 7-12, 1960.

[9] M. D. Paez and T. H. Glisson, "Minimum mean-squared-error quantization in speech PCM and DPCM systems," *IEEE Trans. Commun.,* vol. COM-20, pp. 225-230, Apr. 1972.

[10] P. Fleischer, "Sufficient conditions for achieving minimum distortion in a quantizer," in *IEEE Int. Conv. Rec.,* 1964, pp. 104-111.

[11] P. F. Panter and W. Dite, "Quantizing distortion in pulse-count modulation with nonuniform spacing of levels," *Proc. IRE,* vol. 39, pp. 44-48, 1951.

[12] B. Smith, "Instantaneous companding of quantized signals," *Bell Syst. Tech. J.,* vol. 27, pp. 446-472, 1948.

[13] G. M. Roe, "Quantizing for minimum distortion," *IEEE Trans. Inform. Theory,* vol. IT-10, pp. 384-385, 1964.

[14] V. R. Algazi, "Useful approximation to optimum quantization," *IEEE Trans. Commun. Technol.,* vol. COM-14, pp. 297-301, 1966.

[15] T. Berger, *Rate Distortion Theory.* Englewood Cliffs, NJ: Prentice-Hall, 1971.

[16] H. Gish and J. N. Pierce, "Asymptotically efficient quantization," *IEEE Trans. Inform. Theory,* vol. IT-14, pp. 676-681, 1968.

[17] T. J. Goblick and J. L. Holsinger, "Analog source digitization: A comparison of theory and practice," *IEEE Trans. Inform. Theory,* vol. IT-13, pp. 323-326, Apr. 1967.

Allen Gersho (S'58—M'64) was born in Canada in 1940. He received the B.S. degree from M.I.T. in 1960 and the Ph.D. degree from Cornell University in 1963.

Since 1963 he has been a member of the Technical Staff at Bell Laboratories, Murray Hill, NJ, where his research work has been primarily related to theoretical aspects of communication systems with occasional digressions to circuit theory, social systems modeling, physics, and biophysics. He has recently returned from an academic year (1976-1977) at the System Science Department of UCLA where he taught linear systems and probability theory, as well as giving a series of seminar lectures on the digital coding of analog sources.

Dr. Gersho is a member of the Communication Theory Committee of the IEEE Communications Society.

DIGITAL TELEVISION TRANSMISSION USING BANDWIDTH COMPRESSION TECHNIQUES *

Hisashi Kaneko and Tatsuo Ishiguro

DIGITAL transmission and signal processing techniques have long been viewed as promising and powerful means to achieve efficient television transmission by combining bandwidth compression with digital transmission systems. Recent progress in LSI and digital technologies have made complicated signal processing a technically feasible reality and have led to reasonable hardware size and cost. High-capacity and low-cost MOS memories have made it possible to store an entire television picture frame. This in turn has led to progress in digital television encoding through the development of interframe coding, by which the transmission bit rate can be greatly reduced.

At the same time, progress in digital transmission systems using multiphase modulation has allowed the use of digital transmission formats in microwave links and satellites. These trends have generated enthusiastic efforts to develop and use digital approaches in actual television networks.

In addition, the transmission of video teleconferencing services is growing rapidly. Teleconferencing has long been anticipated as an alternative to travel and many trials and evaluation tests have and are being conducted. Visual communications is a key to teleconferencing. Full motion video transmission is thought to be very helpful and useful in accomplishing interactive communications. However, transmitting a full television signal to accomplish video teleconferencing would be very expensive, requiring a thousand times wider bandwidth than a voice telephone channel. Bandwidth compression is, therefore, required to provide economical teleconferencing systems with compression ratios of 1:10 or even less.

Thus, the growth of broadcast television as well as teleconferencing services will rely more and more on sophisticated video processing techniques. These have already produced reasonable cost/performance results and have made digital television transmission a present reality in actual field applications.

WHY DIGITAL TELEVISION TRANSMISSION

Digital transmission is considered to be advantageous from the channel capacity point of view, particularly when it employs an efficient encoding scheme [6],[7]. Existing television transmission links are generally analog FM. They carry a single network quality television signal per radio channel. Although one might double the channel capacity by restricting the modulation bandwidth [8] or by field interpolation multiplex [9], the resultant transmission quality would not meet network television standards. Furthermore, with analog techniques, high-compression ratios such as 1:10 could never be achieved.

Fig. 1 shows the relationship between transmitter power and RF bandwidth of a radio link. For analog FM, the transmitter power increases significantly as RF bandwidth is reduced (at an SNR of 54 dB weighted). On the other hand, digital transmission using quadraphase PSK modulation, requires smaller RF bandwidth and smaller transmitter power as the transmission bit rate reduces. At equivalent picture quality, digital technology offers obvious advantages over analog methods at the bit rate below 60 Mbits/s.

Furthermore, it carries with it various advantageous features of digital transmission. Picture quality is almost solely determined by the terminal equipment and can be made almost independent of transmission line impairments including those of digital terrestrial links. This implies that equal quality television service can be

The authors are with the Nippon Electric Company Ltd., Kawasaki, Japan.

*Reprinted from *IEEE Communications Society Magazine,* July 1980, Vol. 18, No. 4, pp. 14–22.

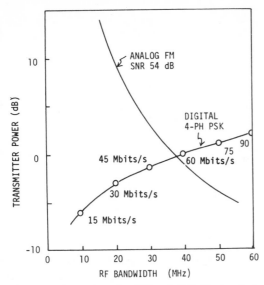

Fig. 1.   A comparison of analog FM and digital transmission in terms of RF bandwidth and transmitter.

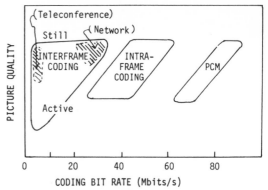

Fig. 2.   Coding bit rate versus picture quality for typical coding methods.

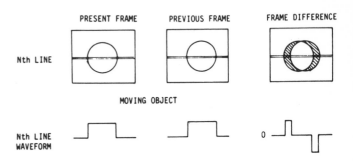

Fig. 3.   Principle of basic interframe coding.

achieved uniformly over a wide service network, irrespective of distance. Sophisticated digital processings can be employed such that error control techniques improve the tolerance of the system to bit error rate, digital television signals can be easily encrypted to protect communication privacy, and adaptive bit sharing techniques can further reduce the required bit rate by sharing some of the total bit rate among plural channels.

WHAT IS INTERFRAME CODING?

Many investigations have been made to realize efficient digital coding [11],[12]. Digital television encoding schemes are generally categorized into three classes: 1) conventional PCM, 2) intraframe coding and 3) interframe coding, as shown in Fig. 2. For NTSC color television signals, conventional PCM or straight A-to-D conversion provides high-quality encoding with 7- or 8-bit encoding at about a 10 MHz sampling, resulting in a 75 through 86 Mbit/s transmission rate [10]. Intraframe coding is well known as a technique to reduce the transmission bit rate by intraframe processing of the signal such as higher order differential PCM (HO-DPCM) or orthogonal transform coding. By these intraframe coding methods, the transmission bit rate can be reduced to about 30 through 60 Mbits/s depending on quality requirements and technique employed. There are also certain tradeoffs between picture quality, bit rate, and hardware complexity.

Much greater reduction in transmission bit rate can be achievable by use of interframe coding, in which the difference signal between two successive frames is encoded and transmitted. The general concept of bit rate reduction by interframe coding can be better illustrated by a simplified model as shown in Fig. 3. Suppose that a soccer ball is crossing a television screen. On looking at

two successive television frames, the soccer ball in the present frame is positioned slightly differently from that of the previous frame, and the difference corresponds to the movement of the soccer ball during the time period: one frame. Instead of transmitting the entire information, if the difference information of the two successive frames is transmitted, the amount of information to be transmitted can be greatly reduced. In fact, if the movement is zero, theoretically, no information need be transmitted. The information to be transmitted is dependent on picture object movement: the more active the movement, the greater the information to be transmitted becomes.

The basic configuration of the interframe coder is shown in Fig. 4. The digitized television signal, converted by an A-to-D converter, is encoded by an interframe coder, in which essentially the difference signal between the present and the previous frame is encoded. The previous frame signal is obtained from frame storage built in the interframe coder. The output of the encoder is again processed through a variable length coder to remove the redundancy contained in the frame differential signal. As mentioned earlier, the information rate is directly dependent on the movement of picture objects. It should be smoothed out to obtain a constant transmission bit rate. This function is accomplished through buffer memory and feedback control to the interframe

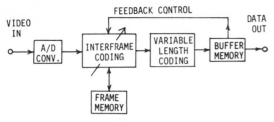

Fig. 4. A basic configuration of interframe encoder.

coder to suppress the excessive generation of information. Since the control of the information generation rate is made by changing the quantizing step size, the signal-to-noise ratio is decreased as the amount of motion increases. When the information generation rate exceeds the transmission rate, picture quality starts to degrade. The encoded picture quality thus varies according to the video source materials.

As will be stated later, for network television signals where much more active motion is encountered, excellent picture quality can be transmitted at a bit rate around 20 through 30 Mbits/s [24],[29]. Relatively still pictures such as those encountered in conference room scenes can be transmitted at 6 Mbits/s, 3 Mbits/s, or even lower bit rates [13]-[22].

It should be noted here that the picture quality measure for interframe coding is different from the existing analog evaluation measure. Even at 3 Mbits/s, the signal-to-noise performance for a still picture is as good as that of 60 Mbit/s PCM. Picture quality impairments occur only when the picture moves actively, and the type of impairment is very much different from simple noise impairment. Presently, no objective evaluation standard has been established. Picture quality can only be subjectively evaluated.

INTERFRAME CODING FOR NETWORK QUALITY TELEVISION

The encoder/decoder arrangements for high-quality transmission of broadcast quality is shown in Fig. 5

[24],[29]. An input composite NTSC color television signal is first converted by an A-to-D converter into a 8 bit PCM signal. The PCM signal is then encoded into a reduced bit rate format of 2-3 bits/sample on average through digital signal processing. In the network television application, the transmission system is required to be transparent to the composite NTSC color video signal. Therefore, the codec is designed so that the composite signal is directly encoded and the waveform of the input signal is preserved except for quantization error.

In an NTSC color television signal, the color subcarrier phase is different by 180° between successive frames. This is undesirable because it generates a large frame difference even when the picture is perfectly still. To solve this, phase inversion is made by a preprocessing circuit in which the composite signal is separated into luminance and chrominance components by reversible transformation and the chrominance component polarity is inverted frame by frame. After phase correction the signal is encoded by an interframe predictive encoder. Adaptive intraframe/interframe coding is used by selecting either an intraframe prediction or an interframe prediction, whichever is appropriate. For quantization of the prediction error signal, eight kinds of quantization characteristics, Q0-Q7, are prepared and one of them is adaptively chosen through feedback control depending on buffer memory occupancy value. Quantization steps for the small amplitude region are, 1(Q0), 1.5(Q1), 2.5(Q3), 4.5(Q4), etc. respectively, where the magnitude 1 corresponds to one quantization step size of the 8 bit A/D conversion. The quantized prediction error is then coded with a variable length coder to reduce a bit rate. The compressed data are stored in a buffer memory with a capacity of about 1 Mbit, and is then transmitted to a line. Buffer memory occupancy value (BMO) is fed back to the adaptive quantizer to control the encoded data generation rate in order to prevent buffer overflow.

The performance of the interframe coder is better

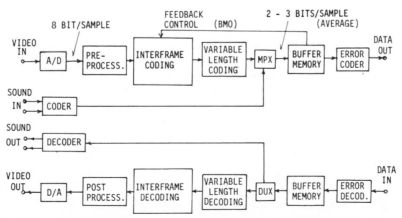

Fig. 5. An interframe encoder and decoder block diagram.

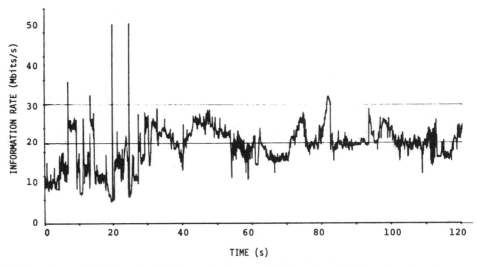

Fig. 6.   Variation of encoded data rate with time. Signal source is "Superbowl '79," most active picture example.

illustrated by way of example in Fig. 6. The example is taken from "Superbowl 79," a most violent picture example, for a period of 2 min. The waveform shows the information generated from the interframe coder without feedback control. The amount of information here indicates more or less original source information. As is seen, the amount of information varies tremendously according to the source scenes. Then, by adding feedback control, the system operates to allow coarse quantization for the portions of excessive source information, and as a result, keeps the output bit rate constant. This is the case for active motion, although generally television programs yield less source information. Fig. 7 shows the statistical data of source information for various source scenes. The dotted line shows the probability of source information rate for various source materials taken from broadcast television pro-

grams for 36 h [29]. It is seen that the average rate is around 15 Mbits/s, and as is seen on the solid line for the cumulative probability, 93 percent of the scenes are handled at 20 Mbits/s and 99 percent at 30 Mbits/s without appreciable noise impairments.

FURTHER IMPROVEMENT IS ACHIEVED BY ADAPTIVE BIT SHARING THROUGH PLURAL CHANNELS

From what we have observed through these statistics, excellent picture quality can be obtained by 20 Mbit/s coding most of the time. Buffer fill occurs with very small probability. According to actual measurements, buffer fill probability decreases very rapidly as the transmission bit rate is increased. The concept of sharing the transmission bit rate among plural television channels becomes quite effective in reducing the probabilities of buffer fill, and in improving picture quality for extremely active motion.

Adaptive bit sharing is a concept quite similar to that of TASI (time assignment speech interpolation) [27], using the advantage of statistical difference among plural channels. When a channel is transmitting a rapidly moving picture, other channels may be transmitting relatively quiet pictures, because the probability of occurrence of rapid picture motion is generally very small. Therefore, we can assign a larger bit rate to the rapidly moving channel, and a smaller bit rate to the other relatively quiet channels, keeping the total bit rate constant. In fact, for three channel transmission with total 60 Mbit/s rate, the bit rate for each channel can be adaptively varied within the 17 to 27 Mbit/s range, with an average of 20 Mbits/s.

Interframe coding performance improvement by adaptive bit sharing is measured [29] by using a real hardware system shown in Fig. 8. Three different broad-

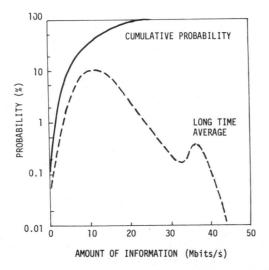

Fig. 7.   Statistics of the encoded data rate for actual broadcast television signals for 36 h in total.

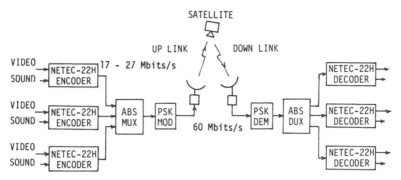

Fig. 8.   A three channel television transmission system.

cast television program signals are supplied to the three encoders, and the adaptive bit rate assignment is performed by an ABS-MUX (adaptive bit sharing multiplexer) with the total bit rate kept constant at 60 Mbits/s.

ERROR CONTROL

This system has a forward acting error control circuit with 239/255 double error correcting BCH code. This error control has excellent error correction capability as shown in Fig. 9 [28]. The calculated mean error free time is longer than 1 h at a line bit error rate of 10^{-5}, and is about 5 s even at 10^{-4} which is generally considered the digital link threshold. The actual observed mean error free time, obtained from a satellite transmission experiment, is plotted in the figure. These data coincide very well with calculated error control performance. In normal satellite conditions, the bit error rate performance is generally better than 10^{-7}, and the mean error free time gets to be much longer than a year.

In addition to forward acting error control, in the decoder, an erroneous line is automatically replaced by

the previous line, making the error less observable to the human eye.

A typical realization of the interframe encoder/decoder for network television is the NETEC-22H developed by Nippon Electric Company, Ltd., as shown in Fig. 10. This configuration consists of three encoders (from the left), an ABS-MUX/DUX bay, a PSK-MODEM, and two receiving decoders. Also each encoder/decoder contains two 15 kHz audio channels.

INTERFRAME CODING FOR VIDEO TELECONFERENCING

So much for the interframe encoding for high-quality network television. Now we will describe the interframe technology for teleconferencing applications where the advantage of interframe coding is more significant in reducing the transmission bit rate down to 6 Mbits/s, 3 Mbits/s, or less. Basic principles of interframe coding are the same as those for network television, but there exist differences in technologies related to achieve high-data compression under conditions of slower picture movement. For this application, transparency is not as

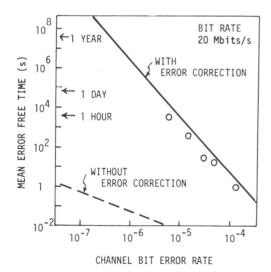

Fig. 9.   Error correction performance of double error correction (DEC) BCH code. The solid line is theoretical and the circles are the field test results.

Fig. 10.   Photograph of the interframe encoder/decoder NETED-22H and ABS-MUX system.

important as in broadcast network use. Algorithm design is therefore directed to a best compromise of the transmission bit rate and picture quality.

A functional block diagram of the interframe encoder and decoder terminal is shown in Fig. 11 [17],[18]. The terminal encodes color video and audio signals into a digital stream and multiplexes them into a bit stream of 3 through 6 Mbits/s. Video signal input and output are NTSC color television signals which are widely used in conventional video equipment. An audio signal with 7 kHz bandwidth is encoded at a bit rate of 128 kbits/s.

In order to accommodate color information, the composite NTSC signal is converted by a color signal processor, CSP-S, into a time division multiplex (TDM) signal which is a baseband video signal with a time compressed chrominance signal inserted into the horizontal blanking interval as shown in Fig. 12. The interframe predictive coding used is an element-to-element difference of frame-to-frame difference coding [19],[31]. The coding parameter/mode-control to prevent buffer overflow in the codec is made by adaptive quantization, field repeating, subsampling, and frame freezing.

The picture quality of the interframe coding is described in relation to the coding parameter/mode-control. When buffer occupancy is in the lowest level, the encoding is made with the finest quantization step size to provide the highest signal-to-quantization noise ratio.

As buffer occupancy increases, the quantization step size is made coarser. The data generation rate reduces as the step size increases. When occupancy is further raised and exceeds a certain level, a field repeating mode is applied in which only the information of every other field, either odd or even, is transmitted. The deleted field is interpolated from the adjacent fields at the receiving side. By the field repeat mode, the information generation rate can be halved. For further occupancy increase, a subsampling mode is added to the field repeating. The subsampling is to transmit every other sample. If the buffer becomes full in spite of the controls described above, the encoding is stopped until the buffer occupancy decreases to a low level. At the receiving side,

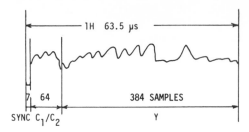

Fig. 12. TDM color TV signal format.

the same picture is repeated, until a new frame picture is received. These mode controls give rise to a tradeoff between the data rate and picture quality and cause impairments as shown in Fig. 13. As the amount of motion increases, the picture quality gradually decreases.

In 6 Mbit/s encoding, a teleconferencing scene with natural motion of attendees can be transmitted within the quantization step-size control. At 3 Mbits/s, the picture is in quantization control most of the time. However, if the picture includes some large motion, field repeating may operate, causing a slight jerkiness. Subsampling and picture freezing occur when the television camera is panned or when television signals are switched from one camera to the other. Decreasing the transmission bit rate results in restricting the range of motion which can be transmitted. However, even allowing such degradation for extreme cases, the interframe coder can handle most teleconference scenes.

Fig. 14 shows a photograph of the NETEC-6/3 terminal. The encoder and decoder are mounted in a single standard 19 in bay with a depth of 24 in.

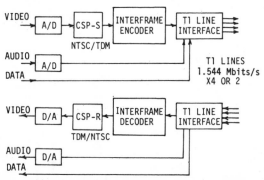

Fig. 11.   A digital television transmission terminal, NETEC-6/3, for teleconferencing use.

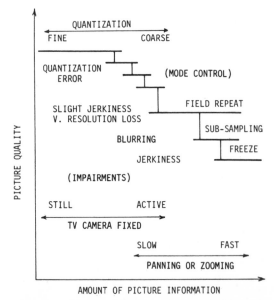

Fig. 13.   Interframe coding performance features.

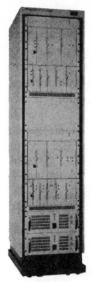

Fig. 14. Photograph of the interframe encoder/decoder, NETEC-6/3.

FUTURE TECHNOLOGIES

As has been stated, the development of interframe coding has reached a point where it is finding real applications in television transmission. However, the future holds still further innovation. LSI technology for high-speed logic circuits will contribute greatly to reducing the size and increasing the economy of hardware.

Further innovation will be realized in more sophistication of interframe coding algorithms by movement compensation. Through movement compensation, more efficient predictions will be possible by tracking the object movement. The concept of movement compensation is not new, but has become a serious area of study in digital video processing [30]-[33]. Several reports have indicated that a doubling or more of the data compression ratio can be expected through these techniques. This is very attractive from an applications point of view. The T1 carrier rate (1.544 Mbits/s) will become sufficient to transmit a video teleconferencing signal. This provides not only a great advantage in transmission cost reduction but also a good match to digital networks widely used now. Also, in broadcast television program transmission, movement compensation will be quite effective particularly for such severe pictures as those taken by camera panning or zooming, and will contribute to the reduction of transmission rate down to 15 through 20 Mbits/s with excellent picture quality.

Efficient modulation techniques are also important for transmission bandwidth compression as well as efficient source coding. A conventional satellite transponder can transmit about 60 Mbits/s over a 36 MHz channel (1.6 bit/Hz). For a terrestrial radio system using eight-phase PSK modulation, the spectral efficiency is in the range of 2.2-2.6 bits/Hz. Extensive studies and developments aimed at higher spectral efficiency (4 bits/Hz) using multilevel QAM, etc. are underway [35]. When such technologies become available, an RF bandwidth of 10 MHz or even narrower will be sufficient to transmit television program signals with a slight increase in transmitter power.

TABLE I
Channel Capacity of Various Schemes over Existing North American Transmission Links

Encoding	PCM	Intra-frame Coding	Interframe Coding Network Quality	Teleconferencing			FM
Bit Rate (Mbits/s)	75-86	32-45	20-30	6	3	1.5	
Satellite (60-64 Mbits/s)	—	1-2	2-3	10	20	40	1-2
Terrestrial (6 GHz, 78 Mbits)	1	2	3	12	24	48	1
(11 GHz, 90 Mbits)	1	2-3	3-4	14	28	56	1-2
DS-3 (45 Mbits)	—	1	2	7	14	28	
DS-2 (6 Mbits)	—	—	—	1	2	4	
DS-1C (3 Mbits)	—	—	—	—	1	2	
DS-1 (1.5 Mbits)	—	—	—	—	—	1	
		HO-DPCM	NETEC-22H	NETEC-6/3		*	

*Movement compensation encoder.

History of Interframe Coding Development

Year	Topic	Authors/Institute/References
1952	Statistics of television signals	Kretzmer, BTL [1]
1959	Frame correlation study	Taki *et al.*, Univ. Tokyo [2]
1965	Frame difference signal statistics	Seyler, APG [3]
1969	Study of bandwidth compression utilizing frame correlation and movement compensation	Rocca, Politec. Milano [30]
1969	Conditional replenishment PCM	Mounts, BTL [4]
1971	Frame difference coding for 1 MHz monochrome picturephone signals	Candy *et al.*, BTL [5]
1974	Interframe coding for 4 MHz color television signals	Iinuma *et al.*, NEC [13],[15]
1975	Adaptive prediction (movement compensation) study	Haskell, BTL [31]
1976	1.5 Mbit/s codec of 1 MHz monochrome videotelephone signals	Yasuda *et al.*, NTT [23]
1976	TRIDEC development	Yasuda *et al.*, NTT [19],[20]
1976	NETEC-22H development	Ishiguro *et al.*, NEC [24]
1977	NETEC-22H satellite transmission experiment and ABS MUX development	Kaneko *et al.*, NEC [28],[26]
1977	1.5 Mbit/s interframe coding of 525 line, monochrome television	Haskell *et al.*, BTL [21]
1978	Movement compensation experiment for broadcast TV signals	Ninomiya, NHK [32]
1979	NETEC-6/3 terminal development	Kaneko, *et al.*, NEC,
1979	TRIDEC commercial service	NTT [18]
1979	30 Mbit/s interfield coding	Hatori *et al.*, KDD [34]
1979	Practical algorithm of movement compensation	Netravali *et al.*, BTL [33]

Thus, future advances in source coding and digital modulation techniques will further enhance the advantage of digital transmission systems.

THIS TECHNOLOGY IS READY FOR APPLICATION

It has been shown that interframe coding techniques provide an effective means to reduce the transmission bit rate of video signals without sacrificing picture quality. In Table I, the channel capacity of various coding schemes over existing digital transmission links is listed. For digital satellite, for example, even a single television channel cannot be transmitted by conventional PCM, whereas one or two channels of television can be transmitted by employing efficient coding techniques. However, much greater advantage is obtained by use of interframe coding. Three network quality television signals can be carried through a transponder.

For teleconferencing applications, a single satellite transponder can manage up to twenty simultaneous conference signals on a TDMA basis, if one uses a 3 Mbit/s interframe coder. The 3 Mbit/s video signal is compatible with the North American digital hierarchy. It can be carried over two T1 lines or a single T1–C line for local distribution and will be compatible with digital toll links multiplexed up to the DS-3 45 Mbit/s rate and beyond.

Our satellite experiment has demonstrated excellent performance for network digital television application. The TRIDEC teleconferencing system has been commercially used by NTT in Japan since 1979 and the NETEC-6/3 interframe encoder is being seriously considered for application in North American teleconferencing applications. During the past decade, interframe coding has grown from theoretical infancy to actual application level. Together with the growth of video transmission demands and nationwide digital transmission over satellite, terrestrial microwave, coaxial cable and optical fiber cables, interframe coding technology will no doubt contribute to new digital television services.

APPENDIX

PREDICTIVE CODING

Data compression of television signals is achieved by redundancy removal of television signals. As is well known, a television signal has strong correlation between picture elements. For example, neighboring picture elements have nearly equal amplitudes in most cases. This leads to the fact that the difference signal has smaller amplitude distribution than the signal itself. Therefore, by transmitting the element difference information instead of the element amplitude, the signal power, or the amount of information to be transmitted can be greatly reduced. This is a simple example of redundancy removal.

A generalized form of such differential technique is "predictive coding." The principle of it is shown in Fig. 15. Input signal X is compared with predicted signal $\hat{X}$. Prediction error signal E given by

$$E = X - \hat{X} \tag{1}$$

is quantized for transmission. The quantized prediction error is added to prediction signal $\hat{X}$ to produce locally decoded signal Y. Prediction signal $\hat{X}$ is obtained through prediction function $P(z)$ as follows:

$$\hat{X} = \sum_{i=1}^{N} a_i \cdot Y_i \tag{2}$$

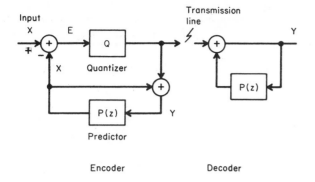

Fig. 15.   Principle of predictive coding.

and

$$P(z) = \sum_{i=1}^{N} a_i \cdot z^{-i} \qquad (3)$$

where a_i's are prediction coefficients, Y_i's are time sequence of locally decoded signal Y, and z is the z-transform operator.

A simplest form of predictive coding is differential PCM (DPCM) which uses only the previous sample in the scan line as the prediction, that is, $N = 1$ and $a_i = 1$ in (2) and (3). For composite NTSC color television signals, efficient prediction is made by using plural samples (for example, $N = 4$) in the same scan line (higher-order DPCM). Two-dimensional prediction uses samples in the previous lines as well as those in the same line, providing slightly higher efficiency than one-dimensional prediction. Interframe coding further uses samples in the previous frame. Movement compensation is considered to be adaptive prediction in which samples Y_i's and coefficients a_i's in interframe coding are adaptively changed depending on the picture movement to minimize the prediction error.

REFERENCES

[1] E. R. Kretzmer, "Statistics of television signals," *Bell Syst. Tech. J.*, vol. 31, pp. 751-763, July 1952.

[2] Y. Taki *et al.*, "Measurements of frame correlation and amplitude distribution of television signals," *NHK Tech. Res.*, vol. 11, pp. 117-126, May 1959.

[3] A. J. Seyler, "Statistics of television frame differences," *Proc. IEEE*, vol. 53, pp. 2127-2128, Dec. 1965.

[4] F. W. Mounts, "A video encoding system using conditional picture-element replenishment," *Bell Syst. Tech. J.*, vol. 48, pp. 2545-2554, Sept. 1969.

[5] J. C. Candy, M. A. Franke, B. G. Haskell, and F. W. Mounts, "Transmitting television as clusters of frame-to-frame differences," *Bell Syst. Tech. J.*, vol. 50, pp. 1889-1919, July-Aug. 1971.

[6] L. S. Golding, "DITEC—A digital television communications system for satellite links," presented at the 2nd Int. Conf. Digital Satellite Commun., Paris, France, Nov. 1972.

[7] W. Zschunke, "Digital transmission of TV by satellite," presented at the Conf. Digital Satellite Commun., Kyoto, Japan, Nov. 1975.

[8] "Transmission performance of television circuits designed for use in international connections," CCIR Rec. 567, 1978.

[9] L. Abbott, "Transmission of four simultaneous television programs via a single satellite channel," *SMPTE J.*, vol. 88, pp. 106-111, Feb. 1979.

[10] A. A. Goldberg, "PCM encoded NTSC color television subjective tests," *SMPTE J.*, vol. 28, pp. 649-653, Aug. 1973.

[11] J. O. Limb, C. B. Rubinstein, and J. E. Thompson, "Digital coding of color video signals—A review," *IEEE Trans. Commun.*, vol. COM-25, pp. 1349-1385, Nov. 1977.

[12] A. N. Netravali and J. O. Limb, "Picture coding: A review," *Proc. IEEE*, vol. 68, pp. 366-406, Mar. 1980.

[13] K. Iinuma *et al.*, "Interframe coding of 4-MHz color television signals," presented at the 1974 Picture Coding Symp., Goslar, West Germany, Aug. 1974.

[14] J. A. Heller, "A real-time Hadamard transform video compression system using frame-to-frame differencing," in *Conf. Rec. 1974 Nat. Telecommun. Conf.*, San Diego, CA, 1974, pp. 77-82

[15] K. Iinuma, Y. Iijima, T. Ishiguro, H. Kaneko, and S. Shigaki, "Interframe coding of 4-MHz color television signals," *IEEE Trans. Commun.*, vol. COM-23, pp. 1461-1466, Dec. 1975.

[16] T. Ishiguro, K. Iinuma, Y. Iijima, T. Koga, and H. Kaneko, "NETEC system: Interframe encoder for NTSC color television signals," presented at the 3rd Int. Conf. Digital Satellite Commun., Kyoto, Japan, Nov. 1975.

[17] K. Iinuma, Y. Iijima, T. Ishiguro, T. Koga, S. Tanaka, and H. Kaneko, "NETEC-6; Interframe encoder for color television signals," *NEC Res. Devel.*, pp. 92-96, Jan. 1977.

[18] H. Kaneko, Y. Iijima, T. Ishiguro, and K. Iinuma, "NETEC-6/3 video transmission equipment for teleconference," in *Intelcom 79 Exposition Proc.*, Dallas, TX, Feb. 1979, pp. 579-582.

[19] H. Yasuda, H. Kuroda, F. Kanaya, and H. Hashimoto, "Transmitting 4-MHz TV signals by combinational difference coding," *IEEE Trans. Commun.*, vol. COM-25, pp. 508-516, May 1977.

[20] H. Kuroda, F. Kanaya, and H. Yasuda, "TRIDEC system configuration," *Rev. Elec. Commun. Labs.*, vol. 25, pp. 1347-1351, Nov.-Dec. 1977.

[21] B. G. Haskell, P. L. Gorden, R. L. Schmidt, and J. V. Scattaglia, "Interframe coding of 525-line, monochrome television at 1.5 Mbits/s," *IEEE Trans. Commun.*, vol. COM-25, pp. 1339-1348, Nov. 1977.

[22] H. Kawanishi, H. Yasuda, and H. Kuroda, "An experimental intermultiframe coder," in *1977 Picture Coding Symp.*, Tokyo, Japan, Aug. 1977, Abstracts S.10-1.

[23] H. Yasuda, F. Kanaya, and H. Kawanishi, "1.544 Mbits/s transmission of TV signals by interframe coding system," *IEEE Trans. Commun.*, vol. COM-24, pp. 1175-1180, Oct. 1976.

[24] T. Ishiguro, K. Iinuma, Y. Iijima, T. Koga, S. Azami, and T. Mune, "Composite interframe coding of NTSC color television signals," in *Proc. Nat. Telecommun. Conf.*, Dallas, TX, Nov. 1976, pp. 6.4.1-6.4.5.

[25] B. G. Haskell, "Buffer and channel sharing by several interframe picturephone coders," *Bell Syst. Tech. J.*, vol. 51, pp. 261-289, Jan. 1972.

[26] H. Kaneko and T. Ishiguro, "Digital transmission of broadcast television with reduced bit rate," in *Proc. Nat. Telecommun. Conf.*, Los Angeles, CA, Dec. 1977, pp. 41.4-41.4.6.

[27] J. M. Fraser *et al.*, "Overall characteristics of a TASI," *Bell Syst. Tech. J.*, vol. 41, July 1962.

[28] H. Kaneko. T. Ishiguro, and M. Sugiyama, "Digital transmission through the satellite," presented at INTELCM 77 Exposition, Atlanta, GA, Oct. 1977.

[29] T. Koga, Y. Iijima, K. Iinuma, and T. Ishiguro, "Statistical analysis of NETEC-22H system performance," in *Proc. Int. Conf. Commun.*, vol. 2, Boston, MA, June 1979, pp. 23.7.1-23.7.5.

[30] F. Rocca, "Television bandwidth compression utilizing frame-to-frame correlation and movement compensation," in *Proc. Symp. Picture Bandwidth Compression*, Apr. 1969. New York: Gorden and Breach, 1972.

[31] B. G. Haskell, "Entropy measurements for nonadaptive and adaptive, frame-to-frame, linear predictive coding of video telephone signals," *Bell Syst. Tech. J.*, vol. 54, pp. 1155-1174, Aug. 1975.

[32] Y. Ninomiya, "Motion correction for interframe coding systems," Tech. Group Image Engineering (IE), IECE Japan, Paper IE78-6, May 1978.

[33] A. N . Netravali and J. D. Robbins, "Motion compensated television coding—Part I," *Bell Syst. Tech. J.*, vol. 58, pp. 631-670, Mar. 1979.

[34] Y. Hatori, H. Murakami, and H. Yamamoto, "30 Mbits/s codec for NTSC-CTV by interfield and intrafield adaptive prediction,"

in *Proc. Int. Conf. Commun.*, Boston, MA, June 1979, pp. 23.7.1-5.

[35] "Special Issue on Digital Radio," *IEEE Trans. Commun.*, vol. COM-27, Dec. 1979.

Hisashi Kaneko received the B.S. degree in electrical engineering in 1956 from the University of Tokyo, Tokyo, Japan, the M.S. degree in electrical engineering in 1962 from the University of California, Berkeley, CA, and the Dr. Eng. degree from the University of Tokyo in 1967.

In 1956 he joined the Central Research Laboratories, Nippon Electric Company where he worked on digital communications, particularly satellite communications, etc. Meanwhile, from 1960 through 1962 he studied at the Electronics Research Laboratory, University of California at Berkeley, as a Research Assistant under Fulbright scholarship. From 1968 through 1970 he worked at Bell Laboratories, Holmdel, NJ, for the formulation of segment companding laws in PCM and the synthesis of digital compandors. Since 1970 he has headed the Communication Research Laboratory, Nippon Electric Company, and devoted his energy to the development of digital communications, including digital coaxial cable transmission, digital signal processings, digital television transmission, digital speech interpolation, etc. He is currently General Manager, Transmission Division, Nippon Electric Company, responsible for the development and manufacturing of transmission equipment.

Dr. Kaneko has published many technical papers including several contributions to the IEEE TRANSACTIONS ON COMMUNICATIONS and technical books. He received the Best Paper Award from the Institute of Electronics and Communication Engineers of Japan in 1967. He is a Senior Member of IEEE, and a member of the Institute of Electronics and Communication Engineers of Japan and the Institute of Electrical Engineers of Japan.

Tatsuo Ishiguro was born in Kyoto, Japan, on January 17, 1940. He received the B.S. degree in electrical engineering from Kyoto University, Kyoto, Japan in 1962.

He joined Nippon Electric Company, Ltd., Kawasaki, Japan in 1962, and is currently Research Manager in the Communication Research Laboratory, Central Research Laboratories. From 1962 to 1967 his work was concerned with ultrasonic amplification in CDS and VHF ultrasonic delay lines. Since 1968, he has been engaged in research on digital encoding and bandwidth compression transmission of television and facsimile signals.

TECHNIQUES FOR DIGITAL SWITCHING *

John C. McDonald, Senior Member, IEEE

Abstract   Digital switching of communication channels is now a practical alternative to analog switching. This application of well-known principles is economically possible due to the availability of low-cost LSI devices and the significant benefits of integrating digital transmission with digital switching. This paper reviews the key elements of a digital switch and summarizes the current debate relating to the design of a local digital switch.

INTRODUCTION

We live in an era where electromechanical and analog technologies are being replaced by digital LSI. It is not surprising, therefore, that electromechanical switching is now being challenged by digital switching. The same technology which allowed the hand-held calculator to obsolete its electromechanical counterpart and the digital watch to threaten its mechanical competition is now being applied to the network and control of telephone switching systems. This challenge has already been successful using LSI components designed primarily for the computer industry. But new digital LSI components are being designed specifically for the communications industry, and the benefits of digital switching will accelerate as these new devices become available.

There are other reasons for the success of digital switching. PCM carrier, introduced in 1961, has found widespread use as an exchange and toll carrier system. PCM

The author is with Vidar, A Division of TRW, Inc., Mountain View, CA 94040.

*Reprinted from *IEEE Communications Society Magazine*, July 1978, Vol. 16, No. 4, pp. 11–19.

technology has also been successfully applied to pair gain systems in the subscriber plant. Digital radio and fiber optic systems will further propagate digital technology. These digital carrier systems will be less expensive in the presence of digital switching since fewer analog-to-digital conversions will be required. Thus, there is an interactive effect between digital carrier and digital switching. An increase in the number of installed digital carrier systems increases the feasibility of digital switching. This in turn increases the feasibility of digital carrier. A trend toward a completely digital network is thereby created with the promise of better subscriber transmission quality, new services, and lower costs.

Digital switching has been in the research stage for many years [1]. The first commercial application to toll and local switching began in France in 1969 [2]. In the United States, the first commercial use occurred in PABX equipment in 1973 [3]. Application to the common carrier toll network followed in 1976 [4], [5]. The first local digital switch in the United States began service in 1977 [6]. Now throughout the world, virtually all switching suppliers are

heavily engaged in product development aimed toward commercial digital switching systems.

This paper reviews the elements of a digital switch and highlights some of the design controversies introduced by this new technology. A series of definitions are presented and the elements required for a digital switch are described. System synthesis is presented to illustrate the design alternates and highlight current debate. Finally, the impact of digital switching on communication network design is described.

DEFINITIONS

There are two major components of a switching system: the switching network and the control. Various commercial systems, step-by-step, crossbar, and electronic switching, have taken their names from either or both of these components. Digital switching takes its name from the switching network since information in the network is represented in digital form.

The following definitions were proposed at NEC'77 [7].

Analog Switch: A means to interconnect two or more circuits whose

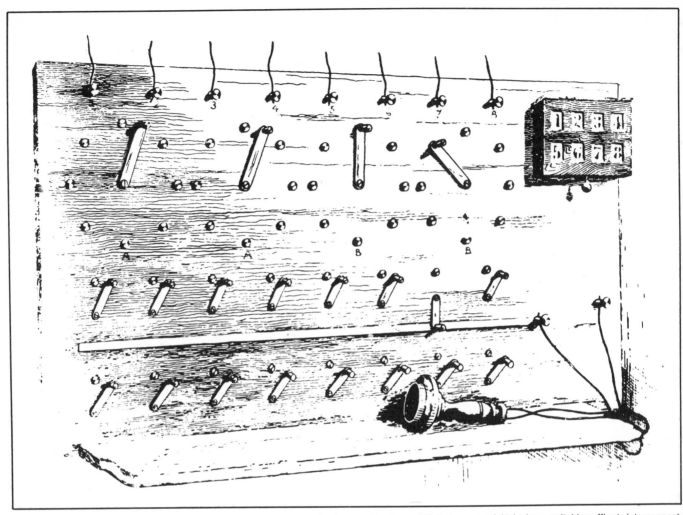

This switchboard, composed of brass strips and screws mounted on a wooden box, enabled the world's first commercial telephone switching office to interconnect up to 8 simultaneous calls among 21 customers. This primitive instrument used a magnetic annunciator or "drop" shown in the upper right to indicate that a connection was requested. Photo courtesy of Bell Laboratories.

information is represented in analog form using a network which may or may not be time divided and may or may not consist of linear elements.

Digital Switch: A means to interconnect two or more circuits whose information is represented in digital form using a time divided network consisting of nonlinear elements.

Merged Technology Switch: A means to interconnect two or more circuits using a series of analog and digital switching stages.

Information in the network of an analog switch is represented in analog form. The network can use relay or electronic crosspoints and the control can be hard-wired or stored-program. Examples of analog switching are given in Table I. Included are examples using pulse amplitude modulation and time-divided techniques. Some analog switches use nonlinear networks with information represented by pulse width modulation.

In a digital switch, information is represented in digital form and the time-divided network uses nonlinear elements. Examples of digital and merged technology switching are also listed in Table I.

DIGITAL SWITCHING ELEMENTS

A digital switch is composed of the elements described in this section.

TABLE I
EXAMPLES OF VARIOUS SWITCHING TECHNOLOGIES

Manufacturer	Analog	Digital	Merged
Automatic Electric	#1 EAX	#3 EAX	
CIT Alcatel	CP400		E10
ITT	TCS5	DSS1	
LM Ericsson	AKE13		AXE10
Northern Telcom	SP1	DMS10	
Nippon Electric	D10		NEAX61S
Stromberg Carlson	ESC1	DCO21	DCO200
TRW/Vidar		ITS5	
Western Electric	#1 ESS	#4 ESS	

These elements are assembled in various ways in the next section to realize a digital switch.

Digital Voice Encoding—Various digital voice encoding techniques have been proposed for use in a digital switch including delta modulation, adaptive delta modulation, PCM, and differential PCM. Most new designs use PCM for the reasons discussed in the next section.

The elements of PCM voice encoding are sampling, quantizing, and coding. All three parameters have been established in the two transmission standards adopted by CCITT, shown in Table II. Each country has chosen one of these standards for PCM transmission in their national telephone network. Conversion must occur between two countries using different standards. Note that both standards sample at 8 kHz and use an 8 bit binary PCM word. However, the quantizing intervals and the multiplexing techniques differ. The North American quantizing standard uses a 15-segment approximation to the μ companding law while the CEPT standard uses a 13-segment approximation to the A-law [8].

Quantizing intervals are chosen to approximate these companding laws and thereby obtain a constant signal to quantizing distortion over a wide range of input levels [8].

Multiplexing—Significant cost savings can be obtained by multiplexing many voice channels together using time-division techniques. This allows time sharing of switching components such as logic gates, interconnecting wires, connector pins, etc. Further efficiencies can be obtained by multiplexing voice and control signaling onto the same path. Multiplexing techniques differ in the number of voice channels per frame, in the technique for deriving control signaling, and in the frame detection technique. The North American multiplexing standard uses "bit stealing" to derive control signaling information as shown in Table I. The CEPT standard uses a dedicated common channel for signaling.

While multiplexing standards exist for trunk carrier equipment, they do not presently exist for subscriber carrier. Many additional signaling and alarm functions are required in a subscriber carrier system which makes standards difficult to establish. Most digital switching systems use combinations of "bit stealing" and common channel techniques for control signaling.

Terminations—A digital switch must provide analog terminations for lines and trunks and digital terminations for multiplexed groups. In a local digital switch, the subscriber loop termination can account for more than half of the total switch cost. This termination must perform the "BORSHT" functions of battery feed, overvoltage protection, ringing the telephone, supervision, hybrid, and test access. Except for the hybrid, similar BORSHT functions are required for analog switching. The digital switch must also provide filtering and digital conversion functions. Design of a subscriber loop termination for a digital switch is described in [9].

A digital termination provides the interface between the digital switch and a multiplexed group of digital channels. While it has been convenient to use "BORSHT" to describe line termination functions, consider the "GAZPACHO" functions of a digital termination: generation of outgoing frame code, alignment of incoming and switch frames, zero string suppression, polar conversion from bipolar to unipolar, alarm processing, clock recovery, hunting during reframe, and office signaling extraction and insertion. Digital terminations must perform additional functions if the switch multiplexing and voice encoding techniques differ from those used in the transmission facility.

Time Divided Networks—Two subscribers served by a digital switch are connected by the time divided network using the principles of time slot interchanging and time multiplexed switching. The operation of a *time slot interchanger* (TSI) is illustrated in Fig. 1. Eight analog terminations are shown in Fig. 1(a) with separate transmit and receive ports. This configuration produces a four-wire circuit between any two terminations.

The switching function is illustrated by connecting terminations 1 and 4. Information, X, transmitted from termination 1 appears on time slot 1 on the transmit multiplex. The time slot interchanger delays X from time slot 1 to time slot 4 on the receive multiplex as shown in Fig. 1(b). X therefore appears on the receive port of termination 4. Similarly, Y is delayed from time slot 4 on the transmit multiplex to time slot 1 on the receive multiplex. Thus, a four-wire transmission path is established between terminations 1 and 4.

The multiplexer and time slot interchanger shown in Fig. 1 are independent. Economies can be realized by making them dependent if we choose the time when X and Y are applied to the transmit multiplex. With a codec per termination sampling can be performed at the time slot when data is required on the receive multiplex, thereby combining the functions of the transmit multiplexer and the time slot interchanger. One can realize a single-stage time-divided network with a desired probability of blocking and without requiring a separate TSI in such a dependent structure.

Fig. 1 illustrates a single-stage

TABLE II
PCM STANDARDS ADOPTED BY CCITT

North American Standard	CEPT Standard
Sampling Rate—8 kHz	Sampling Rate—8 kHz
Modulation—8 Bit PCM	Modulation—8 Bit PCM
Companding—μ255	Companding—A 87.6
Frame—24 Time Slots + 1 Frame Bit	Frame—32 Time Slots
Signaling—Least Significant Bit Stealing every 6 Frames	Signaling—1 Common Channel Containing 2 Words
Bit Rate—1.544 Mbits	Bit Rate—2.048 Mbits
Frame Location—Coded Frame Bit	Frame Location—8 Bit Frame Word
Multiframe Length—12 Frames	Multiframe—16 Frames

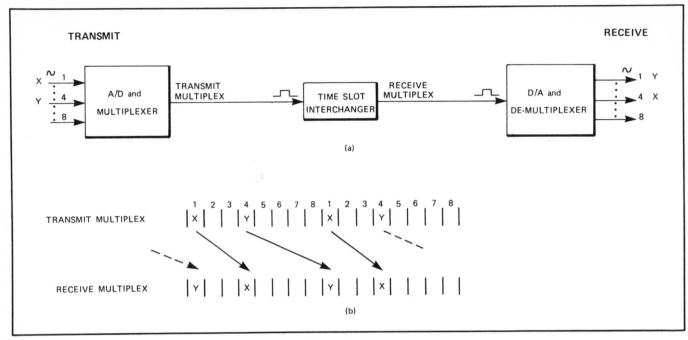

Fig. 1.   Single-stage time-divided network-multiplexer and TSI are independent.

network wherein the number of time slots is equal to the number of terminations. With a constant sampling rate, the time interval for each time slot is inversely proportional to the number of terminations. With 10 000 terminations and an 8 kHz sampling rate, each time slot occupies 12.5 ns. This short interval would place severe requirements on memory access time for the time slot interchanger and tight delay controls would be necessary through gates and cables. Timing requirements can be reduced by using more than one stage in the time-divided network. Since a two-stage network is virtually as complex as three stages [10], most architectures use three or more stages.

The *time–space–time* (T-S-T) network, shown in Fig. 2, will illustrate the operation of a time-multiplexed switching stage. Assume that terminations 1 and 12 are to be connected. Information, X, in termination 1 is transmitted at time slot 1 in the transmit multiplex to TSI-1. This information must appear at the output of TSI-4 at time slot 4 on the receive multiplex, as in Fig. 1(b), to appear at termination 12. A path through the T-S-T network must now be found.

The path-finding algorithm begins by finding an idle time slot through the space stage which can connect TSI-1 with TSI-4. The time slot interchangers can then connect the desired transmit and receive multiplex time slots with the space stage time slot. A question can now be raised regarding the number of time slots in the space stage. This number can differ from the number in the transmit and receive multiplex. To obtain a strictly nonblocking network, the number of space stage time slots must be at least equal to two times the number of time slots in the transmit or receive multiplex minus 1. This can be seen as follows. Assume that the number of time slots in the space stage equals the number of time slots in the multiplex. The path from TSI-1 to TSI-4 requires the connection of crosspoint A during a time slot when B and C are disconnected. However, B could be busy during the first four time slots and C could be busy during the last four time slots. In this case and without rearrangement, the connection between TSI-1 and TSI-4 is blocked. As the number of time slots in the space stage is increased, nonblocking occurs when the number of time slots is twice the number of transmit or receive multiplex time slots minus one. The blocking probability for a single simplex connection with other numbers of

time slots and without rearrangement is given by

$$P = \frac{(M-1)!(M-1)!}{S!(2M-2-S)!} a^S (2-a)^{(2M-2-S)} \quad (1)$$

where

M = Time slots in transmit mux
S = Time slots in space stage
a = Occupancy in Erlangs
$M - 2 < S$.

The path from TSI-1 to TSI-4 is thereby found. The switching process consists of moving X from the transmit multiplex to TSI-1; delaying X until an idle time slot in the space stage; transmitting X through the space stage to TSI-4; delaying X until the proper time slot on the receive multiplex.

A path is now determined between 1 and 12 to transfer information X. The path between 12 and 1 to transfer Y can be established independently of the first path or it can be found using one of many algorithms based on the symmetrical behavior of the switch.

Other combinations of time and space configurations are possible [10]. A comparison between the time–space–time and space–time–space networks is given in [11].

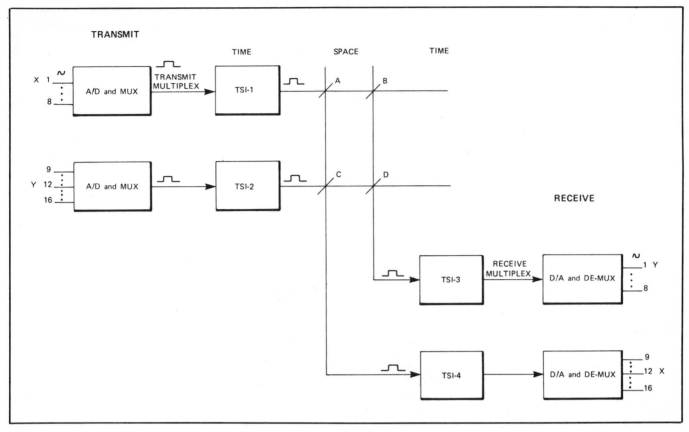

Fig. 2.   Three-stage T-S-T time-divided network—multiplexer and TSI are independent.

Control—Developments in multiprocessor techniques and stored program control are equally suited to analog and digital switching. But the multiplexing techniques used in a digital switch reduce the cost of distributing control functions to obtain a more linear cost versus size profile. Since control signals can be multiplexed with voice channels, many interesting designs are possible. Remoting portions of the digital switch is especially interesting since it can be established without significant change to the call processing software.

Distributed processing using microprocessors is a low-cost and efficient approach. Time-consuming tasks such as line supervision, digit collection, and path finding can be performed by microprocessors. It is common to extract signaling and control information before the multiplex enters the time-divided network. Similarly, signaling is reinserted after the network. For these designs, control information does not propagate through the network connection.

Digital switching with multiplexing and microprocessor technology is stimulating new design approaches for the control function. This approach to control is cost effective over wide size ranges and provides the flexibility to allow new services to be implemented.

Digital Signal Processing—A digital switch presents many opportunities for applying digital signal processing techniques since information in the switch is represented in digital form. Read-only memories (ROM) can be used to synthesize the following progress and signaling tones: dial tone, reorder tone, ringback tone, busy tone, multifrequency (MF) sending tones, idle code, and a milliwatt tone. These ROM functions need only appear once in a switch since broadcast techniques can be used in the switching network to provide simultaneous access to all terminations. ROM can also be used to provide look-up table functions to insert loss, convert from companded to linear code and change companding codes.

Random access memory (RAM) can be used for simple "hit" integration and for dialed digit accumulation. More complex functions employing RAM, adders, and other digital elements include receivers for MF and dual tone multifrequency (DTMF) and conference circuits. Because of the speed of current logic, circuits to implement these functions can be time shared among many functions and terminations.

SYSTEM SYNTHESIS

This section describes the synthesis of digital switching systems using the elements described above. Design alternatives are presented which are currently the subject of debate.

Digital Voice Encoding—The choice of a digital voice encoding technique has been a hotly debated issue for all switching environments but toll. Some designers argue that lines outnumber trunks by about 10 to 1 and that the encoding technique for lines should not be frozen by trunk compatibility requirements.

Low-cost encoding, it is claimed, should be used for lines and format conversion devices should be applied at trunks to establish compatibility. Others argue that PCM is the preferred form of digital modulation for the following reasons.

First, PCM is the only internationally standardized digital modulation technique. While research continues in other encoding schemes, standards are yet to be established. Standardization on PCM is giving the semiconductor industry the necessary volume potential to make commercial LSI devices economically feasible. The second reason for adopting PCM lies in the widespread use of PCM carrier systems. If the switch and carrier modulation techniques are not identical, conversion devices are required at all interfaces. This would add cost and contribute additional quantizing distortion to a digital network. Maintenance costs would be lower if a common technology is used since craftsperson training is unified.

BORSHT location—Where should the BORSHT elements be located? Should they be associated with each subscriber line per the "digital switch" definition or should they be divided with portions shared over many subscriber lines per the "merged technology" definition? This debate was recently aired at the Zurich Conference on Digital Communications [12].

Consider the case where BORSHT with voice encoding is on a per-line basis. A future digital interface per line is then possible by removing the BORSHT and thereby providing a direct digital interface to the subscriber loop. This approach can accommodate a future digital instrument or a mixture of voice and data. Individual terminations, when BORSHT is provided on a per-line basis, can be easily tailored to provide battery boost, special hybrid terminations, etc. This approach also eliminates the need for analog concentration with its architectural and maintenance complication. With digital concentration, intraconcentrator links can be obtained at essentially no cost. With this configuration, groups of lines can be combined and remoted over a digital span line to

realize simple and cost effective pair gain systems. Finally, current silicon technology forecasts favor and possibly require this approach.

Other designers argue that placing the BORSHT on a per-line basis is simply too expensive. Since this interface represents between 50 and 80 percent of the total switch cost, architectures must be established, it is argued, to minimize cost even if restricting future flexibility. These designers conclude that an analog switching network should be used to share portions of the BORSHT functions over many lines. This debate is far from over.

The No. 4 ESS, long-distance traffic "super switch," can handle 500 000 calls per hour. It began service in Chicago and three other areas in 1976. Photo courtesy Bell Laboratories.

Time-Divided Network—A switching network consists of concentration, distribution, and expansion stages. In a digital switch, blocking can be designed with a degree of freedom not possible in analog switching. The distribution stage is usually strictly or essentially nonblocking. Where should analog line and trunk terminations appear? Some designs place both termination types at the concentrator and expander stages where blocking is introduced. Both types compete for a limited number of links to the distribution stage. Other designs segregate lines and trunks since their traffic occupancy is considerably different. In this case, lines terminate at concentration and expansion stages and trunks are applied at nonblocking terminations at the distribution stage.

A significant design decision lies in the network structure. Many combinations of time and space stages have appeared. Indeed, one of the

most interesting aspects of digital switching lies in the large number of variations in the time-divided network. Since the cost of the network is generally less than 10 percent of the overall switch cost, ease in network control and expandability usually dictate the structure.

Multiplexing Format—One of the most interesting debates in North America involves selection of a multiplexing format. Some designers have chosen the North American PCM standard bit rate and multiplexing format for their time-divided network multiplexing scheme. They argue that a less costly and more efficient T-carrier interface is thereby realized. Others have chosen the CEPT bit rate and multiplexing standard which uses a binary number for both channels and bits per frame. They argue that each element of the network should be individually optimized and that conversion devices from one multiplexing scheme to the other are inexpensive.

The ideal switch should interface either transmission standard with no cost penalty. So the debate rages on.

DIGITAL NETWORKS

Digital switching must be compatible with existing analog and digital facilities as it is introduced into the toll and local network. Digital islands will be formed during the transition from an all-analog to an all-digital network. Today's network standards are based on analog technology, but new standards have been established to accommodate the unique properties of digital technology. Performance standards for a digital network are discussed in this section.

Digital Impairments—Digital networks suffer from three information impairments: errors, misframes, and slips. Objectives have been established in the United States for each impairment [13]. The bit error rate objective is 1 in 10^6 for 95 percent of the calls. The misframe rate objective is 4 per trunk per day. The slip rate objective is one slip per trunk per 20 hours.

The slip rate objective is of particular interest since digital network synchronization requirements follow from this objective. Slips occur when the average frequency of the clock in the switch is slightly different from the incoming carrier bit rate. A slip occurs every time this relative drift changes 125 μs. A slip causes a frame to be repeated or deleted depending on the relative drift direction. If the frequency difference is 50 Hz, the incoming PCM frame will move relative to the switch frame at a rate of 32 μs per second. The time to drift one frame, 125 μs, is 3.9 seconds.

The effect of a slip on various services depends on the service. Speech coded with PCM has a low sensitivity to slips. A slip will produce a noise impulse which is often inaudible. Data signals have significantly less redundancy than speech and are therefore less tolerant to slips. Error detection and retransmission is used to eliminate message errors and modems must resynchronize. Therefore, frequent slips can affect channel throughput. Real-time facsimile can be severely impacted by slips. A slip will displace a line being scanned, and will therefore cause distortion in the picture. In this case, one slip can potentially destroy a whole picture.

Techniques must be established in the network to prevent transmission quality deterioration due to slips. Two approaches have been suggested *plesiochronous* and *synchronous* operation. In a plesiochronous network, clocks at each switch are independent, but their frequency is controlled within tight limits. In a synchronous network, clocks are locked with other clocks in the network. A synchronous network has been established for the United States using a master–slave structure [13].

Transmission Loss—The effects of loss, noise, and echo in a telephone channel are well known [14]. Echo is controlled in an analog network by adding loss in proportion to round-trip delay. This has led to the via net loss (VNL) transmission plan. Echo suppressors are used when the round-trip delay exceeds approximately 45 ms.

In a digital network, there are fewer two-to-four wire conversions (sources of echo) and adding loss in small lumps, which is the current VNL practice, would affect quantizing noise. Therefore, the Fixed Loss Plan has been established for a digital network [14]. This plan calls for a net 6 dB loss for toll-connecting trunks. Except for echo suppressors, 6 dB is the maximum loss inserted in trunks in an all-digital toll network.

The local switching entity is a significant contribution to overall network transmission performance. Unlike its analog counterpart, a digital switch requires a four-wire switching path. Four-wire switching is common in toll switches, but is uncommon in local switching. Echo in a four-wire path must be controlled to acceptable "talker" and "listener" levels. If echo is strong, the communication channel will sound hollow, like talking in a barrel. Still stronger echo will cause oscillations (singing).

Listener option tests have been conducted to determine echo requirements for satisfactory transmission. Preliminary results indicate that between 8 and 10 dB of singing margin[1] is required for satisfactory transmission [15],[16]. Work is continuing to further determine the required singing margin. A digital switch also inserts delay into the four-wire path due to filtering, time slot interchanging, cable propagation, etc. Singing margin as a function of four-wire path delay is shown in Fig. 3 (from [15]).

Loss around a four-wire path is heavily influenced by transhybrid loss.[2] This loss in a local digital switch is a function of subscriber loop impedance and the hybrid termination impedance. For an ideal hybrid, the loss equation has the

---

[1]The gain when inserted in a four-wire path which just sustains oscillations.

[2]The loss between the two four-wire points of a hybrid.

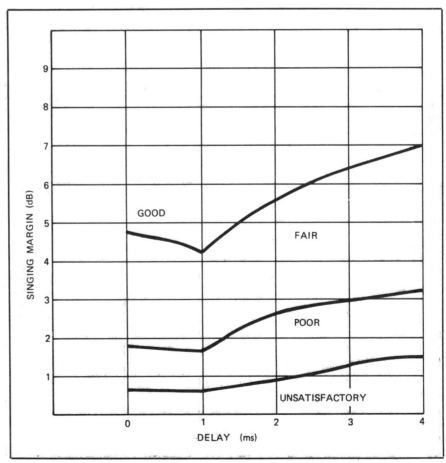

Fig. 3. Subjective evaluation of singing margin as a function of round-trip delay.

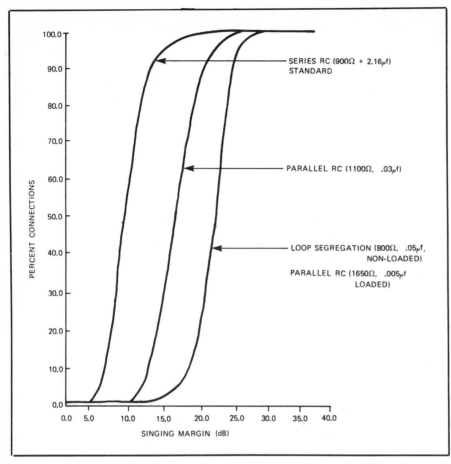

Fig. 4.   Singing margin cumulative distribution function.

mum margin of 8 dB can be established with the standard network by adding 1.5 dB loss (3 dB margin improvement). Loop segregation with new termination values provides a minimum of 13 dB margin with lossless switch operation.

A debate over the proper means to establish adequate singing margin is now in process. Many telephone administrations feel that digital switching with an additional 2 dB loss for intraoffice calls will improve service when compared with existing equipment. They also fear that a facilities segregation scheme would be impossible to administer. Others feel that 0 dB loss must be maintained through any local switch. The impact of adding 1.5 to 2 dB loss for intraoffice calls has been studied, and results are shown in Fig. 5 (from [16]). An improved transmission grade of service is shown for short loops, but a 4 percent degradation is predicted for long loops. These results are currently under additional study.

There are many misleading statements in the literature regarding this issue. Some claim that adding 2 dB loss in the switch is equivalent to

same form as the return loss equation as shown in (2):

$$loss = \frac{Z_T + Z_L}{Z_T - Z_L} \qquad (2)$$

where
    Z_T = hybrid termination
          impedance
    Z_L = subscriber loop impedance.

Listener opinion is plotted versus singing margin and hybrid termination network value for a lossless switch in Fig. 4 (from [16]). Almost 30 percent of the loops tested will not meet an 8 dB margin with the standard hybrid termination network. Therefore, a method must be established to provide additional singing margin.

The techniques currently in use to obtain additional singing margin are: 1) added loss for intraoffice calls with the standard hybrid termination, and 2) subscriber loop segregation into loaded and nonloaded categories with two hybrid terminations. It can be seen from Fig. 4 that a mini-

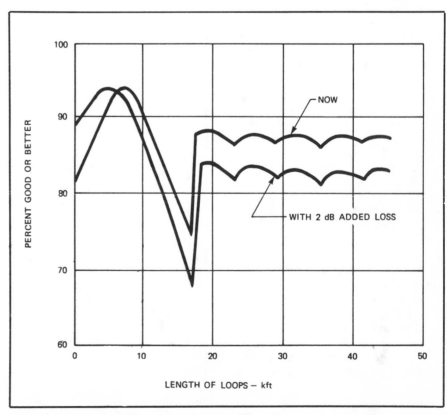

Fig. 5.   Local grade of service with resistance design loops.

adding an extra 2 dB loss in the resistance design of the subscriber loop. They conclude that millions of dollars spent on upgrading the subscriber loop loss from 10 to 8 dB would be lost. This is misleading since loss in a digital switch can be controlled through switchable digital pads. A 2 dB loss can be used for intraoffice calls and 0 dB loss used for interoffice calls. The 2 dB switchable loss is obviously not the same as adding 2 dB loss to the subscriber loop. But the 2 dB pads must be switchable since it has been shown that adding 4 dB loss to the intertoll network would cause an 11 percent reduction in the grade of service for all subscribers. The extra 2 dB switch loss is not required for interoffice trunks since these trunks are designed to have loss as described above. A switchable loss therefore meets the requirements for both intra- and interoffice calls.

SUMMARY

This paper has presented the elements of a digital switch, and current design issues have been summarized. The benefits of digital switching over other available technologies are substantial for many telephone administrations. Resolution of the design issues raised in this paper is not a prerequisite to successful field installation. Digital networks are now being planned where voice is digitized at the subscriber drop by a digital subscriber terminal, is transmitted to a digital central office by a digital repeatered line, is switched to a digital trunk by a digital central office, is transmitted to a digital toll switch over a digital radio or digital cable facility, and is switched by the digital toll switch into the intertoll network. The planning is now complete for an integrated digital network which will set new standards in performance, reduce costs, and provide opportunities for new services.

REFERENCES

[1] H. E. Vaughn, "Research model for time separation integrated communication," *Bell Syst. Tech. J.*, vol. 38, 909, July 1959.

[2] A. E. Pinet, "Introduction of integrated PCM switching in the French telecommunication network," in *Int. Switching Symp. Rec.*, 1972, p. 470.

[3] H. A. Strobel and G. Collins, "The D1201 Digital PABX Delta Modulation Switching System," in *Nat. Telecommun. Conf. Rec.*, vol. I, 1975, p. 1124.

[4] H. E. Vaughn, "An introduction to No. 4 ESS," in *Int. Switching Symp. Rec.*, 1972, p. 19.

[5] J. C. McDonald and J. R. Baichtal, "A new integrated digital switching system," in *Nat. Telecommun. Conf. Rec.*, 1976, p. 3.2-1.

[6] T. H. McKinney and H. W. Stewart, "Digital central office hardware architecture," in *Nat. Telecommun. Conf. Rec.*, 1977, p. 15.4-1.

[7] J. C. McDonald, "Glossary of digital transmission and switching terms," in *Nat. Electron. Conf. Rec.*, 1977, p. 43.

[8] A. Gersho, "Quantization," *IEEE Communications Society Magazine*, vol. 15, p. 16–29, Sept. 1977. See also *Transmission Systems for Communication*, ch. 25, p. 566. Bell Laboratories, Inc., 1970.

[9] J. R. Sergo, Jr., "DSS quad line circuit," in *Int. Symp. on Subscriber Loops and Services Rec.*, 1978, p. 182.

[10] A. W. Kobylar, "Methodology for isolating a set of near optimum PCM digital network configurations," in *Int. Conf. Commun. Rec.*, 1974, p. 34E-1.

[11] S. G. Pitroda, "Selection of an optimum digital PCM switching configuration based on a set of system considerations," in *Int. Conf. Commun Rec.*, 1974, p. 34D-1.

[12] G. White, "Options and progress in digital local switching," in *Proc. Int. Zurich Seminar on Digital Commun.*, 1978, p. A2.1.

[13] J. E. Abate *et al.*, "The switched digital network plan," *Bell Syst. Tech. J.*, vol. 56, p. 1297, Sept. 1977.

[14] L. S. DiBiaso, "Transmission considerations for local switched digital network," *Telephony*, p. 40, Oct. 24, 1977.

[15] R. L. Bunker *et al.*, "Line matching networks to support zero loss operation in digital class 5 offices," in *Int. Symp. on Subscriber Loops and Services Rec.*, 1978, p. 166.

[16] Minutes of USITA/ATT Equipment Compatibility Committee, Meeting MFL-76-12, Dec. 9, 1976, *USITA*, Washington, DC.

John C. McDonald (S'56–M'59–SM'71) is Vice President for Research and Engineering at the Vidar Division of TRW, Inc. He received the Bachelor of Science degree in Electrical Engineering with honors from Stanford University in 1957. He continued his studies while teaching at Stanford and received the Master of Science degree in 1959. He returned to Stanford and received the degree of Engineer in 1964. His thesis research was in the field of solid-state physics.

Mr. McDonald is a member of Tau Beta Pi and Sigma Xi.

INTEGRATED CIRCUITS FOR LOCAL DIGITAL SWITCHING LINE INTERFACES *

Paul R. Gray and David G. Messerschmitt

Technological advances make exploitation of the most advanced ideas in telecommunications systems economically feasible.

The technology of telephony is undeniably evolving at the fastest rate in its history. This is due only in part to current conceptual advances in communication theory and technique. In fact, many of the system concepts currently being implemented were invented and understood to have many advantages two decades ago or more, but were not commercially developed at that time because they were too complicated to be realized economically or reliably. The primary driving force behind the current rapid evolution in telecommunications is integrated circuit technology, which has increased by orders of magnitude the system complexity which can be realized economically. As a direct result of integrated circuitry, our ability to economically implement telecommunications systems is finally catching up with our most advanced ideas as to how they should be realized.

There is no better example of the foregoing than integrated digital transmission and switching, which is one of the strongest trends in the current evolution of telephony. The technical advantages of digital pulse-code modulation (PCM) were understood by communications engineers in the 1940's [1], but the first PCM system was not commercially available until 1962 [2]. Digital switching of PCM streams was in an experimental state in the 1950's [3], but was not realized until the early 1970's [4]. In both cases, the delay from conception to realization was due to a lack of available devices to make exploitation sufficiently economical and reliable.

In the more recent case of digital switching, the first application was to local switches (interconnecting subscribers) in France [4], and the first application in the United States was to toll switches [5] (interconnecting other switches). The toll application is easier to prove-in economically, since the trunk interface is relatively simple and economical as compared to the interface to a subscriber line.

To illustrate this fact, Fig. 1 shows a simplified version of an analog telephone connection in which the long-haul toll switching is done on a four-wire basis (two directions of transmission are separated on different facilities). On the other hand, the local switch and short subscriber line are two-wire (both directions share the same wire pair) to save copper. Conversion from two- to four-wire is performed by a device called a hybrid, to be discussed later. Replacement of this analog toll switch with a digital switch simply involves A/D and D/A conversion at the switch interface in the case of an analog transmission trunk, and more complicated but nevertheless inexpensive functions such as synchronization and framing in the case of a digital transmission trunk.

Replacement of the local switch by one which switches signals in a digital PCM format is a more radical change, as shown in Fig. 2. One reason is that a digital switch must inherently separate the two directions of trans-

The authors are with the Department of Electrical Engineering and Computer Sciences, University of California, Berkeley, CA 94720.

*Reprinted from *IEEE Communications Society Magazine*, May 1980, Vol. 18, No. 3, pp. 12–23.

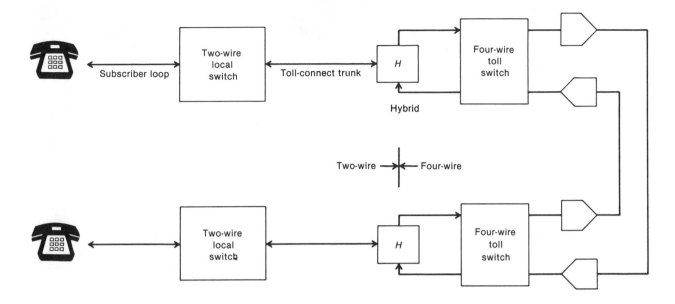

Fig. 1. Typical analog telephone connection.

mission, that is, it is four-wire. Thus, the hybrid moves to the subscriber side of the local switch, and must be implemented on a per subscriber line basis. A second reason is that many interface functions, such as ringing and testing, are needed to interface a subscriber line. These involve high voltages and cannot be performed by a small number of service circuits operating through the switch network itself, as was the case with metallic cross-point analog switches.

A simplified block diagram of the transmission interface between a local digital switch and a subscriber line is

shown in Fig. 3. The line interface performs the hybrid function, provides dc power to the phone, and other functions. The transmit low-pass filter prevents aliasing distortion resulting from any frequency components above 4 kHz, half the subsequent sampling rate of 8 kHz. The A/D converter then converts individual samples to the digital format used internally in the switch. On the receive side, the D/A converter generates analog samples and the low-pass filter reconstructs the analog speech waveform. The A/D and D/A converters together are called a codec (for coder-decoder).

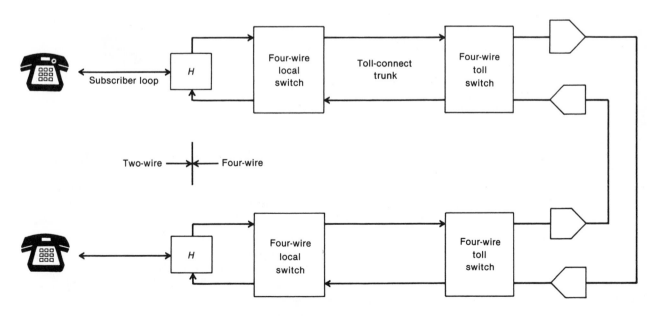

Fig. 2. Typical digital telephone connection.

Virtually none of the functions in Fig. 3 is required at the subscriber-line interface to a metallic crosspoint analog switch, since these functions are associated with A/D conversion and four-wire to two-wire conversion. Therefore, they must be implemented very inexpensively to result in an economically competitive switch. Fortunately, there are offsetting savings in the local digital switch. The switching function itself is more economically realized, and the interface to digital PCM transmission trunks and subscriber line multiplex systems are much cheaper. Because these other functions are so cheap, the total cost of the digital switch is, in fact, dominated by the cost of the functions shown in Fig. 3, together with the related functions of battery feed, protection, and ringing and test access.

Because the functions on Fig. 3 must be realized inexpensively, and because they are replicated on every subscriber line[1] and therefore will be produced in high volume, they are a natural application for special-purpose integrated circuits. This is being pursued aggressively by the vertically integrated telephone equipment manufacturers, and because of the potentially vast market, by many independent semiconductor manufacturers.

The purpose of this paper is to review the functional and interface requirements, as well as circuit and technology alternatives for integrated circuit codecs, filters, and line interfaces.

_____

[1]Another approach to reducing the cost of these functions, an analog crosspoint concentrator switch between subscriber lines and the digital switch line interfaces, will not be discussed here.

CODEC REQUIREMENTS

The codecs utilized in telephony employ nonuniform quantization; that is, their step size increases as the magnitude of the signal sample increases. The so-called $\mu = 255$ law [6] is standard in North America and Japan; Europe uses a slightly different standard, the A-law [6]. These laws provide a wide dynamic range over which the signal-to-quantization noise ratio is approximately constant and a very small step size near the origin.

The $\mu = 255$ coder generates an 8-bit sample, so that 256 intervals of signal level must be distinguished. The total signal range is first divided into 16 intervals (called chords), 8 for positive signals and 8 for negative. These chords double in length as they move away from the origin, as illustrated in Fig. 4. Finally, each chord is divided into 16 equal intervals to yield the 256 total intervals. There is one sign bit, three bits to specify one of eight chords, and four bits to specify the interval (out of 16) on a chord.

There are four types of requirements which are relevant to codec design:
1) idle channel noise
2) crosstalk coupling
3) signal-to-quantizing noise ratio (SNR)
4) gain tracking.

The first two items relate to the overall system design, rather than the coder specifically. However, relative to the design of the coder, the idle channel noise specification cannot be reliably met unless the very small step size on the lowest chord is reasonably accurate. Further, the crosstalk requirements will not be reliably achieved unless the bias of the coder for an idle channel is accurately maintained. For this reason, so-called auto

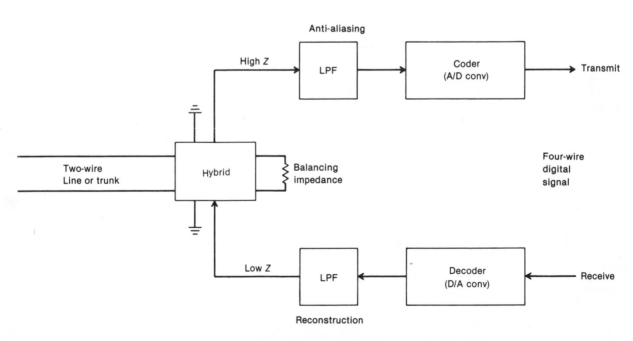

Fig. 3. Block diagram of a line interface circuit.

zero-set circuitry is often incorporated in the codec to maintain accurate idle channel bias. It works by adjusting the bias in a feedback loop to force a positive and negative sign bit with equal frequency.

The last two requirements relate to the accuracy of the A/D thresholds and D/A outputs. They are specified by simple sine-wave measurements, which are readily reproduced in both a laboratory as well as field environment. SNR is measured by transmitting a sine wave (usually at 1 kHz) and measuring the total noise at the output (sine-wave notch filtered out). Gain tracking refers to the tracking between the input and output signal levels as the input level is varied.

INTEGRATED CIRCUIT CODECS

Monolithic voice encoder/decoders (codecs) have become commercially available from a number of inte-

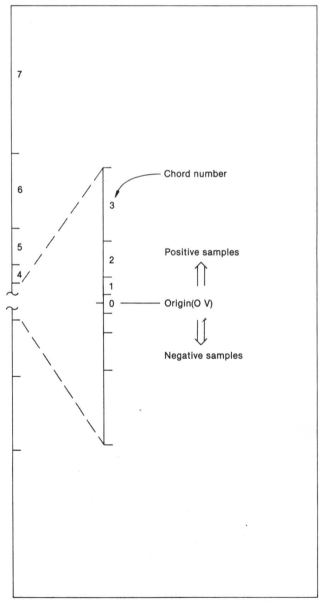

Fig. 4.   μ255 chord structure.

NMOS, CMOS, AND I²L INTEGRATED CIRCUITS

In the early days of IC's, prior to the late 1960's, bipolar technology was supreme. The integrated bipolar transistor, an extension of the discrete transistor, consists of a three-layer sandwich of emitter, base, and collector formed by diffusion of impurities into a silicon substrate. This technology has a relatively high speed of operation because that speed is determined by the thickness of the base, which is controlled by the diffusion rather than the relatively crude (by today's standards) optical photomasking.

Bipolar technology consumes a relatively large die area because of the need for isolation regions around each transistor. MOS technology was developed for digital logic, memory, and microprocessor chips because the MOS transistor does not require isolation, and hence consumes less die area. In an MOS transistor, a current parallel to the silicon surface is controlled by the voltage on a metal electrode (gate). Unfortunately, the speed of an MOS transistor is dependent on the dimensions as defined by the photomask, and hence MOS was initially a slower technology than bipolar. However, the gap is narrowing rapidly with advances in mask resolution.

Not to be outdone, the bipolar proponents developed I²L, in which the bipolar transistor is inverted, eliminating the need for isolation in a common-emitter configuration. The result is a bipolar compatible digital technology with densities close to MOS.

An important variation on MOS is complementary MOS (CMOS). By putting transistors of both polarities on the same chip, power dissipation can be eliminated except during logic transitions, resulting in very low overall power dissipation. However, the die area is about double that of MOS, all else being equal.

For logic, memory, and microprocessor applications, NMOS is the technology of choice for high density, CMOS for low power dissipation, and bipolar for speed. The gap between NMOS and bipolar in speed is rapidly narrowing. For analog functions, bipolar was pervasive until recently. NMOS and CMOS analog circuit elements such as operational amplifiers are now viable and, as explained in the text, for analog filtering functions, NMOS and CMOS are the technologies of choice. In analog applications not requiring filtering, bipolar and I²L are still the most widely used.

grated circuit manufacturers in the past two years. The recent heavy product development activity in this area has resulted from two key developments. The first is rapid progress in the design and manufacture of LSI circuits, such as codecs, which contain both analog and digital circuitry on the same chip. Previously, the most economical realization of such complex functions was to partition them into a digital part realized with MOS technology and an analog part realized with bipolar technology. The resulting inability to fully integrate the codec in one monolithic chip meant that manufacturing costs were too high to be competitive in digital switches.

To realize the codec function in a single chip, some manufacturers utilize CMOS or NMOS technology with sophisticated circuit design techniques to realize the analog functions in this digitally oriented technology, while others add high-density I²L LSI logic to their bipolar analog technology.

The second development which has made monolithic codecs possible is circuit realizations which are relatively insensitive to component mismatches in the passive elements, making it possible to achieve the required transmission performance without any trimming or other adjustment of internal components.

Beyond these common factors, the commercially available devices vary widely in their technological and circuit implementation, as well as in the peripheral features provided. In the following discussion, we survey the circuit techniques and technology used in these devices.

BASIC CODEC ARCHITECTURES

Perhaps the most fundamental architectural feature of the codec is whether one codec is used for each subscriber line as in Fig. 3, or a high-speed coder is shared over many lines using an analog multiplexer. Shared coders result in hardware savings, but present difficult design problems in a completely monolithic realization, such as difficulty in achieving a low level of line-to-line crosstalk and in achieving the high-speed operation required. As a result, shared codecs tend to be implemented on several chips, which reduces but does not negate their potential economic advantages. A second drawback of shared coders is the larger number of lines lost upon a failure (although reasonable reliability requirements can still be met). A third drawback is that analog time-division multiplexing is more difficult and less flexible than digital multiplexing. Despite these drawbacks, shared coders will continue to receive attention in the future due to their economic advantage.

Most of the new product development effort in the recent past has been devoted to per-channel coders, based on the belief that a standard per-line coder can be manufactured in higher volume, resulting in a cost as low as for the shared approach. Although many algorithms have been proposed, the pervasive monolithic codec realization in the devices produced commercially is the traditional successive-approximation technique illus-

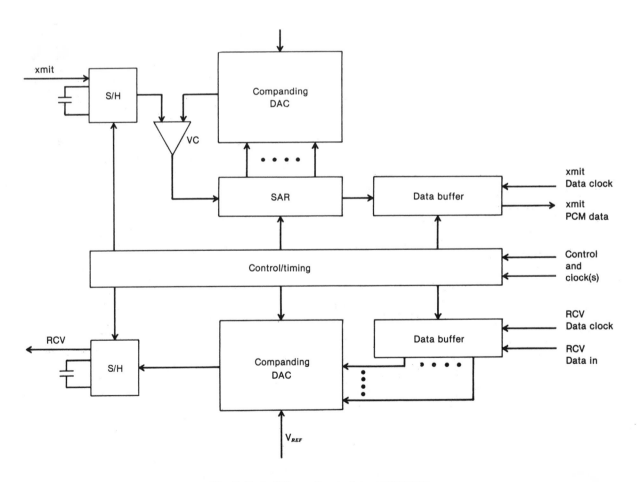

Fig. 5. Basic S/A per line codec components.

trated in Fig. 5. In the encode direction, the four-wire signal from the subscriber line, hybrid, and anti-aliasing low-pass filter (Fig. 3) is sampled in the input sample/hold amplifier, and encoded using the companding DAC, voltage comparator, and successive-approximation register. The resulting data are loaded in the transmit data buffer and subsequently shifted out serially under the control of the transmit data clock. In the receive direction, the data to be decoded are shifted in serially under the control of the receive data clock, and subsequently the decode D/A converter produces an analog voltage which is held in the output sample/hold amplifier. This signal is then connected to the reconstruction low-pass filter (Fig. 3). In some codecs, a single DAC is time shared between the encode and decode directions (which requires complex logic in asynchronous applications), and in other codecs two DAC's are used (eliminating this logic and the decoding sample/hold function).

The principle analog components in the codec are the D/A converter, the sample/hold amplifier, and the comparator. The implementation of the remaining logic circuitry is straightforward in NMOS, CMOS, or I²L technology. The different commercially available codecs are distinct primarily in the technique and technology used to implement the DAC. We now discuss briefly the three primary approaches to the realization of DAC's: current switching, resistor string, and charge redistribution.

DAC IMPLEMENTATION

In the current switching DAC, binary weighted current sources are routed either to the output or to ground through current switches under the control of the digital input to the DAC, as shown in Fig. 6. This technique, which has been used for many years in uniform step-size codecs (monolithic, hybrid and discrete), is well suited to bipolar technology. An important advance in the development of monolithic current-switched codecs was the development of circuit techniques in which the transmission properties of a nonuniform step-size codec are relatively insensitive to the mismatch in the resistors controlling the current sources. The current switched DAC is capable of high-speed operation, but also requires a relatively large silicon area. It has been implemented in both I²L [7] and CMOS [8] technologies.

Much of the early work on codecs was directed towards realizing the DAC in MOS technology so that the dense MOS logic could be used for the digital portions. For the implementation of analog circuits, the MOS transistor is inherently more suited to voltage and charge switching than is the bipolar transistor, has essentially infinite input impedance when used as an amplifier, and MOS technology inherently produces capacitors of very high linearity and stability. These capabilities have led to the use of two nontraditional techniques to implement DAC's in MOS technology: resistor string voltage

switched DAC's and capacitive charge-redistribution DAC's.

An example of a resistor string DAC is shown in Fig. 7 [9], [10]. In this case, the voltage at each junction of the resistor ladder is one-half that at the junction above it. Each of the resistors is divided into 16 segments, and the taps between these segments are connected to a tree of MOS transistors, through which any of the 256 taps of the string can be connected to the ouput.

A second MOS-compatible approach, the charge-redistribution codec [11]-[14], utilizes capacitors, rather than resistors, as precision elements. A simplified schematic diagram of a charge-redistribution coder is shown in Fig. 8. The input voltage is first stored on the top array of capacitors (hence, this type of coder has an inherent sample-and-hold capability), and by shuffling charges around among the capacitors using the MOS switches, the successive bits are generated in eight comparison/decision cycles.

The primary advantages of this type of coder are the inherent sample/hold capability, the low power dissipation because no current flows in the array, and good behavior through the origin. A disadvantage is that two DAC's are required for a complete codec.

FILTERS

Both the transmit and receive filters are basically low-pass filters which reject frequencies above 3 kHz. However, the transmit filter must also attenuate any 60 Hz input component (such as could be induced into the subscriber loop from nearby power lines), and thus becomes in reality a bandpass filter from 300 to 3000 Hz. As with most filters the performance specifications are in terms of passband and stopband attenuation. In this case, the passband ripple specification is quite tight because of concern about transmission quality when a number of these filters are placed in tandem (they appear in PCM digital transmission systems as well). The stopband attenuation requirements relate to frequencies above 4 kHz, which would alias back in-band due to sampling.

In addition to attenuation requirements, the filters will also contribute to crosstalk and idle channel noise. If more than one filter is put in a single monolithic device, care must be exercised to ensure a large crosstalk attenuation between them and little crosstalk coupling through the power supply.

One additional filter requirement relates to the phase response, which is important in voiceband data transmission. The phase response must meet certain linearity requirements, which are usually specified by PAR (peak-to-average ratio) [15].

MONOLITHIC FILTERS

Great progress has been made in the past several years in the area of monolithic analog filters, which as late

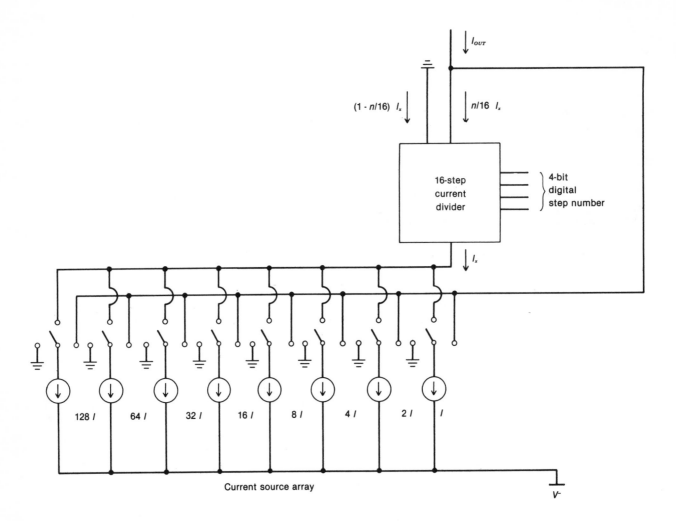

Current source array

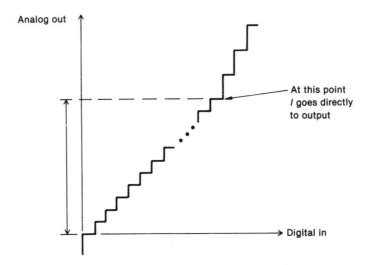

Fig. 6.   Current-switched companding DAC.

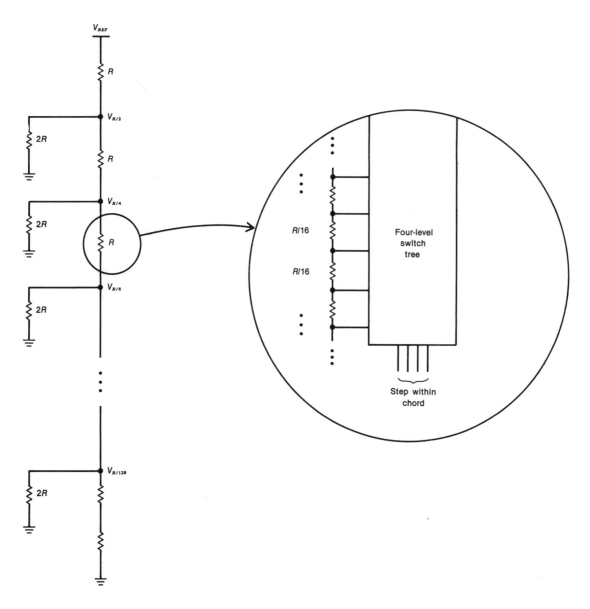

Fig. 7.   Resistor-string companding DAC's.

as 1975 did not appear to be feasible. CCD (charge-coupled device) transversal filters had been under development since 1969, but since they were NMOS devices, there was no way to include the peripheral signal extraction circuitry and a number of peripheral operational amplifiers, etc., were required to implement a complete filter function.

Three developments have changed this picture. First, techniques have evolved for the design of NMOS operational amplifiers [16], [17] and other analog functions, so that these can now be included on the CCD chip in a completely self-contained filter. Second, two new monolithic filtering techniques evolved, including the MOS-compatible switched capacitor filter and the bipolar-

compatible frequency-locked filter [18]. The latter appears to be significantly inferior to CCD and switched capacitor filters in terms of die area and so will not be discussed further here.

We next review the basic principles of operation of CCD and switched capacitor filters.

CCD TRANSVERSAL FILTERS

A CCD consists of a series of potential wells under which charges can be moved under the control of electrodes on the surface of the silicon. The device can realize a transversal filter function by injecting charge packets at one end proportional to the input signal, and

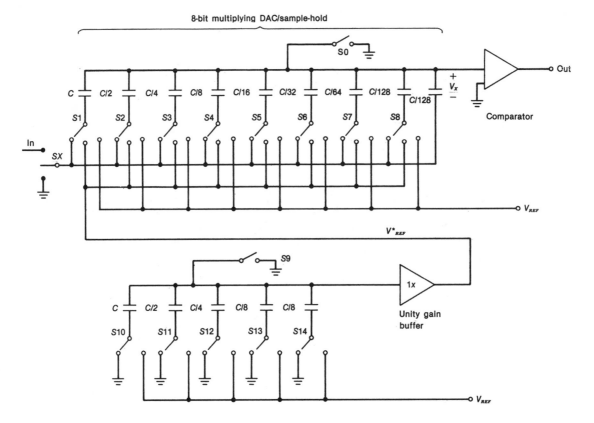

Fig. 8.   Charge redistribution coder.

using surface electrodes along the channel to sense the charges and develop a weighted sum of the delayed samples of the input signal.

An important problem in CCD filters has been the fact that the output signal sensing amplifier had to sense small differences in charge output in the presence of large common-mode charges, resulting in severe requirements on the MOS-compatible sensing amplifier and an inadequate dynamic range. This problem has been solved using the double-split electrode structure [19] in which a large center section of the CCD electrodes is electrically grounded. This plus the development of NMOS operational amplifiers for on-chip signal extraction circuitry has made the self-contained CCD filter a viable competitor to the traditional hybrid and discrete component approaches.

SWITCHED CAPACITOR FILTERS

The switched capacitor filter utilizes a small capacitor switched at a high rate to emulate the behavior of a resistor, and permits most classical active RC filtering techniques to be realized in completely monolithic form.

The basic concept of the switched capacitor "resistor" is illustrated in Fig. 9 [20], [21]. If the capacitor is first charged to voltage $V1$, and subsequently discharged to voltage $V2$, then an amount of charge equal to $C(V1-V2)$

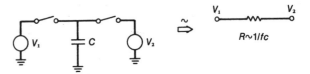

Fig. 9.   Switched-capacitor resistor.

will have flowed from source $V1$ to source $V2$. If the switching operation is performed at a rate f, then the average current is simply $fC(V1-V2)$. Thus, the switched capacitor behaves like a resistance $1/fC$. The delay inherent in the switching process must also be taken into account in the detailed design process.

The application of the concept in a simple single-time-constant circuit is illustrated in Fig. 10. Here the continuous resistor is replaced by the switched capacitor resistor to yield a single-pole filter whose cutoff frequency depends only on a capacitor ratio and the externally supplied clock frequency rather than the RC product as in the continuous case. Capacitor ratios can be accurately controlled in MOS technology without trimming or other adjustments.

The transmission performance requirements on PCM filters generally dictate that they be realized with a five-pole, four-zero elliptic configuration. A doubly termi-

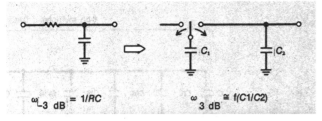

Fig. 10. Switched-capacitor filter elements.

nated passive *LC* network realization has excellent sensitivity properties and can be transformed into an active ladder equivalent [22], [23] and then to a switched capacitor equivalent (by simply replacing the analog integrators with switched capacitor integrators).

Switched-capacitor filters which are commercially available [24] include both transmit and receive filters, internal clock generation, continuous anti-aliasing filters, the 50/60 Hz high-pass hum rejection filter, and a transformer driver to interface directly with a hybrid transformer.

LINE INTERFACES

The seven primary functions which a line interface must provide are

1) battery feed (powering the telephone)
2) line supervision (determining when the telephone is off-hook)
3) ringing access
4) test access
5) overvoltage protection
6) two-wire to four-wire conversion (hybrid)
7) balanced to unbalanced transmission conversion (for immunity to noise on the subscriber loop).

Items 3) and 4) are generally provided by interrupting the line with a relay. The battery feed circuit must provide a relatively low impedance at dc and simultaneously a high impedance at voice frequencies to avoid shorting out the signal path (traditionally done with a large series inductance). Supervision is defined as the detection of dc current flowing in the loop, indicating an off-hook telephone (and also used to detect dial pulses, which consist of a periodic interruption of dc continuity). Overvoltages which must be protected against include induced 60 Hz from power lines and the occasional power line cross or lightning strike (limited by external protective devices to about 1000 V).

The purpose of the hybrid can be seen in Fig. 11. The four-wire path constitutes a closed feedback path through the two hybrids on each end of a connection. The hybrid must present a high loss from the four-wire port to four-wire port, thereby preventing oscillation and the two types of echo shown, while simultaneously presenting a low loss on the talker speech path from the two-wire port to both four-wire ports. It can be implemented with transformer windings or with a single can-

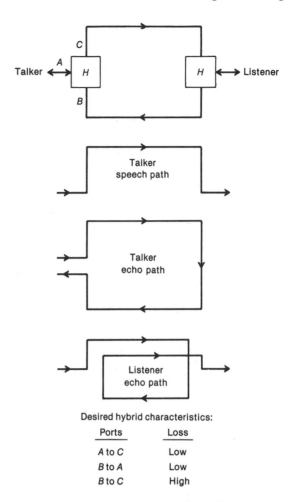

Desired hybrid characteristics:

Ports	Loss
A to C	Low
B to A	Low
B to C	High

Fig. 11. Hybrid and echo paths.

celling filter, and in both cases requires knowledge of the subscriber loop impedance to achieve balance.

One problem which has arisen in local digital switching is that when the switch desirably has no insertion loss between two-wire ports, the loss around the four-wire path is not always adequate when a single compromise balance impedance is used [25], [26]. However, studies have shown that two balance impedances, one for each of two common types of subscriber loops, would be adequate. Techniques for selecting the appropriate balance network automatically are being developed.

The final line interface requirement is the balanced-to-unbalanced conversion. Induced 60 Hz, noise, and crosstalk from the subscriber loop are common mode signals, appearing equally on both wires (in telephony, a common-mode signal is called a longitudinal signal). Thus, by inputting to the filter only the difference between the voltages on the two wires (called the metallic signal), these unwanted longitudinal signals are rejected.

MONOLITHIC LINE INTERFACE CIRCUITS

Considerable effort is currently underway in the semiconductor industry and at telephone industry laboratories to provide a low-cost all-silicon replacement for the line interface function which is now performed with a transformer. Parallel efforts are underway to provide the same function at lower cost by using an LSI device in conjunction with an inexpensive transformer.

The basic functional configuration for an all-silicon approach is shown in block diagram form in Fig. 12. The transmit and receive four-wire signals to and from the filter are shown on the left, and the two leads of the subscriber line are shown on the right. The two other essential connections are battery ground and the −48 to −56 V battery voltage. The overriding consideration is that voltage surges on the order of 1000 V can occur, but the maximum voltage difference that can be accommodated between any two leads on integrated circuits with low-cost production technologies available today is about 100 V. This means that an external series element must be placed in the subscriber line leads to provide the remaining 900 V or so of voltage drop, and some form of diode clamp must be provided to clamp the terminals of the IC so that they do not exceed the battery voltage. Perhaps the most straightforward way of achieving this is shown in Fig. 12.

Once this basic configuration is adopted, a second problem is that normal operation must be maintained in the presence of large common-mode (longitudinal) voltages. One possibility would be to have the T' and R' terminals of the IC behave like a floating voltage source for common-mode signals so that no current would flow as a result of these common-mode signals. Unfortunately, this would cause the protection diodes to forward bias, making normal operation impossible. The only remaining option is to make the line circuit look like a short circuit to common-mode signals, so that the T' and R' terminals maintain a constant dc voltage with respect to each other and with respect to ground when the common-mode signals are applied. This implies large longitudinal currents that must be absorbed in the monolithic circuit, resulting in substantially increased power dissipation as compared to a transformer interface.

In addition, the line interface must provide the basically inductive impedance variation with frequency (low impedance at dc and high impedance at voice frequencies), which implies internal active circuitry. This active circuitry can take two forms: current drive with voltage sensing, and voltage drive with current sensing. In the current drive configuration, two controlled equal current sources are used to drive the line. The voltage across the line is sensed by a differential amplifier, and a control circuit is used to provide an active feedback path to give the desired relation between line voltage and line current. Since the desired ac and dc impedances are different, the control function contains a frequency selective filter. The advantage of this circuit is that it appears to be possible to implement it rather simply using a current mirror [27]. The basic voltage drive configuration, which requires less power dissipation in the presence of large longitudinal currents, consists of two voltage amplifiers connected as a bridge output amplifier. The loop current is sensed by two series resistors and converted to a single-ended signal. This signal is then used to make the loop voltage have the desired dependence on loop current so as to achieve the correct impedance.

The hybrid function can be realized without a transformer by inserting a voltage divider to model the voltage divider formed by the output impedance of the line interface unit and the impedance presented by the line, and using the output of this voltage divider to cancel the unwanted feedthrough from the four-wire path to the four-wire path. Typically the required balance network is a very simple RC network, which can conveniently be realized using the switched capacitor techniques described earlier.

Several manufacturers have under development fully integrated line interface circuitry which utilizes current drive with voltage sensing with either external or internal high current transistors to realize the current sources.

Perhaps the most difficult problem facing the solid-state line circuits is reliability. The devices will have to operate in an environment where they are thermally cycled, experience high voltage, and experience large surges in current and voltage. Whether or not such devices can achieve the stringent reliability requirements under such conditions remains to be seen. In the next few years, it should become apparent whether the optimum solution to the problem is an all-silicon line interface or one which incorporates a transformer.

As can now be better appreciated, the transformer is a simple device beautifully suited to the task of providing simultaneously isolation from foreign potentials, the correct impedance levels, battery feed, and longitudinal signal rejection. Its primary problem is that, in the traditional configuration, it must carry the large dc battery feed current, which requires a bulky and expensive core to avoid magnetic saturation. However, at least two methods have been investigated to essentially eliminate this dc magnetic flux, yielding a smaller and less expensive transformer. One method is to drive a secondary winding with a current source so as to null the dc flux.

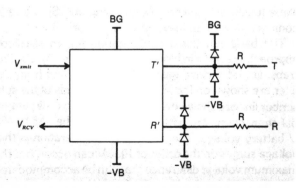

Fig. 12.   Basic considerations in SLIC's.

The primary current can be sensed (partially negating the isolation advantages of the transformer) or the magnetic flux can be sensed directly, as with a Hall effect device. The disadvantage of this approach, its additional power consumption, is overcome by the other approach, which is to use a floating battery feed circuit on the primary side of the transformer, with a capacitor blocking the current from flowing in the transformer itself [28]. Because the circuit is at a floating potential with respect to ground, it is inherently insusceptible to longitudinal foreign potentials, and longitudinal balance is inherently very good. It can also be designed to have a lower power dissipation than traditional battery feed circuits.

CONCLUSIONS

It is clear that the devices which have been described here have already made a major impact in the telephone industry. The prime example is local digital switching, which is economical largely as a result of these devices, and is being actively pursued by every major telephone switching equipment manufacturer.

FUTURE TRENDS IN TELECOM IC's

Advances in both circuit techniques and in technology will result in the combination of the codec and filter functions on a single chip within the next two years, substantially reducing the cost of these functions. In fact, one such combination chip has reached the production stage [19]. The first of these combination chips will almost certainly utilize NMOS technology because of the higher density achievable in the digital portions of the circuit. Beyond this development, higher circuit density may allow the inclusion of the filtering and coding functions for two, four, or eight lines on one chip.

Another significant development is the heavy emphasis in the semiconductor industry at the present time on the development of high-density silicon-gate CMOS LSI technologies for low-power microprocessors and memories. The result will likely be production CMOS technologies which are very dense and can be applied to telecommunications products to narrow the gap in cost between CMOS and NMOS implementations of codecs and filters. Because of the importance of power dissipation in many system applications, this development could result in a long-term shift towards CMOS for these types of products.

In the line interface area, the next few years should make it clear whether an all-solid state interface or one involving a transformer is more cost effective in PBX and/or central office applications.

For the longer term, it is clear that there are other opportunities for application of special-purpose integrated circuits in telephony. Two which are already looming on the horizon are digital transmission on the subscriber loop all the way to the customer premises, and techniques for the reduction of bit rate for speech transmission, of interest in subscriber loop and long-haul applications. Digital subscriber loop transmission requires full-duplex 64 kbit/s data transmission on a pair of wires, and will presumably require measures for compensating for a wide range of line characteristics automatically. Putting codec and filter at the subscriber instrument requires low power dissipation (suggesting CMOS) in order to be powered over the loop, but also will relax their transmission performance requirements (for example, the anti-aliasing low-pass filter will probably be much simpler because there is no crosstalk and reduced noise). Bit rate reduction for speech would reduce the 64 kbits/s, which is difficult on longer subscriber loops, and would basically be an extension of the present codec technology.

The only limitation in the use of LSI components is their availability. At present, the demand for the components is greater than their supply.

REFERENCES

[1] B.M. Oliver, J.R. Pierce, and C.E. Shannon, "Philosophy of PCM," *Proc. IRE*, vol. 36, p. 1324, 1948.

[2] D.F. Hoth, "T1 carrier system," *Bell Lab. Rec.*, vol. 40, p. 358, 1962.

[3] H.E. Vaughn, "Research model for time separation integrated communications," *Bell Syst. Tech. J.*, vol. 38, p. 909, July 1959.

[4] J. Bourbao and J.B. Jacob, "New development in E-10 digital switching systems," in *Int. Switching Symp. Rec.*, 1976, p. 421.

[5] H.E. Vaughn, "An introduction to No. 4 ESS," in *Int. Switching Symp. Rec.*, 1972, p. 19.

[6] H. Kaneko, "A unified formulation of segment companding laws and synthesis codecs and digital compandors," *Bell Syst. Tech. J.*, vol. 49, p. 1555, Sept. 1970.

[7] R.A. Blauschil *et al.*, "A single chip I²L PCM codec," *IEEE J. Solid-State Circuits*, vol. SC-14, Feb. 1979.

[8] S. Kelley and D. Ulmer, "A single-chip CMOS PCM codec," *IEEE J. Solid-State Circuits*, vol. SC-14, Feb. 1979.

[9] A.J. Cecil *et al.*, "A two-chip PCM codec for per-channel applications," in *Dig. Tech. Papers, 1979 Int. Solid-State Circuits Conf.*, San Francisco, CA, Feb. 1979.

[10] M.E. Hoff, J. Huggins, and B.M. Warren, "An NMOS telephone codec for transmission and switching applications," *IEEE J. Solid-State Circuits*, vol. SC-24, Feb. 1979.

[11] Y.P. Tsividis, P.R. Gray, D.A. Hodges, and J. Chacko, "A segmented μ-255 law PCM voice encoder using NMOS technology," *IEEE J. Solid-State Circuits*, vol. SC-10, Dec. 1975.

[12] G.F. Landsburgh and G.F. Smarandious, "A two-chip PCM codec," in *Dig. Tech. Papers, 1978 Int. Solid-State Circuits Conf.*, Feb. 1978.

[13] J.T. Daves *et al.*, "A PCM voice codec with on-chip filters," *IEEE J. Solid-State Circuits*, vol. SC-14, Feb. 1979.

[14] K.B. Ohri and M.J. Callahan, "Integrated PCM codec," *IEEE J. Solid-State Circuits*, vol. SC-14, Feb. 1979.

[15] L.W. Campbell, "The PAR meter: Characteristics of a new voiceband rating system," *IEEE Trans. Commun. Technol.*, vol. COM-18, p. 147, Apr. 1970.

[16] Y.P. Tsividis and P.R. Gray, "An integrated NMOS operational amplifier with internal compensation," *IEEE J. Solid-State Circuits*, vol. SC-11, Dec. 1976.

[17] D. Senderowics, D.A. Hodges, and P.R. Gray, "High-performance NMOS operational amplifier," *IEEE J. Solid-State Circuits*, vol. SC-13, Dec. 1978.

[18] K.S. Tan and P.R. Gray, "Fully integrated analog filters using bipolar JFET technology," *IEEE J. Solid-State Circuits*, vol. SC-13, Dec. 1978.

[19] A.A. Inrahim, G.J. Hupe, and T.G. Foxall, "Double split electrode transversal filter for telecommunications applications," *IEEE J. Solid-State Circuits*, vol. SC-14, Feb. 1979.

[20] B.J. Hostica, R.W. Brodersen, and P.R. Gray, "MOS sampled data recursive filters using state switched capacitor integrators," *IEEE J. Solid-State Circuits*, vol. SC-12, Dec. 1977.

[21] J.T. Caves, M. Copeland, C.F. Rahim, and S.D. Rosenbaum, "Sampled data analog filtering using switched capacitors as resistors equivalents," *IEEE J. Solid-State Circuits*, vol. SC-12, Dec. 1977.

[22] D.J. Allstott, R.W. Brodersen, and P.R. Gray, "MOS switched capacitor ladder filters," *IEEE J. Solid-State Circuits*, vol. SC-14, Dec. 1978.

[23] G.M. Jacobs, D.J. Allstott, R.W. Brodersen, and P.R. Gray, "Design techniques for MOS switched capacitor ladder filters," *IEEE Trans. Circuits Syst.*, Dec. 1978.

[24] P.R. Gray, D. Senderowics, H. Ohara, and B.W. Warren, "A single chip NMOS dual channel filter for PCM telephony applications," in *Dig. Tech. Papers, 1979 Int. Solid-State Circuits Conf.*, Feb. 1979.

[25] L.S. DiBiaso, "Transmission considerations for the local switched digital network," *Telephony*, p. 40, Oct. 23, 1977.

[26] R.L. Bunker, F.J. Scida, and R.P. McCabe, "Line matching networks to support zero-loss operation in digital class 5 offices," presented at the Int. Conf. on Subscriber Loops and Syst., Atlanta, GA, Mar. 1978.

[27] F. Boxall, "Hybrid circuit," U.S. Patent 4 004 109, Jan. 18, 1977.

[28] H.E. Mussman and D.F. Smith, "Design techniques which reduce the size and power of the subscriber interface to a local exchange," in *Conf. Rec., 1978 Zurich Seminar on Digital Commun.*

Paul R. Gray was born in Jonesboro, AR, on December 8, 1942. He received the B.S., M.S., and Ph.D. degrees in 1963, 1965, and 1969, respectively, from the University of Arizona, Tucson.

From 1969 to 1971 he was a member of the Technical Staff at Fairchild Research and Development Laboratories, where he was involved in the development of new circuit techniques and technologies for use in analog integrated circuits. In 1971 he became a Visiting Lecturer in the Department of Electrical Engineering and Computer Sciences, University of California, Berkeley, where he is now a Professor. His research activity at Berkeley has involved new circuit techniques and computer design aids for MOS and bipolar integrated circuits, and he has coauthored a book on the subject.

Dr. Gray has served as Consultant to a number of semiconductor manufacturers. During the academic year 1977-1978 he took a leave of absence from Berkeley to serve as Project Manager for PCM filter development with Intel Corporation. Dr. Gray is a member of Eta Kappa Nu and Sigma Xi, and served as Editor of the IEEE JOURNAL OF SOLID-STATE CIRCUITS from 1977 through 1979.

David G. Messerschmitt has been an Associate Professor of Electrical Engineering and Computer Sciences at the University of California, Berkeley, since 1977. From 1968 to 1974 he was a member of the Technical Staff and from 1974 to 1977 a Supervisor at Bell Laboratories, Holmdel, NJ, where he did systems engineering, development, and research on digital transmission lines and terminals, digital speech interpolation, and digital signal processing, particularly as it relates to both low and high bit rate encoding of speech. His current research interests are analog and digital signal processing, with applications to voice encoding; digital transmission; phase-locked loops, and adaptive filtering. He has published 22 papers and has 8 patents issued or pending in these fields. Since 1977 he has served as a Consultant to industry, including, among other companies, TRW VIDAR, Hughes Aircraft, Acumenics, and Intel.

Dr. Messerschmitt received the B.S. degree from the University of Colorado in 1967, and the M.S. and Ph.D. degrees from the University of Michigan in 1968 and 1971, respectively. He is a Senior Member of IEEE and is a member of Eta Kappa Nu, Tau Beta Pi, and Sigma Xi. He is currently serving as Editor for Transmission Systems for the IEEE TRANSACTIONS ON COMMUNICATIONS.

Section 2

Data and Modulation Techniques

EDITORS' COMMENTS

To efficiently and reliably transmit information over an actual channel, many operations on the raw data are usually performed. One of the most important is that of modulation. The desired result of any operation on the data is typically to implement a system that uses either as small a bandwidth or as small a signal-to-noise ratio as possible and still transmits a given amount of information over the channel with a minimum prescribed degree of fidelity (or reliability).

Fundamental to being able to accomplish this task is the need to understand the system implications of whatever definition of bandwidth is being used, as well as the invariable tradeoff that arises between power and bandwidth. Once these ideas are understood, various modulation formats can be considered, some being more power efficient (at the expense of bandwidth), and others being more bandwidth efficient (at the expense of power).

In recent years, many new digital modulation techniques have been both analyzed and implemented. While the two most classical techniques, namely phase-shift keying (PSK) and frequency-shift keying (FSK), are still extensively used, many other techniques, in particular higher order alphabet techniques, have become popular. These include minimum shift keying (MSK) and various forms of combined amplitude and phase modulation such as 16-ary quadrature amplitude modulation (sometimes referred to as quadrature-amplitude-shift keying). The motivation for considering new modulation techniques varies. In the case of satellite communications, it is typically to find a scheme which performs well when used over a nonlinear channel. In the case of mobile radio, it is to find a technique which is as bandwidth efficient as possible and which still performs well when used over a multipath channel.

It is clear then that there is interest in the judicious choice of a transmitted waveform relative to the appropriate channel characteristic. This requires a sufficient knowledge of the channel so that its effect on the transmitted signal can be accurately determined. Examples of common communication channels in addition to satellite links are high frequency (HF) fading channels and telephone links.

Finally, because of the tremendous demand to maximize the information rate across a bandlimited channel, a phenomenon known as intersymbol interference (ISI) arises and is one of the key sources of degradation in a digital communication system. When a time-limited pulse is passed through a bandlimited channel, the pulse is dispersed in time so that it occupies a time interval greater than (and possibly much greater than) the time interval to which it was initially limited. Any time a contiguous stream of time-limited pulses is used over a bandlimited channel, this time dispersion causes some energy of each pulse to overlap into the time intervals of adjacent pulses. This is known as intersymbol interference. To reduce the effect of intersymbol interference on the performance of a digital communication system, equalization techniques are often employed at the receiver. There are a variety of different equalizers that can be designed, but they all attempt to compensate, in some sense, for the time-dispersion caused by the system.

In this section, there are eight papers which discuss different aspects of the problem of reliable data transmission. An overview of the way in which data communications came to be transmitted over links that were originally designed for voice is provided in the first paper by E. R. Kretzmer. This paper, entitled "The Evolution of Techniques for Data Communication Over Voiceband Channels," provides a somewhat historical summary of the advances made in voiceband data.

The next two papers are concerned with the basic concepts of bandwidth and its relation to system performance. The first paper is by F. Amoroso and is entitled "The Bandwidth of Digital Data Signals." It presents many different definitions of bandwidth, and illustrates their use by presenting comparisons of bandwidth occupancy for several modulation techniques. "Spectrum Conservation by Efficient Channel Utilization" by E. Bedrosian is the second of these two papers, and this one takes an information theoretic look at the relationship between power and bandwidth in a digital communication system.

The next three papers discuss bandwidth efficient modulation techniques. The first two, entitled "Correlative Coding: A Bandwidth-Efficient Signaling Scheme" and "Minimum Shift Keying: A Spectrally Efficient Modulation," are both by S. Pasupathy. The paper on correlative coding describes a technique, known at times as partial-response signaling, for transmitting signals over a bandwidth-limited channel without suffering a large degradation in system performance due to ISI. This is accomplished by introducing a controlled amount of ISI. The paper on minimum shift keying describes an MSK signal and compares its performance and characteristics to those of quadrature phase shift keying (QPSK) and offset QPSK. The third paper on modulation techniques is by N. F. Dinn and is entitled "Digital Radio: Its Time Has Come." This paper describes the application of certain multilevel modulation schemes such as octal or 8 phase PSK and 16-ary quadrature amplitude modulation to digital radio links.

The last two papers in this section treat the topic of equalization of digital communication links. The first paper is by J. F. Hayes and is entitled "The Viterbi Algorithm Applied to Digital Data Transmission." It provides a description of how the Viterbi algorithm, originally developed to decode convolutional codes, can be used to equalize a system that has been degraded by the introduction of ISI. A different approach to equalization is provided by P. Monsen in the last paper of the section. This one is entitled "Fading Channel Communications." In addition to providing a description of the dominant characteristics of HF channels, it shows how adaptive equalizers can be used to mitigate the effects of the ISI that are produced by most multipath channels.

COMMUNICATION TECHNOLOGY: 25 YEARS IN RETROSPECT, PART IV, THE EVOLUTION OF TECHNIQUES FOR DATA COMMUNICATION OVER VOICEBAND CHANNELS *

E. R. Kretzmer, Fellow, IEEE

Abstract—Data communication has grown dramatically in the past two decades, both in technical sophistication and in usage. While channels and networks designed expressly for data use have emerged, the voiceband channels of the telephone network continue to be the major transmission medium; data sets, or modems, play a role analogous to that of the telephone in voice communication. Fundamental developments such as adaptive equalization along with bandwidth-conserving signal formats have allowed the modem to better match the characteristics of the analog channel, resulting in increased available throughput. The quest for further improvements, along with elegant implementations and attractive user features, provide a continuing challenge to communication engineers.

When the need arose to communicate digital data at transmission rates substantially higher than telegraph speeds, first for defense about mid-century and then for industry in the late 1950's, all the ingredients required for successful implementation were at hand. Decades earlier, Nyquist had formulated the filtering or band shaping requirements to allow the independent transmission of a sequence of signal samples. Just prior to mid-century, Shannon had published his celebrated information theory, which showed engineers the maximum rate at which they could signal through a channel if only they were clever enough. The telephone network, especially in the United States, had reached a high state of development and widespread accessibility; it seemed like an ideal vehicle to carry the new data communication traffic.

By around 1960, data communication was off to a solid start. The ones and zeros were transformed into signals well suited to the telephone lines by what was then called data subsets ("sub" for subscriber), later to be called data sets, and still later, modems. The line signals were typically binary frequency-modulated signals, so-called FSK or frequency shift-keyed signals, where a one was represented by one frequency and a zero by another. Alternatively, they were phase shift-keyed signals in which, for example, four different possible phase shifts represented the four possible combinations of pairs of ones and zeros. Such modulation schemes are still in use today, augmented and supplemented by additional schemes of greater complexity. Thus, we are now able to signal through a given channel at a higher bit rate.

Even now, some 30 years after Shannon's work, new insights and understanding are pushing the maximum data rates to creep up toward the theoretical channel capacity—roughly 20 kbits/s for a voiceband telephone channel. Rates as high as 80 percent of that capacity have been attained experimentally—with error rates low enough for encoded speech and low enough to have negligible effect on the Shannon information rate. The error rates are not low enough to be of use in data communication, however. Thus, the highest commercially available data rate is currently about half of the theoretical capacity. To make even this much possible has required a host of innovations and has stimulated notable work in such areas as error control, scrambling of data, timing acquisition, carrier recovery, automatic equalization, adaptive equalization, and echo cancellation.

Over the last decade, commerce and industry have become so dependent on efficient and reliable data communication that it is now as indispensable to many operations as is voice communication in everyday life. For the most part, the voiceband channels of the telephone network are used as the transmission medium.This medium has been gradually improved to enhance its effectiveness for both voice and data communications. The improvement has included the control of certain channel characteristics important to data transmission, but previously unknown and unnoticed inasmuch as they had little effect on voice transmission. Much work has been done over the years to characterize the voice transmission channels with respect to all the impairments that became known as affecting data transmission performance [1]. In addition, a great deal of effort has been and continues to be spent on the establishment of standards for interfacing the data

The author is with Bell Laboratories, Holmdel, NJ 07733.

*Reprinted from *IEEE Communications Society Magazine*, January 1978, Vol. 16, No. 1, pp. 10–14.

communications channels with computers and user terminals.

When the first modems were put into service around 1960, not only did the designers have the benefit of the work of Nyquist and Shannon, but they also had the beginnings of modern solid-state circuitry to work with. Since then, modems have evolved from discrete transistor circuitry to designs using integrated circuits—first small-scale, then medium-scale, and now large-scale integrated circuits. Even microprocessors have found their way into data communication. However, while these technological improvements have been tremendously important, the last few years have been the years of the user; designers have made ever stronger efforts to understand the real needs of the users of data communication services and have designed their equipment accordingly. The sophisticated user is now able to continuously monitor his data communication network and locate bottlenecks by built-in diagnostics.

Thus, in the space of less than 20 years a new industry was born, grew up, and has now matured. Its growth has been phenomenal and is even now greater, percentagewise, than that of almost any other segment of communication.

THE DEVELOPMENT OF SIGNALING AND EQUALIZATION TECHNIQUES

Probably first among the insights and innovations which have made possible the progress described above was the application of the FSK signal. This digital version of FM was correctly identified over two decades ago as the way to achieve modest rates at low expense and with high immunity to noise or interference. The rates have crept from 600 bits/s to mostly 1200 and some 1800 bit/s usage. Along the way there have been many valuable analytical results [2], [3].

The differentially phase shift-keyed (DPSK) signal with four possible phase shifts, also close to 20 years old, was an early ingenious choice; it is the charter member of the class of signaling formats now recognized as the most effective, having a signal constellation with well-spaced points in the two-dimensional amplitude-phase space. The idea of differential detection, which uses the preceding signal element as a reference in deciding on the present signal element or symbol, not only led to a simple receiver design [4], but also a welcome insensitivity to low-frequency phase jitter, one of the transmission impairments found on voice channels. Four-phase DPSK is universally used for 2400 bits/s.

The quest for higher speeds in the early sixties led to the exploration of minimum-bandwidth systems of the vestigial sideband (VSB) type. These systems are capable of nearly twice the symbol rate for a given bandwidth, and, combined with multilevel coding, e.g., eight levels, triple the bit rate relative to the symbol rate. It was recognized at the outset that such high-rate signals require rather precise compensation for any amplitude and phase distortion introduced by the channel. VSB entails a linear translation of the spectrum from baseband to passband, allowing the compensation, or equalization, to be done in the baseband, following demodulation; this helped to pave the way for the first automatic equalizer.

Automatic equalization was the crucial element in the realization of a multifold increase in the highest attainable rate—up to 9600 bits/s [5], [6]. The process was initially based on the observation at the receiver of a series of individually transmitted impulses, which become distorted in transmission (i.e., measurement of system impulse response). The departures of the received signal from the ideal (nondistorted) waveform were used to adjust the taps of a transversal filter in an iterative fashion until the intersymbol interference had been "forced to zero." Subsequent improvements replaced the single-impulse "training" by a sort of on-the-job training whereby iterative corrections were done during data reception; this was aptly called adaptive equalization. Adaptive equalization is based on "decision-directed" operation, in which the iterative corrections are derived from final (not necessarily correct) output decisions in the receiver [7]. This mode of operation has come into universal usage in the steady state. However, the short-message traffic typical of polling networks requires very rapid start-up, for which a predetermined pseudorandom training sequence is used. Concurrent with this evolution was the utilization of the mean-squared error minimization criterion in place of the zero-forcing algorithm [8], [9].

Meanwhile, back on the modulation or signal format front,

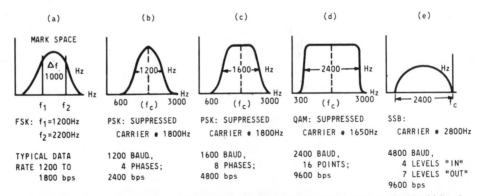

Data signal spectra. Spectra associated with 1200 to 9600 bit/s modems; all but (e) use double-sideband modulation; single-sideband and vestigial sideband (not shown) are no longer favored by most modem designers. Transmission is synchronous for (b)-(e), with baud or symbol rate equal to the 6 dB bandwidth for (b)-(d) and twice the total bandwidth for (e).

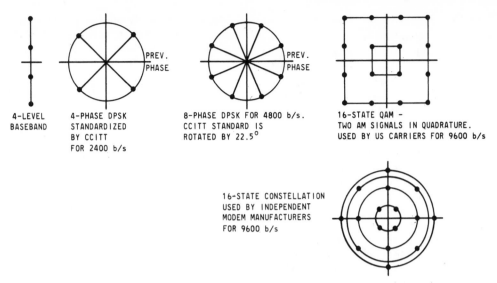

4-LEVEL BASEBAND

4-PHASE DPSK STANDARDIZED BY CCITT FOR 2400 b/s

8-PHASE DPSK FOR 4800 b/s. CCITT STANDARD IS ROTATED BY 22.5°

16-STATE QAM – TWO AM SIGNALS IN QUADRATURE. USED BY US CARRIERS FOR 9600 b/s

16-STATE CONSTELLATION USED BY INDEPENDENT MODEM MANUFACTURERS FOR 9600 b/s

Signal constellations. "Constellations" characterizing alternative data signal formats. Number of allowed states, N, ranges from 4 to 16. Bit rate equals $\log_2 N$ times baud or symbol rate.

efforts continued to reduce the required bandwidth to permit more "bits per buck." Single sideband transmission is even more bandwidth conserving than vestigial sideband transmission. However, the vestige can be dispensed with only if the very low frequency components normally present in a data signal are somehow eliminated. The duobinary scheme [10] pointed the way to a generalization termed correlative or partial-response signaling [11], [12], one form of which made feasible true single sideband multilevel data

transmission. This allows a rate of fully 2 baud (signal elements or symbols per second) per hertz of absolute bandwidth, with true zeros at the band edges, equal to Nyquist's theoretical limit. However, a price is paid for this: the receiving detector must be able to discern three levels, even though the original signal had two levels. More generally, it must be able to discern $2n-1$ levels if n-level coding is used. Such schemes are in use today on highly linear and quiet media such as cables and microwave links.

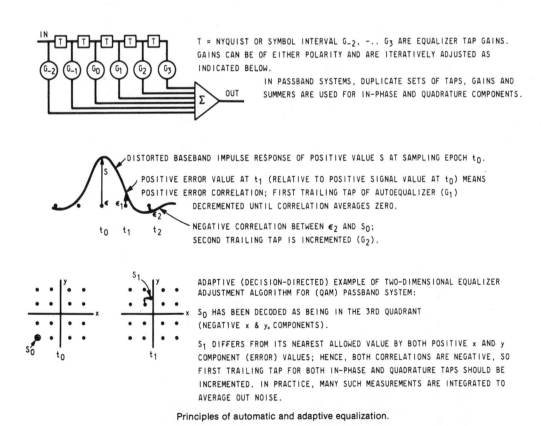

T = NYQUIST OR SYMBOL INTERVAL G_{-2}, $-..$ G_3 ARE EQUALIZER TAP GAINS. GAINS CAN BE OF EITHER POLARITY AND ARE ITERATIVELY ADJUSTED AS INDICATED BELOW.

IN PASSBAND SYSTEMS, DUPLICATE SETS OF TAPS, GAINS AND SUMMERS ARE USED FOR IN-PHASE AND QUADRATURE COMPONENTS.

DISTORTED BASEBAND IMPULSE RESPONSE OF POSITIVE VALUE S AT SAMPLING EPOCH t_0.

POSITIVE ERROR VALUE AT t_1 (RELATIVE TO POSITIVE SIGNAL VALUE AT t_0) MEANS POSITIVE ERROR CORRELATION; FIRST TRAILING TAP OF AUTOEQUALIZER (G_1) DECREMENTED UNTIL CORRELATION AVERAGES ZERO.

NEGATIVE CORRELATION BETWEEN ϵ_2 AND S_0; SECOND TRAILING TAP IS INCREMENTED (G_2).

ADAPTIVE (DECISION-DIRECTED) EXAMPLE OF TWO-DIMENSIONAL EQUALIZER ADJUSTMENT ALGORITHM FOR (QAM) PASSBAND SYSTEM:

S_0 HAS BEEN DECODED AS BEING IN THE 3RD QUADRANT (NEGATIVE x & y_* COMPONENTS).

S_1 DIFFERS FROM ITS NEAREST ALLOWED VALUE BY BOTH POSITIVE x AND y COMPONENT (ERROR) VALUES; HENCE, BOTH CORRELATIONS ARE NEGATIVE, SO FIRST TRAILING TAP FOR BOTH IN-PHASE AND QUADRATURE TAPS SHOULD BE INCREMENTED. IN PRACTICE, MANY SUCH MEASUREMENTS ARE INTEGRATED TO AVERAGE OUT NOISE.

Principles of automatic and adaptive equalization.

SIGNAL SHAPINGS AND CONSTELLATIONS

Data modem design has evolved further, building on the early four-phase signal constellation and utilizing the wide range of spectral shapings allowable under Nyquist's criteria for avoiding intersymbol interference. For baseband operation, such shapings, with the response down to half-amplitude at frequency F, allow signaling at $2F$ baud. So-called full cosine roll-off gives 100 percent excess bandwidth, since the true band edge is at $2F$. This is allowable, even advantageous, for low speeds such as 2400 bits/s. For the highest speeds, however, excess bandwidths approaching the order of only 10 percent have come into general use—facilitated by the autoequalizer's ability to compensate for departures from perfect shaping and by other new technology developments. Thus, the signaling rate approaches 1.8 baud/Hz at baseband, or 0.9 baud/Hz at passband using double-sideband modulation such as phase-reversal keying. Overlapping two such passband signals in quadrature (90° out of phase, so they can be separated perfectly) yields essentially the same signal format as four-phase modulation. Allowing two amplitude levels besides four phases results in a 16-point constellation. This signal format is generally called quadrature amplitude modulation (QAM), although its structure is similar to that of an amplitude/phase modulated (AM/PM) signal. QAM, commonly used to provide 9600 bit/s service (about 3.6 bits/s/Hz), provides near-optimum signal/noise performance. Another 16-point constellation in common use for 9600 bits/s has its points spaced at 90° on four concentric circles, yielding very low sensitivity to phase jitter. At 4800 bits/s, eight-phase modulation is in common use, with the eight points spaced at 45° around a circle. Constellations having up to 64 points have been implemented in order to test the feasibility of 16 kbits/s—more than 5 bits/s/Hz. There is practically no limit to the number of different constellations one can design [13], each with its characteristic pros and cons; but the interest of compatibility, of course, argues for a single choice per speed.

The use of these sophisticated constellations to attain high-speed service could not have been realized without important inventions in the area of carrier recovery and autoequalization. Optimum signal/noise performance demands that demodulation be coherent, necessitating the recovery of a carrier which is suppressed in the transmitted signal. Furthermore, if the channel impairments have imparted phase jitter to that signal, the recovered carrier should "track" that jitter. The use of decision-directed phase-locked loops has been a key development here [14]-[16].

With two-dimensional constellations, the desire to track phase jitter with the smallest possible delay militates in favor of placing the autoequalizer in the passband rather than baseband. This has led to highly refined structures [16], [17], generally comprising in-phase and quadrature tap weights. These equalizers are capable of compensating for carrier phase and even small amounts of frequency offset. Importantly, they can also train in a time span comparable to the time spanned by their tapped delay elements [18].

THE FUTURE FOR MODEMS

The references cited in this paper represent only a small fraction of the many excellent contributions to the impressive progress in the field of data communication. The basic problem of bettering analog channel utilization for digital communication continues to challenge some of the best minds.

In concert with the increasing functional sophistication of data modems, the implementations have made good use of new technology. Signal processing within the modem has gone from largely analog to largely digital. Two principal trends are now clearly discernible: 1) large-scale integration of major modem functions, and 2) high-speed microprocessor implementation of all the modem functions. Both approaches permit the inclusion of many complex features in remarkably small volume.

Data communication, of course, entails a great deal more than modems and transmission lines. A rapid evolution of computer communication networks is underway, in which the data channels provide the basic foundation. Also, there have long been efforts toward the use of digital channels instead of analog voiceband channels. These trends have led many to ask whether the time has come when data traffic will go largely onto new networks built for that purpose instead of onto the telephone network. The answer, most likely, is that such a change will be a gradual, evolutionary one—tempered by the unrivaled accessibility of the voice network.   □

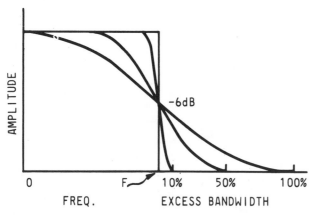

Baseband Nyquist shapings for zero intersymbol interference at signaling rate 2F baud. For double-sideband passband operation, simply add a mirror image to the left and increment all frequencies by carrier frequency.

REFERENCES

[1] M. D. Balkovic, H. W. Klancer, S. W. Kiare, and W. G. McGruther, "High-speed voiceband data transmission performance on the switched telecommunications network," *Bell Syst. Tech. J.*, vol. 50, pp. 1349-1384, Apr. 1971.

[2] W. R. Bennett and J. R. Davey, *Data Transmission.* New York: McGraw-Hill, 1965. ch. 9.

[3] R. W. Lucky, J. Salz, and E. J. Weldon, Jr., *Principles of Data Communications.* New York: McGraw-Hill, 1963, ch. 8.

[4] P. A. Baker, "Phase modulation data sets for serial transmission at 2000 and 2400 bits per second, Part 1," *AIEE Trans. (Commun. and Electron.)*, July 1962.

[5] F. K. Becker, L. N. Holzman, R. W. Lucky, and E. Port, "Automatic equalization for digital communication," *Proc. IEEE*, vol. 53, pp. 96-97, Jan. 1965.

[6] R. W. Lucky, "Automatic equalization for digital communication," *Bell Syst. Tech. J.*, vol. 44, pp. 547-588, Apr. 1965.

[7] ——, "Techniques for adaptive equalization of digital com-

munication," *Bell Syst. Tech. J.*, vol. 45, pp. 255-286, February, 1966.

[8] B. Widrow and M. E. Hoff, Jr., "Adaptive switching circuits," in *IRE Wescon Conv. Rec.*, part 4, Aug. 1960, pp. 96-104.

[9] A. Gersho, "Adaptive equalization of highly dispersive channels for data transmission," *Bell Syst. Tech. J.*, pp. 55-70, Jan. 1969.

[10] A. Lender, "The duobinary technique for high-speed data transmission," *IEEE Trans. Commun. Electron.*, vol. 82, pp. 214-218, May 1963.

[11] E. R. Kretzmer, "Binary data communication by partial response transmission," in *1965 ICC Conf. Rec.*, pp. 451-455; also, "Generalization of a technique for binary data communication," *IEEE Trans. Commun. Technol.*, pp. 67-68, Feb. 1966.

[12] A. M. Gerrish and R. D. Howson, "Multilevel partial response signaling," in *1967 ICC Conf. Rec.*, p. 186.

[13] G. J. Foschini, R. D. Gitlin, and S. B. Weinstein, "On the selection of a two-dimensional signal constellation in the presence of phase jitter and Gaussian noise," *Bell Syst. Tech. J.*, vol. 52, pp. 927-965, July-Aug. 1973.

[14] H. Kobayashi, "Simultaneous adaptive estimation and decision algorithm for carrier modulated data transmission systems," *IEEE Trans. Commun. Technol.*, vol. COM-19, pp. 268-280, June 1971.

[15] R. Matyas and P. J. McLane, "Decision-aided tracking loops for channels with phase jitter and intersymbol interference," *IEEE Trans. Commun.*, vol. COM-22, pp. 1014-1023, Aug. 1974.

[16] D. D. Falconer, "Jointly adaptive equalization and carrier recovery in two-dimensional digital communication systems," *Bell Syst. Tech. J.*, vol. 55, pp. 317-334, Mar. 1976.

[17] R. D. Gitlin, E. Y. Ho, and J. E. Mazo, "Passband equalization of differentially phase-modulated data signals," *Bell Syst. Tech. J.*, vol. 52, pp. 219-238, Feb. 1973.

[18] K. H. Mueller and D. A. Spaulding, "Cyclic equalization—A new rapidly converging equalization technique for synchronous data communication," *Bell Syst. Tech. J.*, pp. 369-406, Feb. 1975.

THE BANDWITH OF DIGITAL DATA SIGNALS *

Frank Amoroso

The spectrum is finite, yet as technology manipulates it, a dazzling increase in bandwidth capability emerges.

THE engineering of data communications systems invites involvement with a number of rather refined parametric concepts, such as bit error rate, antenna gain, radiated power, communication efficiency, and bandwidth. Of these, none has been the subject of more lively discussion and revision than bandwidth. The implications of bandwidth can vary considerably from context to context, as the profusion of definitions of bandwidth will attest. The purpose of the present article is to explore the subject of data transmission bandwidth through an examination of its various definitions.

WHY DISCUSS DATA BANDWIDTH?

Most of the current attention given to data bandwidth centers on the problem of spectrum allocation in an increasingly crowded radio frequency spectrum. But the perennial interest in band occupancy springs from a much broader base of concerns, many outside the domain of radio transmission. Not only may it be important to know how much bandwidth a signal occupies, but also the extent to which a given "band-limited" medium may be exploited for data transmission.

The early experience with open wire telephone lines soon led to the physical realization of sharply band-limited channels. Reactance inherent in the lines combined with the cascading of many equalizers and repeater amplifiers quickly produced very pronounced frequency cutoff characteristics. Nyquist's telegraph transmission theory [1] in 1928 dealt very rigorously with the theory of data signals over such circuits, which were modeled as strictly band-limited channels. More recently [2] it has been shown that the CATV

*Reprinted from *IEEE Communications Society Magazine*, November 1980, Vol. 18, No. 6, pp. 13–24.

transportation trunk, a cascade of properly terminated coaxial cable sections and repeater amplifiers, each with gradual cutoff characteristics, presents a sharply band-limited end-to-end response characteristic.

Even the record-playback characteristic of a digital magnetic tape station [3] has been modeled as a data communication channel with limited bandwidth capability. To this fact should be added the common knowledge that many microwave system components, antennas, output power devices, and voltage tunable oscillators impose bandwidth constraints of their own.

In HF radio transmission the bandwidth may not be limited so much by allocation as by a multipath interference effect, which can constrain the usable bandwidth to as little as 100 Hz. Satellite downlink communication is best performed in the rather restricted band from 2 to 4 GHz. Below 2 GHz, galactic noise becomes a significant degradation, while above 4 GHz, the noise due to oxygen absorption is combined with possible fading from precipitation to make the link less attractive.

The needs of the spread-spectrum communication system place a rather contrasting premium on *maximum* bandwidth occupancy. To decrease the power spectral density of the signal without reducing the transmitter power, to reduce the effectiveness of enemy jammers, to multiplex many signals occupying the very same band, and to increase the precision of timing information derived from a signal all imply greater system bandwidth per data rate.

NO SINGLE DEFINITION SUFFICES

The availability of a single universal definition of bandwidth would decidedly simplify the specification and structuring of a great variety of systems. However, some reflection on the subject will lead to the conclusion that no universally satisfying definition of bandwidth can exist.

The most relevant and widely cited theory for data transmission systems bandwidth is contained in the sampling theorem of Nyquist and in Shannon's theorem on channel capacity [4]. Nyquist says that any signal of bandwidth W

hertz can be completely characterized by analog samples taken once every $1/2W$ seconds. Shannon says that a noisy channel of bandwidth W will theoretically support errorless data transmission at a channel capacity given by

$$C = W \log_2 \left(1 + \frac{S}{N}\right)$$

bits per second, where S/N is the ratio of signal power to noise power.

The theories of Nyquist and Shannon, however concise and fundamental, suffer the practical disadvantage of being based on the assumption of strict band limitation. This means that no signal power whatever is allowed outside the defined band. Unfortunately, a strictly band-limited signal would imply infinite time delay in the channel and either infinitely complex signal synthesis or infinitely complex channel filters. On the other hand, the possible generalization of the fundamental theorems to cover nonstrictly band-limited signals continues to pose perplexing questions [5] even for the preeminent investigator.

The practical interest in strict, or nearly strict, band limitation is most often an indirect expression of concern for adjacent channel interference in a frequency division multiplexed (FDM) system. Here almost any definition of occupied bandwidth only imprecisely measures the interference. In fact, a recent investigation showed [6] that total interchannel crosstalk can be minimized by optimization of the data modulation, and without any direct reference to bandwidth. Definitions of bandwidth, when used, serve the needs of the regulatory agency or the solicitor of a technical proposal who must seek to control interference while allowing the designer freedom to choose his own modulation scheme. Definitions of bandwidth arising in these circumstances are bound to be somewhat arbitrary.

In the case of spread-spectrum communication, the measure of bandwidth is more likely to ask "is the signal spread out enough" rather than "within what bounds is it confined." This is like saying that a gallon of paint should be applied in a layer not to exceed a certain thickness rather than insisting that the paint never splatter beyond certain boundaries. The definitions of bandwidth for spread-spectrum systems can be very precise and practical in their own right, as further discussion will show.

THE SIGNIFICANCE OF POWER SPECTRAL DENSITY

Before proceeding to discuss various types of bandwidth, it will be good to discuss the underlying frequency function on which they all depend. This is called the power spectral density and is familiar as the typical display on a swept spectrum analyzer [7]. The general feeling is that for every data modulation scheme there exists a precisely known power spectral density from which the bandwidth can be deduced or computed according to any given definition of bandwidth.

The power spectral densities generally given for data signals are based on critical assumptions of random data sequences and long averaging times. There may be circumstances under which one or both of those assumptions will fail to apply. For instance, the power spectral density usually given for 180° binary phase shift keying takes the form

$$S_n(f) = \left[\frac{\sin \pi (f - f_c) T}{\pi (f - f_c) T}\right]^2 \qquad (1)$$

where f_c is the carrier frequency and T is the bit duration. During the transmission of simple idling patterns, such as all 1's or alternating 1's and 0's, the spectrum will not remotely resemble that given in (1), but rather will consist of discrete spectral lines, i.e., infinitely tall spikes of spectral density. In a more complex example, random data, when convolutionally encoded for error correction, lose some of their randomness. Interdependency between bits is introduced, and the spectrum [8] will depart from (1).

Even if the criterion of randomness has been satisfied, the spectrum will not conform to (1) if the observation time is short. Fig. 1 shows the power spectral density of a 32-bit burst of random data, with the spectrum (jagged curve) expressed in decibels. The smooth curve represents (1), and the bit pattern is inset. Notice that the burst spectrum conforms very broadly to the smooth curve, but for the most part departs markedly from it. The theoretically satisfying spectra often observed on swept analyzers will be seen to conform with (1) for sweep rates set very slow with respect to the bit rate and with very long random sequences.

It is worth mentioning that a number of recent developments tend to favor the use of random data streams. A U.S. Commerce Department requirement [9] for the protection of "unclassified information transmitted by and between government agencies and contractors that would be useful to an adversary" has spurred inquiry into bulk encryption. This will have the effect of converting nonrandom data into random data. Police radio nets are being forced to convert to encrypted digitized speech [10] to prevent eavesdropping curiosity-seekers from congregating at the scenes of hostage situations and other sensitive operations. Of course, such data sources as digitized speech and digitized video are notably nonrandom in their unscrambled form. Scrambling is often added to facilitate bit timing extraction even if security and privacy are not necessarily desired.

A second factor that encourages the use of precise mathematical formulas such as (1) for digital data spectra is that the formulas are often rather simply derived. For many of the common modulation techniques, such as binary phase shift keying (BPSK), quaternary phase shift keying (QPSK), and minimum shift keying (MSK), the spectrum of the (random) data stream is just the spectrum of the individual data pulse, and that expression is mathematically simple. The spectra of higher order phase and amplitude shift keyed modulation, such as 8 PSK 16 PSK, and quadrature amplitude modulation used in digital radios, will take the same form as BPSK and QPSK but with the frequency variable rescaled to reflect higher bandwidth efficiency.

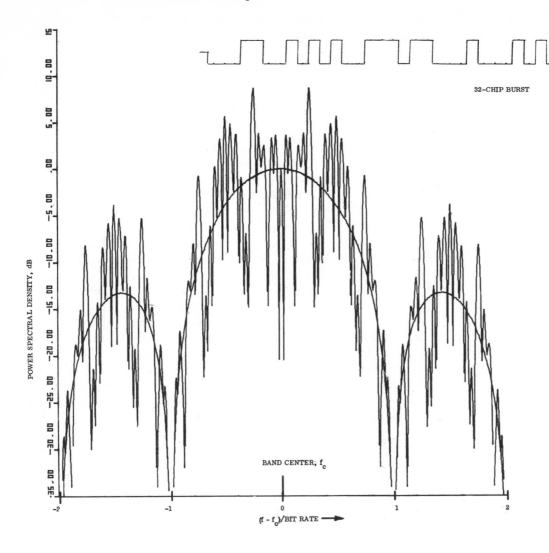

Fig. 1.
Spectrum of
32-chip burst.

Such other modulation types as sinusoidal frequency shift keying (SFSK)[11], phase continuous FSK with arbitrary modulation index [12], or the quasi-band-limited variant [13] on MSK required more complex computation to arrive at the power spectral density. The computational results are, however, precise enough to enable the bandwidth computation for any definition of bandwidth in current use.

Recently the level of interest in constant carrier envelope modulation has grown, owing largely to the desire for hard limiter operation in satellite relays. All of the phase and frequency modulation types mentioned thus far provide constant carrier envelope. However, the use of spectrum shaping filters following the power output stage or the use of amplitude modulation, as in QAM, will generally destroy the constant envelope property of a wave.

The following discussion of bandwidth will focus on several constant carrier envelope modulation schemes. These by no means exhaust the available possibilities, but their power spectra are varied enough to reveal the implications of the various definitions of bandwidth. The modulation types are BPSK, QPSK, MSK, SFSK, and the quasi-band-limited

keying mentioned previously. The latter type contains hard limiting to achieve constant carrier envelope.

DEFINITIONS OF BANDWIDTH

The power spectral densities of the five modulation types under consideration are shown in Fig. 2(a) and (b). None is strictly band limited. In fact, no constant envelope wave can be strictly band limited, save for the case of an unmodulated carrier. This is readily established through the squared envelope theorem [14] of analytic signal theory. The definitions of bandwidth to be considered are all amenable to constant envelope signaling and represent a fairly complete catalog.

Fractional Power Containment Bandwidth

The first type of bandwidth to be discussed has been adopted by the Federal Communications Commission (FCC Rules and Regulations Section 2.202) and states that the occupied bandwidth is the band which leaves exactly 0.5 percent of the signal power above the upper band limit and exactly 0.5 percent of the signal power below the lower band

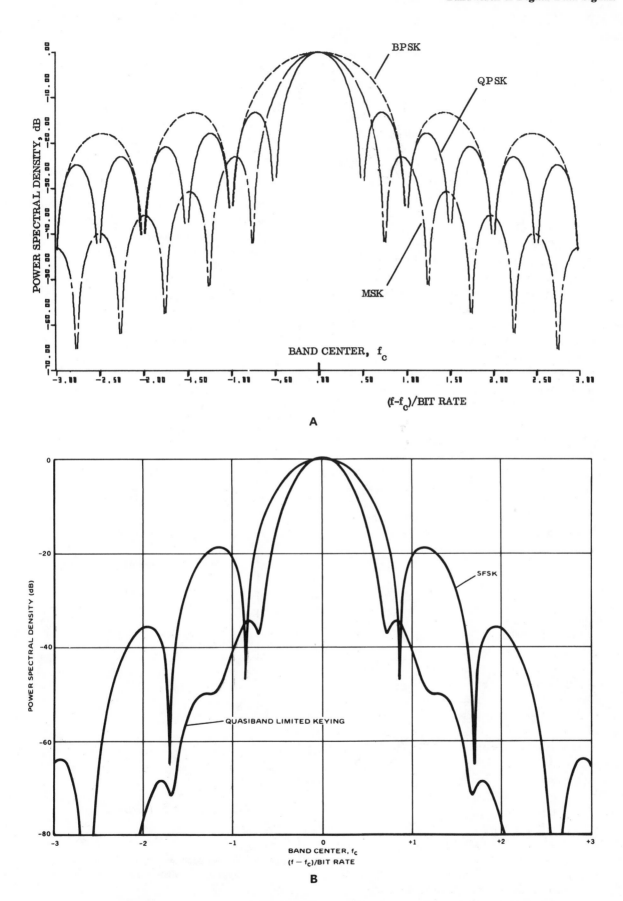

Fig. 2.   (a) Power spectral densities for BPSK, QPSK, and MSK. (b) Power spectral densities for SFSK, quasi-band-limited keying.

limit. Thus 99 percent of the signal power is inside the occupied band.

An aid to applying the given definition to the modulation types at hand is provided in Fig. 3. This is a plot of fractional power outside a band centered on the carrier, plotted versus bandwidth in units of bit rate. When fractional power is −20 dB, then 1 percent of the power is outside the band. Since the data spectra are symmetrical, the FCC definition is satisfied at the −20 dB level. For instance, the 99 percent bandwidth of MSK is 1.18 Hz/bits/s. The plots of the type in Fig. 3 are generally simpler to use than those of Fig. 2, as they do not exhibit the complex lobed behavior found in the power spectral density.

A lower bound (solid curve) is also given in Fig. 3, as it represents the minimum achievable bandwidth over a continuum of power containment percentages [15]. The lower bound applies to all modulation schemes whose basic data pulses last for 2 bit durations or less. With the exception of quasi-band-limited keying, this category includes all modulation schemes under consideration. Of these, MSK comes very close to having the minimum achievable bandwidth at the −20 dB level. For this and other reasons to be stated shortly the MSK technique has been termed a bandwidth efficient modulation method. Note that the communication efficiency[1] of BPSK, QPSK, MSK, and SFSK are all identical.

The quasi-band-limited modulation scheme differs from the others in two important respects. First, its basic pulse lasts 4 bit durations, and second its communication efficiency is about 0.64 dB worse than the other methods when the simplest demodulation and detection is employed.

Inspection of Fig. 3 will show that if the above bandwidth criterion were adjusted to be either much less stringent or

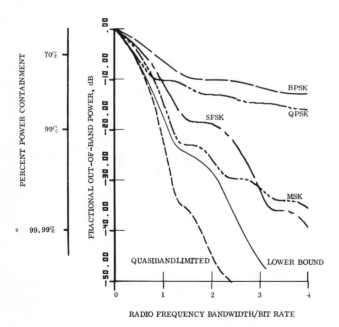

Fig. 3.   Fractional out-of-band power
for various modulation schemes.

COMMUNICATIONS EFFICIENCY COMPARISON

Communications efficiency measures the effectiveness with which signal to noise ratio is used to achieve low bit error rates. The figure below shows the bit error rate performance of the modulation schemes under consideration. Coherent detection is assumed, which means that a perfect reference carrier is available at the receiver for demodulation. The performance of BPSK, QPSK, MSK, and SFSK are identical. The performance curve of quasi-band-limited keying lies about 0.64 dB to the right of the others, so that the quasi-band-limited scheme requires about 0.64 dB more signal to noise ratio to achieve the same error rates. In this sense it may be said that the communications efficiency of quasi-band-limited keying is 0.64 dB worse than the other modulation methods.

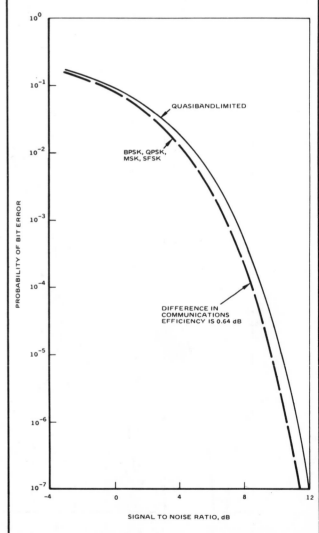

**Relative communications efficiency of quasi-band-limited
keying and other modulation methods.**

much more stringent, then MSK would no longer be as attractive. For instance, if only 70 percent of the power were required to be within the band, then QPSK would occupy less bandwidth than MSK and would nearly attain the lower bound bandwidth. If 99.99 percent of the energy were required, then SFSK would be more bandwidth efficient than MSK. The numerical value of bandwidth in units of bit rate would, of course increase as the criterion were made more stringent.

The relevant bandwidths[2] per bit rate for the 99 percent FCC criterion are

Modulation Type	99 Percent Energy Containment Bandwidth
BPSK	20.56
QPSK	10.28
MSK	1.18
SFSK	2.20
Quasi-band-limited	0.95.

The exceptionally high values for BPSK and QPSK owe to the very slow rate of their spectral rolloff, hence high levels of power far from the carrier.

Null-to-Null Bandwidth

A simpler and more popular measure of bandwidth, honored by time and wide acceptance [16] if not by profound theoretical implications, is called null-to-null bandwidth. This is a reference to the lobed behavior of the power spectral density observed in Fig. 1. The null-to-null bandwidth is just the width of the main spectral lobe. The assumption, clearly, is that the power spectral density possesses a main lobe bounded by well-defined spectral nulls, i.e., frequencies where the power spectral density is nil. Such is not the case for quasi-band-limited keying, nor is it the case for a number of other modulation schemes recently developed. Therefore null-to-null bandwidth lacks complete generality. For the modulation types considered here the null-to-null bandwidths are

Modulation Type	Null-to-Null Bandwidth
BPSK	2.00
QPSK	1.00
MSK	1.50
SFSK	1.72
Quasi-band-limited	no well defined nulls.

The null-to-null bandwidth also contains most of the signal power, if rough criteria can be accepted. Fig. 3 gives the relevant quantitative information. The degree to which the spectrum is generally spread is also roughly measured by the null-to-null bandwidth. Therefore this mode of specifying

bandwidth is often found in spread spectrum system descriptions.

Bounded Power Spectral Density

Another widely used method of specifying bandwidth is to state that everywhere outside the specified band the power spectral density must have fallen at least to a certain stated level below that found at the band center. Typical attenuation levels might be 35 or 50 dB, although greater levels of attenuation have been used in system specifications.

The 35 and 50 dB bandwidths for the modulation types under consideration can, in part, be found in Fig. 2, and are completely tabulated below.

Modulation Type	35 dB Bandwidth	50 dB Bandwidth
BPSK	35.12	201.04
QPSK	17.56	100.52
MSK	3.24	8.18
SFSK	3.20	4.71
Quasi-band-limited	1.68	2.38

Here again, the enormous bandwidths registered for BPSK and QPSK are evidence of their very slow rate of spectral rolloff. Also, the comparative bandwidths of MSK and SFSK are seen to be sensitive to the particular setting of the spectrum bound. Their 35 dB bandwidths are nearly equal, but the SFSK 50 dB bandwidth is only half that of MSK. The reason is clear in Fig. 2, as the spectral density of SFSK quickly drops below −50 dB at frequencies much beyond twice the bit rate, while the MSK spectrum continues to roll off slowly.

It must be noted that a bound on power spectral density does not necessarily imply a bound on total power outside the band; it only has the effect of bounding the total power in some finite interval of frequency outside the band. Vice versa, a limit on the total power outside the band does not necessarily imply a bound on the power spectral density in some highly localized region of frequency outside the band. For instance, a periodic signal component, at whatever infinitesimal power level, will always present infinite power spectral density at one or more frequencies.

The idea of defining bandwidth purely by bounding the spectral level outside the authorized band is found in Electronics Industries Association Standard RS-152-B, on Land Mobile Communication in the 25-470 MHz band. The bounded power spectral density method of specifying bandwidth also appears frequently in specifications concerned with adjacent channel interference.

Noise Bandwidth

Now attention is shifted to measures of bandwidth that focus on the general dispersion of the spectrum rather than on the details of spectral sidelobe structure. The first of these measures is called noise bandwidth. Conceived initially to permit rapid computation of the noise power out of an ampli-

---

[1] Communication efficiency is measured by bit error rate versus signal to noise ratio, as explained in the inset at right.

[2] Bandwidth here is always in units of bit rate.

fier with a wide-band noise input, the noise bandwidth concept can now profitably be applied to the evaluation of data link performance in the face of intentional interference, or jamming.

Noise bandwidth of a signal is defined as the value of bandwidth which satisfies the relation

$$W_N \, S \, (f_c) = P$$

where

W_N = noise bandwidth

$S \, (f_c)$ = power spectral density at the band center, assumed to be the maximum value of spectral density over all frequencies.

P = total signal power over all frequencies.

Fig. 4 illustrates the noise bandwidth concept for BPSK. The power spectral density is plotted on a linear rather than logarithmic amplitude scale; hence, its appearance differs from the representation in Figs. 1 and 2.

The area under the power spectral density curve over all frequencies is, by definition, the total signal power. This is the shaded area. The area of the rectangle is, by the proper setting of W_N, also equal to the total power. The noise bandwidth W_N is the width of the rectangle, or 1 Hz/bit/s.

The significance of noise bandwidth in combatting jammers relates to a potential strategy used by jammers to degrade system performance. It is well known that optimum link performance in white Gaussian noise is attained by the use of a receiver matched filter. The frequency response of the matched filter matches the power spectral density of the data and therefore has peak response at the band center. This being the case, the jammer will want to concentrate his power toward the band center, where he will produce the strongest spurious response at the detector which follows the matched filter.

One way to counter the jammer is to pick a modulation format which leads to a minimum of peak matched filter frequency response, hence a minimum value of $S(f_c)$ per radiated signal power P. But $S(f_c)/P$ is just the inverse of the noise bandwidth, from (2). The signal design strategy is then simply a maximization of noise bandwidth. The noise bandwidths for the modulation types under consideration are tabulated below.

Modulated Type	Noise Bandwidth
BPSK	1.00
QPSK	0.50
MSK	0.62
SFSK	0.73
Quasi-band-limited	0.48

The most jam resistant modulation type is BPSK, according to the jamming strategy outlined. The signal design strategy just discussed applies equally well to the direct sequence pseudonoise (DSPN) spread-spectrum systems [17]; only the nomenclature need be changed. "Chip rate" replaces "bit rate" and "chip matched filter" replaces "bit matched filter." Curiously, the concept of noise bandwidth has not yet gained great popularity in the specification of DSPN spread spectrum systems. Many alert engineers, however, recognize its value.

Half-Power Bandwidth

A very popular measure of bandwidth which, like noise bandwidth, gauges the degree of dispersion of the spectrum is

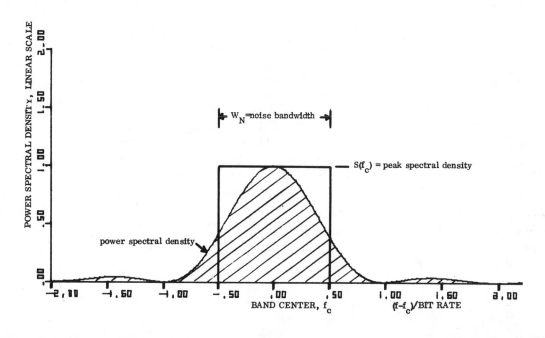

Fig. 4.
Definition of
noise bandwidth
applied to BPSK.

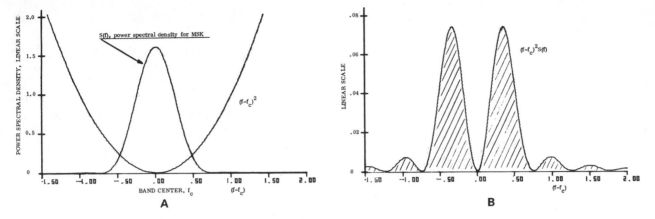

Fig. 5. Frequency domain definition of Gabor bandwidth for MSK, unit bit rate.

the half-power bandwidth. This is just the interval between frequencies at which the power spectral density has dropped to half power, or 3 dB below the peak value. Hence half-power bandwidth is also known as 3 dB bandwidth. The values are tabulated below.

Modulated Type	Half-Power Bandwidth
BPSK	0.88
QPSK	0.44
MSK	0.59
SFSK	0.70
Quasi-band-limited	0.47

The half-power bandwidth is especially interesting to designers of frequency hopped (FH) spread-spectrum systems. The spectra at individual hop frequencies are overlapped at the half-power points to produce a long-term average spectrum that is very nearly flat.

Half-power bandwidth is usually slightly smaller than the noise bandwidth, depending on the details of the spectral rolloff characteristic. The steeper the spectral rolloff the more closely the half-power bandwidth will approximate the noise bandwidth.

Gabor Bandwidth

The final measure of bandwidth is again a general indicator of spectral dispersion, its rather curious definition now

bearing the name of its originator [18]. Gabor bandwidth has a dual definition, one in the frequency domain and one in the time domain [19]. Either definition is sufficient by itself, and both definitions give the same numerical result. The two definitions are linked by the powerful Parseval relation of analytic signal theory. In both definitions the signal power is assumed, for simplicity, to be unity.

First, the frequency domain definition is illustrated in Fig. 5, for MSK signaling at unit bit rate. In Fig. 5(a), power spectral density $S(f)$ is shown on a linear scale together with a plot of $(f - f_c)^2$. In Fig. 5(b), the product $(f - f_c)^2 S(f)$ is shown. The area under that product curve over all frequency is called A. Now the Gabor bandwidth is just $\sqrt{A}$. This method of computing bandwidth is akin to the computation of the standard deviation of a probability density function, which is a legitimate measure of the width of the function. We prefer to tabulate twice the Gabor bandwidth as it compares more directly with the passband bandwidths discussed previously. The "passband Gabor bandwidth" of MSK is exactly 0.5 Hz/bit/s.

Next the Gabor bandwidth calculation is attempted for BPSK. The $(f - f_c)^2 S(f)$ product is plotted in Fig. 6. As is evident, the plotted function fails to decay with increasing distance from the band center. Therefore the area under the curve over all frequencies is . . . infinite! Both BPSK and QPSK possess infinite Gabor bandwidth.

Fig. 6. Gabor bandwidth calculation for BPSK.

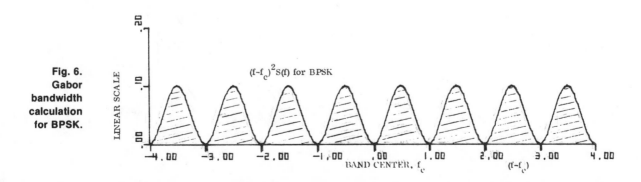

A further insight into the meaning of this situation is gained by considering the time domain definition. Fig. 7(a) shows the typical bit pulse for MSK. Fig. 7(b) shows the slope of that pulse, and Fig. 7(c) shows the square of the slope. The shaded area in Fig. 7(c), over the duration of the pulse is called B. Now the passband Gabor bandwidth is $2\sqrt{B/4\pi^2}$, $= 0.5$ Hz as before. To give a further interpretation, notice that the pulse in Fig. 7(a) is one-half cycle of a sine wave at 0.25 Hz. When such a sine wave modulates a carrier, two spectral lines 0.5 Hz apart are produced. Hence the passband Gabor bandwidth of 0.5 Hz is correct. In fact, if the pulse in Fig. 7(a) had been any number of cycles of a sine wave then the passband Gabor bandwidth would have been correctly computed according to the present definition.

Generally the Gabor bandwidth, being based on the squared slope of the bit pulse, is a measure of "wiggliness" of the bit pulse. This is perhaps an intuitively satisfying way to think of bandwidth. Because the basic bit pulses of BPSK and

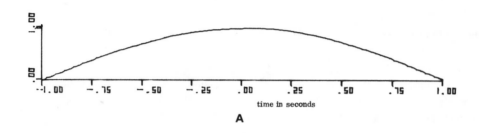

A

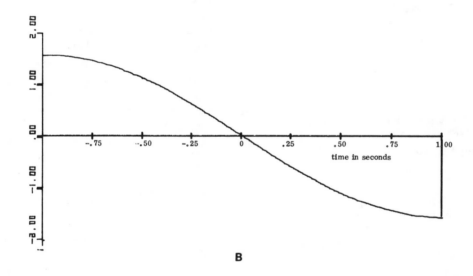

B

**Fig. 7.
Time domain
definition of
Gabor bandwidth
for MSK,
unit bit rate.
(a) Basic pulse
for MSK, unit
bit rate.
(b) Slope of
basic pulse.
(c) (Slope of
basic pulse)².**

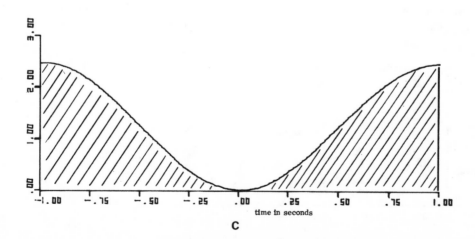

C

QPSK are both perfectly rectangular, they have infinite slope at their edges; hence, they are infinitely wiggly. This accounts for their infinite Gabor bandwidth.

The passband Gabor bandwidths for modulation types under consideration are

Modulation Type	Passband Gabor Bandwidth
BPSK	infinite
QPSK	infinite
MSK	0.50
SFSK	0.61
Quasi-band-limited	0.37.

The results in [19] lead directly to the conclusion that MSK gives the minimum possible Gabor bandwidth of all modulation types whose basic pulses are limited to 2 bit duration. This adds more substance to the claim that MSK is a bandwidth efficient modulation scheme.

FCC Spectrum Envelope

While the FCC spectrum envelope (FCC Rules and Regulations Section 21.106) does not strictly qualify as a definition of bandwidth, it ought to be mentioned, as it serves to set the upper limit on data rate in certain frequency bands.

The rule states that "the mean power of emissions shall be attenuated below the mean output power of the transmitter in accordance with the following schedule."

"For operating frequencies below 15 GHz, in any 4 kHz band, the center frequency of which is removed from the assigned frequency by more than 50 percent up to and including 250 percent of the authorized bandwidth, as specified by the following equation but in no event less than 50 dB: $A = 35 + 0.8 \, (P - 50) + 10 \log_{10} B$ (attenuation greater than 80 dB is not required) where

A = attenuation (in decibels) below the mean output power level;
P = percent removed from the carrier frequency
B = authorized bandwidth in megahertz.

The equation for attenuation A involves authorized bandwidth B as a parameter, and B takes on different values in different frequency bands. For instance $B = 30$ MHz at 6 GHz, 40 MHz at 11 GHz, etc.

In digital microwave radio, the data rates will be many megabits per second, so power spectral density will be practically unchanged over any particular 4 kHz band. Therefore, the power in any 4 kHz band will simply be given by the power spectral density multiplied by 4 kHz.

Fig. 8 depicts the expression for A in two cases, with 30 MHz authorized bandwidth and with 40 MHz authorized

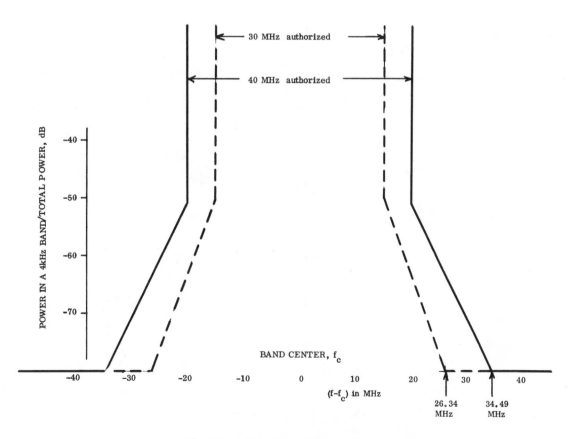

Fig. 8. FCC envelopes for 30 and 40 MHz bandwidths.

APPLICATION OF THE FCC SPECTRUM ENVELOPE—AN EXAMPLE

The problem in this example is to find the maximum data rate permitted under the FCC spectrum envelope when the authorized bandwidth is 30 MHz and the modulation method is quasi-band-limited keying.

The power spectral density of quasi-band-limited keying as given in Fig. 2(b) is expressed in decibels and is normalized to 0 dB at $f = f_c$. To calculate the power contained in a 4 kHz band, as referred to in the FCC specification, it is necessary to account for the bit rate and the noise bandwidth of the modulation. From Table I, the noise bandwidth is 0.48 times the bit rate, or

$$W_N = 0.48R = \frac{P}{S(f_c)}$$

where R is the bit rate, P is total power, and $S(f_c)$ is the true (unnormalized) value of power spectral density at the band center. So

$$S(f_c) = \frac{P}{0.48R}.$$

Now if $S_n(f)$ is the normalized power spectral density,

$$S(f) = \frac{P}{0.48R}\ S_n(f)$$

where Fig. 2(b) shows $10\log_{10} S_n(f)$.

Now the power in a 4 kHz band, at the data rates of interest, is closely approximated by the following expression in dBW:

$$\text{power in 4 kHz} = 10\log_{10} 4000 + 10\log_{10} \frac{P}{0.48R}$$
$$+ 10\log_{10} S_n(f)$$

and the power in a 4 kHz band per total power P, expressed in decibels, becomes

$$A(f) = 39.21 - 10\log_{10}R + 10\log_{10} S_n(f).$$

This is plotted in the figure below for $R = 25$ Mbits/s, together with the FCC envelope for 30 MHz authorized bandwidth. The quasi-band-limited spectrum is just tangent to the FCC envelope at the -80 dB level. Any increase in bit rate would cause the envelope to be violated. Therefore 25 Mbits/s is the maximum allowable data rate with 30 MHz authorized bandwidth and quasi-band-limited keying.

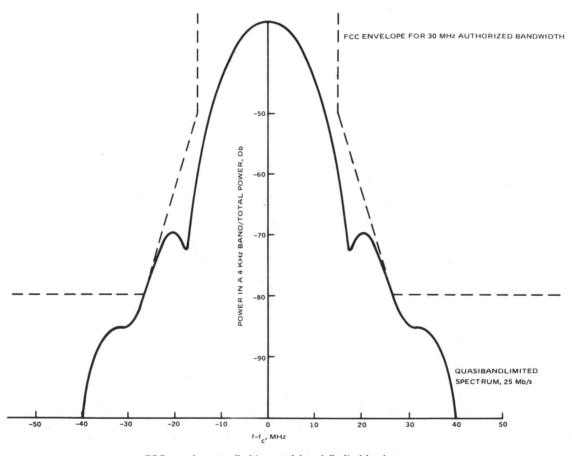

FCC envelope applied to quasi-band-limited keying.

TABLE I
Bandwidths for Digital Data Modulation, Units of Bit Rate

BW Type Modulation type	99% Energy Containment BW	Null-to-Null BW	35 dB BW	50 dB BW	Noise BW	Half-Power BW	Passband Gabor BW
BPSK	20.56	2.00	35.12	201.04	1.00	0.88	infinite
QPSK	10.28	1.00	17.56	100.52	0.50	0.44	infinite
MSK	1.18	1.50	3.24	8.18	0.62	0.59	0.50
SFSK	2.20	1.72	3.20	4.71	0.73	0.70	0.61
Quasi-band-limited	0.95	No well-defined nulls	1.68	2.38	0.48	0.47	0.37

bandwidth. It is interesting to ask whether this pair of spectrum envelopes can constitute a single consistent definition of bandwidth. In other words, if a given data spectrum just satisfies the 40 MHz envelope, then does it also just satisfy the 30 MHz envelope when the data rate is decreased by the factor ¾? The answer is no. In explaining why, it is legitimate to assume that the data signal power is fixed, since A is a ratio with signal power in the denominator.

First observe that the 40 MHz envelope first reaches −80 dB at $f_c \pm 34.49$ MHz. The 30 MHz envelope first reaches −80 dB at $f_c \pm 26.34$ MHz. Say that the data spectrum, multiplied by 4 kHz, experiences a sidelobe peak of exactly −80 dB amplitude at exactly $f_c \pm 35.12$ MHz. The 40 MHz envelope is just satisfied there. Now decrease the data rate by the factor ¾. This causes two things to happen. First, the sidelobe peak moves to $f_c \pm ¾ \times 35.12$ MHz = $f_c \pm 26.34$ MHz. Second, the entire power spectral density *rises* by the ratio $10 \log_{10} 4/3$ or 1.25 dB, in order to keep the total power constant. Now the sidelobe peak has value −78.75 dB and therefore violates the −80 dB bound at $f_c \pm 26.34$ MHz for 30 MHz bandwidth. It is clear that the two envelopes shown in Fig. 8 do not imply a single consistent definition of bandwidth, hence the FCC Rule does not constitute a definition of bandwidth.

For any given authorized bandwidth it should be possible to calculate the maximum permissible data rate for any given modulation scheme as long as the power spectral density is known. The inset, which examines the FCC spectrum envelope, gives an example.

SUMMARY

Table I summarizes the bandwidths of the modulation schemes considered. A very wide range of values is represented. Perhaps this will give a hint of the implications of the strict application of bandwidth definitions to some of the classical modulation types.

If in certain instances, BPSK and QPSK seem to be giving rise to unreasonably large bandwidths, it must be remem-

bered that the ideal waveform assumptions have been rigidly adhered to. In practice, the inherent bandwidth limitations of system components will likely modify the spectrum at large multiples of the data rate. This will change the spectrum and reduce the bandwidth to an extent dependent heavily on the specific application at hand.

Recent work [20] has shown that offset QPSK can be filtered and then hard limited to produce a spectrum comparable with the quasi-band-limited keying discussed here. The resulting spectrum is known only through Monte Carlo simulation and has much of the raggedness characteristic of Fig. 1 of the present article. From such simulation results it is difficult to compute bandwidth under the many definitions given here.

Although a great deal of effort has been bent toward the maximization of bandwidth efficiency in various contexts, very little attention has been given to the optimization of spread-spectrum DSPN modulation. With the substantially different criteria, this field presents an interesting challenge of its own.

REFERENCES

[1] H. Nyquist, "Certain topics in telegraph transmission theory," *Trans. AIEE*, vol. 47, pp. 617-644, Apr. 1928.
[2] A. S. Taylor and L. H. Janes, "Field testing the performance of a cable TV system," *Proc. IEEE*, vol. 58, pp. 1086-1102, July 1970.
[3] H. M. Sierra, "Design of a pulse narrowing network," *ElectroTechnol.*, vol. 72, pp. 38-42, Sept. 1963.
[4] W. R. Bennett and J. R. Davey, *Data Transmission*. New York: McGraw-Hill, 1965, pp. 305-306.
[5] D. Slepian, "On bandwidth," *Proc. IEEE*, vol. 64, pp. 292-300, Mar 1976.
[6] R. E. Eaves and S. M. Wheatley, "Optimization of quadrature-carrier modulation for low crosstalk and close packing of users," *IEEE Trans. Commun.*, vol. COM-27, pp. 176-185, Jan. 1979.
[7] M. Engleson and L. Garret, "Digitial radio measurements using the spectrum analyzer," *Microwave J.*, vol. 23, pp. 35-48, Apr. 1980.
[8] D. Divsalar and M. Simon, "Spectral characteristics of convolutionally encoded digital signals," *IEEE Trans. Commun.*, vol. COM-28, pp. 173-186, Feb. 1980.
[9] D. Kahn, "Cryptology goes public," *IEEE Commun. Mag.*, vol. 18, pp. 19-28, Mar. 1980.
[10] D. Williams, "OK Police, Fire Department use of digitized voice systems," *Electron. News*, p. 22, Jan. 30, 1978.

[11] F. Amoroso, "Pulse and spectrum manipulation in the minimum (frequency) shift keying (MSK) format," *IEEE Trans. Commun.*, vol. COM-24, pp. 381-384, Mar. 1976.

[12] M. G. Pelchat, "The autocorrelation function and power spectrum of PCM/FM with random binary modulating waveforms," *IEEE Trans. Space Electron. Telem.*, vol. SET-10, pp. 39-44, Mar. 1964.

[13] F. Amoroso, "The use of quasi-bandlimited pulses in MSK transmission," *IEEE Trans. Commun.*, vol. COM-27, pp. 1616-1624, Oct. 1979.

[14] G. B. Lockhart, "A spectral theory for hybrid modulation," *IEEE Trans. Commun.*, vol. COM-21, pp. 790-800, July 1973.

[15] P. H. Wittke and G. S. Deshpande, "Investigation of bandwidth efficient modulation schemes," Queen's University at Kingston, Ont., Canada, Dep. Elec. Eng., Res. Rep. 79-2, Mar. 1979.

[16] C. L. Cuccia, J. L. Spilker, and D. T. Magill, "Digital communications at gigahertz data rates," *Microwave J.*, vol. 13, pp. 89 ff, Jan. 1970.

[17] R. C. Dixon, *Spread Spectrum Systems.* New York: Wiley, 1976, Section 2.1

[18] D. Gabor, "Theory of communication," *J. Inst. Elec. Eng.*, Pt. 3, vol. 93, pp. 429-457, 1946.

[19] A. H. Nuttall and F. Amoroso, "Minimum Gabor bandwidth of *M* orthogonal signals," *IEEE Trans. Inform. Theory*, vol. IT-11, pp. 440-444, July 1965.

[20] S. A. Rhodes, "Effects of hardlimiting on bandlimited transmissions with conventional and offset QPSK modulation," in *Conf. Rec., 1972 Nat. Telecommun. Conf.*, Dec. 4-6, 1972, IEEE Publ. 72 CHO 601-5-NTC, pp. 20F-1-20F-7.

Frank Amoroso was born in Providence, RI, on July 31, 1935. He received the B.S. and M.S. degrees in electrical engineering from the Massachusetts Institute of Technology in 1958. From 1958 to 1960 he pursued further graduate studies in electrical engineering at Purdue University. From 1964 through 1966 he studied mathematical analysis under F. G. Tricomi at the University of Turin, Italy, as a guest of the Italian government. He has served on the technical staffs of Edgerton, Germeshausen, and Grier, Inc., M.I.T. Instrumentation Laboratory, Melpar Applied Science Division, Litton Systems Advanced Development Laboratory, RCA Laboratories (David Sarnoff Research Center), Mitre Corporation, Collins Radio Company, Rockwell International Autonetics and Space Divisions, and the Northrop Corporation Electronics Division. Since 1972 he has been with the Hughes Aircraft Company in Fullerton, CA.

He served to the rank of First Lieutenant at the Institute for Exploratory Research, U.S. Army Electronics Command, Ft. Monmouth, NJ, from 1961 to 1962, and held an instructorship in electrical engineering at Purdue University from 1958 to 1960.

He received the RCA Laboratories Award in 1964 for "research leading to improved digital magnetic recording," and is now a Registered Professional Engineer in California. He currently appears in *Who's Who in Technology Today*, *Who's Who in the West*, and *Who's Who in California*.

SPECTRUM CONSERVATION BY EFFICIENT CHANNEL UTILIZATION *

Edward Bedrosian, Senior Member, IEEE

Abstract   The growth of the communication industry has resulted in ever-increasing demands on the radio spectrum. This, in turn, has spurred the exploitation of previously unused frequency bands and the development of techniques for making better use of those already available. In this paper, spectral utilization is examined using the information-theory techniques of Shannon. The power-bandwidth tradeoff for the ideal channel is established and the performance of practical analog and digital modulation techniques is compared with this ideal. It is shown that well-designed systems tend to operate near the "knee" of the power-bandwidth tradeoff curve for the ideal channel and that they are frequently within 10 dB or less of the performance of the ideal.

INTRODUCTION

The growing demand for communications, particularly between widely separated points, has resulted in an ever-increasing proliferation of radio circuits. This growth has been most dramatic in the area of satellite communications where a number of civilian and military systems employing a variety of repeaters currently provide telephone, television, and data circuits, frequently on a global basis. As might be expected, the problem of maintaining mutual interference at acceptable levels is becoming a matter of increasing concern.

A number of things can be done to improve our capability of satisfying this growing demand for communications. These include redundancy removal to reduce the information rate generated by the sources, the use of efficient transmission techniques to make the best use of existing channels, and the development of additional channels. The second of these is considered in this paper.

The term channel, as used here, denotes a characteristic or distinguishable segment of the radio spectrum that is characterized by its bandwidth W. The channel involves an electromagnetic propagation path that joins the transmitter and receiver as illustrated in the block diagram of Fig. 1. The source is assumed to generate information, which may be either discrete, consisting of digital data produced at a specified rate, or continuous, consisting of band-limited analog functions (i.e., those whose spectra lie within a specified band). Equipment used to remove redundancy from the raw information is assumed to be contained in the source; the corresponding equipment to reintroduce the redundancy, if desired, is assumed to be contained in the destination.

The transmitter comprises the equipment to modify the source information in a suitable manner and to impose it on a carrier that is appropriate to the channel—this process is known as modulation. The power output of the transmitter is denoted by P and the spectrum of the signal it radiates is assumed to be contained in the bandwidth W of the channel. The attenuation or spreading loss associated with a conventional radio link and all frequency-dependent effects are disregarded in the communication model. Thus, the channel is simply a unity gain device that uniformly conveys the transmitter output to the receiver. Its important attribute is that it is assumed to contribute an additive Gaussian noise having a uniform power spectral density N_0 to the signal. Under these conditions, the signal power at the input to the receiver is P and the noise power at the same point is

$$N = N_0 W. \tag{1}$$

The receiver, which is assumed to be noiseless (or whose noise can be suitably attributed to the channel), demodulates the carrier to recover the source information.

It is clear that the noise added to the signal in the channel degrades the quality of the information being transmitted. If

This paper is based in part on research for the Defense Advanced Research Projects Agency under Contract DAHC15-73-C-0181.

The author is with The RAND Corporation, Santa Monica, CA 90406.

*Reprinted from *IEEE Communications Society Magazine*, March 1977, Vol. 15, No. 2, pp. 20–27.

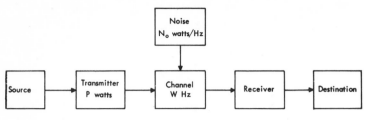

Fig. 1.   Block diagram of radio communication system.

the signal is to be transmitted with a specified quality and if the additive noise density N_0 is taken as a characteristic of the channel, the communication problem becomes one of selecting or devising modulation techniques that make the best use of the transmitter power and the channel bandwidth. Inasmuch as there are direct costs associated with generating power and the radio spectrum is a limited commodity, there is a strong incentive to reduce the use of each as much as possible. Not surprisingly, it develops that for a given rate of information transmission, the reduction of one usually requires an increase in the other, i.e., there is a power-bandwidth tradeoff at work. This leads to two basic questions: 1) How are the power and bandwidth related theoretically for ideal systems? 2) How do practical systems compare with this ideal?

The approach here will be to use the results developed in modern information theory to answer the first question and to summarize the results available in the technical literature of communication theory to describe the performance of practical systems in an attempt to answer the second. It should be noted that only the single-user case is considered—practical systems must operate in the presence of other users with the result that there is an inevitable mutual interference that may dominate the effects of simple channel noise. Future theoretical studies should include a consideration of the effects of mutual interference.

THE THEORETICAL LIMIT

Although the basic limitations to the communication of information had been understood in an approximate way for some time, it was not until the work of Shannon in 1948 [1] that the concepts were formulated rigorously. The principal result (Theorem 17) states that the capacity C of a channel of bandwidth W perturbed by white thermal noise of power N when the average transmitter power is limited to P is given by

$$C = W \log \left(1 + \frac{P}{N} \right). \tag{2}$$

When W is in Hz and the base 2 is used for the logarithm, the capacity is given in bits/s. It is possible to transmit digital information over such a channel at a rate R, where $R \leqslant C$, with an arbitrarily small frequency of errors by using a sufficiently complicated coding system. It is not possible to transmit information at a rate $R > C$ without a finite error rate.

Shannon [2] used the formula for channel capacity to exhibit graphically the bound for the performance achievable by practical systems. This plot, which is shown in Fig. 2, gives

the normalized channel capacity C/W, in (bits/s)/Hz, as a function of the signal-to-noise ratio in the channel. A plot of its reciprocal, i.e., the normalized channel bandwidth W/C in Hz/(bits/s) as a function of the signal-to-noise ratio in the channel, is sometimes used [3] to illustrate the power-bandwidth tradeoff inherent in the ideal channel. However, such a plot is not a pure tradeoff because it can be seen from (1) that the noise power N is proportional to the bandwidth W. Substituting in (2) and rearranging leads to

$$\frac{P}{N_0 C} = \frac{W}{C} \left[2^{C/W} - 1 \right] \tag{3}$$

which is a true power-bandwidth tradeoff. In (3), both sides have been divided by C to yield dimensionless ratios and the base 2 logarithm has been used.

The ratio P/C in (3) is the energy per bit. To show this, note that a signal occupying a channel of width W Hz has, according to the sampling theorem [4], $2W$ independent samples/s and that when the channel is operating at capacity it can support an information rate of C bits/s. Therefore, the channel can convey up to $C/2W$ bits/sample. On the other hand, a source of power P W, operating in a channel of width W Hz, and therefore generating $2W$ independent samples/s, is transmitting energy at a rate $(P/2W)$ (W/[sample/s]) or J/sample. Thus the ratio P/C is simply the energy in J/bit, E_b, and (3) can be written

$$\frac{E_b}{N_0} = \frac{W}{C} \left[2^{C/W} - 1 \right]. \tag{4}$$

The normalized channel capacity C/W, in (bits/s)/Hz, is plotted as a function of the bit energy-to-noise density ratio E_b/N_0 in Fig. 3. Like Fig. 2, it sets the limiting performance that can be achieved by practical systems. The true power-bandwidth tradeoff appears in Figs. 7 to 9, in which the

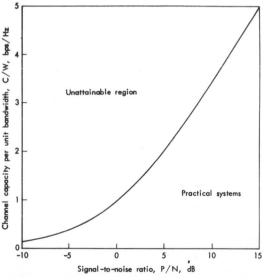

Fig. 2.   Channel capacity normalized to channel bandwidth versus channel signal-to-noise ratio.

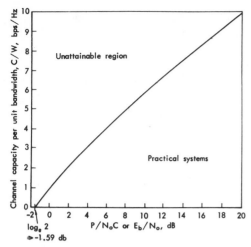

Fig. 3. Channel capacity normalized to channel bandwidth versus normalized channel power-to-noise density ratio or bit energy-to-noise density ratio.

normalized channel bandwidth W/C, in Hz/(bits/s), is plotted as a function of the bit energy-to-noise density ratio E_B/N_0.

INFORMATION RATE

It should be noted that the plot in Fig. 3 exhibits a finite limit [5]. From (4) it can be shown that

$$\lim_{\substack{W \to \infty \\ \text{or} \\ C \to 0}} \frac{E_b}{N_0} = \log_e 2 = 0.693 = -1.59 \text{ dB}. \qquad (5)$$

The existence of a limiting value of E_b/N_0 below which no error-free information rate can be supported per unit of bandwidth (Fig. 3) or below which even an infinite bandwidth cannot support a finite information rate (Figs. 7 to 9) is in seeming contradiction with the traditional statement of capacity as given by (2), which holds that there is always a nonvanishing information per unit bandwidth (or a finite bandwidth that will support a given information rate) regardless of how small the signal-to-noise ratio P/N becomes. That there is no inconsistency can be seen by noting from Fig. 2 that, for a fixed bandwidth, the capacity of the ideal channel (i.e., its information rate) goes to zero as the power is decreased to zero. Thus, the duration of a bit of information (i.e., $1/C$) becomes arbitrarily long so that the energy per bit is indeterminate ($P \cdot 1/C = 0 \cdot \infty$); equation (5) shows that the energy per bit approaches $\log_e 2$. Alternatively, it can be seen that, for a fixed capacity, the bandwidth required gets arbitrarily large as the signal-to-noise ratio is reduced to zero. Thus, the noise power in the widening channel is also getting arbitrarily large so that the signal power is indeterminate ($P/N \cdot N = 0 \cdot \infty$) as the signal-to-noise ratio goes to zero; the energy per bit is again shown by (5) to be the ratio of the finite limiting power to the assumed fixed capacity.

Another apparent contradiction is observed when bit error probabilities are calculated for practical systems. Plots of E_b/N_0 for such systems invariably display a smooth increase of error probability as E_b/N_0 is decreased. For example, the bit-error probability plot for binary phase-shift keying (PSK)

in Fig. 4 shows the bit error probability tending to 0.5 in the limit as E_b/N_0 approaches zero [6]. Thus, there is apparently always a nonvanishing information rate regardless of how small E_b/N_0 becomes. The trouble is that the value of E_b traditionally used in such calculations for practical systems is invariably the signal energy per *transmitted bit*. However, the meaning of E_b in (4) is the signal energy per bit of *received information*. Thus, the information loss caused by the noisy channel must be taken into account; this can be done by calculating the equivocation [1]. The rate of actual transmission R is equal to the rate of production $H(x)$, i.e., the entropy of the source, less the average rate of conditional entropy.

$$R = H(x) - H_y(x). \qquad (6)$$

For the binary PSK case, the rate of production is 1 bit/digit. When, for example, the value of E_b/N_0 per transmitted bit is 0.1 (−10 dB), the error probability is 0.32 (from Fig. 4) and the actual rate becomes

$$R = 1 + 0.32 \log_2 0.32 + 0.68 \log_2 0.68$$
$$= 0.0956 \text{ bits/ digit}.$$

Hence,

$$\frac{E_b}{N_0} = \frac{(E_b/N_0) \ (\text{J/digit}) \ / \ (\text{W/Hz})}{R \qquad \text{bits/digit}}$$

$$= 1.046 \frac{\text{J/bit}}{\text{W/Hz}} = + 0.19 \text{ dB}.$$

Thus, the apparent value of -10 dB for E_b/N_0 is actually equal to 0.19 dB/received bit, which is well above the limit of −1.59 dB.

A similar information-rate reduction is shown in Fig. 5 for pulse-code modulation (PCM), where the output information rate per unit bandwidth is shown as a function of the output signal-to-noise ratio [7]. When the signal-to-noise ratio in the

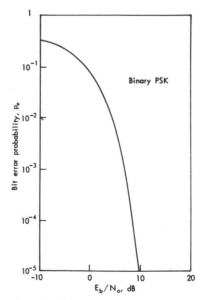

Fig. 4. Bit error probability as a function of energy per transmitted bit-to-noise density ratio.

channel is large, the output signal-to-noise ratio is high because few errors are made and the information rate approaches a limiting value of 2 (bits/s)/Hz (when binary digits are being sent). As the channel signal-to-noise ratio decreases, the increased rate of error production reduces both the output signal-to-noise ratio and the rate of information receipt in the manner illustrated in Fig. 5.

The information rate per unit bandwidth for 7-digit PCM systems is shown in Fig. 6 as a function of channel signal-to-noise [8]. It illustrates the connection between output information rate and channel signal-to-noise ratio alluded to in the discussion of Fig. 5 and, further, introduces the concept of a fidelity criterion for continuous sources. The rate relative to an equivocation criterion is calculated using an error probability that reflects the digital character of PCM. The rate relative to an equivocation criterion is calculated by recognizing that the basic information source is continuous and that the true measure of information transmission is determined by the fidelity with which the continuous signal is reproduced at the destination. Shannon has shown [1, theorem 23] that the information rate for any continuous source of average power P and band W relative to an rms measure of fidelity is bounded by

$$W \log \frac{P_1}{N} \leqslant R \leqslant W \log \frac{P}{N} \qquad (7)$$

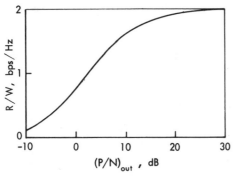

Fig. 5.   Information rate per unit bandwidth for PCM systems versus output signal-to-noise ratio.

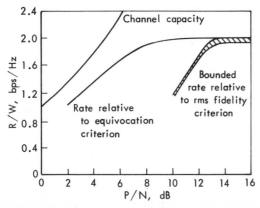

Fig. 6.   Information rate per unit bandwidth versus channel signal-to-noise ratio for 7-digit systems.

where P_1 is the entropy power of the source and N is allowed mean-square error between the original and recovered messages.

The entropy power of a source is defined as the average power of a normally distributed source having the same entropy as the original source. For PCM systems having uniform distributions of amplitudes (or with quantization steps adjusted to approximate a uniform distribution), it can be determined that

$$P_1 = \frac{6}{\pi e} P \qquad (8)$$

which leads to

$$\log \frac{6}{\pi e} \left(\frac{P}{N}\right)_{out} \leqslant \frac{R}{W} \leqslant \log \left(\frac{P}{N}\right)_{out} \qquad (9)$$

These bounds are plotted in Fig. 6, which takes into account the relationship between input and output signal-to-noise ratios for 7-digit systems.

It should be noted that the upper bound in (8) is achieved for a source having a Gaussian distribution of amplitudes [1, theorem 22]. Since, for a single sideband (SSB) signal, the channel and information bandwidths are the same and the output signal-to-noise ratio equals the channel signal-to-noise ratio, it follows that the information rate for a Gaussian SSB signal is

$$R = W \log \frac{P}{N} \qquad (10)$$

which differs trivially from the channel capacity, (2), at moderate signal-to-noise ratios. Interest in rate-distortion theory, as this subject is called, is increasing [9] - [11].

BANDWIDTH AS A PARAMETER

In continuous systems, two bandwidths are apparent. One of these is the bandwidth of the channel, which, with the exception of the reference to SSB in the preceding section, has been the only bandwidth of interest so far; the other is the bandwidth of the source. The source signal is usually a smooth time function of voltage or current that is the analog of speech sound waves or television light intensity, for example. Such functions can be filtered or otherwise constrained in frequency content so as to be said to be contained within a frequency band often referred to as the source or information bandwidth. Information rate is meaningless for such a signal because the signal has independent samples at intervals equal to the reciprocal of twice its bandwidth and each of these samples represents a point on a continuum of values. In effect, the information rate is infinite. Only by degrading its quality, as was done by introducing the concept of a fidelity criterion in connection with (7), can an information rate be ascribed to a continuous source.

The process of modulation, as applied to continuous signals, is one of transforming the information-bearing signal so that it (or a function of it) appears as either the amplitude and/or phase (or, equivalently. the frequency) of a carrier. This results in the familiar SSB or double sideband (DSB) signals, with or without the carrier, phase modulation (PM), frequency modulation (FM), and hybrid systems such as single sideband FM (SSBFM) [12] which uses both amplitude and phase modulation simultaneously. The consequence is always a modulated signal that has a bandwidth equal to or greater than that of the information signal. For example, the channel bandwidth is equal to the source bandwidth for SSB as already noted. is doubled for DSB, and is doubled (at least) or more for PM and FM. Early speculation that small-deviation FM had a reduced bandwidth was dispelled by Carson [13]. There is at present no known method for reducing the bandwidth of a continuous signal without removing redundancy or degrading its quality.

Digital systems are in decided contrast to continuous systems in that they are characterized by a data rate rather than by a bandwidth. Only by assigning waveforms to the data streams so that they assume specified levels of amplitude or phase at regular sampling intervals can the data stream be said to have a bandwidth. But inasmuch as the sampling intervals can be put as far apart as desired, for a given data rate, by sending more information in each sample, it is clear that the bandwidth can be made as small as desired. (Of course, the higher power required to permit distinguishing the larger number of states possible in each sample is the price paid for the bandwidth reduction. This is the simplest manifestation of the power-bandwidth tradeoff.)

The bandwidth of a digital waveform can also be increased as much as desired, first by reducing the number of bits per sample (and, hence, decreasing the interval between samples) until only binary states are being transmitted and, second, by introducing redundancy into the data stream by suitable encoding.

It can be seen from the foregoing that the entire range of the power-bandwidth tradeoff can be exploited when digital data are being transmitted whereas, for continuous sources, the channel bandwidth must be equal to or greater than the source bandwidth unless the source quality is degraded. The latter is usually done by digitizing the continuous source.

THE POWER-BANDWIDTH TRADEOFF

The normalized channel bandwidth plots in Figs. 7 to 9 illustrate the basic power-bandwidth tradeoff that can be effected in the limit for an ideal communication system; as noted earlier, only the plot against E_b/N_0 is a pure tradeoff. A characteristic "knee" is in evidence—the knee would be even sharper if the abscissa were plotted in linear units rather than in decibels.

The existence of this knee means that systems seeking to reduce the channel bandwidth they occupy or to reduce the signal level they use must make an increasingly unfavorable exchange in the other parameter. For example, from Fig. 7, an ideal system operating at an E_b/N_0 of 1.8 dB and using a normalized bandwidth of 0.5 Hz/(bits/s) would have to increase E_b/N_0 to 20 dB to reduce the bandwidth occupancy to 0.1 Hz/(bits/s).

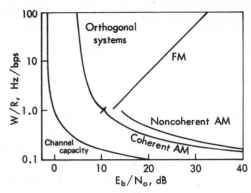

Fig. 7. Power-bandwidth tradeoffs for orthogonal and digital-equivalent continuous AM and FM systems for a bit error probability of 10^{-6}.

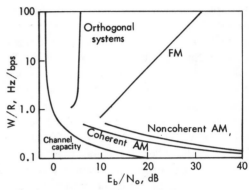

Fig. 8. Power-bandwidth tradeoffs for orthogonal and digital-equivalent continuous AM and FM systems when the quantizing and error noise powers are equal.

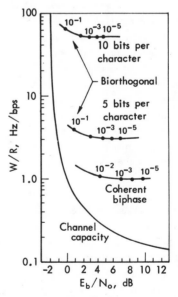

Fig. 9. Power-bandwidth tradeoffs for coherent biphase and biorthogonal coding.

Thus, a 66-fold increase in power is required to achieve a 5-fold reduction in bandwidth. Tradeoffs in the other direction are even more frustrating because even an arbitrarily large increase in bandwidth cannot reduce the signal level by more than about 3.4 dB, which is a factor of only about 2.2. For example, from Fig. 9, quadrupling the bandwidth to 2

Hz/(bits/s) reduces the E_b/N_0 required by about 2.6 dB, a factor of about only 1.8.

Practical systems display power-bandwidth tradeoffs that have a variety of shapes, though many agree with that of the ideal channel in a general way. Power-bandwidth tradeoff plots are shown in Figs. 7 to 9 for a variety of digital and digitized-analog modulation techniques. Orthogonal and digital-equivalent continuous systems using amplitude modulation (AM) with coherent and noncoherent reception and frequency modulation (FM) are shown in Fig. 7 for the case when the error probability is 10^{-6} and in Fig. 8 for the case when the quantizing and error noise powers are equal [5]. The AM systems reduce bandwidth by sending more information per sample. When more information per sample is sent in FM systems the frequency deviation is increased, thereby increasing bandwidth. The orthogonal systems, which use signaling waveforms that are uncorrelated with respect to one another, increase bandwidth by introducing redundancy. Thus, the systems shown typify the bandwidth contraction and expansion techniques just described. Except for the FM systems, they are seen roughly to parallel the performance of the ideal channel, showing a displacement of about 7 to 10 dB for orthogonal systems and from about 10 to 20 dB for the AM systems. These values reflect the degree to which they fail to achieve the ideal. Wide-band FM systems are seen to be very inefficient in their use of the channel even though they do permit a power reduction in absolute terms.

Another power-bandwidth tradeoff for digital systems is portrayed in Fig. 9 for coherent biphase and two biorthogonal codes [14]; these are systems that use orthogonal codes and their negatives to increase the alphabet size. Bit error probability is given as a parameter on these plots to indicate how, in the spirit of the discussion relating to (6), the equivocation caused by errors reduces the information per digit as E_b/N_0 is decreased. If lines of equal error probability were drawn through these points, they would be seen to yield power-bandwidth tradeoff curves similar to the one in Fig. 7 for orthogonal systems. For an error probability of 10^{-5}, the displacement from the ideal is again seen to be about 7 to 10 dB.

In the coherent biphase system, the transmission consists of a reference carrier phase or its opposite. This amounts to transmitting the carrier or its negative, which corresponds to suppressed-carrier DSB AM. Inasmuch as there are 2 samples/s/Hz and 1 bit/sample when using binary digits, information is being generated at the rate of 2 (bits/s)/Hz—the use of a DSB modulation system then results in a bandwidth utilization of 1 Hz/(bit/s), which is the limiting value for large E_b/N_0 in Fig. 9. The two multidimensional biorthogonal systems use larger bandwidths and less energy/bit for the same error rate; it is apparent that a considerable bandwidth expansion must be used to achieve only moderate power savings.

SYSTEM COMPARISONS

The normalized channel-capacity plot shown in Fig. 3 is closely related to the normalized channel-bandwidth plots of Figs. 7 to 9 that served as the basis for the power-bandwidth tradeoffs illustrated previously for various systems. In this Section comparisons are presented for similar systems using the form of Fig. 2 as a basis. The principal difference is that a power-bandwidth tradeoff illustrates how power and bandwidth may be exchanged whereas the normalized channel-capacity plots emphasize the way in which increasing power increases information rate. In both cases, the comparison with the ideal is displayed.

The performance of PCM and quantized pulse-position modulation (PPM) systems [2] is shown in Fig. 10 for an error probability of 10^{-5}. The parameter for the PCM systems is the number of bits per sample (so that the point 5 denotes a PCM system with 32 levels of quantization). The dashed line joining the points has no significance but serves to illustrate how the set of points parallels the capacity of the ideal channel. The practical PCM systems are seen to require about 8 dB more power than the ideal system; at large signal-to-noise ratios multilevel PCM systems are within 4.5 dB of the ideal [15]. The parameter for the PPM systems is the number of discrete positions the pulse is allowed to assume (thus, the 4-position system transmits 2 bits/sample). The PPM systems with only a few pulse positions fall up to 10 to 12 dB short of the ideal. At small signal-to-noise ratios, the multiposition systems approach to within 3 dB of the channel capacity.

FM systems and, again, PCM are shown [16] in Fig. 11. The PCM computations are based on [4] and the resulting curve is similar to the one in Fig. 10. The parameter α on the FM curves is the ratio of the channel bandwidth W to the message bandwidth B, i.e.,

$$\alpha = \frac{W}{B} \qquad (11)$$

so it is actually the bandwidth-expansion factor. It is related, roughly, to the FM modulation index

$$m = \frac{\Delta f}{B} \qquad (12)$$

(where Δf is the peak frequency deviation) by Carson's rule

$$W = 2B \, (1 + m). \qquad (13)$$

The normalized information rate for the FM systems is not calculated as it was in Figs. 7 and 8. Here, the information rate is found by calculating the capacity of the output (after demodulation) using the channel-capacity formula given by (2); it is then normalized with respect to the channel bandwidth to form R/W. This approach is approximately valid with respect to an equivocation criterion to the extent that the information rate of the output can be said to be given by (10). In general, in agreement with Figs. 7 and 8, wide-band FM systems are seen to make very inefficient use of bandwidth.

A number of digitized AM, FM, and PM systems are compared in Fig. 12 for a 10^{-4} error probability [17]; the parameters on the plots refer to the number of levels of quantization that are used. It can be seen that polar baseband (i.e., having zero average value), vestigial side-band suppressed

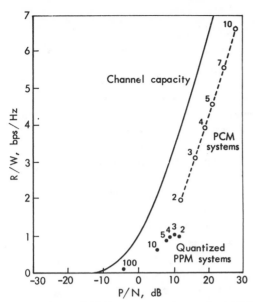

Fig. 10. Comparison of PCM systems and quantized PPM systems with channel capacity for 10^{-5} error probability.

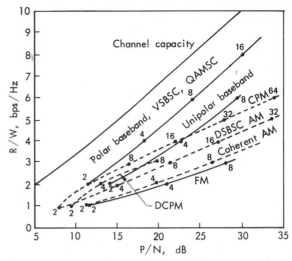

Fig. 12. Comparison of AM, FM, and PM systems with channel capacity for a 10^{-4} error probability.

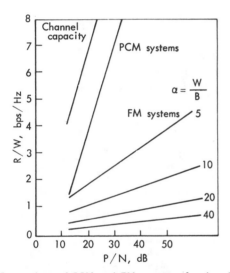

Fig. 11. Comparison of PCM and FM systems of various bandwidth expansion factors with channel capacity.

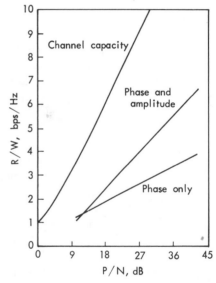

Fig. 13. Comparison of phase only and phase and amplitude modulation with channel capacity.

carrier (VSBSC), and quadrature AM suppressed carrier (QAMSC) are the most efficient; unipolar baseband is less efficient because of the presence of a dc component. Conventional AM with suppressed carrier (DSBSC) is better than full-carrier AM, even with coherent detection; in fact, the latter is hardly better than FM. PM with coherent detection (CPM) is roughly comparable to unipolar baseband and is particularly desirable at low signal-to-noise ratios using binary signaling. Using differentially coherent detection (DCPM) results in somewhat reduced efficiency (a difference of 3 dB is typical).

The plots in Fig. 12 suggest that there is a possibility for improved performance by combining PM and AM. This is indeed the case, as can be seen from Fig. 13, which compares the performance of phase-only and combined phase and ampli-

tude modulation using coherent detection [18]. The number of states for the two systems are not indicated but, typically, they increase with the channel signal-to-noise ratio as they do in Fig. 12. The combined phase and amplitude system is optimized in the sense that the number of states in phase and in amplitude and the threshold levels are adjusted to maximize the information rate for a given signal-to-noise ratio. In general, the phase and amplitude states are not independent because it is clear that the number of equally discernible phase states increases with the amplitude level. At the crossover point, the system performances are equal at a signal-to-noise ratio of 11 dB using 8 states each. At higher signal-to-noise ratios, where more states can be accommodated, combined phase and amplitude modulation makes more efficient use of the channel.

CONCLUSION

The power-bandwidth tradeoff inherent in the ideal communication channel specifies the manner in which these quantities can be exchanged for one another in an optimum sense. Typically, the tradeoff curve displays a characteristic knee in the vicinity of which power-bandwidth trades are equally favorable in either direction. At values of bandwidth or power that are removed in either direction from the knee, an increasingly unfavorable tradeoff is required of one variable to effect a given change in the other. The power-bandwidth tradeoffs for practical systems generally display a behavior similar to that of the ideal channel and, except for some specialized wide-band systems, tend to be not far removed from the knee.

The efficiency of the ideal channel is measured by its capacity per unit of bandwidth. It increases with signal-to-noise ratio and provides the limiting performance against which to compare practical systems. Well designed systems, which are often near the knee of the power-bandwidth tradeoff in practical implementations, tend to be within 10 dB or so of the ideal. SSB and similar analog and digital systems are typical of such efficient systems. FM, particularly when wideband, makes inefficient use of the spectrum although it does offer a convenient mechanism for effecting a power-bandwidth tradeoff. Combined phase and amplitude systems offer promise for improved efficiency but they may suffer from the effects of amplitude fluctuations and interference from other systems.

The search for efficient modulation schemes is complicated greatly by factors beyond the efficiency as measured in an information-theoretic sense. Since these factors tend to be economic in character they introduce considerations that can be cited here but are not evaluated.

One of these is the efficient use of amplifiers—FM, for example, uses amplifiers driven to saturation and therefore capable of delivering a greater power output for a given power input than they could if they were required to transmit amplitude information as well. Also, there is the complexity associated with the modulation and demodulation equipment and, in some systems, in the encoding and decoding equipment. Synchronization, which is either desirable because it improves detection efficiency or is necessary because detection cannot be achieved without it, can be difficult to achieve in some circumstances.

Interference effects may, in fact, dominate efficiency considerations. Information-theoretic considerations based on the additive Gaussian channel may be relevant in many applications but when these considerations are used to make best use of the spectrum, the inevitable mutual interference between practical channels make the model less applicable. The inherent resistance of a given modulation technique to adjacent and cochannel interference may be of greater importance than the efficiency with which it uses its channel in the absence of interference. Frequency modulation is a good example of technique with enhanced interference resistance.

The potential for improvement in spectrum utilization can, in light of the foregoing, be expressed only in the most general terms. Broadly speaking, improvements can be obtained by exploiting one or both of the following system aspects.

1) *The degree to which well designed systems fail to achieve the theoretical limit:* Improvements in this area are very difficult to achieve because they usually result from the development of new and perhaps revolutionary techniques. A sophisticated coding technique such as convolutional encoding and sequential decoding is an example. The use of combined phase and amplitude modulation is another.

2) *The extent to which the power-bandwidth tradeoff can be exploited to increase total system capacity at the expense of power:* A complicated interplay of technological and practical factors results when this is attempted in actual systems involving many users.

It is likely that only small improvements in spectrum utilization can be realized along either of these lines. The first offer little potential because well designed systems are already so near the ideal and the second because it would require the unwelcome relaxation of economic constraints.

REFERENCES

[1] C. E. Shannon, "A mathematical theory of communication," *Bell Syst. Tech. J.,* vol. 27, pp. 379-423, July 1948, and vol. 27, pp. 623-656, Oct. 1948; also C. E. Shannon and W. Weaver, *The Mathematical Theory of Communication.*   Urbana, IL: Univ. of Illinois Press, 1949.

[2] C. E. Shannon, "Communication in the presence of noise," *Proc. IRE,* vol. 37, pp. 10-21, Jan. 1949.

[3] D. A. Bell, *Information Theory.*   London, England: Pitman and Sons, 1953.

[4] B. M. Oliver, J. R. Pierce, and C. E. Shannon, "The philosophy of PCM," *Proc. IRE,* vol. 36, pp. 1324-1331, Nov. 1948.

[5] R. W. Sanders, "Communication efficiency comparison of several communication systems," *Proc. IRE,* vol. 48, pp. 575-588, Apr. 1960.

[6] W. C. Lindsey and M. K. Simon, *Telecommunication Systems Engineering.*   Englewood Cliffs, NJ: Prentice-Hall, 1973.

[7] H. F. Mayer, "Principles of pulse code modulation," in *Advances in Electronics,* vol. III, L. Marton, Ed.   New York: Academic, 1951.

[8] E. Bedrosian, "Weighted PCM," *IRE Trans. Inform. Theory,* vol. IT-4, pp. 45-49, Mar. 1958.

[9] T. J. Goblick, Jr., and J. L. Holsinger, "Analog source digization: A comparison of theory and practice," *IEEE Trans. Inform. Theory,* vol. IT-13, pp. 323-326, Apr. 1967.

[10] F. Jelinek, "Evaluation of distortion rate functions for low distortions," *Proc. IEEE,* vol. 55, pp. 2067-2068, Nov. 1967.

[11] H. H. Tan and K. Yao, "Evaluation of rate-distortion functions for a class of independent identically distributed sources under an absolute magnitude criterion," *IEEE Trans. Inform. Theory,* vol. IT-21, pp. 59-64, Jan. 1975.

[12] E. Bedrosian, "The analytic signal representation of modulated waveforms," *Proc. IRE,* vol. 50, pp. 2071-2076, Oct. 1962.

[13] J. R. Carson, "Notes on the theory of modulation," *Proc. IRE,* vol. 10, pp. 57-64, Feb. 1922.

[14] R. C. Hansen, "Comparison of digital modulation method in active satellite communication systems," in *Space Radio Communication,* G. M. Brown, Ed.   Amsterdam, The Netherlands: Elsevier, 1962.

[15] H. S. Black, *Modulation Theory.*   Princeton, NJ: Van Nostrand, 1953.

[16] W. L. Wright, "Choice of optimum modulation method in active satellite communication systems," in *Space Radio Communication,* G. M. Brown, Ed.   Amsterdam, The Netherlands: Elsevier, 1962.

[17] W. R. Bennett and J. R. Davey, *Data Transmission.* New York: McGraw-Hill, 1965.

[18] J. C. Hancock and R. W. Lucky, "Performance of combined amplitude and phase-modulated communication systems," *IRE Trans. Commun. Syst.,* vol. CS-8, pp. 232-237, Dec. 1960.

Edward Bedrosian (S'51-A'53-M'54-SM'56-F'77) was born in Chicago, IL, on May 22, 1922. He received the B.Sc., M.S., and Ph.D. degrees from Northwestern University, Evanston, IL, in 1949, 1950, and 1953, respectively.

After graduation he joined the Research Staff of Motorola, Inc., in Chicago and, later, in Riverside, CA, where he was engaged in studies of antennas and communication systems. Since 1959 he has been with The RAND Corporation, Santa Monica, CA, conducting studies of military and civilian communication and satellite systems.

Dr. Bedrosian is a member of Sigma Xi.

CORRELATIVE CODING:
A BANDWITH-EFFICIENT
SIGNALING SCHEME *

Subbarayan Pasupathy

Abstract   In most transmission channels, bandwidth is at a premium, and an important attribute of any good digital signaling scheme is its ability to make efficient use of the bandwidth. Conventional Nyquist-type pulse amplitude modulation signaling schemes, which are designed to eliminate intersymbol interference, achieve high data rates only at the expense of a large number of signal levels. In many applications, correlative coding or partial response signaling, which introduces intersymbol interference in a controlled way, is able to achieve high data rates with fewer levels and hence with better error rate performance. In addition to higher data rates, correlative schemes achieve convenient spectral shapes and have error-detecting capabilities without introducing redundancy into the data stream. This paper explains how correlative schemes work and why they are advantageous.

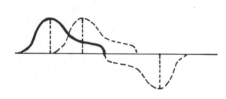

INTRODUCTION

The correlative coding technique, also known as partial response signaling, originated in the 1960's and was initially applied to data transmission over private and switched telephone circuits [1]. Recently, and for the first time, the method has been successfully employed in voice communication where the analog voice signal is transmitted as digital pulses by means of pulse-code modulation (PCM). A new electronic repeater system (duobinary) that allows 48 separate telephone conversations to be transmitted where only 24 were sent before was announced by GTE Lenkurt Inc. during 1976 [2]; this effectively doubles the capacity of most existing T1-type telephone cables. This recent application illustrates the key feature and the main advantage of a correlative signaling scheme—namely, its efficient use of available bandwidth. The purpose of this paper is to present a simplistic look at some of the major aspects of correlative schemes.

SYNCHRONOUS DATA TRANSMISSION

The model of a baseband synchronous pulse amplitude modulation (PAM) data transmission system is shown in Fig.

The author is with the Department of Electrical Engineering, University of Toronto, Toronto, Ont., Canada.

Reprinted from IEEE Communications Society Magazine. July 1977, Vol. 15, No. 4, pp. 4–11.

1. Digital transmission is carried out by transmitting a pulse train at the rate of R pulses/s ($R = 1/T$ where T is the pulse spacing). In binary communication, the pulse amplitudes are ± 1 in some convenient units. At the receiving end, a properly synchronized sampler (sampling every T seconds) is, ideally, supposed to recover the pulse amplitude values.

In Fig. 1(a), the amplitudes $\{x_k\}$, represented by a train of impulses, are fed to a transmitting filter which suitably shapes the pulse waveform, producing pulses of reasonable bandwidth. The channel, due to its imperfect amplitude and phase characteristics, distorts the pulse shape, and the random noise which is inevitably present in any communication system adds its own share of interference. The receiving filter tries to undo the damage done by the channel, and the sampler output is typically compared with a predetermined threshold to decide whether +1 or -1 was sent. For example, as shown in Fig. 1(a), the threshold can be set at zero; the transmitted symbol is taken to be +1 if the sampler output is positive and -1 if the output is negative.

Fig. 1(b) shows a convenient representation of the system in Fig. 1(a) from an analysis point of view, where all the transmitting, channel, and receiving filters have been lumped into a single equivalent channel with transfer function $G(f)$. The noise is shown as being added in following $G(f)$. Since the typical channel (e.g., telephone line) is bandlimited, $G(f)$ can be assumed to be zero for frequencies above some W Hz. The

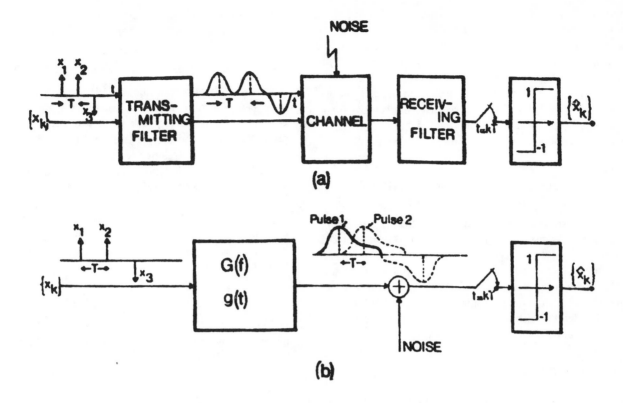

Fig. 1. Synchronous data communication system.

train of pulses will all have the same shape $g(t)$, the impulse response of this equivalent channel.

PULSE SHAPE FOR ZERO-INTERSYMBOL INTERFERENCE

Even in the absence of noise, we see from Fig. 1(b) that symbol detection will, in general, suffer from intersymbol interference, i.e., the tail of one pulse spills over into adjacent symbol intervals so as to interfere with the correct interpretation of the pulse amplitudes at the sampling instants. Nyquist [3] considered the problem of designing the pulse shape $g(t)$ such that no intersymbol interference occurs at the sampling instants. He showed that the minimum bandwidth needed in order to transmit R symbols/s without intersymbol interference is $R/2$ Hz. This occurs when the equivalent channel transfer function is made rectangular and the corresponding pulse shape is

$$g(t) = \frac{\sin(\pi R t)}{\pi R t} \qquad (1)$$

Consequently, $R/2$ is called the "Nyquist bandwidth." Fig. 2 shows this $G(f)$ (in heavy line) and $g(t)$. Even though $g(t)$ has a long tail, note how it passes through zero at the other sampling points, thus generating no interference between successive symbols. When the transmitted symbols are independent, and the noise is uncorrelated at the sampling instants, then each symbol can be recovered without resorting to the past history

of the waveform. Consequently, such systems can be referred to as zero-memory systems.

Nyquist's result can be stated another way; given a channel with bandwidth W Hz, the maximum theoretical rate at which

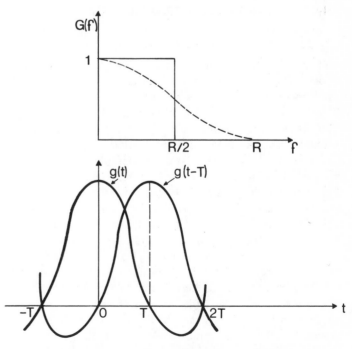

Fig. 2. Pulse shaping in zero-memory systems.

symbols can be sent without interference is $2W$ symbols/s. Thus the maximum symbol rate packing (the symbol rate per hertz of available bandwidth) is 2 symbols/s/Hz, using the pulse shape in Fig. 2. Unfortunately, the pulse shape of (1) is theoretically impossible to achieve without infinite delay. Furthermore, it is impractical to approximate because its very slowly decreasing tail will cause excessive intersymbol interference if any perturbations from the ideal situation occur, such as, for example, a small timing offset in the sampler.

SYMBOL RATE PACKING IN PRACTICAL ZERO-MEMORY SYSTEMS

Nyquist discussed more "practical" shapes for $G(f)$ which have smoother transitions at the spectral band edge, yet also create no intersymbol interference. To maintain a symbol rate of R symbols/s, all these practical filters require more bandwidth than the Nyquist bandwidth of $R/2$. One common shape frequently used is the raised-cosine spectrum with 100 percent "excess bandwidth" (shown in dotted lines in Fig. 2), which occupies twice the Nyquist bandwidth and is given by

$$G(f) = \cos^2\left(\frac{\pi f}{2R}\right), \qquad |f| \leqslant R. \qquad (2)$$

This yields a symbol rate packing of 1 symbol/s/Hz. Thus, one can conclude that under Nyquist's ground rules, even though the theoretical maximum is 2 symbols/s/Hz, the practical symbol rate packing in zero-memory systems is more like 1 symbol/s/Hz. (Typical values of excess bandwidth today range from about 15 to 100 percent.)

DATA RATE INCREASE BY MULTILEVEL TRANSMISSION

A symbol rate of R gives, for binary transmission, a data rate of R bits/s. A logical extension of binary transmission, leading to higher data rate, is the multilevel system where the pulse amplitudes $\{x_k\}$ can take on one of M levels ($M=2$ corresponds to binary). Generally, M is a power of 2, i.e., $M = 2^k$ and thus each symbol or pulse represents k bits of information. Thus, by using M-ary transmission, one can achieve a practical data rate packing of k bits/s/Hz and thus k times the data rate capability of binary systems. This, of course, is at the expense of an increased number of levels, which implies, for fixed signal power, greater sensitivity to noise and poorer error rate performance (i.e., smaller noise levels can now cause one signal level to be confused with an adjacent signal level.) Moreover, multilevel systems require complex equipment. Thus, from the view point of achieving higher data speeds with good error performance, a good signaling scheme should achieve high data rate packing with as few levels as possible. Within the ground rules set by Nyquist, there is little that can be done to improve data rate packing without suffering the effects of a large number of levels. Even though multilevel zero-memory systems are adequate for many applications, for high data rate packing of 4-6 bits/s/Hz, the number of levels required in a zero-memory system becomes prohibitive.

THE CONCEPT OF CORRELATIVE CODING

It was at this time that Lender [4] showed that the ground rules laid by Nyquist were unnecessarily restrictive. One of the key assumptions in Nyquist's work was that the transmitted pulse amplitudes be selected independently. By introducing dependencies or correlation between the amplitudes and by changing the detection procedure, Lender could achieve the symbol rate of $2W$ symbols/s in a bandwidth of W Hz. Thus, the correlative schemes conceived by Lender, and later generalized by Kretzmer [5], could be regarded as a practical means of achieving the theoretical maximum symbol rate packing of 2 symbols/s/Hz postulated by Nyquist, using realizable and perturbation-tolerant filters.

DUOBINARY SCHEME

The basic ideas behind correlative schemes will now be illustrated by considering the example of the duobinary scheme (or "class 1 partial response") for binary transmission. The generalization to other schemes and M-ary transmission is straightforward.

Assume that a sequence $\{x_k\}$ of binary symbols is to be transmitted at the rate of R symbols/s over the ideal rectangular lowpass channel of bandwidth $R/2$ Hz. Let the digits first be passed through a simple digital filter as shown in Fig. 3. This discrete linear filter simply adds to the present digit the value of the previous digit. For every impulse at the input to the digital filter, we get two impulses at the output. Thus, every digit is bound to interfere with the next digit to be transmitted. The symbol sequence $\{y_k\}$ at the channel input is the sum of x_k and the previous symbol value x_{k-1}, i.e.,

$$y_k = x_k + x_{k-1}. \qquad (3)$$

Hence, unlike the zero-memory systems, the input amplitudes $\{y_k\}$ are no longer independent, and this dependency or correlation between transmitted levels can also be thought of as artificially introducing intersymbol interference. Since the ideal Nyquist filter transmits all input amplitudes without distortion, in the absence of noise, we get at the sampler output once again the sequence $\{y_k\}$. Note, however, that the intersymbol interference introduced by the digital filter is of a controlled amount, i.e., the interference in determining $\{x_k\}$ comes only from the preceding symbol $\{x_{k-1}\}$. The controlled amount of interference is the key behind correlative or partial-response signaling. Whereas signaling in zero-memory systems is based on eliminating intersymbol interference, signaling in correlative or finite-memory systems is based on allowing a controlled amount of interference. The beneficial aspects of this interference will be explored in the rest of the paper; an analogy that comes to mind is that of semiconductors and the benefits which arose from the use of a controlled amount of impurity.

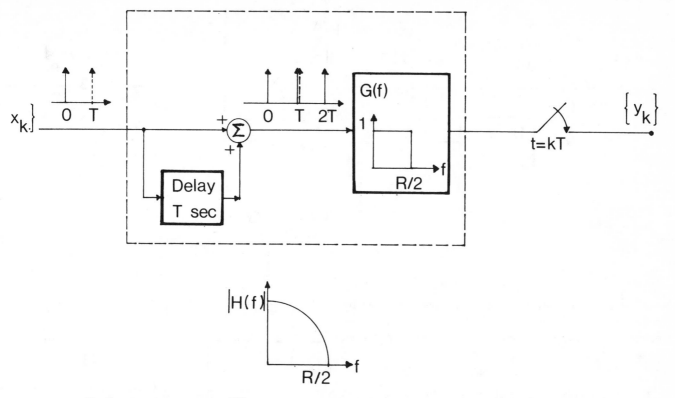

Fig. 3.   Duobinary signaling. $H(f)$ is the equivalent transfer function due to cascade of digital filter and $G(f)$.

DATA DETECTION AND ERROR PROPAGATION

The cascade of the digital filter and the ideal channel $G(f)$ in Fig. 3 can be shown to be equivalent to

$$|H(f)| = 2 \cos (\pi f/R), \qquad |f| \leqslant R/2. \qquad (4)$$

This is because a delay element has a transfer function $e^{-j2\pi fT}$, so the digital filter has the transfer function $1 + e^{-j2\pi fT}$, and hence

$$\begin{aligned} H(f) &= G(f) \, (1 + e^{-j2\pi fT}) \\ &= G(f) \, (e^{j\pi fT} + e^{-j\pi fT})e^{-j\pi fT} \\ &= 2G(f) \cos \pi fT \cdot e^{-j\pi fT}. \end{aligned}$$

Thus, $H(f)$ which has a gradual rolloff to the band edge, can also be implemented by practical and realizable analog filtering and no separate digital filter need be used. The corresponding impulse response $h(t)$ is shown in Fig. 4 and one can see that there are only two nonzero samples.

If x_k is ± 1, y_k is seen from (3) to have three values: -2, 0, and 2. Thus, the binary input is converted into a three-level output and, in general, for M-ary transmission, we get 2M-1 levels. However, the data can still be detected correctly using (3). If x_{k-1} has been decided, its effect on y_k can be eliminated by subtraction and x_k can be decided, etc. This demodulation technique (known at present as nonlinear decision feedback equalization) is essentially an inverse of the operation of the digital filter at the transmitter. However, one immediately apparent drawback of this demodulation technique is that errors, once made, tend to propagate. This is due to the fact that a decision on the present digit x_k depends on the correctness of the decision on the previous digit x_{k-1}. A

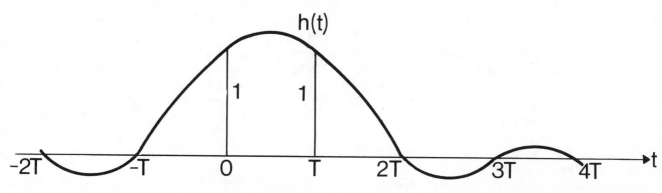

Fig. 4.   Duobinary pulse shape.

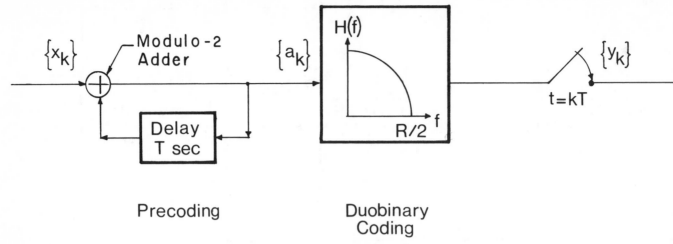

Fig. 5.   A precoded duobinary scheme.

means of avoiding this error progapation was introduced by Lender [4] and is known as "precoding." Unlike the *linear* operation of duobinary coding, this is a *nonlinear* operation.

PRECODING

The precoding operation performed on the data sequence x_k converts it into another binary sequence $\{a_k\}$ (see Fig. 5). Reverting to original symbol values $x_k = 0,1$ we can form the a_k's by

$$a_k = x_k \oplus a_{k-1}. \qquad (5)$$

($\oplus$ represents modulo 2 summation of the binary digits, equivalent to a logical "exclusive—or".) The (0,1) sequence $\{a_k\}$ is converted into a (-1,1) sequence $\{b_k\}$ by the rule

$$b_k = 2a_k - 1. \qquad (6)$$

Then, the transmitted pulse in the nth symbol interval is $\pm h(t-nT)$, corresponding to $b_n = \pm 1$. From (3) with precoded data, then, the received symbols are given by

$$y_k = b_k + b_{k-1}. \qquad (7)$$

Using (5)-(7) we can use an example input sequence $\{x_k\}$ to set up the table in Fig. 6: In the absence of noise, we can decode by the rule

$$x_k = 0, \qquad \text{if } y_k = \pm 2$$
$$x_k = 1, \qquad \text{if } y_k = 0.$$

In the presence of noise, the received signal will be rectified and the data sequence will be obtained according to the rule

$$x_k = \begin{cases} 0, \text{if} & |y_k| \geqslant 1 \\ 1, \text{if} & |y_k| \leqslant 1. \end{cases}$$

Since no knowledge of any sample other than y_k is involved in deciding x_y, error propagation cannot occur.

The generalization of binary to M-ary transmission is straightforward. The precoding will involve mod-M addition,

x_k		0	0	1	0	1	1
a_k	1	1	1	0	0	1	0
b_k	1	1	1	-1	-1	1	-1
y_k		2	2	0	-2	0	0

Fig. 6.   Data sequences in duobinary scheme.

and a mod-M operation must be taken prior to the M-ary decision process at the receiver.

OTHER PARTIAL RESPONSE SIGNALING SCHEMES

Generalization of duobinary scheme to other schemes, collectively known as "partial response" schemes, has been done [5]. The duobinary scheme gives rise to intersymbol interference from only one previous pulse amplitude. By choosing various combinations of integer values for the coefficients $\{f_k\}$ in Fig. 7, one can obtain different useful shapes $H(f)$ having a controlled amount of interference over a span of a few digits. For example, if $f_0=1$ and $f_2=-1$ are the only two nonzero coefficients, the digital filter subtracts the second previous symbol from the present symbol. This scheme is widely used and is known as class 4-partial response or modified duobinary scheme. The overall spectral shape $H(f)$ is a half cycle sinusoid and is given by

$$|H(f)| = \sin (2\pi f/R), \qquad f \leqslant R/2.$$

Unlike the duobinary, the modified duobinary has no dc component and is therefore easily adaptable to single sideband

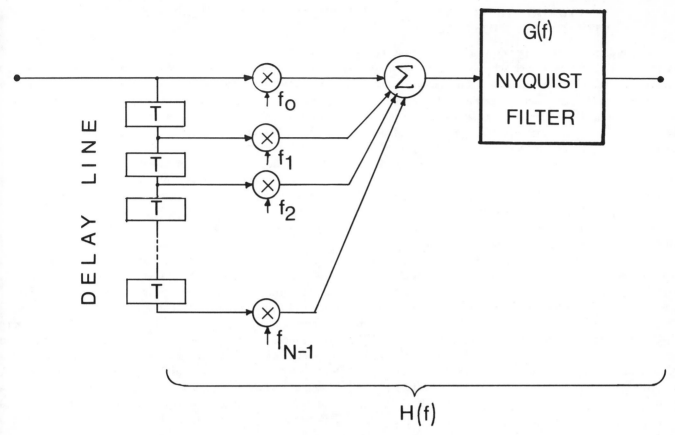

H(f)

Fig. 7.   Generalized partial response coding.

(SSB) modulation. (Very sharp cutoff filters would not be required since there is little signal power near the carrier frequency.)

CHOICE OF PARTIAL RESPONSE SIGNALING SCHEME

The choice of $H(f)$ for any particular application is dictated mainly by some of the following considerations.

1) *Choice of G(f):* In order to maximize data rate in the available bandwidth, many partial response systems are designed to occupy the minimum Nyquist bandwidth of $R/2$, i.e., $G(f)$ in Fig. 7 is chosen to be the idea rectangular filter. Other choices of $G(f)$ (such as raised cosine shape) occupying larger bandwidth are possible and used often. The choice depends on the actual bandwidth available and the data rate desired.

2) *Spectral nulls:* Reduced low frequency components in the spectrum are desirable in systems such as transformer coupled circuits, dc powered cables, SSB modems, and carrier systems with carrier pilot tones. Thus a spectral null at $f=0$ is often a desirable feature for $H(f)$. Similarly, a null at $f=R/2$ is also quite useful—a pilot tone inserted at this point can be used for clock recovery. In addition, a general requirement will be that there be no nulls or severe ripples in the middle of the passband.

3) *Output levels:* The number of output levels for a prac-

tical correlative system is a compromise between the complexity of instrumentation and the data rate desired. In addition, there is a tendency for the error performance at a given signal-to-noise ratio to degrade with a large number of output levels. Five output levels for binary input can be considered as a practical upper bound.

Using the above criteria, a systematic search of various $H(f)$ shapes can be made, and nine of the more useful systems have been analyzed and their performances compared [6]. All of these schemes have three or five output levels for binary input. The two systems which stand out because of their simplicity, useful spectral shapes, and performance measures are the duobinary [see (4)] and the modified duobinary [see (9)] schemes. These are also the ones that are widely used in practice. For example, in the recently developed GTE Lenkurt repeater 9148A, the modified duobinary scheme is used (even though, for brevity, it is referred to as duobinary repeater). Besides the obvious advantages of the spectral nulls at $f=0$ and $f=R/2$, the shape of the spectral density of modified duobinary can be shown to be the same as that of the bipolar code used in T-1 systems. Also, fault-locating patterns and error-detecting procedures are similar. Thus, the modified duobinary system is compatible with the existing T1-type systems; compatibility with existing systems may also be an important factor in choosing a partial response scheme.

HIGH DATA RATE PACKING
WITH FEWER OUTPUT LEVELS

The high data rate packing of correlative schemes is probably the major advantage of such schemes over multilevel zero memory schemes. For channels with gradual cutoff, a characteristic of most physical channels, correlative techniques often permit signaling with fewer levels than zero-memory schemes for the same data rate packing. Fewer levels implies better signal-to-noise ratio, and hence better error performance. If the equivalent channel $G(f)$ in a zero-memory system is assumed to be a raised cosine filter with bandwidth of R Hz, an $M(=2^k)$ level $\{x_k\}$ sequence in Fig. 1 would achieve a data rate packing of k bits/s/Hz. On the other hand, if a duobinary or modified duobinary scheme is used on a channel with the same bandwidth of R Hz, a $2^{k/2}$-level x_k sequence in Fig. 5 would achieve the same k bits/s/Hz packing with only $2(2^{k/2})-1$ output levels. For example, a data rate packing of 4 bits/s/Hz would require 16 levels in a zero-memory system and only 7 levels in a correlative system. This is one of the main reasons for choosing a 7-level modified duobinary scheme to transmit a 1.544 Mbit/s data stream in the approximately 500 kHz available in the baseband of the existing microwave radio channel normally used for carrying voice channels. This technique, making efficient use of the normally unused frequency band under voice channels, is called a Data Under Voice scheme [7].

SPEED TOLERANCE

Another advantage of correlative schemes is that the high data rate packing is achieved by means of practical filtering which is tolerant to timing perturbations. Speed tolerance is a measure commonly used to measure the sensitivity of a partial response system to changes in the signaling rate. It is convenient to define the concept of an eye pattern and then define speed tolerance in terms of the eye pattern. An eye pattern for a random input pulse train is formed by superimposing the received pulse train waveform (before the sampler in Fig. 5) over a 1-symbol interval. This is done experimentally by observing the received waveform on an oscilloscope synchronized externally with a clock running at R Hz. As a result, a pattern resembling a human eye is formed.

The eye pattern for a duobinary scheme, in the absence of noise, is shown in Fig. 8. Note that at the best sampling time (where the eye opening is maximum), there are only 3 levels $(0, \pm2)$. This is because the duobinary waveform $h(t)$ (Fig. 4) has only two nonzero samples of equal value at the desired sampling rate of R pulses/s. But if the signaling rate (and hence the synchronized sampling rate) is higher than R, there will be many more nonzero sample values and the intersymbol interference due to these will cause many more levels to spread out around the 3 levels in Fig. 8. If the decision thresholds are placed midway in the gaps between the levels, the system will not make any errors until the levels overlap in one of the eyes

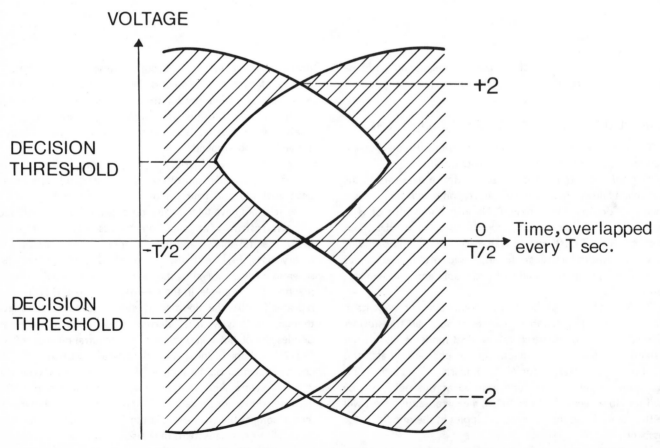

Fig. 8.   Duobinary eye pattern for binary transmission.

(i.e., the eye closes). Speed tolerance can be defined as the increase in transmission rate at which the smallest eye closes. Above this rate, some data sequences will cause errors even in the absence of noise.

The duobinary and modified duobinary schemes have speed tolerances of 43 and 16 percent, respectively, over and above the Nyquist rate. These can be contrasted with the ideal rectangular filter zero-memory system which has zero speed tolerance. The speed tolerance figures for correlative schemes also suggest that if some intersymbol interference can be tolerated, a rate even higher than R symbols/s in a bandwidth of $R/2$ Hz can be employed.

ERROR DETECTION AND SYSTEM MONITORING

Another unique characteristic of correlative schemes is that errors can be detected without the transmitter having to introduce any redundancy into the original data stream. Error detection provides a convenient way of monitoring the transmission line condition.

The principle of error detection can be explained in the 3-level duobinary scheme considered before. From the table in Fig. 6, we can see that because of the correlation between digits, only certain types of transitions are allowed. The following constraints apply to the 3-level duobinary scheme:

1) A positive (negative) level at one sampling time may not be followed by a negative (positive) level at the next sampling instant.

2) If a positive (negative) peak is followed by a negative (positive) peak, they must be separated by an odd number of center samples.

3) If a positive (negative) peak is followed by another positive (negative) peak, they must be separated by an even number of center samples.

The duobinary waveform can be constantly checked by a level violation monitor, so that any violation of these rules results in an error indication. Although certain combinations of bit errors will be compensatory in nature and will not show up as 3-level violations, the 3-level error rate will be a good indication of the actual error rate. Similar predetermined rules can be laid out for other partial response formats, and violations of these patterns can be used for error detection. Other schemes of error control have also been proposed [8], [9].

OTHER FEATURES AND CONCLUSIONS

While the high data rate packing with perturbation-tolerant filtering, convenient spectral shape, and error-detection capability can be termed as the three major attractive features of correlative or partial response schemes, there are other aspects that this paper has not gone into. Partial response schemes have been applied to various modulation systems such as FM [10] and SSB [11] and have been considered for various line-of sight propagation applications [12]. A generalized form of partial response precoding has been introduced where the coefficients f_k in Fig. 7 need not be integers [13]. Correlative coding has also been compared to other coding schemes [14]. Maximum likelihood sequence decoding, making full use of the correlation between signal levels, has been considered [15], [16]. Special features for equalization of partial response systems have been examined [17]. It is safe to predict that since bandwidth is usually at a premium in any transmission medium, correlative schemes will continue to be used because of their high data rate packing and simple implementation.

ACKNOWLEDGMENT

The author is sincerely grateful to A. Lender for his advice and suggestions in compiling this paper.

REFERENCES

[1] A. Lender, "Correlative level coding for binary data transmission," *IEEE Spectrum*, vol. 3, pp. 104-115, Feb. 1966.
[2] *IEEE Spectrum*, vol. 14, pp. 46-47, Jan. 1977.
[3] H. Nyquist, "Certain topics on telegraph transmission theory," *Trans. AIEE*, vol. 47, pp. 617-644, Apr. 1928.
[4] A. Lender, "The duobinary technique for high speed data transmission," *IEEE Trans. Commun. Electron.*, vol. 82, pp. 214-218, May 1963.
[5] E.R. Kretzmer, "Generalization of a technique for binary data communication," *IEEE Trans. Commun. Technol.*, vol. COM-14, pp. 67-68, Feb. 1966.
[6] P. Kabal and S. Pasupathy, "Partial-response signaling," *IEEE Trans. Commun.*, vol. COM-23, pp. 921-934, Sept. 1975.
[7] K.L. Seastrand and L.L. Sheets, "Digital transmission over analog microwave systems," in *Conf. Rec., IEEE Int. Conf. Commun.*, 1972, pp. 29-1 - 29-5.
[8] J.W. Smith, "Error control in duobinary data systems by means of null-zone detection," *IEEE Trans. Commun. Technol.*, vol. COM-16, pp. 825-830, Dec. 1968.
[9] H. Kobayashi, and D.T. Tang, "On decoding of correlative level coding systems with ambiguity zone detection," *IEEE Trans. Commun. Technol.*, vol. COM-19, pp. 467-477, Aug. 1971.
[10] T.L. Swartz, "Performance analysis of a three-level modified duobinary digital FM microwave system," in *Conf. Rec., IEEE Int. Conf. Commun.*, 1974, pp. 5D-1 - 5D-4.
[11] F.K. Becker, E.R. Kretzmer, and J.R. Sheehan, "A new signal format for efficient data transmission," *Bell Syst. Tech. J.*, vol. 45, pp. 755-758, May-June 1966.
[12] P.A. Bello et al., "Line-of-sight techniques investigation," *Rome Air Development Center, Rep.* AD/A-006 104, Jan. 1975.
[13] H. Harashima and H. Miyakawa, "Matched-transmission technique for channels with intersymbol interference," *IEEE Trans. Commun.*, vol. COM-20, pp. 774-780, Aug. 1972.
[14] H. Kobayashi, "A survey of coding schemes for transmission or recording of digital data," *IEEE Trans. Commun. Technol.*, vol. COM-19, pp. 1087-1100, Dec. 1971.
[15] ——, "Correlative level coding and maximum-likelihood decoding," *IEEE Trans. Inform. Theory*, vol. IT-17, pp. 586-594, Sept. 1971.
[16] G.D. Forney, Jr., "Maximum likelihood sequence estimation of digital sequences in the presence of intersymbol interference," *IEEE Trans. Inform. Theory*, pp. 363-378, May 1972.
[17] K.H. Mueller, "A new, fast-converging mean-square algorithm for adaptive equalizers with partial-response signaling," *Bell Syst. Tech. J.*, vol. 54, pp. 143-153, Jan. 1975.

Subbarayan Pasupathy (M'73) was born in Madras, India, on September 21, 1940. He received the B.E. degree in telecommunications from the University of Madras, India, in 1963, the M. Tech. degree in electrical engineering from the Indian Institute of Technology, Madras, in 1966, and the M. Phil. and Ph.D. degrees in engineering and applied science from Yale University, New Haven, CT, in 1970 and 1972, respectively.

During 1965 to 1967 he was a Research Scholar at the Indian Institute of Technology, Madras, and worked as a Teaching Assistant at Yale University during 1968 to 1971. During 1972 to 1973 he was a Post-Doctoral Fellow at the University of Toronto, Toronto, Ont., Canada, where he is now an Assistant Professor of Electrical Engineering. His research interests include digital communication, statistical communication theory, and signal processing.

MINIMUM SHIFT KEYING: A SPECTRALLY EFFICIENT MODULATION *

Subbarayan Pasupathy

Compact power spectrum, good error rate performance, and easy synchronization make MSK an attractive digital modulation technique.

The ever increasing demand for digital transmission channels in the radio frequency (RF) band presents a potentially serious problem of spectral congestion and is likely to cause severe adjacent and cochannel interference problems. This has, in recent years, led to the investigation of a wide variety of techniques for solving the problem of spectral congestion. Some solutions to this problem include: 1) new allocations at high frequencies; 2) better management of existing allocations; 3) the use of frequency-reuse techniques such as the use of narrow-beam antennas and dual polarizing systems; 4) the use of efficient source encoding techniques; and 5) the use of spectrally efficient modulation techniques [1]. This article will consider the last approach and analyze, in particular, a modulation scheme known as minimum shift keying (MSK). The MSK signal format will be explained and its relation to other schemes such as quadrature phase shift keying (QPSK), offset QPSK (OQPSK), and frequency shift keying (FSK) pointed out. The main attributes of MSK, such as constant envelope, spectral efficiency, error rate performance of binary PSK, and self-synchronizing capability will all be explained on the basis of the modulation format.

The author is with the Department of Electrical Engineering, University of Toronto, Toronto, Ont., Canada.

*Reprinted from *IEEE Communications Society Magazine*, July 1979, Vol. 17, No. 4, pp. 14–22.

SPECTRAL EFFICIENCY AND MSK

In any communication system, the two primary communication resources are the transmitted power and channel bandwidth. A general system-design objective would be to use these two resources as efficiently as possible. In many communication channels, one of the resources may be more precious than the other and hence most channels can be classified primarily as power-limited or band-limited. (The voice-grade telephone circuit, with approximately 3 kHz bandwidth, is a typical band-limited channel, whereas space communication links are typically power limited). In power-limited channels, coding schemes would be generally used to save power at the expense of bandwidth, whereas in band-limited channels "spectrally efficient modulation" techniques would be used to save bandwidth.

The primary objective of spectrally efficient modulation is to maximize the bandwidth efficiency, measured in bits/s/Hz.

The primary objective of spectrally efficient modulation is to maximize the *bandwidth efficiency*, defined as the ratio of data rate to channel bandwidth (in units of bits/s/Hz). A secondary objective of such modulation schemes may be to achieve this bandwidth efficiency at a prescribed average bit error rate with minimum expenditure of signal power. Some channels may have other restrictions and limitations which may force other constraints on the modulation techniques. For example, communication systems using certain types of nonlinear channels call for an additional feature, namely, a constant envelope, which makes the modulation impervious to such impairments. This is needed because a memoryless nonlinearity produces extraneous side-bands when passing a signal with amplitude fluctuations. Such sidebands introduce out-of-band interference with other communication systems.

A typical example where such considerations are appropriate is in time-division multiple access (TDMA) satellite communication, where the traveling wave tube (TWT) amplifiers are operated near power saturation for high efficiency. For frequency division multiple access (FDMA) application also, constant envelope properties are useful at each ground terminal, if the high power amplifier is operated near power saturation like a class "C" device, where the response would be nonlinear. In this article, we will be concerned with these and similar applications which call for a constant envelope, bandwidth-efficient, digital-modulation technique. Recent investigations into signaling schemes for such applications have centered upon MSK.

Modulation studies during the late 1960's led to the development of MSK [2], [3]. MSK was used by the Data Transmission Co. (Datran) for its proposed data network in 1972 [4]. Other applications which have considered and/or used MSK since then include a proposed AT&T domestic satellite system [5],[6], military tactical radio [7], extremely low frequency (ELF) underwater communication systems [8], and Canadian communications technology satellite (CTS) experiments [9]. We will examine the major attributes of MSK which make it a suitable candidate for such applications. We begin with a brief review of some related modulation techniques such as FSK, PSK, QPSK, and OQPSK.

FSK and PSK

The constraint of a constant envelope feature for the modulation scheme narrows the search to two major signaling techniques, namely, FSK and PSK. Consider binary communication—transmitting a pulse every T seconds (at the signaling rate of $1/T$ baud) to denote one of two equally likely information symbols, $+1$ or -1. FSK denotes the two states by transmitting a sinusoidal carrier at one of two possible frequencies, whereas binary PSK (BPSK) uses the two opposite phases of the carrier (i.e., 0 and 180°). Fig. 1 shows typical signals in the two types of modulation. Note that BPSK is also equivalent to amplitude modulating the carrier with the information bit stream, i.e., multiplication with $+1$ or -1. The two schemes can be compared on the basis of their bit error rate (BER) performance (i.e., the average number of errors in transmitting a long bit stream)

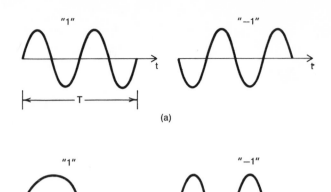

Fig. 1.   (a) Binary phase shift keying; (b) Frequency shift keying.

through an ideal channel. The ideal channel is taken to be a linear all-pass channel, corrupted only by additive white Gaussian noise with a constant (one-sided) power spectral density of N_0 W/Hz. The required ratio, E_b/N_0, of signal energy per bit (E_b) and noise level N_0 to achieve a given BER (such as 1 error in 10^5 bits) is the quantity of interest.

It can be shown that optimum receivers for binary signals in such channels call for matched filters (also known as correlator-detectors) with perfect carrier phase reference available at the receiver. For such coherent receivers, which base their bit decisions after observing the signal over T seconds, there exists a class of signals, of which PSK is one, which turns out to be optimum in the sense of requiring the minimum amount of E_b/N_0 for a specified BER. This optimum class of signals is called "antipodal," i.e., the two signals denoting the two possible information symbols have exactly the same shape but opposite polarity. On the other hand, from the viewpoint of simpler receiver implementation, noncoherent detection schemes, which do not make use of the carrier phase reference, can also be used. For example, a noncoherent detector for FSK can use two bandpass filters tuned to the two frequencies, followed by envelope detectors and bit-rate samplers, and base the binary decision on which of the two sampled envelopes is the larger. It can be shown that for such noncoherent detection, the optimum class of signals is the class of noncoherently orthogonal signals. (Orthogonal signals are those that do not interfere with one another in the process of detection. If the demodulation process is coherent, they can be called coherently orthogonal and if noncoherent detection is employed, noncoherently orthogonal. For example, in noncoherent FSK signaling using envelope detectors, the two FSK signals at frequencies f_1 and f_2 are said to be noncoherently orthogonal, if, when a tone at f_1 is transmitted, the sampled envelope of the output of the receiving filter tuned to f_2 is zero, i.e., no cross talk.) In the case of FSK, the minimum separation between the two frequencies equals the signaling rate $1/T$ for noncoherent orthogonality. The performance of such a noncoherently orthogonal FSK scheme is much poorer than the coherent antipodal PSK scheme. Even if coherently orthogonal FSK is used and detected by coherent methods in an effort to improve performance, it is still poorer by 3-dB (in terms of E_b/N_0) than PSK.

The poorer performance of FSK has been responsible for restricting the use of FSK mainly to low-data-rate low-efficiency applications, while PSK has been the preferred scheme for efficient higher-data-rate applications. We will show later that this is not a fair assessment of FSK. In fact, MSK, a coherently orthogonal FSK modulation scheme, requires only $1/2T$ Hz frequency separation and achieves a performance equivalent to that of BPSK, when a coherent receiver bases its decision after observing the signal over 2 bit periods ($2T$ seconds) rather than just one. Before we view MSK as a particular case of FSK, it is helpful to understand several other PSK schemes, such as QPSK and OQPSK, and to view MSK as a particular variation of OQPSK signaling.

QPSK and OQPSK

The optimum E_b/N_0 performance achievable with BPSK led to a search for mechanisms to improve the bandwidth efficiency of PSK schemes without any loss of performance. It was found that since $\cos 2\pi f_c t$ and $\sin 2\pi f_c t$ (where f_c is the carrier frequency) are coherently orthogonal signals, two binary bit streams modulating the two carrier signals in quadrature can be demodulated separately. (In analog communication, this idea has been used for a long time to multiplex two signals on the same carrier, so as to occupy the same bandwidth, e.g., the two chrominance signals in color television are

modulated onto the color subcarrier this way). Such a modulation scheme, increasing the bandwidth efficiency of binary PSK by two, is known as QPSK and is shown in Fig. 2 (a).

The input binary bit stream $\{a_k\}$, $(a_k = \pm 1)$ $k = 0,1,2,$ $\cdots$ arrives at a rate of $1/T$ baud and is separated into two streams $a_I(t)$ and $a_Q(t)$ consisting of even and odd bits, respectively, as shown in the example waveforms of Fig. 2(a). The two pulse trains modulate the inphase and quadrature components of the carrier and the sum $s(t)$, the modulated QPSK signal, can be represented as

$$s(t) = \frac{1}{\sqrt{2}} a_I(t) \cos (2\pi f_c t + \frac{\pi}{4}) + \frac{1}{\sqrt{2}} a_Q(t) \sin (2\pi f_c t + \frac{\pi}{4}). \quad (1)$$

The two terms in (1) represent two binary PSK signals and can be detected independently due to the orthogonality of $\cos(2\pi f_c t + \frac{\pi}{4})$ and $\sin(2\pi f_c t + \frac{\pi}{4})$. Using a well-known trigonometric identity, (1) can also be written as

$$s(t) = \cos(2\pi f_c t + \Theta(t)) \quad (2)$$

where, as shown in Fig. 3, $\Theta(t) = 0°, \pm 90°$ or $180°$ corresponding to the four combinations of $a_I(t)$ and $a_Q(t)$.

The OQPSK signaling can also be represented by (1) and (2) and the difference between the two modulation techniques is only in the alignment of the two bit streams. The odd and even bit streams, transmitted at the rate of $1/2T$ baud, are synchronously aligned in QPSK [as shown in Fig. 2(a)] such that their transitions coincide. OQPSK modulation is obtained by a shift or offset in the relative alignments of $a_I(t)$ and $a_Q(t)$ data streams by an amount equal to T. Fig. 2(b) shows the offset. (OQPSK is also sometimes referred to as staggered QPSK.)

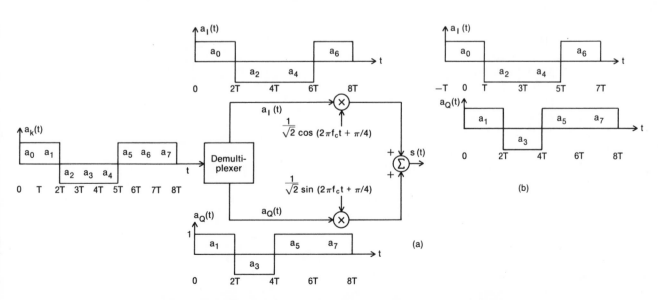

Fig. 2. (a) QPSK modulator; (b) Staggering of data streams in OQPSK.

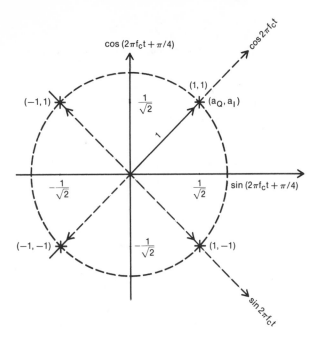

Fig. 3.   Signal space diagram for QPSK and OQPSK.

The difference in time alignment in the bit streams does not change the power spectral density and hence both QPSK and OQPSK spectra have the same $(\sin 2\pi fT/2\pi fT)^2$ shape, associated with the rectangular pulse used for signaling. However, the two modulations respond differently when they undergo bandlimiting and hardlimiting operations, encountered in applications such as satellite communications. The difference in the behavior of the two modulations can be understood by a study of the phase changes in the carrier in the two modulations.

In QPSK, due to the coincident alignment of $a_I(t)$ and $a_Q(t)$ streams, the carrier phase can change only once every $2T$. The carrier phase over any $2T$ interval is any one of the four phases shown in Fig. 3 depending on the values of $\{a_Q(t), a_I(t)\}$. In the next $2T$ interval, if neither bit stream changes sign, the carrier phase remains the same. If one component ($a_I(t)$ or $a_Q(t)$) changes sign, a phase shift of $\pm 90°$ occurs. A change in both components results in a phase shift of $180°$. Fig. 4(a) shows a typical QPSK signal waveform, for the sample sequence $a_I(t)$ and $a_Q(t)$ shown in Fig. 2(a). In a satellite communication system, the modulated signal shown in Fig. 4(a) is bandlimited by a bandpass filter so as to conform to out-of-band spectral emission standards. The bandlimited QPSK will no longer have constant envelope and in fact, the occasional phase shifts of π rad in the carrier will make the envelope go to zero [10]. At the satellite repeater, this signal will undergo hardlimiting which, while restoring the constant envelope to the signal, will at the same time restore essentially all the frequency sidelobes back to their original level prior to filtering. These undesired sidebands negate the

bandlimiting of the QPSK signal carried out at the transmitter and introduce out-of-band radiation on the satellite downlink that may interfere with other communication systems. On the other hand, bandlimiting and hardlimiting operations do not seem to produce the same deleterious effect on an OQPSK signal.

In OQPSK, the binary components cannot change states simultaneously. One component has transitions in the middle of the other symbol and hence only one component can switch at a time. This eliminates the possibility of $180°$ phase changes and phase changes are limited to $0°$, $\pm 90°$ every T seconds. Fig. 4(b) shows a typical OQPSK waveform for the example bit streams in Fig. 2(b). When a OQPSK signal undergoes bandlimiting, the resulting intersymbol interference (smearing of adjacent pulses on one another) causes the envelope to droop slightly in the region of $\pi/2$ rad phase transitions. Since phase shifts of $180°$ have been avoided, the envelope does not go to zero as it does in the bandlimited QPSK case. When the bandlimited OQPSK goes through a hard limiter, the slight envelope droop is removed by the limiting process. However, limiting affects only the envelope and the phase is preserved. Consequently, the absence of rapid phase shifts (and hence high frequency content) in the region of a $\pi/2$ phase change means that limiting will not regenerate the high frequency components originally removed by the bandlimiting filter. Thus, out-of-band interference is avoided.

Tests [10] indicate that unlike QPSK, the spectrum of OQPSK after limiting remains essentially unchanged and seems to retain its bandlimited nature in almost its entirety. OQPSK signals also seem to perform better than QPSK in the presence of phase jitter associated with noisy reference carriers [11]. Furthermore, the offset of T s between the 2 bit streams in OQPSK has been shown to be optimum in terms of phase jitter

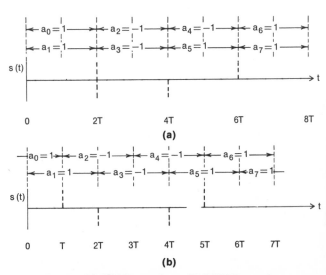

Fig. 4.   (a) QPSK waveform; (b) OQPSK waveform.

immunity in the presence of additive Gaussian noise [12]. All these advantages possessed by OQPSK stem mainly from the fact that OQPSK avoids the large phase change of 180° associated with the QPSK format. This suggests that further suppression of out-of-band interference in bandlimiting-hardlimiting applications can be obtained, if the OQPSK signal format can be modified to avoid phase transitions altogether. This can be thought of as an obvious motivation for designing constant envelope modulation schemes with continuous phase. MSK is one such scheme, as will be discussed next.

MSK can be viewed as a form of offset quadrature phase shift keying with a half-sinusoidal rather than rectangular weighting.

MSK

MSK can be thought of as a special case of OQPSK with sinusoidal pulse weighting [13],[14]. Consider the OQPSK signal, with the bit streams offset as shown in Fig. 2(b). If sinusoidal pulses are employed instead of rectangular shapes, the modified signal can be defined as MSK and equals

$$s(t) = a_I(t)\cos(\frac{\pi t}{2T})\cos 2\pi f_c t + a_Q(t)\sin(\frac{\pi t}{2T})\sin 2\pi f_c t.$$

$$(3)$$

Fig. 5 shows the various components of the MSK signal defined by (3). Fig. 5(a) shows the modified in-phase bit stream waveform, for the sample $\{a_I\}$ stream shown in Fig. 2(b). The corresponding values of the even-bits are shown as ± 1 inside the waveform. The in-phase carrier [the first term in (3)], obtained by multiplying the waveform in Fig. 5(a) by $\cos 2\pi f_c t$, is shown in Fig. 5(b). Similarly, the sinusoidally shaped odd-bit stream and the quadrature carrier are shown in Fig. 5(c) and (d). The composite MSK signal $s(t)$, the addition of Fig. 5(b) and (d), is shown in Fig. 5(e). The waveform in Fig. 5(e) can be better understood if we use a well-known trigonometric identity to rewrite (3) as

$$s(t) = \cos[2\pi f_c t + b_k(t)\frac{\pi t}{2T} + \phi_k]$$

$$(4)$$

where b_k is $+1$ when a_I and a_Q have opposite signs and b_k is -1 when a_I and a_Q have the same sign and ϕ_k is 0 or π corresponding to $a_I = 1$ or -1. Note that $b_k(t)$ can also be written as $-a_I(t)a_Q(t)$.

From Fig. 5(e) and (4), we deduce the following properties of MSK.

1) It has constant envelope.

2) There is phase continuity in the RF carrier at the bit transition instants.

3) The signal is an FSK signal with signaling frequencies $f_+ = f_c + 1/4T$ and $f_- = f_c - 1/4T$. Hence the frequency deviation equals half the bit rate, i.e., $\Delta f = f_+ - f_- = 1/2T$.

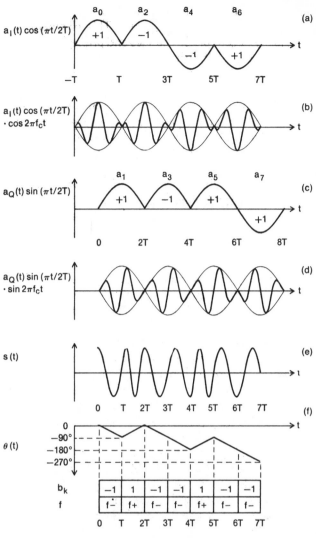

Fig. 5. MSK waveforms.

This is the minimum frequency spacing which allows the two FSK signals to be coherently orthogonal, in the sense discussed in the section on FSK and PSK; hence the name "minimum shift" keying. Since the frequency spacing is only half as much as the conventional $1/T$ spacing used in noncoherent detection of FSK signals, MSK is also referred to as Fast FSK [3].[1]

4) The excess phase of the MSK signal, referenced to the carrier phase, is given by the term

$$\Theta(t) = b_k(t)\frac{\pi t}{2T} = \pm\frac{\pi t}{2T}$$

in (4), which increases or decreases linearly during each bit period of T seconds. A bit b_k of $+1$ corresponds to an increase of the carrier phase by 90° and corresponds to an FSK signal at the higher frequency f_+. Similarly,

---

[1]There are several forms of MSK described in the literature; however, all of them are spectrally equivalent to the version described here. Most of them differ in details, such as bit-to-symbol precoding and the use of differential encoding.

$b_k = -1$ implies a linear decrease of phase by 90° over T s, corresponding to the lower frequency f_-. (In order to make the phase continuous at bit transitions, the carrier frequency f_c should be chosen such that f_c is an integral multiple of $1/4T$, one-fourth the bit rate.) The excess phase $\Theta(t)$ is shown in Fig. 5(f), with the corresponding frequencies and values of b_k shown below.

Thus MSK can be viewed either as an OQPSK signal with sinusoidal pulse weighting or as a continuous phase (CPFSK) signal with a frequency separation equal to one-half the bit rate.

PULSE SHAPING AND POWER SPECTRA

The power spectra of QPSK, OQPSK, and MSK (shifted to baseband) can all be expressed by the magnitude squared of $P(f)$, the Fourier transform of the symbol shaping function $p(t)$. Thus, for QPSK and OQPSK [see (1)]

$$p(t) = \begin{cases} \dfrac{1}{\sqrt{2}} & |t| \leqslant T \\ 0 & \text{elsewhere.} \end{cases} \tag{5}$$

and for MSK [see (3)]

$$p(t) = \begin{cases} \cos\left(\dfrac{\pi t}{2T}\right) & |t| \leqslant T. \\ 0 & \text{elsewhere.} \end{cases} \tag{6}$$

Thus the spectral density $G(f)$ for QPSK and OQPSK (normalized to have the same power) is given by

$$\frac{G(f)}{T} = 2\left(\frac{\sin 2\pi fT}{2\pi fT}\right)^2 \tag{7}$$

and for MSK is given by

$$\frac{G(f)}{T} = \frac{16}{\pi^2}\left(\frac{\cos 2\pi fT}{1-16f^2T^2}\right)^2. \tag{8}$$

The spectra are sketched in Fig. 6.

The difference in the rates of falloff of these spectra can be explained on the basis of the smoothness of the pulse shape $p(t)$. The smoother the pulse shape, the faster is the drop of spectral tails to zero. Thus, MSK, having a smoother pulse, has lower sidelobes than QPSK and OQPSK. A measure of the compactness of a modulation spectrum is the bandwidth B which contains 99 percent of the total power. For MSK, $B \cong (1.2/T)$ while for QPSK and OQPSK, $B \cong (8/T)$ [13]. This indicates that in relatively wide-band satellite links (where, for example, filtering is not used after the nonlinearities), MSK may be spectrally more efficient than QPSK or OQPSK. However, as can be seen from Fig. 6, the MSK spectrum has a wider mainlobe than QPSK and OQPSK. This suggests that in narrow-band satellite links, MSK may not be the preferred method. Computer simulations [15], which take into account all

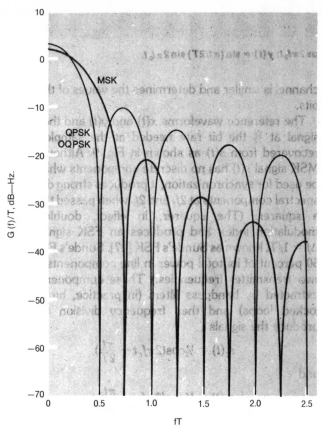

Fig. 6.   Spectral density of QPSK, OQPSK, and MSK.

relevant parts of typical wide-band and narrow-band TDMA satellite links, tend to support the above conclusions. Assuming the transmitter and satellite amplifiers to be operating at power saturation, results [15] show that MSK gives superior performance to QPSK when α, the product of channel spacing and symbol duration exceeds 1.8 and to OQPSK only when α exceeds 2.3. OQPSK is shown to have superior performance to QPSK except when α is less than 1.4. However, it should be pointed out that in realistic system applications [13], [15], the difference in the required (E_b/N_0) for the three schemes seems to be less than 1 dB and the choice of modulation method depends on other, less obvious, criteria.

MSK has some excellent special properties that make it an attractive alternative when other channel constraints require bandwidth efficiencies below 1.0 bit/s/Hz. For example, the continuous phase nature of MSK makes it highly desirable for high-power transmitters driving highly reactive loads [8]. Since intersymbol switching occurs when the instantaneous amplitude of $p(t)$ is zero, the finite rise and fall times and data asymmetry inevitably present in practical situations have a minimal effect on the MSK performance [16]. In addition, as we shall see below, MSK has simple demodulation and synchronization circuits.

MSK TRANSMITTER AND RECEIVER

A typical MSK modulator is shown in Fig. 7 [6],[9],[16]. The multiplier produces two phase coherent signals at frequencies f_+ and f_-. The advantage of this method of forming the binary FSK signals is that the signal coherence and the deviation ratio are largely unaffected by variations in the data rate [9]. The binary FSK signals, after being separated by means of narrow bandpass filters, are properly combined to form the in-phase and quadrature carrier components. These carriers are multiplied with the odd and even bit streams $a_I(t)$ and $a_Q(t)$ [which are offset by T, as in Fig. 2(b)] to produce the MSK modulated signal $s(t)$ [as defined in (3)].

The block diagram of a typical MSK receiver is shown in Fig. 8. The received signal (equal to $s(t)$ of (3) in the absence of noise and intersymbol interference) is multiplied by the respective in-phase and quadrature carriers $x(t)$ and $y(t)$ followed by integrate and dump circuits. The multiplier-integrator constitutes correlation detection or matched filtering, an optimum coherent receiver in the absence of intersymbol interference. Note the integration interval of $2T$ s. The demodulation works as follows: if $s(t)$ of (3) is multiplied by $x(t)$ $(= \cos \pi t / 2T \cos 2\pi f_c t)$, the low frequency component of the result (the output of the integrator) equals $a_I(t)(1 + \cos \pi t / T)/4$ and hence the polarity of the sampler output determines the value of $a_I(t)$. The operation on the quadrature channel is similar and determines the values of the odd-bits.

The reference waveforms $x(t)$ and $y(t)$ and the clock signal at ½ the bit rate needed at the samplers are recovered from $s(t)$ as shown in Fig. 9. Although the MSK signal $s(t)$ has no discrete components which can be used for synchronization, it produces strong discrete spectral components at $2f_+$ and $2f_-$ when passed through a squarer. (The squarer, in effect, doubles the modulation index and produces an FSK signal with $\Delta f = 1/T$, known as Sunde's FSK [17]. Sunde's FSK has 50 percent of its total power in line components at the two transmitter frequencies.) These components are

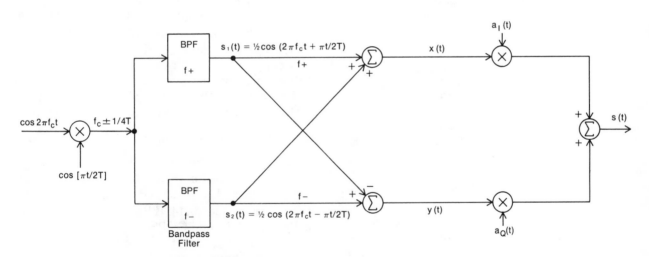

Fig. 7. MSK modulator: $x(t) = \cos(\pi t / 2T) \cos 2\pi f_c t; \; y(t) = \sin(\pi t / 2T) \sin 2\pi f_c t.$

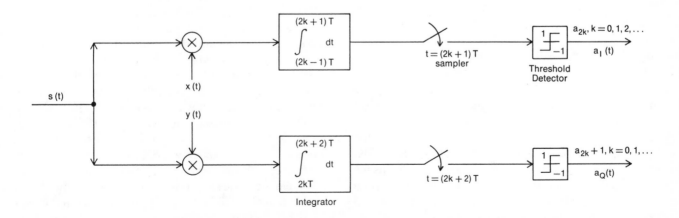

Fig. 8. MSK receiver: $x(t) = \cos(\pi t / 2T) \cos 2\pi f_c t; \; y(t) = \sin(\pi t / 2T) \sin 2\pi f_c t.$

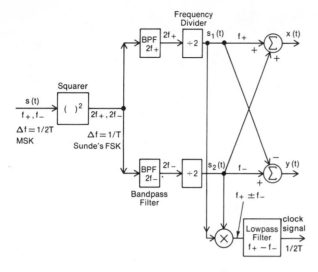

Fig. 9.   Synchronization circuits for MSK: $f_+ = f_c + 1/4T$;
$f_- = f_c - 1/4T$.

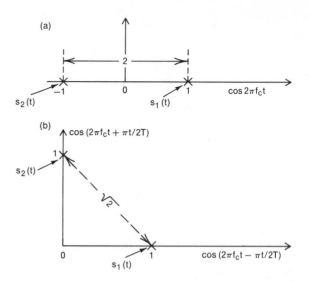

**Fig. 10.   Signal space diagrams for (a) BPSK (antipodal);
and (b) FSK (orthogonal).**

extracted by bandpass filters (in practice, by phase-locked loops) and then frequency division circuits produce the signals

$$s_1(t) = \tfrac{1}{2}\cos(2\pi f_c t + \frac{\pi t}{2T}) \qquad (9)$$

and

$$s_2(t) = \tfrac{1}{2}\cos(2\pi f_c t - \frac{\pi t}{2T}), \qquad (10)$$

respectively. The sum and difference $s_1 + s_2$ and $s_2 - s_1$ produce the reference carriers $x(t)$ and $y(t)$, respectively. If $s_1(t)$ and $s_2(t)$ are multiplied and low-pass filtered, the output is $\tfrac{1}{4}\cos 2\pi t/2T$ (a signal at $\tfrac{1}{2}$ the bit rate) which is the desired timing waveform. Thus the MSK format lends itself to very easy self-synchronization.

The natural 180° ambiguity produced by the divide-by-two circuits [i.e., whether the outputs are $\pm s_1(t)$, $\pm s_2(t)$] can be removed by differential encoding and other techniques [14],[18]. (Differential encoding is a technique in which changes or transitions in bit streams, rather than the absolute values themselves, are encoded as +1 and −1. At the receiver, a differential decoder will be employed to generate the original bit stream.)

ERROR RATE PERFORMANCE

In addition to its bandwidth efficiency and self-synchronizing capabilities, the error rate performance of MSK in an ideal channel, as defined in the section on FSK and PSK is also of interest. It is relevant at this point to summarize the performance capabilities of related binary and quadrature modulations. Binary PSK (BPSK) with $\pm \cos 2\pi f_c t$ representing "1" and "−1," is, as we mentioned before, an example of antipodal signaling and can be represented as shown in Fig. 10(a) with a normalized distance of two between them. In QPSK, the two carriers used ($\cos(2\pi f_c t + \pi/4)$ and $\sin(2\pi f_c t + \pi/$

4)) are orthogonal and hence the two bit streams $a_I(t)$ and $a_Q(t)$ can be demodulated independently. Hence for the same (E_b/N_0), the error probabilities of coherently detected BPSK and QPSK are the same. Since staggering the bit streams does not change the orthogonality of the carriers, OQPSK has the same performance as BPSK and QPSK.

MSK (3) uses antipodal symbol shapes ($\pm \cos \frac{\pi t}{2T}$ and $\pm \sin \frac{\pi t}{2T}$) over $2T$ to modulate the two orthogonal carriers just as was the case in QPSK. Also, the energy in the two shapes $p(t)$ in (5) and (6) are the same. Thus, when matched filtering is used to recover the data (as was the case in Fig. 8), MSK has the same performance as BPSK, QPSK, and OQPSK.

Note, however, that it is the detection of the MSK signal on the basis of observation over $2T$ seconds that results in this good performance. If MSK is coherently detected as an FSK signal with bit decision made over an observation interval of T seconds, MSK would be poorer than BPSK by 3 dB. This is because the coherent performance of equal energy binary signals in white Gaussian noise depends only on the "distance" between the two signals in the signal space—the larger the distance, the less the probability of error. (This is intuitively appealing since the larger the distance, the less the possibility of mistaking one signal for the other.) As Fig. 10(b) shows, MSK, viewed as an example of orthogonal FSK signaling, has only a distance of $\sqrt{2}$ between the two signals. This decrease in the distance in FSK as compared to BPSK translates into an (E_b/N_0) increase of 3 dB needed to maintain the same error rate as in a BPSK scheme. This 3 dB disadvantage of FSK over BPSK vanishes in the case of MSK, where bit decisions are made after observing the waveform over two bit periods.

Even though T is the duration of one bit and one decision per transmitted bit is needed, better performance may be obtained by observing the received waveform over a longer period, thus giving us more knowledge about the underlying signal and/or noise process. In MSK, the phases are related over 2 bit periods and hence by observing over $2T$ seconds and using the continuous phase nature of MSK, we know more about the signal. In fact, as is evident from the OQPSK format of MSK (3), we know that over $2T$ seconds, it is one of two antipodal signals which has been transmitted and hence the equivalence in the performance of MSK and BPSK.

Since MSK is a type of FSK, it can also be noncoherently detected (e.g., by means of a discriminator) whereas QPSK systems require either a fully coherent or differentially coherent detection scheme. This possibility of noncoherent detection permits inexpensive demodulation of MSK when the received (E_b/N_0) is adequate and provides a low-cost flexibility feature in some systems.

EXTENSIONS AND GENERALIZATIONS

MSK or continuous phase FSK (CPFSK) may be generalized to include other values of Δf, the frequency separation, and a longer bit memory before the decision has to be made. For larger observation intervals such as $3T$ or $5T$, a maximum improvement of 0.8 dB has been reported for $\Delta f = 0.715/T$ [18]-[20]. However, the complexity of the circuits involved does not seem to favor these schemes over the simple yet efficient MSK modulation.

Similarly, while retaining the advantage of good bit error rate performance, the spectral properties of MSK can be improved by shaping the data pulses further. Note that in MSK, the symbol pulse shape $p(t)$ is $\cos[\pi t g(t)/2T]$ where $g(t) = 1$, $0 \leqslant t \leqslant T$. Other choices of $g(t)$ are possible with the spectral falloff rate depending on the end-point behavior of the shape chosen [21],[22]. For example, a function such as $g(t) = \sin(2\pi t/T)/(2\pi t/T)$ (known as sinusoidal frequency shift keying [23]) results in a much smoother $p(t)$ and produces an asymptotic spectral falloff that is twice as fast as in MSK. Unfortunately, all these generalizations tend to produce a broader (main lobe) spectrum than MSK thus worsening the performance at low bandwidth/bit rate values. MSK has been extended to multiple level pulses, known as multiple amplitude MSK (MAMSK) [24]. Other recent works [25],[26] indicate that application of an efficient baseband coding scheme such as correlative coding [27] to MSK may be the answer to further spectral economy and good performance.

SUMMARY

The MSK scheme was shown to be a special case of continuous phase FSK signaling with frequency deviation equal to ½ the bit rate. MSK can also be viewed as a form of offset QPSK signaling in which the symbol pulse is a half-cycle sinusoid rather than the usual rectangular form. It combines in one modulation format many attractive attributes such as constant envelope, compact spectrum, the error rate performance of BPSK, and simple demodulation and synchronization circuits.

MSK is an excellent modulation technique for digital links when bandwidth conservation and the use of efficient amplitude-saturating transmitters are important requirements.

These features make MSK an excellent modulation technique for digital links in which bandwidth conservation and the use of efficient transmitters with nonlinear (amplitude saturated) devices are important design criteria.

REFERENCES

[1] J. G. Smith, "Spectrally efficient modulation," in *Proc. IEEE Int. Conf. Commun.* (ICC '77), June 1977, pp. 3.1-37-3.1.41.

[2] M. L. Doelz and E. H. Heald, Collins Radio Co., "Minimum-shift data communication system," U.S. Patent 2 977 417, Mar. 28, 1961.

[3] R. DeBuda, "Coherent demodulation of frequency shift keying with low deviation ratio," *IEEE Trans. Commun.*, vol. COM-20, pp. 429-435, June 1972.

[4] W. A. Sullivan, "High capacity microwave system for digital data transmission," *IEEE Trans. Commun.*, vol. COM-20, pp. 466-470, June 1972.

[5] D. M. Brady, "FM-CPSK: Narrowband digital FM with coherent phase detection," in *Proc. IEEE Conf. Commun.* (ICC '72), June 1972, pp. 44-12-44-16.

[6] ——, "A constant envelope digital modulation technique for milli-meter-wave satellite system," in *Proc. IEEE Int. Conf. Commun.* (ICC '74), Minneapolis, MN, June 1974.

[7] E. J. Sass and J. R. Hannum, "Minimum shift keying modem for digitized voice communications," *J. RCA*, pp. 80-84, 1974.

[8] S. L. Bernstein et al., "A signaling scheme for extremely low frequency (ELF) communication," *IEEE Trans. Commun.*, vol. COM-22, pp. 508-528, Apr. 1974.

[9] D. P. Taylor et al., "A high speed digital modem for experimental work on the communications technology satellite," *Can. Elec. Eng. J.*, vol. 2, no. 1, pp. 21-30, 1977.

[10] S. A. Rhodes, "Effects of hardlimiting on bandlimited transmissions with conventional and offset QPSK modulation," in *Proc. Nat. Telecommun. Conf.*, Houston, TX, 1972, pp. 20F/1-20F/7.

[11] ——, "Effect of noisy phase reference on coherent detection of offset QPSK signals," *IEEE Trans. Commun.*, vol. COM-22, pp. 1046-1055, Aug. 1974.

[12] R. D. Gitlin and E. H. Ho, "The performance of staggered quadrature amplitude modulation in the presence of phase jitter," *IEEE Trans. Commun.*, vol. COM-23, pp. 348-352, Mar. 1975.

[13] S. A. Gronemeyer and A. L. McBride, "MSK and offset QPSK modulation," *IEEE Trans. Commun.*, vol. COM-24, pp. 809-820, Aug. 1976.

[14] H. R. Mathwich, J. F. Balcewicz, and M. Hecht, "The effect of tandem band and amplitude limiting on the E_b/N_0 performance of minimum (frequency) shift keying (MSK)," *IEEE Trans. Commun.*, vol. COM-22, pp. 1525-1540, Oct. 1974.

[15] M. Atobe, Y. Matsumoto, and Y. Tagashira, "One solution for constant envelope modulation," in *Proc. 4th Int. Conf. Digital Satellite Commun.*, Montreal, P.Q., Canada, Oct. 1978, pp. 45-50.

[16] R. M. Fielding et al., "Performance characterization of a high data rate MSK and QPSK channel," in *Proc. IEEE Int. Conf. Commun.* (ICC '77), June 1977, pp. 3.2-42-3.2-46.

[17] W. R. Bennett and S. O. Rice, "Spectral density and autocorrela-

tion functions associated with binary frequency shift keying," *Bell Syst. Tech. J.*, vol. 42, pp. 2355-2385, Sept. 1963.

[18] R. DeBuda, "Fast FSK signals and their demodulation," *Can. Elec. Eng. J.*, vol. 1, no. 1, pp. 28-34, 1976.

[19] W. B. Osborne and M. B. Luntz, "Coherent and noncoherent detection of CPFSK," *IEEE Trans. Commun.*, vol. COM-22, pp. 1023-1036, Aug. 1974.

[20] R. DeBuda, "About optimal properties of fast frequency-shift keying," *IEEE Trans. Commun.* (Corresp.), vol. COM-22, pp. 1726-1728, Oct. 1974.

[21] M. K. Simon, "A generalization of minimum-shift-keying (MSK)-type signaling based upon input data symbol pulse shaping," *IEEE Trans. Commun.*, vol. COM-24, pp. 845-856, Aug. 1976.

[22] M. Rabzel and S. Pasupathy, "Spectral shaping in MSK-type signals," *IEEE Trans. Commun.* vol. COM-26, pp. 189-195, Jan. 1978.

[23] F. Amoroso, "Pulse and spectrum manipulation in the minimum (frequency) shift keying (MSK) format," *IEEE Trans. Commun.* (Corresp.), vol. COM-24, pp. 381-384, Mar. 1976.

[24] W. J. Weber, III, *et al.*, "A bandwidth compressive modulation system using multiamplitude minimum shift-keying (MAMSK)," *IEEE Trans. Commun.*, vol. COM-26, pp. 543-551, May 1978.

[25] G. J. Garrison, "A power spectral density analysis for digital FM," *IEEE Trans. Commun.*, vol. COM-23, pp. 1228-1243, Nov. 1975.

[26] F. de Jager and C. B. Dekker, "Tamed frequency modulation, A novel method to achieve spectrum economy in digital transmission," *IEEE Trans. Commun.*, vol. COM-26, pp. 534-542, May 1978.

[27] S. Pasupathy, "Correlative coding; A bandwidth efficient signaling scheme," *IEEE Commun. Soc. Mag.*, vol. 17, pp. 4-11, July 1977.

Subbarayan Pasupathy was born in Madras, India. He received the B.E. degree in telecommunications from the University of Madras, India, in 1963, the M. Tech. degree in electrical engineering from the Indian Institute of Technology, Madras, India, in 1966, and the M. Phil. and Ph.D. degrees in engineering and applied science from Yale University, New Haven, CT, in 1970 and 1972, respectively.

From 1965-1967, he was a Research Scholar and part-time Lecturer at the Indian Institute of Technology, Madras, India, and he worked as a Teaching Assistant at Yale University, CT, from 1968-1971. From 1972-1973, he was a Postdoctoral Fellow at the University of Toronto, Toronto, Ont., Canada, working in the area of array processing of sonar signals. Since 1973, he has been with the Department of Electrical Engineering, University of Toronto, Toronto, Ont., Canada, where he is now an Associate Professor and Chairman of the Communications Group. His current research interests lie in the areas of digital communications, sonar-radar systems, and communication theory.

Dr. Pasupathy is a registered Professional Engineer in the Province of Ontario, Canada, and a member of the IEEE.

DIGITAL RADIO: ITS TIME HAS COME*

Recent advances in digital transmission radically increase the feasibility of using it for long distance.

DIGITAL transmission in metropolitan areas (*T* Carrier) has been with us for almost 20 years. Until recently however, efforts to extend digital technology into the intercity area have met with, at best, limited success. This is due primarily to the cost of laying cable and deploying repeaters. With the advent of digital radio the ability to "go digital" has been economically extended out to several hundred miles. For distances longer than that, analog transmission is still the economic choice.

Interestingly enough, analog radio line haul costs are actually less expensive than digital radio line haul costs per voice channel mile. However, digital radio permits the use of digital banks and digital multiplex equipment which are indeed less expensive than their analog counterparts. This can be seen in the relative transmission costs shown in Fig. 1, where the digital termination costs (ordinate intercept) offset the somewhat higher line haul cost of digital radio out to a crossover of about 300 mi. The economic prove-in of digital radio for intercity applications has made it the facility choice for route relief in many areas of the country. In addition, with digital switching becoming more common, digital radio facilitates interconnection of these digital switches by providing a high-capacity digital route for the *T* Carrier signals terminating on these switches.

The increasing proliferation of digital radio during the last several years may well portend the coming of the long-awaited, much-heralded digital transmission revolution. For years, communication oracles have been predicting the future demise of analog facilities, but each year the imminent demise is postponed by enhanced analog capabilities developed by

*Reprinted from *IEEE Communications Society Magazine,* November 1980, Vol. 18, No. 6, pp. 6–12.

analog system designers. For example, they have provided increased channel capacity, time assignment speech interpolation (TASI), and single-sideband radio. Now however, with digital radio on the scene, the area for economic application of digital transmission is greatly extended. No longer are "bits" confined to metropolitan areas or small cross-sectional outstate applications. Digital radio has been found to be a valuable addition to the repertoire of intercity facility designers.

With all the interest in digital radio, it is appropriate to provide additional background on its genealogy, its operation, its limitations and its applications. In this article we will review the evolution of digital radio, examine some of the modulation schemes, discuss the impact of propagation variation induced stresses, and identify the approaches taken to ameliorate them. The performance of digital radio will be examined along with some of the benefits accruing from its use, and finally some long-term prognostications of its future will be made.

Although this article focuses primarily on 6 GHz digital radio, the reader should recognize that 2 and 11 GHz digital radio have also found widespread use. 2 GHz is used primarily for small cross sections, often finding use as a feeder for 6 and 11 GHz. Both 6 and 11 GHz are used for short-haul applications up to several hundred miles. 11 GHz digital radio was introduced first because the wider bandwidth available in that band eased some of the design difficulties. 6 GHz now would normally be the facility selected because of the longer hop length over which it can be engineered. This reduces the number of repeaters required and hence reduces cost. However, due to frequency congestion, 11 GHz must often be used instead. There are, in addition, 18 GHz radio systems. These systems tend to have unique applications and consequently have not found widespread acceptance in the United States.

DIGITAL RADIO ROOTS

As we stand at the beginning of a tremendous growth in the deployment of digital radio, it is informative to review how

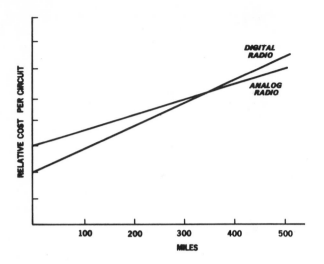

Fig. 1.   Relative transmission costs.

we got here. Contrary to popular opinion, digital radio is not a new application of radio. In fact, radio was used as a vehicle for digital transmission long before it was used to carry voice traffic. As far back as 1894, Sir Oliver Lodge demonstrated a wireless transmitter and receiver. Marconi followed up with a demonstration of radio telegraphy across the English Channel in 1899. By 1901 he had transmitted signals across the Atlantic Ocean between Poldhu, England, and St. Johns, Newfoundland. DeForest's "audion," a three-electrode tube, became available in 1907 and quickly became the detector of choice because of its greatly increased sensitivity. It was at that point that radiotelegraphy really came of age.

At that time, coding was typically done by "keying" the signal on and off, although some frequency shift keying was accomplished by shorting out portions of a tuning coil. With these relatively crude approaches, the transmitted information

rate was quite limited, at least by today's standards. It did not take engineers long, however, to begin increasing the information rate. One approach was simply to automate the transmittal and decoding process so the information content per unit time could be increased. Other approaches utilized carriers at different frequencies, each of which was keyed as rapidly as possible. All of these applications were of course aimed at telegraphy. Digitized voice transmission on an economic basis was still a long way off. However, radiotelegraphy did provide the capability for ship-to-shore transmission and thus found widespread use prior to and during World War I.

During this period, emphasis was shifting from telegraphy ("digital radio") to telephony ("analog radio"). Although interest continued in increasing telegraph transmission rates, radiotelephony on an analog basis became the focus. In fact, by 1927 commercial radiotelephony from New York to London was instituted on an analog basis. Digital radio for telephony did not really take hold until almost 50 years later.

DIGITAL RADIO DESCRIPTION

The term digital radio encompasses any of the modulation schemes which alter the carrier characteristics in a discrete fashion such that the modulated signal can assume only a finite number of unique changes in state. These states may be in frequency, phase, amplitude, or some combination.

The ways of implementing digital radio are quite varied. Typically the underlying concept is accomplished by changing the carrier phase and/or amplitude according to the state of a digital input signal. Perhaps the easiest scheme to conceptualize is the one used initially for radio telegraphy where, in a single-step modulation process, the carrier was turned on or off. The duration of the signal could of course be readily interpreted as a dot or dash in Morse code. This allowed existing equipment and telegraphers to function as

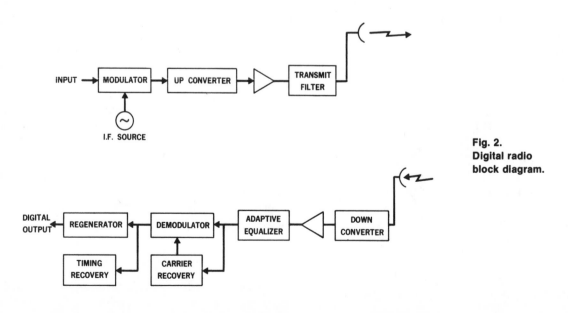

**Fig. 2.
Digital radio
block diagram.**

before without being concerned about whether the signal was transmitted over wire or radio. This on-off keying (OOK) was, of course, simply a binary amplitude modulation of the carrier. As indicated before, when first implemented, this rudimentary approach was effective, but far from efficient in terms of the amount of transmitted information per unit of time.

The approach often taken now involves modulating an intermediate frequency (IF) carrier, typically 70 MHz, according to the input signal and then upconverting, amplifying, and transmitting the resulting signal. At the receiver, the complementary processing takes place. The signal is downconverted, adjusted in level, perhaps amplitude equalized, and then the demodulation or detection process takes place resulting in a regenerated baseband digital signal. Thus the key to the digital radio concept lies in two items:

1) a finite set of states represented by the carrier at the transmitter.

2) a regenerated digital signal at the receiver.

See Fig. 2 for a representative block diagram.

The concepts are apparently straightforward; what is the problem? Indeed, if bandwidth were readily available and there were no minimum voice circuit loading requirement, digital radio design might be readily accomplished. However, radio spectrum is a valuable resource and the FCC, in Docket 19311, has decreed a minimum channel capacity of 1152 voice channels at 4, 6, and 11 GHz. Since existing digital banks digitize voice signals on the basis of 24 channels per 1.544 Mbit/s signal (DS-1 signal), the 1152 voice channel requirement implies a minimum of 48 DS-1 signals, or approximately 75 Mbits/s. Although some digital radio in the field does carry only 1152 voice channels at 78 Mbits/s, most designs at 6 and 11 GHz achieve a capacity of 1344 voice channels, requiring approximately 90 Mbits/s. This capacity was not chosen lightly. There exists a hierarchy of digital transmission rates, and the 90 Mbit/s capacity can accommodate two DS-3 signals. Each DS-3 signal, at a 44.736 Mbit/s rate, corresponds to 672 digitized voice signals. The 90 Mbit/s rate facilitates the handling of larger digital cross sections, namely, DS-3 signals; it thus reduces multiplex requirements and thereby reduces cost.

Providing 90 Mbits/s over existing radio channelization does not come easy. At 11 GHz the bandwidth is 40 MHz, implying a capacity of 90/40 or 2.25 bits/s/Hz of bandwidth. At 6 GHz the bandwidth is only 30 MHz and thus a capacity of 3 bits/s/Hz is required. At 4 GHz the bandwidth is only 20 MHz and thus would require 4.5 bits/s/Hz. Current digital radio technology has not matured sufficiently to allow more than about 3 bits/s/Hz.

8PSK DIGITAL RADIO

As indicated above, accommodating two DS-3 signals in the 30 MHz channel at 6 GHz requires 3 bits/s/Hz of bandwidth. To accomplish this an eight-level signal is required to be transmitted at 30 Mbaud. With this requirement, one of the first schemes to be used was 8PSK, i.e., phase shift keying with eight possible states. In this scheme the carrier phase is altered 30×10^6 times per second, shifting it to one of eight positions relative to its previous position (see Fig. 3).

Since the input consists of two DS-3 ($\sim$45 Mbits/s each) signals, some processing of the input is required. This is done by recoding the signals into three 30 Mbit/s signals. The state of each of the three input signals is then examined each time a new modulation state is to be generated. It is then straightforward, conceptually at least, to interpret the three states as comprising a binary number with value zero through seven. That number is then mapped into one of the eight selectable carrier phase changes. This is the essence of the 8PSK approach.

It should be noted that the information is mapped into the *changes* in the carrier phase as opposed to the absolute phase. This precludes the need for establishing and maintaining an absolute phase reference at the receiver.

Along with specifying voice occupancy limits, the FCC also has limits on the in-band energy and the adjacent channel interference which can be tolerated. Since digital signal spectra have significant energy over a wide frequency band (see Fig. 4), additional processing is required at the transmitter to limit it. Along with controlling the pulse shape of the modulating baseband signal, typically two steps are used: one before and one after modulation takes place.

Control of the in-band energy, i.e., tone suppression, is accomplished by scrambling the input signal. This effectively randomizes the signal so that the energy spectrum, both before and after modulation, is essentially flat across the band.

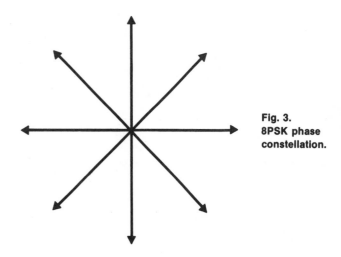

**Fig. 3.
8PSK phase
constellation.**

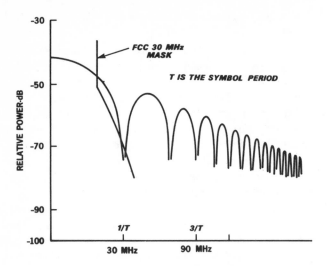

Fig. 4.   Unfiltered 8PSK spectrum.

Following the amplifier[1] it is necessary to address the adjacent channel energy constraint. This is done by filtering the signal in such a way that the signal energy levels meet the FCC mask (see Fig. 5). In addition to satisfying the FCC criteria, the filters play an important role in determining the amount of margin that exists. By controlling the combined characteristics of the transmit and receive filters so that they satisfy Nyquist's criterion, intersymbol interference at the sampling instant can be minimized, thus maximizing the available noise margin.

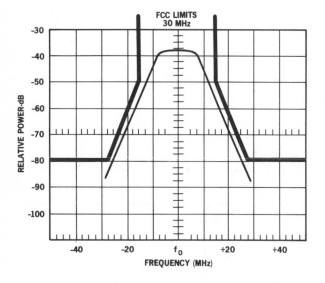

Fig. 5.   Transmitted spectrum relative to FCC mask.

---

[1]It should be noted that 8PSK, of course, results in a virtually constant envelope signal. This greatly reduces the linearity requirements on amplifiers, hence facilitating their design.

At the receiver, it is necessary to obtain timing and relative carrier phase information. This is best accomplished when the received signal contains many state transitions. The scrambler introduced at the transmitter serves this function well. Once the receiver establishes its timing and phase reference, it need only detect the change in phase to demodulate correctly. Thus there is no need for an absolute phase reference. Just as the transmitter coded a 3 bit word into one of eight phase changes, so the receiver maps each of the phase changes, back into a 3 bit word. This processing would indeed be straightforward if the signal had only been transmitted through an ideal channel, but there are impairments to contend with. First there is additive noise. In general the margin against thermal noise is in excess of 40 dB, and for the 11 GHz systems it is not unusual to have 50 dB. However, noise is not the only problem. At 11 GHz, rain fading generally dominates and indeed the 50 dB of flat fade margin can be utilized. At 6 GHz however the limiting factor is typically multipath fading. This results not only in decreased received power, it also results in amplitude distortion across the band (Fig. 6).

In particular, with 3 bits per symbol and a signaling rate of 30 Mbaud, over a 30 MHz channel, there is little margin against channel dispersion. Consequently, along with gain control, it has been found necessary to introduce adaptive amplitude equalization to compensate for time varying multipath fade distortion.

One additional difficulty associated with 8PSK digital radio at 6 GHz has been the adjacent channel interference. Although various filtering approaches have been somewhat successful, the desired margin has not been universally achieved. In the hope of overcoming the pitfalls of 8PSK, an alternative modulation scheme, 16 QAM (quadrature amplitude modulation with 16 states), has been finding favor in the digital radio arena.

16 QAM DIGITAL RADIO

16 QAM is a combined amplitude and phase modulation scheme with, as the name implies, 16 different states. As can

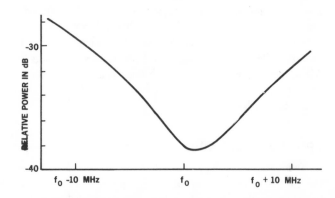

Fig. 6.   Channel amplitude shape during multipath fade.

be seen in Fig. 7, the resultant carrier can be viewed as having two components, the in-phase *I*, and the quadrature *Q*, each of which may take one of four possible values. From the figure it is clear that the resultant signal is not constant in amplitude; in fact, three separate amplitude levels exist. Thus it is necessary that the power amplifiers be quite linear to avoid distortion. Note that there is approximately a 10 dB difference in power between the minimum and maximum carrier levels.

The benefit in 16 QAM arises from the fact that with 16 possible states, each symbol contains 4 bits of information. Thus to transmit 90 Mbits/s it is only necessary to transmit at 90/4 Mbaud, i.e., 22.5 Mbaud. This reduced symbol rate implies a reduction in the required bandwidth thereby lessening the filter design problem. A more gradual transition between the filter's passband and stopband can be accommodated while still providing sufficent out-of-band suppression to meet adjacent channel interference resistrictions. In addition, by virtue of its reduced bandwidth occupancy, the 16 QAM signal is itself more immune to adjacent channel interference. One should not lose sight of the fact that we did not get something for nothing. Going to a 16-level signal, among other things, places an additional constraint on the amplifiers. They must now be linear. In addition, rather than detecting one of eight signals or phases, it is now necessary to detect which of 16 signals has been sent, with the added complexity of detecting both amplitude and phase. As might be expected, the 16 QAM signal is also sensitive to the in-band dispersion caused by multipath fading. Thus, just as with the 8PSK system, the receiver must include adaptive amplitude equalization.

The implementation of a 16 QAM digital radio system can be summarized as follows.

The two DS-3 input signals are scrambled, combined, and then converted into four parallel 22.5 Mbit/s bit streams. Alternatively, each DS-3 signal may be scrambled and converted into two 22.5 Mbit/s bit streams, such that each DS-3 signal maintains its identity throughout the process. In either event, the parallel bit streams then go through a differential encoding process so that it is the change in the signal which carries the information. At that point, each rail is filtered and applied to a balanced modulator. The modulated signal is amplified, upconverted, and transmitted over the air. At the receiver, the complementary processing takes place. Timing and phase references are established, the downconverted signal is gain adjusted, equalized, and demodulated.

SPACE DIVERSITY

Up to this point little has been said about system performance other than to indicate the need for adaptive equalization. It would be totally misleading at this point to leave the reader with the assumption that this was all that was necessary. In fact, if 6 GHz digital radio (either 8PSK or 16 QAM) were implemented with adaptive equalization and installed on most existing radio hops, it would not operate satisfactorily when fading occurs. Currently available equalizer technology has been unable to cope adequately with the selective fade induced channel misalignment. Thus a means of reducing the exposure to fading is also required.

Frequency diversity provides very little benefit at the relatively shallow fade depths, typically less than 30 dB, where dispersion often becomes significant. As a result, for 6 GHz applications, it is necessary that most hops in an average fading environment be equipped not only with equalizers, but also with space diversity. Adaptive equalization and space diversity operate together so that the net result is better than the product of their individual improvement capabilities.

Either adaptive equalization or space diversity would individually reduce outage, periods when the error rate is worse than 10^{-3}, by typically less than a factor of 10. However, when used together they work synergistically. The equalizer reduces the number of times when space diversity is required by compensating for many dispersive fades. In essence, this improves the effective fade margin. At these deeper fades, from 33 to 36 dB, the fading on the two antennas becomes more decorrelated and hence the benefit of space diversity is further enhanced. As a result, the combined effect of space diversity and equalization can reduce outage by greater than a factor of 100.

The manner in which space diversity is implemented varies. In some cases, the signals from the two antennas are gain adjusted, phased, and then combined (Fig. 8). This approach is referred to as continuous combining and may be implemented at either IF or RF. The benefits of combining are that there are no switch transients to introduce errors and there is a potential for more powerful equalization schemes if the selective fades on the two signals are different, as is typically the case.

A second approach to space diversity involves switching

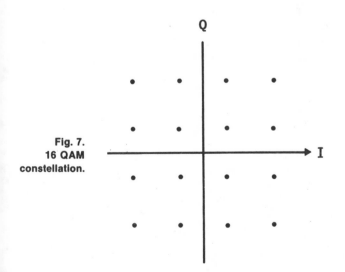

**Fig. 7.
16 QAM
constellation.**

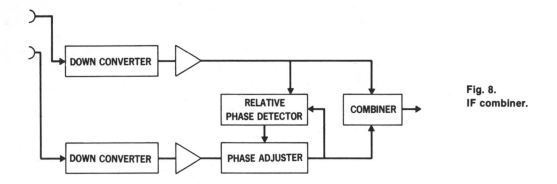

Fig. 8.
IF combiner.

from one signal to the other (see Fig. 9). Although not universally true, most digital radio installations using a switching approach do so at baseband. Typically the two signals are adaptively phased in time so that the switch from one signal to the other is done hitlessly. In this way, if the receiver can detect an erosion of margin, the switch to diversity can be accomplished before errors actually occur. In addition, if the signal paths have been properly "DADE'd",[2] i.e., matched in absolute delay, the switch events do not themselves cause errors.

We have discussed digital radio implementation and how the effects of multipath fading can be reduced. Now we will look at what the net results are in terms of error rate and outage. The cited results are an amalgamation of many experiments on different systems and thus, while representative, are not to be construed as being applicable to any specific system.

PERFORMANCE

Digital radio performance is easily measured and presented; it is only the interpretation that is difficult. In general, the user is primarily interested in two measures:

1) the amount of time the system is unavailable,

[2]Differential Absolute Delay Equalization

2) the quality of performance (bit error rate or BER as a function of time) during its availability.

The first item is the outage, and if that is interpreted to mean all time for which the error rate exceeds 10^{-3}, then a single plot of the time the BER exceeds various levels will satisfy both performance measures. A typical plot is given in Fig. 10, but what does it mean? If the indicated performance occurred in a nonfading environment, then the system performance might be interpreted as being poor. Conversely, if the indicated performance had been achieved in an environment where, say, the signal on the primary antenna had experienced 6000 s faded below 30 dB, then the system performance would be interpreted as quite good. Thus any interpretation of performance requires that one know the environment in which the performance was measured.

As noted earlier, the prime controller of digital radio performance is the amplitude distortion remaining after adaptive equalization. However, since the relative frequency of occurrence of various dispersive fades are not as well known as the single frequency fade behavior, digital radio systems are generally engineered according to the amount of anticipated time the total power is faded below 30 dB. For this reason a performance factor for voice (PFV) has been developed. The PFV normalizes the performance characteristics according to the fade activity. Specifically, the time for which the error rate exceeds 10^{-3} (outage level) divided by

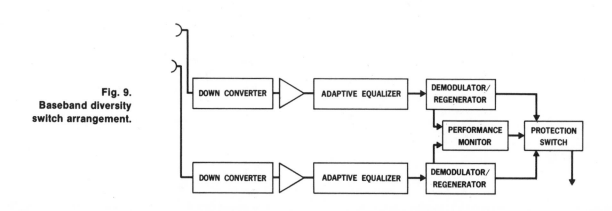

Fig. 9.
Baseband diversity
switch arrangement.

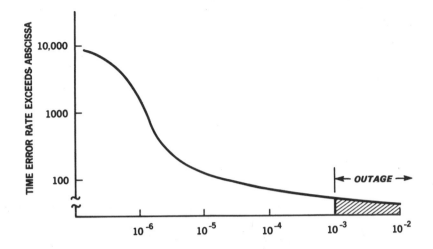

Fig. 10.
Error
performance.

the amount of time the fade on the primary antenna exceeds 30 dB is defined as the PFV. The PFV is not a unique indication of the digital radio system's intrinsic capabilities, but rather how well the specific system being measured is performing. For example, if considerable space diversity improvement had been engineered into the system (say 150 ft separation and full-size diversity antenna) the PFV would be better, i.e., smaller, than if 30 ft separation with reduced gain antennas had been used. In addition, the equalizers and the switching or combining algorithms all have an impact on the PFV. When equipped with space diversity and adaptive equalization, the values of PFV for digital radio range from about 0.05 to less than 0.01 with a typical value being about 0.02.

This implies that for every 1000 s the total channel power faded below 30 dB on the primary antenna, a system engineered to have a PFV equal to 0.02 would experience outage, i.e., BER $> 10^{-3}$, for about 20 s. It has been found that when routes are properly engineered and maintained, 6 GHz digital radio can meet one-hop one-way outage levels of less than 1 min for 3000 s of time faded below 30 dB on the primary channel. This is well within the current short-haul radio outage objectives.

DIGITAL RADIO ATTRIBUTES

There are many reasons why digital radio is currently so attractive. It provides a means of digitally interconnecting metropolitan areas which are themselves hotbeds of digital transmission activity. Along with providing digital connectivity, it reduces the cost of intercity trunking, and it provides a facility directly compatible with data transmission and other digital services. In many cases it can utilize existing radio towers and antennas thus shortening the interval for providing service. The technology, although still immature, is already cost competitive. Thus one would expect the relative cost differences to become even more favorable for digital radio as the technology matures. Finally, one would expect future enhancements to further increase the capacity.

CONCLUSIONS

Since digital radio is such an economically attractive alternative to most cable systems and analog radio, the number of applications will continue to increase in the foreseeable future. Although equalization and diversity schemes have yet to be pushed to their limits, the outage performance capability is already within acceptable short-haul limits. The performance is still somewhat fragile in the sense that relatively small (~ 2 dB) residual (i.e., after adaptive equalization) amplitude misalignment is sufficient to cause errors, but improved technology should compensate for this and provide a bright future. If the past bears any relation to the future, then new equalization techniques, higher efficiency in bits/s/Hz, and long-haul 4 GHz digital radio can all be anticipated over the next ten years.

REFERENCES

Rather than my enumerating a large number of fertile sources of information on digital radio, I feel the interested reader should obtain a copy of the Special Digital Radio issue of the IEEE TRANSACTIONS ON COMMUNICATIONS, vol. COM-27, December 1979, which contains a wealth of technical information itself, and, in addition, has extensive reference lists.

Neil F. Dinn is Supervisor of the Radio Characterization Studies Group, part of the Transmission Facilities Engineering Laboratory at Bell Laboratories in Holmdel, NJ. His current interests focus on radio propagation experiments, including the design of the specialized microprocessor based measurement and data acquisition equipment needed to implement the experiments.

Upon joining Bell Labs in 1967, Mr. Dinn worked on timing control and adaptive equalization for digital systems. Subsequently, he worked on the systems engineering for the T1C transmission system. He has been involved in satellite studies, development of criteria for single-sideband transmission, and a computerized planning program for outstate applications.

Mr. Dinn received the B.S.E.E. degree from Northeastern University and the M.S.E.E. degree from the Massachusetts Institute of Technology. He is a member of Phi Kappa Phi, Sigma Xi, and Eta Kappa Nu.

THE VITERBI ALGORITHM
APPLIED TO DIGITAL
DATA TRANSMISSION *

J. F. Hayes

INTRODUCTION

Since its development by Richard Bellman in the 1940's, dynamic programming has found wide application in control and circuit theory. [1] It is only recently that dynamic programming, in the form of the Viterbi algorithm, has been applied to communication problems. The first application was to the decoding of convolutional codes [3]. Subsequently, the technique was extended to the detection of data signals transmitted over linear models of voiceband channels [4]-[7]. Earlier work on optimum sequential detection of data signals was done by Chang and Hancock [8].

This paper describes, through systematic derivations and numerical examples, what the Viterbi algorithm is and how it works.[2] The reader is presumed to have a good background in basic communication theory but no specialized knowledge of data transmission or dynamic programming.

INTERSYMBOL INTERFERENCE

A major impairment encountered in the high-speed transmission of digital data over voice frequency lines is intersymbol interference. Consider the situation depicted in Fig. 1(a), in which a single pulse is transmitted over a relatively narrow-band channel, resulting in the pulse being smeared in time at the output. A sequence of pulses, such as in pulse amplitude modulation systems, suffers intersymbol interference when the energy from one pulse spills over into adjacent symbol intervals so as to interfere with the detection of these adjacent pulses [see Fig. 1(b)]. Thus, a sample at the center of a symbol interval is a weighted sum of amplitudes of pulses in several adjacent intervals. This effect, combined with random noise, leads to error.

The current practice is to minimize the effect of intersymbol interference by channel equalization [9], which adjusts the pulse shape so that it does not interfere with neighboring pulses at pulse centers [see Fig. 1(a) and (b)]. Although this approach is effective in many cases, minimizing the effect of intersymbol interference in this way is inherently suboptimum, since even the interference contains information about the symbols that were transmitted. In

theory, when the channel causes a time dispersion of signal energy, the whole received signal rather than center values should be used to detect any symbol or group of symbols. Heretofore, the obstacle to optimum detection of a whole sequence of pulses has been computational complexity. The number of computations required in a straightforward approach grows exponentially with the length of the transmitted sequence. Furthermore, computation cannot begin until the entire sequence has been received.

The significance of the dynamic programming approach is that the number of computations required for optimum detection grows only linearly with the length of the transmitted sequence, and hence computations can be carried out while the sequence is being received. Although this approach reduces computational complexity, it is still, in most cases of practical interest, beyond the capability of present day processors. This difficulty will, perhaps, be overcome with the growth of computer technology. Moreover, there are suboptimum implementations which may yield performance close to the optimum.

EXAMPLE OF DYNAMIC PROGRAMMING

Dynamic programming is essentially a computational procedure for finding an optimum path or trajectory. The following rather pedestrian example[3] will serve to illustrate its basic principles. A certain Professor X walks each day from his office in the EE building to the faculty lounge for lunch (see Fig. 2). Between the two buildings lie two small streams, christened by some campus wag as the Publish and the Perish. Each stream runs north to south and is crossed by two foot bridges. In our example, these bridges shall be designated by the stream they cross and by the appellation north or south. One day our scholarly friend decides to find the shortest path to the faculty lounge. He could, of course, simply calculate the length of all possible paths and choose the shortest. However, sensing that a general principle is involved, Professor X eschews the brute force approach. He first writes down the distances from his office to the two bridges across the Publish. He then *postulates* that the optimum path is via the north bridge across the Perish. Under this assumption, he calculates the minimum path from his office *to this bridge* by comparing the two paths over the Publish. The same procedure is repeated for the south bridge across the Perish. At this point, the professor notes that for the purpose of further calculations, he need only keep track of the *shortest* path to the north bridge on the Perish and the *shortest* path to the south bridge on the Perish. In observing

The author is with Bell Laboratories, Holmdel, N.J. 07733.

[1] A number of texts on the subject have been written; see, for example, [1] and [2].

[2] For a treatment of the Viterbi algorithm from a different point of view, we recommend [21].

*Reprinted from *IEEE Communications Society Magazine*, March 1975, Vol. 13, No. 2, pp. 15-20.

[3] For a somewhat more complex example, see [2, ch. 1].

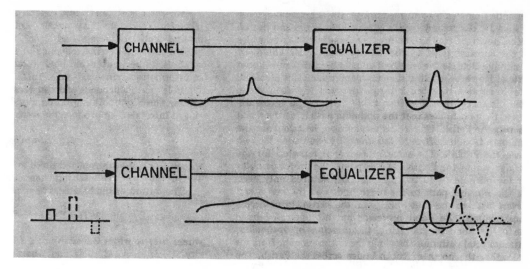

Fig. 1.   (a) Single pulse. (b) Pulse train.

this simplification, the good professor has hit upon the basic principle of dynamic programming, the *principle of optimality*. The optimum total path must lie along the optimum path from his office to either the north or south bridge across the Perish. The final step is a comparison of the total distances to the lounge via the north and south bridge across the Perish. The step-by-step procedure followed by Professor X is shown in Table I (note the distances in Fig. 2).

In carrying out these calculations, six additions are necessary. Brute force enumeration would require eight additions. Now, if Professor X had to cross N streams, each with two bridges, dynamic programming would require $4(N-1) + 2$ additions, whereas straight enumeration would require $(N-1)2^N$ additions. Notice the difference between linear and exponential growth with N here.

This example illustrates *forward* dynamic programming since the computation proceeds from the starting point of the journey. The computation could just as well have been carried out from the

Faculty Lounge working backward, illustrating *backward* dynamic programming.[4] The principle of optimality applies to both; the optimum total path must lie along an optimum subpath from the beginning or end to any intermediate point. This principle, applied to finite dimensional problems, gives rise to systematic and efficient algorithms for calculating optimum paths.

BASEBAND SIGNAL MODEL

We shall now relate this general mathematical theory to the reception of digital data signals. Let us first consider a mathematical model of a baseband signal, i.e., a signal not modulated onto a carrier.[5] Let the sequence of numbers, called information symbols, to be transmitted be denoted $a_1, \cdots, a_N$. N, the number of information symbols, is large but finite. It is assumed that these symbols are independent and can each assume L equally probable values. These symbols amplitude modulate a train of pulses occurring at intervals T to produce the transmitted waveform

$$s(t) = \sum_{i=1}^{N} a_i p(t-iT) \qquad (1)$$

where $p(t)$ is the transmitted pulse and the symbol rate is $1/T$ Bd. The bit rate over the baseband channel is $(1/T)\log_2 L$ bits/s.

The output of the baseband channel is written

$$y(t) = \sum_{i=1}^{N} a_i h(t-iT) + n(t) \qquad (2)$$

where $n(t)$ is white Gaussian noise with double-sided power density spectrum $N_0/2$ W/Hz and where $h(t)$ is the convolution of $p(t)$ with the impulse response of the baseband channel. In the following derivation, we shall, for simplicity, refer to $h(t)$ as the channel impulse response.

A key assumption is that $h(t)$ has finite duration mT (as suggested in Fig. 1). This assumption has two consequences. First, all elements of the N symbol sequence are received in the finite interval $0 \leq t \leq \tau$, $(N+m)T < \tau < \infty$. The second consequence bears upon a term that arises in the sequel. We make the definition

$$r_{i-j} \triangleq \int_0^\tau h(t-iT)h(t-jT)dt . \qquad (3a)$$

TABLE I

Minimum Distance from Office to	Publish	Perish	Faculty Lounge
via North bridge	0.5	1.2	1.4
via South bridge	0.8	1.0	1.3

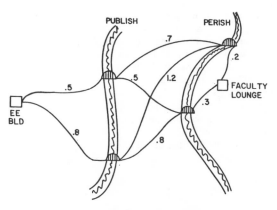

Fig. 2.   A pedestrian example.

---

[4]For our example, the distinction between backward and forward is trivial. However, there are problems that naturally fit one or the other. We shall see one shortly.

[5]The derivation of the Viterbi algorithm presented in the sequel is due to Ungerboeck [10], who also considered the passband case.

Now the finite memory of the channel implies that

$$r_{i-j} = 0, \qquad \text{for } |i-j| > m . \tag{3b}$$

We shall refer to m as the *memory* of the channel in units of T.

MAXIMUM LIKELIHOOD SEQUENCE ESTIMATION

Our objective is to operate on the received signal $y(t)$, $0 \leq t \leq \tau$ so as to produce an estimate $a_1^*, a_2^*, \cdots, a_N^*$ of the sequence of transmitted symbols $a_1, a_2, \cdots, a_N$. Given that $y(t)$ is perturbed by additive noise, we cannot reproduce the transmitted sequence with certainty. Rather, we seek to minimize the probability of sequence error,[6] i.e., the probability that $a_1^*, a_2^*, \cdots, a_N^*$ is different from $a_1, a_2, \cdots, a_N$. Under our assumptions, on the transmitted sequence, maximum likelihood sequence estimation (MLSE) produces this minimum error probability.

In order to define MLSE, first define the probability density functional

$$P[y(t), \ 0 \leq t \leq \tau \mid a_1 = \hat{a}_1, a_2 = \hat{a}_2, \cdots, a_N = \hat{a}_N]$$

as the probability that $y(t)$, $0 \leq t \leq \tau$ is received under the *assumption* that the transmitted symbols are $\hat{a}_1, \hat{a}_2, \cdots, \hat{a}_N$. Notice that for a particular received signal, there are L^N values of this quantity since there are L^N sequences $\hat{a}_1, \cdots, \hat{a}_N$. In MLSE we estimate the transmitted sequence to be the sequence which maximizes this likelihood.[7] Ostensibly, L^N calculations are required to find this maximum. The virtue of the Viterbi algorithm is that the number of calculations necessary for MLSE grows linearly with N rather than exponentially.

We now derive an expression for the likelihood which shows explicitly the calculations that are necessary. From (2), if $a_1, a_2, \cdots, a_N$ are assumed to have particular values $\hat{a}_1, \hat{a}_2, \cdots, \hat{a}_N$, then $y(t)$ is a Gaussian process with mean $\sum_{i=1}^{N} \hat{a}_i h(t-iT)$. A straightforward derivation[8] allows us to write

$$P[y(t); 0 \leq t \leq \tau \mid a_1 = \hat{a}_1, a_2 = \hat{a}_2, \cdots, a_N = \hat{a}_N]$$

$$= K \exp\left[-\frac{1}{N_0} \int_0^\tau [y(t) - \sum_{i=1}^{N} \hat{a}_i h(t-iT)]^2 \right] dt \tag{4}$$

where K is a constant independent of $y(t)$ and the sequence $\hat{a}_1, \cdots, \hat{a}_N$. In finding the optimum sequence, we are interested not in absolute values of the likelihood, but in relative values for different sets of $\hat{a}_1, \hat{a}_2, \cdots, \hat{a}_N$. This observation allows the problem to be simplified quite apart from the Viterbi algorithm. The likelihood is maximized by choosing the set $\hat{a}_1, \hat{a}_2, \cdots, \hat{a}_N$ which minimizes the integral

$$\int_0^\tau \left[y(t) - \sum_{i=1}^{N} \hat{a}_i h(t-iT) \right]^2 dt .$$

We now expand the quadratic term under the integral sign, resulting in three terms, one of which is $\int_0^\tau y^2(t)dt$. This energy term is independent of $\hat{a}_1, \cdots, \hat{a}_N$ and may be ignored in making comparisons. The *objective function* to be minimized is then

$$D = -2 \sum_{i=1}^{N} \hat{a}_i Z_i + \sum_{i=1}^{N} \sum_{j=1}^{N} \hat{a}_i \hat{a}_j r_{i-j} \tag{5}$$

[6]In a later section, the distinction between sequence error and bit error is discussed.

[7]For a discussion of decision rules, see [11].

[8]The random process $y(t)$ is approximated by a sequence of Karhunen-Loeve expansions. All of the coefficients in these expansions are Gaussian random variables. A limit of these expansions yields (4). For details, see [12] and [13].

where $Z_i \triangleq \int_0^\tau y(t)h(t-iT)dt$. [Recall that r_{i-j} was defined in (3).]

The two terms in (5) embody the two kinds of information at our disposal. The terms $Z_i = \int_0^\tau y(t)h(t-iT)dt$; $i = 1,2,\cdots,N$ can be viewed as the sampled output of a filter matched to the channel impulse response $h(t)$. All that we need to know about the received signal $y(t)$ is in these samples. The term $\sum_{i=1}^{N} \sum_{j=1}^{N} \hat{a}_i \hat{a}_j r_{i-j}$ indicates our knowledge of the memory in the channels [see (3b)].

THE VITERBI ALGORITHM

Although (5) reflects considerable simplification, a brute force approach to its minimization requires L^N calculations. Like Professor X, we shall eschew this approach. By suitable definitions, the problem can be cast as that of optimum path selection. By use of the principle of optimality, the optimum path can be found.

We now begin a series of manipulations that lead to the desired result. First we decompose the objective function. From (5) we have

$$D = -2 \sum_{i=1}^{N} \hat{a}_i Z_i + \sum_{i=1}^{N} \sum_{j=1}^{N} \hat{a}_i \hat{a}_j r_{i-j}$$

$$= -2 \sum_{i=1}^{N-1} \hat{a}_i Z_i + \sum_{i=1}^{N-1} \sum_{j=1}^{N-1} \hat{a}_i \hat{a}_j r_{i-j}$$

$$-2\hat{a}_N Z_N + 2\hat{a}_N \sum_{i=N-m}^{N-1} \hat{a}_i r_{N-i} + \hat{a}_N^2 r_0 . \tag{6}$$

Notice that there is a replication in (6) in that the first two terms on the RHS are similar in form to the LHS. We shall continue this replication in the course of developing the algorithm. Notice also that the last three terms of (6) are a function only of $\{ \hat{a}_{N-m}, \hat{a}_{N-m+1}, \cdots, \hat{a}_N \}$ and not of the rest of the possible transmitted sequence. Also, these terms depend only upon one output of the matched filter Z_N.

By the use of appropriate definitions, the objective function can be put in a more suitable form. We define the set of *state vectors*

$$\sigma_k \triangleq \{ \hat{a}_{k-m+1}, \hat{a}_{k-m+2}, \cdots, \hat{a}_k \}, \qquad k = m, m+1, \cdots, N. \tag{7}$$

Note that σ_k contains all the data symbols, except for $\hat{a}_{k+1}$, that will determine y_{k+1}. There is a one-to-one correspondence between a sequence of state vectors $\sigma_m, \sigma_{m+1}, \cdots, \sigma_N$ and an estimated sequence of transmitted symbols $\hat{a}_1, \hat{a}_2, \cdots, \hat{a}_N$, although it is apparent that the set of state vectors has much redundancy. The problem of choosing an optimum sequence from the set $a_1, \cdots, a_N$ can therefore be recast as that of choosing an optimum $\sigma_m, \cdots, \sigma_N$. Estimating the optimum sequence of states $\sigma_m, \sigma_{m+1}, \cdots, \sigma_N$ can be viewed as optimum path selection through a *lattice* representing the states. In Fig. 3, such a lattice is shown for the case $L = 2$ and $m = 3$. The dotted lines indicate the transitions that can be made from one state to another. For example, state $\sigma_6 = \{ \hat{a}_4 \hat{a}_5 \hat{a}_6 \} = \{ -1, +1, -1 \}$ can only have predecessor states $\sigma_5 = \{ -1, -1, +1 \}$ or $\sigma_5 = \{ +1, -1, +1 \}$, i.e., states with the same a_4 and a_5. In order to formulate the problem as optimum path selection, we must now derive suitable distance measures.

We proceed by defining the quantities

$$U(Z_1, \cdots, Z_k; \sigma_m, \cdots, \sigma_k) \triangleq -2 \sum_{i=1}^{k} \hat{a}_i Z_i + \sum_{i=1}^{k} \sum_{j=1}^{k} \hat{a}_i \hat{a}_j r_{i-j} \tag{8a}$$

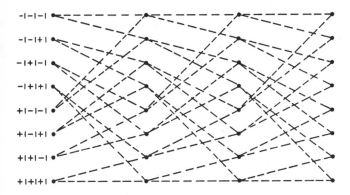

Fig. 3.   Lattice diagram. $L = 2$, $M = 3$.

and

$$V(Z_k; \sigma_{k-1}, \sigma_k) \overset{\Delta}{=} -2\hat{a}_k Z_k + 2\hat{a}_k \sum_{i=k-m}^{k-1} \hat{a}_i r_{k-i} + (\hat{a}_k)^2 r_0,$$

$$k = m+1, \cdots, N. \quad (8b)$$

In terms of these definitions, we can write

$$U(Z_1, \cdots, Z_k; \sigma_m, \cdots, \sigma_k) = U(Z_1, \cdots, Z_{k-1}; \sigma_m, \cdots, \sigma_{k-1})$$

$$+ V(Z_k, \sigma_{k-1}, \sigma_k). \quad (9)$$

The problem of finding the set of states that minimizes the objective function (5) can be succinctly written $\min_{\sigma_m, \cdots, \sigma_N} U(Z_1, \cdots, Z_N; \sigma_m, \cdots, \sigma_N)$. Expanding upon this notation, we write

$$I(Z_1, \cdots, Z_N) \overset{\Delta}{=} \min_{\sigma_m, \cdots, \sigma_N} U(Z_1, \cdots, Z_N; \sigma_m, \cdots, \sigma_N)$$

$$= \min_{\sigma_N} \min_{\sigma_m, \cdots, \sigma_{N-1} | \sigma_N} U(Z_1, \cdots, Z_N; \sigma_m, \cdots, \sigma_N). \quad (10)$$

Equation (10) indicates that the minimization is carried out in *two steps:* 1) with σ_N held fixed to one of its L^m values, minimize over $\sigma_m, \cdots, \sigma_{N-1}$ (L^{N-1} comparisons, since $(N-1)$ *data estimates* go into this set of state vectors); 2) with the values obtained in 1), minimize over σ_N (L comparisons). This merely specifies the order of a brute force approach. The next step breaks the minimization into three operations and introduces the definitions of (8).

$$I(Z_1, \cdots, Z_N) = \min_{\sigma_N} \min_{\sigma_{N-1} | \sigma_N} \min_{\sigma_m, \cdots, \sigma_{N-2} | \sigma_{N-1}, \sigma_N}$$

$$[U(Z_1, \cdots, Z_{N-1}; \sigma_1, \cdots, \sigma_{N-1}) + V(Z_N, \sigma_{N-1}, \sigma_N)]. \quad (11)$$

Up to this point, we have merely manipulated the problem of finding the set $\hat{a}_1, \cdots, a_N$ which maximizes $P_r[y(t)\ 0 \leq t \leq \tau | a_1 = \hat{a}_1, a_2, = \hat{a}_2, \cdots, a_N = \hat{a}_N]$ into an entirely equivalent problem as stated in (11). As we have noted earlier, there is a one-to-one mapping between the state set $\sigma_m, \cdots, \sigma_N$ and the symbol set

$\hat{a}_1, \hat{a}_2, \cdots, \hat{a}_N$. In connection with the quantities in (11), we make two observations which are crucial to the derivation of the algorithm.

$$\min_{\sigma_m, \cdots, \sigma_{N-2} | \sigma_{N-1}, \sigma_N} V(Z_N, \sigma_{N-1}, \sigma_N)$$

$$= V(Z_N, \sigma_{N-1}, \sigma_N). \quad (12a)$$

This simply says that if σ_{N-1} and σ_N are fixed, no variation of the states $\sigma_m, \cdots, \sigma_{N-2}$ can change the value of $V(Z_N, \sigma_{N-1}, \sigma_N)$.

$$\min_{\sigma_m, \cdots, \sigma_{N-2} | \sigma_{N-1}, \sigma_N} U(Z_1, \cdots, Z_{N-1}; \sigma_m, \cdots, \sigma_{N-1})$$

$$= \min_{\sigma_m, \cdots, \sigma_{N-2} | \sigma_{N-1}} U(Z_1, \cdots, Z_{N-1}; \sigma_m, \cdots, \sigma_{N-1}).$$

$$(12b)$$

If σ_{N-1} is fixed, fixing σ_N also has no bearing on the value of $U(Z_1, \cdots, Z_{N-1}; \sigma_m, \cdots, \sigma_{N-1})$. Combining (10), (11), and (12) yields

$$\min_{\sigma_N} \min_{\sigma_m, \cdots, \sigma_{N-1} | \sigma_N} U(Z_1, \cdots, Z_N; \sigma_m, \cdots, \sigma_N)$$

$$= \min_{\sigma_N} \min_{\sigma_{N-1} | \sigma_N} [V(Z_N; \sigma_{N-1}, \sigma_N) \quad (13)$$

$$+ \min_{\sigma_m, \cdots, \sigma_{N-2} | \sigma_{N-1}} U(Z_1, \cdots, Z_{N-1}; \sigma_m, \cdots, \sigma_{N-1})].$$

By the same steps that led to (13), we can decompose the second term on the RHS of (13). We find

$$\min_{\sigma_m, \cdots, \sigma_{N-2} | \sigma_{N-1}} U(Z_1, \cdots, Z_{N-1}\ \sigma_m, \cdots, \sigma_{N-1})$$

$$= \min_{\sigma_{N-2} | \sigma_{N-1}} [V(Z_N; \sigma_{N-2}, \sigma_{N-1})$$

$$+ \min_{\sigma_m, \cdots, \sigma_{N-3} | \sigma_{N-2}} U(Z_1, \cdots, Z_{N-2}; \sigma_m, \cdots, \sigma_{N-2})].$$

This decomposition can be continued. It is easiest to express this with some additional notation. We write

$$F(\sigma_{k+1}) = \min_{\sigma_k | \sigma_{k+1}} [V(Z_{k+1}; \sigma_k, \sigma_{k+1}) + F(\sigma_k)],$$

$$k = m, \cdots, N-1 \quad (14)$$

where

$$F(\sigma_m) \overset{\Delta}{=} U(A_1, \cdots, Z_m; \sigma_m)$$

and

$$F(\sigma_k) \overset{\Delta}{=} \min_{\sigma_m, \cdots, \sigma_{k-1} | \sigma_k} U(Z_1, \cdots, Z_k, \sigma_m, \cdots, \sigma_k),$$

$$k = m + 1, \cdots, N-1.$$

Although not indicated explicitly, $F(\sigma_k)$ is still dependent on $Z_1, \cdots, Z_k$.

Equations (13) and (14) embody the Viterbi algorithm. One begins by computing $F(\sigma_m)$ for each of the L^m values of the state σ_m [see (8a)]. Then employing (14), L^m values of $F(\sigma_{m+1})$ are found. Recall that in connection with Fig. 3 we said that σ_{m+1} had L possible predecessor states so that $\min\limits_{\sigma_m \mid \sigma_{m+1}}$ requires L comparisons for each of the L^m possible values of σ_{m+1}. The quantity $F(\sigma_{m+1})$ is the "minimum distance" to a particular value of state σ_{m+1}. From the principle of optimality, the optimum total path through the state lattice, as in Fig. 3, must lie on one of the paths to each of the L^m possible realizations of state σ_{m+1}. In the same way that $F(\sigma_{m+1})$ is computed, so are the successive quantities $F(\sigma_{m+2}), F(\sigma_{m+3}), \cdots, F(\sigma_N)$. We end with L^m possible paths through the lattice, each ending with a different value of σ_N. The final step is choosing the path for which $F(\sigma_N)$ is minimum.

It should be emphasized that L^m is related to the number of levels per transmitted symbol and the memory of the channel, but not to N, the number of symbols transmitted. Hence, we have shown that the Viterbi algorithm requires a fixed number of computations per symbol independent of the number of the symbols received.

EXAMPLE

In order to illustrate the Viterbi algorithm, we go through an example step-by-step. Let a system with the characteristics $L = 2$ (data values ± 1), $m = 2$, $r_0 = 1$, $r_{\pm 1} = 0.5$, and $r_{\pm 2} = -0.25$ be given. We consider a sequence seven symbols long with the successive outputs of the matched filter being $Z_1 = 1.5$, $Z_2 = 2.0$, $Z_3 = 0.5$, $Z_4 = 1.0$, $Z_5 = -1.5$, $Z_6 = -3.0$, and $Z_7 = 0.5$. The steps required to detect the transmitted sequence are as follows.

1) For each of the $L^m = 4$ states, $\sigma_2 = \{\hat{a}_1, \hat{a}_2\}$, calculate $U(Z_1, Z_2, \sigma_2)$ according to (8a). These values, along with those obtained in the succeeding steps of the algorithm, are shown in the lattice diagram of Fig. 4.

2) Apply (14) to find the optimum path to each of the four states $\sigma_3 = \{\hat{a}_2, \hat{a}_3\}$. For example, let $\sigma_3 = \{-1, +1\}$. As shown in Fig. 4, the predecessor states are $\{\pm 1, -1\}$. From $\{-1, -1\}$ the distance is $+11.5$ and from $\{+1, -1\}$ it is $+2.5$. Thus the optimum path to $\{-1, +1\}$ is via $\{+1, -1\}$. This is indicated in Fig. 4 by the solid line.

3) Repeat step 2) for each of the succeeding states σ_4, σ_5, σ_6, and σ_7, obtaining four paths through the lattice.

4) Optimize over the final state σ_7. In Fig. 4 we see that the optimum path passes through $\sigma_7 = \{-1, +1\}$. The optimum path is the one which leads through the succession of states $\{+1, +1\}$, $\{+1, -1\}$, $\{-1, +1\}$, $\{+1, -1\}$, $\{-1, -1\}$, $\{-1, +1\}$ indicating the symbol sequence $\hat{a}_1 = +1$, $\hat{a}_2 = +1$, $\hat{a}_3 = -1$, $\hat{a}_4 = +1$, $\hat{a}_5 = -1$, $\hat{a}_6 = -1$, and $\hat{a}_7 = +1$.

MERGES

From the foregoing, we see that it is not necessary to wait until the entire sequence $Z_1, Z_2, \cdots, Z_N$ has been received before making calculations. Equation (14) shows that the quantity $F(\sigma_k)$ is updated by $V(Z_{k+1}, \sigma_k, \sigma_{k+1})$, which is a function of the most recent output Z_{k+1} of the matched filter.

Examination of Fig. 4 also discloses that it is not necessary to wait until the entire output of the matched filter $Z_1, \cdots, Z_N$ has been received before making decisions. Consider the four paths leading to state σ_4. There is a *merge* in that each of these paths passes through the state $\sigma_2 = \{+1, +1\}$. Whatever happens from this point on does not change anything before the merge, so we can immediately make the decision $\hat{a}_1 = +1$, $\hat{a}_2 = +1$. In Fig. 4 it happens that merges take place in the minimum time, i.e., the minimum number of steps required to go from any value of one state to any value of another. This minimum time is equal to the memory, which in this case is $m = 2$. The assumption of a different set of channel characteristics and matched filter outputs could lead to the lattice diagram shown in Fig. 5. Here the merge does

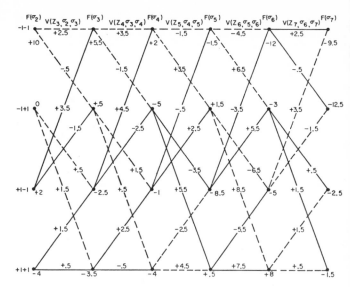

Fig. 4.   Lattice diagram.

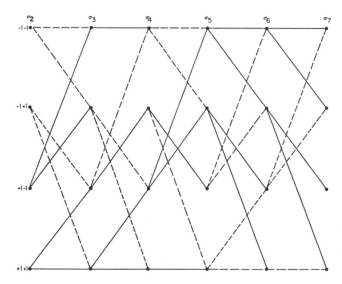

Fig. 5.   Lattice diagram optimum sequence detection.

not take place in minimum time. It is not until state σ_5 has been reached that one can make decisions. Merging is a random phenomenon, and it is possible that, in an unfortunate set of circumstances, no decisions can be made until the end of the entire sequence.

IMPLEMENTATION

The complexity of the Viterbi algorithm can be assessed by computing the amount of memory and the number of arithmetic operations required to implement it. From the foregoing, it is evident that both of these quantities depend directly upon the number of possibilities L^m for the state vector at any given time. In a not unlikely case, $L = 4$ and $m = 6$, implying $1\,048\,576$ possibilities. Clearly, for the dynamic programming technique to be feasible with present technology, the number of states must be reduced. Two techniques for reducing the number of states have been considered in the literature. Qureshi and Newhall [15] propose a prefiltering technique which reduces the spread of the signal pulses. The combined effect of the prefilter and the channel yields a reduction of the memory m. This approach is inherently suboptimum since the additive noise is enhanced. Vermeulen and Hellman [16] study a state reduction technique in which only the most probable states are retained as the algorithm is carried out. Both of these techniques show promising results under certain circumstances. (See also [22].)

PERFORMANCE

From the properties of maximum likelihood estimation, we are assured that the probability of sequence error, i.e., mistaking one sequence for another, is minimized. In his paper on the Viterbi algorithm [7], Forney[9] presented upper and lower bounds on the probability of bit error for large signal-to-noise ratio when maximum likelihood sequence estimation is used.

Minimizing the probability of sequence error corresponds roughly to the minimization of probability of block error in digital communications. Iterative algorithms similar to the foregoing have been developed for minimizing the probability of incorrectly detecting individual symbols [19]. Optimum bit-by-bit detection and optimum sequence detection are not synonymous.[10] In a sense, optimum sequence detection considers all erroneous sequences to be equally bad. Thus for low signal-to-noise situations, errors may lead to detected sequences that are very far from the true sequence. However, for high signal-to-noise ratio, it is conjectured that both techniques would give about the same performance. The comparison of these two techniques is a subject that merits further attention.

CONCLUSION

We have studied the application of the Viterbi algorithm to the detection of digital signals. The Viterbi algorithm is one of several nonlinear approaches to signal detection. We recommend the survey article by Lucky [20] summarizing the work in this area.

Another survey paper by Forney [21] places the Viterbi algorithm in a wider context than digital transmission over narrow-band channels. The Viterbi algorithm is generally applicable to the problem of detecting a discrete-time finite-state Markov process immersed in additive memoryless noise. To date, the Viterbi algorithm has been applied to decoding convolutional codes, FSK, text recognition, and magnetic tape recording. Doubtlessly, more applications will be found as work in this area progresses.

ACKNOWLEDGMENT

The author would like to express his appreciation to R. D. Gitlin and F. R. Magee for many illuminating discussions, and to S. B. Weinstein for many helpful suggestions on the manuscript.

REFERENCES

[1] R. E. Bellman, *Dynamic Programming*, Princeton, N.J.: Princeton Univ. Press, 1957.
[2] S. E. Dreyfus, *Dynamic Programming and the Calculus of Variations*. New York and London: Academic, 1965.
[3] A. J. Viterbi, "Error bounds for convolutional codes and an asymptotically optimum decoding algorithm," *IEEE Trans. Inform. Theory*, vol. IT-13, pp. 260-269, Apr. 1967.
[4] H. Kobayashi, "Application of probabilistic decoding to digital magnetic recording systems," *IBM J. Res. Develop.*, vol. 15, pp. 64-74, Jan. 1971.
[5] ——, "Correlative level coding and maximum-likelihood decoding," *IEEE Trans. Inform. Theory*, vol. IT-17, pp 586-594, Sept. 1971.
[6] J. K. Omura, "Optimal receiver design for convolutional codes and channels with memory via control theoretical concepts," *Inform. Sci.*, vol. 3, pp. 243-266, July 1971.
[7] G. D. Forney, Jr., "Maximum likelihood sequence estimation of digital sequences in the presence of intersymbol interference," *IEEE Trans. Inform. Theory*, vol. IT-18, pp. 363-378, May 1972.
[8] R. W. Chang and J. C. Hancock, "On receiver structures for channels having memory," *IEEE Trans. Inform. Theory*, vol. IT-12, pp. 463-468, Oct. 1966.
[9] R. W. Lucky, J. Salz, and E. J. Weldon, Jr., *Principles of Data Communications*. New York: McGraw-Hill, 1968, ch. 6.
[10] G. Ungerboeck, "Adaptive maximum-likelihood receiver for carrier modulated data-transmission systems," *IEEE Trans. Commun.*, vol. COM-22, pp. 624-636, May 1974.
[11] G. L. Turin, *Notes on Digital Communication*. New York: Van Nostrand Reinhold, 1969, ch. 2.
[12] *ibid.*, ch. 3.
[13] W. B. Davenport, Jr. and W. L. Root, *An Introduction to the Theory of Random Signals and Noise*. New York: McGraw-Hill, 1958.
[14] L. K. Mackechnie, "Receivers for channels with intersymbol interference" (Abstract), presented at the IEEE Int. Symp. Inform. Theory, 1972.
[15] S. U. H. Qureshi and E. E. Newhall, "An adaptive receiver for data transmission over time-dispersive channels," *IEEE Trans. Inform. Theory*, vol. IT-19, pp. 448-451, July 1973.
[16] F. L. Vermeulen and M. E. Hellman, "Reduced state Viterbi decoding for channels with intersymbol interference" (Abstract only), presented at the IEEE Int. Symp. Inform. Theory, 1972.
[17] G. J. Foschini, "Performance bound for maximum likelihood reception of digital data," *IEEE Trans. Inform. Theory*, to be published.
[18] R. R. Anderson and G. J. Foschini, "The minimum distance for MLSE digital data systems of limited complexity," *IEEE Trans. Inform. Theory*, to be published.
[19] K. Abend and B. D. Fritchman, "Statistical detection for communications channels with intersymbol interference," *Proc. IEEE*, vol. 58, pp. 779-785, May 1970.
[20] R. W. Lucky, "A survey of the communication theory literature, 1968-1973," *IEEE Trans. Inform. Theory*, vol. IT-19, pp. 725-739, Nov. 1973.
[21] G. D. Forney, Jr., "The Viterbi algorithm," *Proc. IEEE*, vol. 61, pp. 268-278, Mar. 1973.
[22] D. D. Falconer and F. R. Magee, Jr., "Adaptive channel memory truncation for maximum likelihood sequence estimation," *Bell Syst. Tech. J.*, vol. 52, pp. 1541-1561, Nov. 1973.

[9]In a recent paper, Foschini [17] made further contributions to this aspect of Forney's work. Also, Anderson and Foschini [18] have described how to calculate these bounds in practical situations.

[10]For a summary of work on bit-by-bit detection, see the survey paper by Lucky [20].

Jeremiah F. Hayes (S'65-M'66) received the B.E.E. degree from Manhattan College, New York, N.Y., in 1956, the M.S. degree from New York University, New York, N.Y., in 1961, and the Ph.D. degree from the University of California, Berkeley, in 1966.

Dr. Hayes was a member of the faculty of Purdue University, Lafayette, Ind., from 1966 to 1969. Since 1969 he has been a Member of the Technical Staff of Bell Laboratories. His current work is in the area of computer communication networks.   □

FADING CHANNEL COMMUNICATIONS *

Peter Monsen

Adaptive processing can reduce the effects of fading on beyond-the-horizon digital radio links.

INTRODUCTION

Two radio propagation channels for beyond-the-horizon communications, troposcatter, and HF are currently being reexamined. In the past, transmission over these radio channels had been considered unreliable due to fading effects. Recently, conversion from analog to digital transmission and the use of new adaptive signal processing techniques have offered promise of acceptable network communication quality. In addition, over-the-horizon radio provides economic and/or security advantages relative to satellite, cable, or line-of-sight terrestrial microwave links. There is renewed interest in the use of high frequency (HF) and troposcatter communications for networks carrying digital traffic.

HF radio uses frequencies in the range of 2 to 30 MHz. At these frequencies, communications beyond line-of-sight is achieved through refractive bending of the radio wave in the ionosphere from ionized layers at different elevations. In most cases more than one ionospheric "layer" causes the return of a refracted radio wave to the receiving antenna. The impulse response of such a channel exhibits a discrete multipath structure. The time between the first arrival and the last arrival is the multipath delay spread. Changes in ion density in individual layers due to solar heating cause fluctuations in each multipath return. This time varying multipath characteristic produces alternately destructive and constructive interference. The resulting fading can produce complete loss of signal as those who have heard replays

of Winston Churchill's radio talks during World War II can attest.

Troposcatter radio transmission was discovered only after World War II when it was noted that microwave signals from beyond the horizon radars were much stronger than predicted from diffraction calculations over the earth's surface. A commonly held theory of this phenomenon, developed by Tatarski [1], holds that random fluctuations in the dielectric constant in the troposphere divert some small fraction of the impinging energy back to the receiver. The name tropospheric scatter or troposcatter derives from this concept of random redirection of the incident wave by the troposphere. Significant scatter returns occur from a "common volume" defined by the receiving and transmitting antenna beam patterns. Scatter returns from different points within the common volume have different path delays. Signals scattered from points separated by more than the decorrelation distance of the fluctuations in the dielectric constant are not correlated. Thus, as in the HF example, the impulse response characterizing the channel has a time-varying multipath structure with delay spread but without the discrete layers. Troposcatter systems have been widely used in military applications for beyond line-of-sight communications up to about 600 miles. The frequency range for this application extends from about 400 to 5000 MHz. Reliable communications require redundant transmission paths provided through the use of multiple frequencies, antennas separated in space, or scatter at two different beam angles. The several redundant paths are referred to as diversity paths, and the number of paths is termed the order of diversity.

The multipath delay spread limits the channel capacity which can be achieved in present analog systems. Only

The author is with SIGNATRON, Inc., Lexington, MA 02173.

*Reprinted from *IEEE Communications Society Magazine*, January 1980, Vol. 18, No. 1, pp. 27-36.

transmission bandwidths less than the reciprocal of this multipath delay can be achieved. Signals of larger bandwidths become distorted due to the multipath dispersion. In FM systems this dispersion causes intermodulation noise after detection.

With the introduction of satellite communication systems, which do not suffer from extensive multipath fading, the future of HF and troposcatter systems appeared to be limited. Economic and security factors have altered this assessment, particularly for digital transmission.

With digital signal formats, adaptive methods can exploit multipath structure to improve performance.

With digital signal formats, adaptive methods can be devised to measure the multipath structure and exploit it as an extra form of diversity to improve performance. Unlike the capacity of analog systems, the capacity of digital systems is not restricted by the multipath delay spread. From a network viewpoint, fades in tandem digital links do not have a cumulative effect because the signal can be regenerated at each node.

Adaptive troposcatter systems have been demonstrated which are efficiently able to detect digital signals perturbed by a fading channel medium while tracking the fading variations. If receiver adaptation requires significantly less signal-to-noise ratio per bit than receiver detection of the digital signal, the receiver decisions can be effectively used as the estimate of the transmitted signal to achieve what is referred to as *decision-directed* adaptation. Such systems can be operated without transmission of special pilot or reference signals for channel tracking.

In troposcatter communication, adaptive techniques have increased the digital rate capability by at least an order of magnitude.

In troposcatter communication, adaptive techniques have increased the digital rate capability by at least an order of magnitude. An adaptive equalizer modem [2],[3] developed for military troposcatter links has been successfully field tested at digital rates up to 12.6 Mbits/s in a 15 MHz channel allocation. In HF systems, adaptive signal processing techniques are now being considered with goals of digital rates on the order of 5 kbits/s in a 3 kHz channel. In both these examples, the digital symbol period is of the same order as the channel multipath delay spread.

Applicability of adaptive signal processing techniques is critically dependent on whether the rate of fading is slower than the rate of signaling. As discussed below, both HF and troposcatter radio links can be considered to be slow fading multipath channels.

SLOW-FADING MULTIPATH CHANNELS

For digital communication over beyond-the-horizon radio links, an attempt is made to maintain transmission linearity, i.e., the receiver output should be a linear superposition of the transmitter input plus channel noise. This is accomplished by operation of the power amplifier in a linear region or, with saturating power amplifiers, by using constant-envelope modulation techniques. For linear systems multipath fading can be characterized by a transfer function of the channel $H(f; t)$. This function is the two-dimensional random process in frequency f and time t that is observed as carrier modulation at the output of the channel when sine wave excitation at the carrier frequency is applied to the channel input. For any continuous random process, we can determine the minimum separation required to guarantee decorrelation with respect to each argument.

For the time varying transfer function $H(f; t)$, let t_d and f_d be the decorrelation separations in the time and frequency variables, respectively. If t_d is a measure of the time decorrelation in seconds, then

$$\sigma_t = \frac{1}{2\pi t_d} \text{ Hz}$$

is a measure of the fading rate or bandwidth of the random channel. The quantity σ_t is often referred to as the *Doppler spread* because it is a measure of the width of the received spectrum when a single sine wave is transmitted through the channel. The dual relationship for the frequency decorrelation f_d in hertz suggests that a delay variable

$$\sigma_f = \frac{1}{2\pi f_d} \text{ seconds}$$

defines the extent of the multipath delay. The quantity σ_f is often referred to as the *multipath delay spread* as it is a measure of the width of the received process in the time domain when a single impulse function is transmitted through the channel.

Typical values of these spread factors for HF and troposcatter communication are

HF	Troposcatter
$\sigma_t \sim 0.1$ Hz	$\sigma_t \sim 1$ Hz
$\sigma_f \sim 10^{-3}$ seconds	$\sigma_f \sim 10^{-7}$ seconds

where the symbol $\sim$ denotes "on the order of."

The spreads can be defined precisely as moments of spectra in a channel model [4] which assumes wide sense stationarity (WSS) in the time variable and uncorrelated scattering (US) in a multipath delay variable. This WSSUS model and the assumption of Gaussian statistics for $H(f;t)$ provide a statistical description in terms of a single two-dimensional correlation function of the random process $H(f;t)$.

This characterization has been quite useful and accurate for a variety of radio link applications. However the stationarity and Gaussian assumptions are not necessary for the utilization of adaptive signal processing techniques on these channels. What is necessary is first that sufficient time exists to "learn" the channel characteristics before they change, and second, that decorrelated portions of the frequency band be excited such that a diversity effect can be realized. These conditions are reflected in the following two relationships in terms of the previously defined channel factors, the data rate R, and the bandwidth B:

$$R(\text{bits/s}) \gg \sigma_t \ (\text{Hz}) \qquad \text{Learning Constraint}$$

$$B(\text{Hz}) \gtrsim f_d \ (\text{Hz}) \qquad \text{Diversity Constraint}$$

where the symbol $\gtrsim$ denotes "on the order of or greater than."

The learning constraint insures that sufficient signal-to-noise ratio (SNR) exists for reliable communication at rate R over the channel. Clearly if $R \sim \sigma_t$, the channel would change before significant energy for measurement purposes could be collected. When $R \gg \sigma_t$, the received data symbols can be viewed as the result of a channel sounding signal and appropriate processing can generate estimates of the channel character during that particular stationary epoch. The signal processing techniques in an adaptive receiver do not necessarily need to measure the channel directly in the optimization of the receiver, but the requirements on learning are approximately the same. If only information symbols are used in the sounding signal, the learning mode is referred to as *decision-directed*. When digital symbols known to both the transmitter and receiver are employed, the learning mode is called *reference-directed*. An important advantage of digital systems is that in many adaptive communications applications, adaptation of the receiver with no wasted power for sounding signals can be accomplished using the decision-directed mode. This is possible in digital systems because of the finite number of parameters or levels in the transmitted source symbols and the high likelihood that receiver decisions are correct.

Diversity in fading applications is used to provide redundant communications channels so that when some of the channels fade, communication will still be possible over the others that are not in a fade. Some of the forms of diversity employed are space using multiple antennas, angle of arrival using multiple feedhorns, polarization, frequency, and time. These diversity techniques are sometimes called *explicit diversity* because of their externally visible nature. An alternate form of diversity is termed *implicit diversity* because the channel itself pro-

vides redundancy. In order to capitalize on this implicit diversity for added protection, receiver techniques have to be employed to correctly assess and combine the redundant information. The potential for *implicit fre-*

The learning constraint insures adequate time to measure the channel characteristics before they change. The diversity constraint insures adequate bandwidth to combat deep, frequency-selective fades.

---

quency diversity arises because different parts of the frequency band fade independently. Thus, while one section of the band may be in a deep fade, the remainder can be used for reliable communications. However, if the transmitted bandwidth B is small compared to the frequency decorrelation interval f_d, the entire band will fade and no implicit diversity can result. Thus, the second constraint $B \gtrsim f_d$ must be met if an implicit diversity gain is to be realized. In diversity systems a little decorrelation between alternate signal paths can provide significant diversity gain. Thus it is not necessary for $B \gg f_d$ in order to realize implicit frequency diversity gain although the implicit diversity gain clearly increases with the ratio B/f_d. Note that the condition $R \ll B \gtrsim f_d$ does not preclude the use of implicit diversity because a bandwidth expansion technique can be used in the modulation process to spread the transmitted information over the available bandwidth B. We shall distinguish between these low-data rate and high-data rate conditions because the appropriate receiver structures take on somewhat different forms.

The implicit diversity effect described here results from decorrelation in the frequency domain in a slow-fading $(R \gg \sigma_t)$ application. This implicit frequency diversity can in some circumstances be supplemented by an *implicit time diversity* effect which results from decorrelation in the time domain. In fast fading applications $(R \gtrsim \sigma_t)$ redundant symbols in a coding scheme can be used to provide time diversity provided the code word spans more than one fade epoch. In our slow-fading application this condition of spanning the fade epoch can be realized by interleaving the code words to provide large time gaps between successive symbols in a particular code word. The interleaving process requires the introduction of signal delay longer than the time decorrelation separation t_d. In many practical applications which require transmission of digitized speech, the required time delay is unsatisfactorily long for two-way speech communication. For these reasons there is more emphasis on implicit frequency diversity techniques in practical systems. The receiver structures to be discussed next are applicable to situations where the implicit frequency diversity applies.

ADAPTIVE RECEIVER STRUCTURES

We consider a pulse amplitude modulation system wherein the sample set $\{a_k\}$ is to be communicated over the channel using a modulation technique which forms a one-to-one correspondence between the sample a_k and the amplitude of a transmitted pulse. Independent modulation of quadrature carrier signals (i.e., $\sin 2\pi f_o t$ and $\cos 2\pi f_o t$, f_o = carrier frequency) is included in this class. An important example with optimum detection properties is quadrature phase shift keying (QPSK) which transmits the sample set $\{a_k = +1 + j\}$ by changing the sign of quadrature carrier pulses in accordance with the sign of the real and imaginary parts of the source sequence $\{a_k\}$.

A. Receivers for Channels with Negligible Intersymbol Interference

If each sample a_k can be one of M possible amplitudes ($M=4$ for QPSK), the transmitted data rate is

$$R = (\log_2 (M))/T$$

where $1/T$ is the transmitted symbol rate.

Most terrestrial over-the-horizon channel applications utilize media which are signal-to-noise ratio limited rather than bandwidth limited. In order to maximize signal detectability, only a few amplitudes are usually employed in these applications. When the symbol period T is much greater than the total width of the multipath dispersion of the channel, only a small portion of adjacent symbols interfere with the detection of a particular symbol. For the slow fading application, the diversity constraint requires that the signal bandwidth B be on the order of or larger than the frequency decorrelation interval f_d. Conditions for negligible intersymbol interference (ISI) and adaptive processing to obtain implicit diversity are then

$$T \gg 2\pi\sigma_F = \frac{1}{f_d}$$

$$B \gtrsim f_d.$$

When the number of amplitudes M is small, these conditions imply a low-data rate system relative to the available bandwidth, i.e., a bandwidth expansion system. The low-data rate condition for implicit frequency diversity can be expressed in terms of the data rate and bandwidth as

$$R \ll B \log_2 (M).$$

In the absence of intersymbol interference, it is well known [5] that the optimum detection scheme contains a noise filter and a filter matched to the received pulse shape. The optimum noise filter has a transfer function equal to the reciprocal of the noise power spectrum $K(f)$. When the additive noise at the receiver input is white, i.e., its spectrum is flat over the frequency band of interest, the noise filter can be omitted in the optimum receiver. The optimum receiver for a fixed channel trans-

fer function $H(f)$ then contains a cascade filter with component transfer functions

$$R(f) = \frac{1}{K(f)} H^*(f)$$

Noise Matched
Filter Filter

where the * denotes complex conjugation. In practical applications, signal delay must be introduced in order to make these filters realizable.

In general K and H change with time and the adaptive receiver must track these variations. The tapped-delay-line (TDL) filter is an important filter structure for such channel tracking applications. The TDL filter shown in Fig. 1 consists of a tapped delay line with signal multiplications by the tap weight w_i for each tap. For a bandpass system of bandwidth B, the sampling theorem states that any linear filter can be represented by parallel TDL filters operating on each quadrature carrier component with a tap spacing of $1/B$ or less. The optimum receiver can then be realized by a cascade of two such parallel TDL quadrature filters: one with tap weights adjusted to form the noise filter, the second with tap weights adjusted to form the matched filter. Since the cascade of two band-limited linear filters is another bandlimited filter, in some applications it is more convenient to employ one TDL to realize $R(f)$ directly. In practice, signals cannot be both time and frequency limited so that these TDL filters can only approximate the ideal solution. One advantage of the TDL filter is the convenience in adjusting the tap weight control voltage as a means of tracking the channel and noise spectrum variations.

The optimum receiver requires knowledge of the noise power spectrum $K(f)$ and the channel transfer function $H(f)$. When $K(f)$ is not flat over the band of interest, the input noise process contains correlation which is to be removed by the noise filter. Techniques to reduce *correlated noise effects* include: 1) prediction of future noise values and cancellation of the correlated component; 2) mean square error filtering techniques using an appropriate error criterion; and 3) noise excision techniques where a fast Fourier transform (FFT) is used to identify and excise noise peaks in the frequency domain.

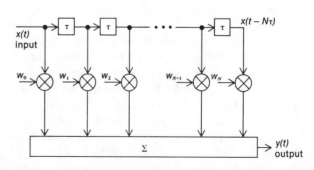

Fig. 1. Tapped-delay line (TDL) filter.

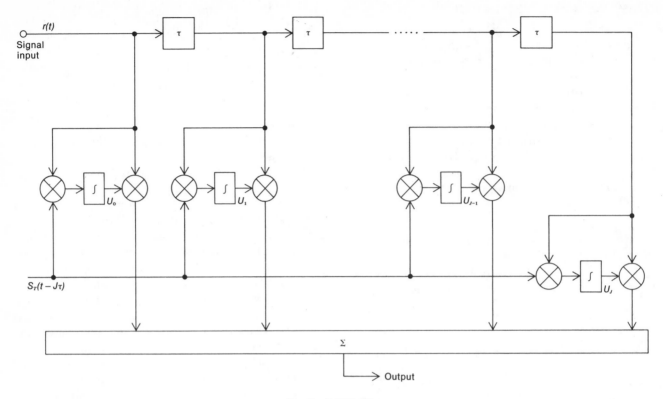

Fig. 2. RAKE filter

The problem of noise filtering is usually important in bandwidth expansion systems because of interference from other users as well as hostile jamming threats. For realization of the matched filter, a RAKE TDL filter using the concepts developed by Price and Green [6] can be used to adaptively derive an approximation to $H^*(f)$. A RAKE filter is so named because it acts to "rake" all the multipath contributions together. This can be accomplished using the TDL filter shown in Fig. 2, where the TDL weights are derived from a correlation of the tap voltages with a *common test sequence*, i.e., $S(t)$. This correlation results in estimates of the equivalent TDL channel tap values. By proper time alignment of the test sequence, the RAKE filter weights become estimates of the channel tap values but in inverse time order as required in a matched filter design. For adaptation of the RAKE filter, the test sequence may be either a known sequence multiplexed with the modulated information or it may be receiver decisions used in a decision-directed adaptation.

An alternate structure for realizing the matched filter is a recirculating delay line which forms an average of the received pulses. This structure was proposed as a means of reducing complexity in a RAKE filter design [7] for a frequency-shift-keying (FSK) system. For a pulse amplitude modulation (PAM) system, the structure would take the form shown in Fig. 3. An inverse modulation operation between the input signal and a local replica of the signal modulation is used to strip the signal modulation from the arriving signal. The recirculating delay line

can then form an average of the received pulse which is used in a correlator to produce the matched filter output. This correlation filter is considerably simpler than the RAKE TDL filter shown in Fig. 2.

In both the RAKE TDL and correlation filter, an averaging process is used to generate estimates of the received signal pulse. Because this signal pulse is imbedded in receiver noise it is necessary that the measurement process realize sufficient signal-to-noise ratio. This fundamental requirement is the basis for the learning constraint

$$R\,(\text{bits/s}) \gg \sigma_t\,(\text{Hz})$$

introduced earlier. If the signal rate R from which adapation is being accomplished is not much greater than the channel rate of change σ_t then the channel will change before the averaging process can build up sufficient signal-to-noise ratio for an accurate measurement. This requirement limits the application of adaptive receiver techniques with implicit frequency diversity gain to slow fading applications relative to the data rate. Fortunately, many channels have fade rates on the order of a few Hertz and data requirements thousands of times larger.

The receiver for this small-ISI example has, in general, a noise filter to accentuate frequencies where noise power is weakest and a matched filter structure which coherently recombines the received signal elements to provide the implicit diversity gain. The implicit diversity can be viewed as a frequency diversity because of the decorrelation of received frequencies. The matched filter in this view is a frequency diversity combiner which combines each frequency coherently according to its received strength. Without the matched filter, incoherent combining of the received frequencies would occur and no implicit diversity effect would be realized.

An important application of this low-data rate system is found in jamming environments where excess bandwidth is used to decrease jamming vulnerability. In more benign environments, however, most communication requirements do not allow for a large bandwidth relative to the data rate and if implicit diversity is to be realized in these applications, the effect of intersymbol interference must be considered.

B. High-Data Rate Receivers

When the transmitted symbol rate is on the order of the frequency decorrelation interval of the channel, the frequencies in the transmitted pulse will undergo different gain and phase variations resulting in reception of a distorted pulse.

Although there may have been no intersymbol interference (ISI) at the transmitter, the pulse distortion from the channel medium will cause interference between adjacent samples of the received signal. In the time domain, ISI can be viewed as a smearing of the transmitted pulse by the multipath thus causing overlap between successive pulses. The condition for ISI can be expressed in the frequency domain as

$$T^{-1} (\text{Hz}) \gtrsim f_d \ (\text{Hz})$$

or in terms of the multipath spread

$$T (\text{seconds}) \lesssim 2\pi\sigma_f \ (\text{seconds}).$$

Since the bandwidth of a PAM signal is at least on the order of the symbol rate T^{-1} Hz, there is no need for bandwidth expansion under ISI conditions in order to provide signal occupancy of decorrelated portions of the frequency band for implicit diversity. However it is not obvious whether the presence of the intersymbol interference can wipe out the available implicit diversity gain. Within the last decade it has been established that adaptive receivers can be used which cope with the intersymbol interference and in most practical cases wind up with a net implicit diversity gain. These receiver structures fall into three general classes: correlation filters with time gating, equalizers, and maximum likelihood detectors.

Adaptive receivers can cope with intersymbol interference and wind up with a net diversity gain.

---

1) *Correlation Filters:* These filters approximate the matched filter portion of the optimum no-ISI receiver. The correlation filter shown in Fig. 3 would fail to operate correctly when there is intersymbol interference between received pulses because the averaging process would add overlapped pulses incoherently. When the multipath spread is less than the symbol interval, this condition can be alleviated by transmitting a time gated pulse whose "off" time is approximately equal to the width of the channel multipath. The multipath causes the gated transmitted pulse to be smeared out over the entire symbol duration but with little or no intersymbol interference. The correlation filter can then be used to match the received pulse and provide implicit diversity [8]. In a configuration with both explicit and implicit diversity, moderate intersymbol interference can be tolerated because the diversity combining adds signal components coherently and ISI components incoherently. Because the off-time of the pulse can not exceed 100 percent, this approach is clearly data rate limited for fixed multipath conditions. In addition, the time gating at the transmitter results in an increased bandwidth which may be undesirable in a bandwidth-limited application. The power loss in peak power limited transmitters due to time gating can be partially offset by using two carrier frequencies with independent data modulation [9].

2) *Adaptive Equalizers:* Adaptive equalizers are linear filter systems with electronically adjustable parameters which are controlled in an attempt to compensate for intersymbol interference. Tapped delay line filters are a common choice for the equalizer structure as the tap weights provide a convenient adjustable parameter set. Adaptive equalizers have been widely employed in telephone channel applications [10] to reduce ISI effects due to channel filtering. In a fading multipath channel application, the equalizer can provide three functions simultaneously: noise filtering, matched filtering for explicit and implicit diversity, and removal of ISI. These functions are accomplished by adapting a tapped delay line equalizer (TDLE) to force error measure to a minimum. By designing the error measure to include the degrada-

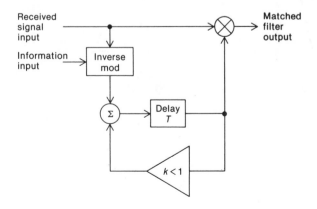

Fig. 3. Correlation filter for PAM system.

tion due to correlated noise, ISI, filtering, and improper diversion combining, the TDLE will minimize their combined effects.

A linear equalizer (LE) is defined as an equalizer which linearly filters each of the N explicit diversity inputs. An improvement to the LE is realized when an additional filtering is performed upon the detected data decisions. Because it uses decisions in a feedback scheme, this equalizer is known as a decision-feedback equalizer (DFE).

The operation of a matched filter receiver, an LE, and a DFE can be compared from examination of the received pulse train example of Fig. 4. The binary modulated pulses have been smeared by the channel medium producing pulse distortion and interference from adjacent pulses. Conventional detection without multipath protection would integrate the process over a symbol period and decide a $+1$ was transmitted if the integrated voltage is positive and -1 if the voltage is negative. The pulse distortion reduces the margin again in that integration process. A matched filter correlates the received waveform with the received pulse replica thus increasing the noise margin. The intersymbol interference arises from both future and past pulses in these radio systems since the multipath contributors near the mean path delay normally have the greatest strength. This ISI can be compensated for in a linear equalizer by using properly weighted time shifted versions of the received signal to cancel future and past interferers. The DFE uses time shifted versions of the received signal only to reduce the future ISI. The past ISI is cancelled by filtering past detected symbols to produce the correct ISI voltage from these interferers. The matched filtering property in both the LE and DFE is realized by spacing the taps on the TDLE at intervals smaller than the symbol period.

The DFE is shown in Fig. 5 for an Nth order explicit diversity system. A forward filter (FF) TDLE is used for each diversity branch to reduce correlated noise effects, provide matched filtering and proper weighting for explicit diversity combining, and reduce ISI effects. After diversity combining, demodulation, and detection, the data decisions are filtered by a backward filter TDLE to eliminate intersymbol interference from previous pulses. Because the backward filter compensates for this "past" ISI, the forward filter need only compensate for "future" ISI.

An automatic gain control (AGC) amplifier is shown for each diversity branch in order to bring the fading signal into the dynamic range of the TDLE. A decision-directed error signal for adaptation of the DFE is shown as the difference between the detector input and output. Qualitatively one can see that if the DFE is well adapted this error signal should be small. Reference-directed adaptation can be accomplished by multiplexing a known bit pattern into the message stream for periodic adaptation.

When error propagation due to detector errors is ignored, the DFE has the same or smaller mean-square error than the LE for all channels [11]. The error propagation mechanism has been examined by a Markov chain analysis [12] and shown to be negligible in practical fading channel applications. Also in an Nth order diversity application, the total number of TDLE taps is generally less for the DFE than for the LE. This follows because the former uses only one backward filter after combining of the diversity channels in the forward filter.

The performance of a DFE on a fading channel can be predicted [16]-[18] using a transformation technique which converts implicit diversity into explicit diversity and which treats the ISI effects as a Gaussian interferer.

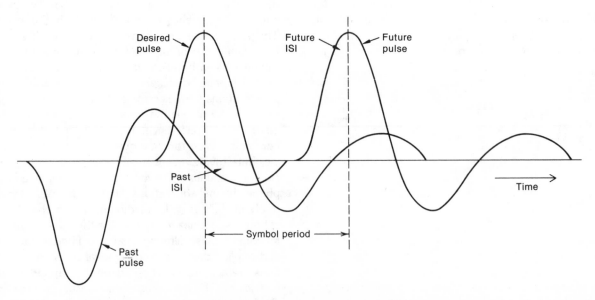

Fig. 4.   Received pulse sequence after channel filtering.

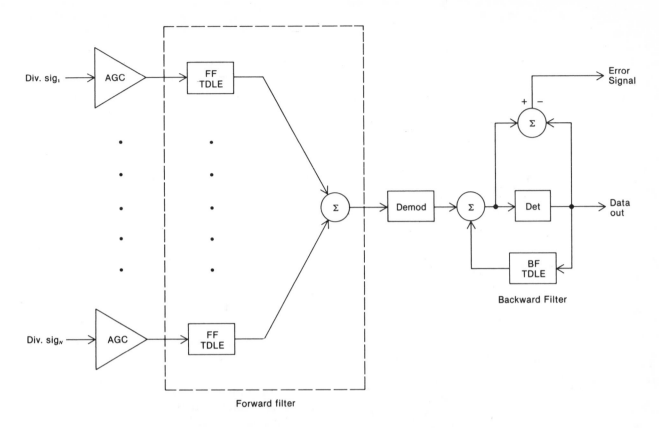

Fig. 5.   Decision-feedback equalizer, *Nth*-order diversity.

As an example, the average probability of error versus the total received bit energy (E_b) relative to the noise spectral density (N_0) is shown in Fig. 6 for a quadruple diversity system. The dashed line represents the zero multipath spread ($\sigma_f = 0$) performance and the solid lines show performance for different DFE configurations (N = number of forward filter taps and Δ = normalized tap spacing) and ISI conditions when the ratio of multipath spread to symbol period T is 0.25. The no-ISI condi-

---

A decision feedback equalizer with a modest number of taps performs almost as well as one with an infinite number.

---

tions are performance bounds determining by setting the ISI components to zero. When $\sigma/T = 0.25$, performance would be to the right of the dashed line if adaptive signal processing were not employed. The equalizer is seen to remove this degradation and also provide an implicit diversity gain which is measured by the difference between the solid line $N = 3$, $\Delta = 0.5$ curve and the dashed line. The difference between the $N = 3$, $\Delta = 0.5$ curve and the next curve labeled "No ISI" is the intersymbol interference penalty. With the filter parameters $N = 3$,

$\Delta = 0.5$, no technique for removing ISI can do better than this curve. The small-ISI penalty in this typical example is a strong argument for the use of the DFE versus more powerful ISI techniques. Finally the leftmost solid line approximates the very best that can be done as results show negligible improvement as the number of taps is increased further. The small difference exhibited shows that a DFE with only a modest number of forward filter taps performs to an ideal DFE with an infinite number of taps.

A DFE modem has been developed [3] with data rates up to 12.5 Mbits/s for application on troposcatter channels with up to four orders of diversity. This DFE modem uses only a three-tap forward filter TDLE and a three-tap backward filter TDLE. Extensive simulator and field tests [3],[16] have shown that implicit diversity gain is realized over a wide range of actual conditions while ISI effects are mostly eliminated. Thus operation at data rates near the frequency decorrelation distance is possible with no large intersymbol interference penalty. Measured results agree well with the predicted performance for which Fig. 6 is a typical example.

3) *Maximum Likelihood Detectors:* Since the DFE minimizes an analog detector voltage, it is unlikely that it is optimum for all channels with respect to bit error probability. By considering intersymbol interference as a conventional code defined on the real line (or complex

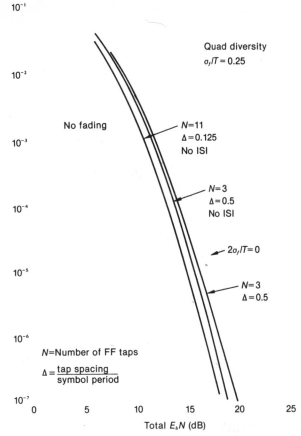

Fig. 6.   DFE performance, quad diversity.

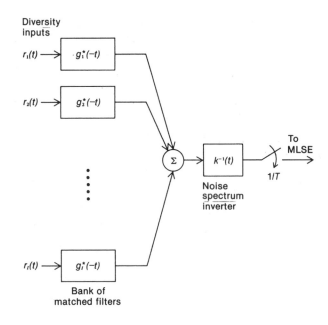

Fig. 7.   Diversity combiner for MLSE receiver.

line for bandpass channels), maximum likelihood sequence estimation algorithms have been derived [13],[14] for the PAM channel. These algorithms provide a decoding procedure for receiver decisions which minimize the probability of sequence error. A maximum likelihood sequence estimator (MLSE) receiver still requires a noise filter and matched filters for each diversity channel. After these filtering and combining operations, a trellis decoding technique is used to find the most likely transmitted sequence. Fig. 7 illustrates the filtering, combining, and sampling functions which precede the MLSE.

The MLSE algorithm works by assigning a state for each intersymbol interference combination. Because of the one-to-one correspondence between the states and the ISI, the maximum likelihood source sequence can be found by determining the trajectory of states.

If some intermediate state is known to be on the optimum path, then the maximum likelihood path originating from that state and ending in the final state will be identical to the optimal path. If at time n, each of the states has associated with it a maximum likelihood path ending in that state, it follows that sufficiently far in the past the path history will not depend on the specific final state to which it belongs. The common path history is the maximum likelihood state trajectory [13].

Since the number of ISI combinations and thus the number of states is an exponential function of the multipath spread, the MLSE algorithm has complexity which grows exponentially with multipath spread. The equalizer structure exhibits a linear growth with multipath spread. Also, the requirement for diversity combining and matched filtering in the MLSE receiver requires about the same circuit and adaptation implementation complexity as an equalizer for this requirement alone. By comparing Figs. 5 and 7 for the DFE and MLSE receiver, the systems are seen to be similar except for the replacement of the backward filter in the DFE by the decoding algorithm in the MLSE receiver. The backward filter is an L tap TDL filter whereas the MLSE decoding algorithm has computational complexity with exponential growth as a function of multipath spread. In return for this additional complexity, the MLSE receiver results in a smaller (sometimes zero) intersymbol interference penalty for channels with isolated and deep frequency selective fades. However in many applications where high orders of diversity are employed, these deep selective frequency fades do not occur frequently enough to significantly affect the average error probability. This result is illustrated in the performance curve given in Fig. 6 which showed only a small-ISI penalty for the DFE with just three taps as compared to the DFE with as many taps when there is no ISI.

NEW AREAS OF RESEARCH

Present adaptive equalizers for fading channel applications use an estimated gradient algorithm for channel tracking which can be quite slow for channels with deep selective frequency fades or in the presence of hostile electronic interference. Algorithms derived from the Kalman estimation equations have been suggested [19] as a means of realizing the full potential of adaptive tracking capability. Faster tracking would provide an impetus for HF equalization where digital data rates are not always many orders of magnitude greater than the channel rate of change. Adaptive receivers using multiple antennas in a fading multipath channel environment can provide antijamming protection from antenna nulling in addition to implicit diversity gain. Faster tracking algorithms would increase system flexibility to a wider range of jamming threats.

REFERENCES

[1] V. I. Tatarski, "The effects of the turbulent atmosphere on wave propagation," Israel Program for Scientific Translation, Jerusalem, Israel, 1971.

[2] P. Monsen, "High speed digital communication receiver," U.S. Patent 3 879 664, Apr. 22, 1975.

[3] D. R. Kern and P. Monsen, "Megabit digital troposcatter subsystem (MDTS)," GTE Sylvania, Needham, MA, and SIGNATRON, Lexington, MA, Final Rep. ECOM-74-0040-F.

[4] P. A. Bello, "Characterization of randomly time-variant linear channels," *IEEE Trans. Commun. Syst.*, vol. CS-11, pp. 360-393, Dec. 1963.

[5] L. A. Wainstein and V. D. Zubakov, *Extraction of Signals from Noise.* Englewood Cliffs, NJ: Prentice-Hall, 1962, ch. 3.

[6] R. Price and P. E. Green, Jr., "A communication technique for multipath channels," *Proc. IRE*, vol. 46, pp. 555-569, Mar. 1958.

[7] S. M. Sussman, "A matched filter communication system for multipath channels," *IRE Trans. Inform. Theory*, vol. IT-6, pp. 367-372, June 1960.

[8] M. Unkauf and O. A. Tagliaferri, "An adaptive matched filter modem for digital troposcatter," *Conf. Rec., Int. Conf. Commun.*, June 1975.

[9] ——, "Tactical digital troposcatter systems," in *Conf. Rec., Nat. Telecommun. Conf.*, vol. 2, Dec. 1978, pp. 17.4.1-17.4.5.

[10] R. W. Lucky, J. Salz, and E. J. Weldon, Jr., *Principles of Data Communication.* New York: McGraw-Hill, 1968, ch. 6.

[11] P. Monsen, "Feedback equalization for fading dispersive channels," *IEEE Trans. Inform. Theory*, vol. IT-17, pp. 56-64, Jan. 1971.

[12] ——, "Adaptive equalization of the slow fading channel," *IEEE Trans. Commun.*, vol. COM-22, Aug. 1974.

[13] G. D. Forney, Jr., "Maximum-likelihood sequence estimation of digital sequences in the presence of intersymbol interference," *IEEE Trans. Inform. Theory*, vol. IT-18, pp. 363-377, May 1972.

[14] G. Ungerboeck, "Adaptive maximum-likelihood receiver for carrier-modulated data transmission systems," *IEEE Trans. Commun.*, vol. COM-22, pp. 624-636, May 1974.

[15] R. Bellman, *Dynamic Programming.* Princeton, NJ: Princeton Univ. Press, 1957.

[16] P. Monsen, "Theoretical and measured performance of a DFE modem on a fading multipath channel," *IEEE Trans. Commun.*, vol. COM-25, pp. 1144-1153, Oct. 1977.

[17] M. Schwartz, W. R. Bennett, and S. Stein, *Communications Systems and Techniques.* New York: McGraw-Hill, 1966, ch. 10.

[18] P. Monsen, "Digital transmission performance on fading dispersive diversity channels," *IEEE Trans. Commun.*, vol. COM-21, pp. 33-39, Jan. 1973.

[19] D. Godard, "Channel equalization using a Kalman filter for fast data transmission," *IBM J. Res. Develop.*, pp. 267-273, May 1974.

Peter Monsen received the B.S. degree in electrical engineering from Northeastern University, Boston, MA, in 1962, the M.S. degree in operations research from the Massachusetts Institute of Technology, Cambridge, MA, in 1963, and the Eng.Sc.D. degree in electrical engineering from Columbia University, New York, NY, in 1970.

During a two year period beginning in January 1964, he served as a Lieutenant in the U.S. Army at the Defense Communication Agency, Arlington, VA, where he was concerned with communication systems engineering. From 1966 to 1972 he was with Bell Laboratories, Holmdel, NJ, where he was Supervisor of a Transmission Studies Group whose work included fading-channel characterization and adaptive equalization. He is presently with SIGNATRON, Inc., where he was responsible for the development of a 12.6 Mbit/s troposcatter modem. He holds U.S. patents on the decision-feedback equalizer and on an adaptive receiver for cross-polarized signals. His current areas of interest include an angle diversity study and test program, the development of a new troposcatter modem, and high-data rate HF techniques.

Dr. Monsen is a member of Eta Kappa Nu, Tau Beta Pi, Sigma Xi, and the IEEE.

Section 3

Computer Communications

EDITORS' COMMENTS

The most profound part of the technological revolution of the 1970s lies in the area of computer communications. The computer has become so powerful and useful that it finds applications in all facets of our lives. The computer is used in such industries as banking, transportation, entertainment, and general time-sharing services, just to name a few. The underlying requirement for successful and efficient use of computer systems is first to set up a computer network consisting of users or terminals, computers or processors, and communication links or channels to carry data. Then we must decide upon an operating protocol such that stable control can be maintained throughout the network.

Within this apparently simple framework for computer communications, there are scores of important topics that must be addressed. For example, in a computer network, one must decide upon a geographic topology. Some common network topologies that might be considered are a star, loop, tree, or mesh. Closely associated with the topology is the data capacity of the communication links connecting the nodes in the network—a non-trivial design problem. Equally important in a computer communication system is the question of routing and flow control procedures. These two latter concerns must insure that data can be transmitted throughout the network without undo delay. If we are dealing with a random access system, where users can randomly access a common channel, such as in packet radio or satellite communications, then the emphasis becomes the protocols that dictate flow control and how well the available channel can be utilized. Throughout, we must be concerned with basic performance criteria such as reliability, transparency, economy, efficiency, convenience, and security, although we will certainly encounter specific requirements on delay time, throughput, and cost. The tutorials found in this section will very substantially fill in the details of the many facets of computer communications.

The set of five papers in this section covers the basics of computer communications beginning with terminology, ending with future trends, and including some of the state-of-the-art distribution protocols. The articles appear in chronological order. The lead paper by I. M. Soi and K. K. Aggarwal is entitled "A Review of Computer-Communication Network Classification Schemes." In a largely nontechnical manner, the authors are able to describe various data network classification schemes. After defining such fundamental terms as a computer network, a node, a host, they turn their attention to seven different points of view by which one can classify a computer network. For example, the designer's view considers techniques for interconnecting computers via circuit, message, packet, or hybrid switching. The manager's view considers only the topological aspects of a network. The most extensive classification is the operational view, where the method of routing the communication entity is of primary importance. This paper lays a good foundation for the rest of the section.

The second paper in this section is written by L. Kleinrock and is entitled "On Resource Sharing in a Distributed Communication Environment." The key point found here is that although there are many poorly utilized resources in an information processing system, if many people share the resources such as in a distributed computer network, then we will arrive at a true cost-effective system. This point is emphasized by examining the average time delay (taken as the figure of merit) of a computer network as a function of message rate and channel capacity. Kleinrock then considers a multiaccess broadcast channel in a distributed environment and the problem of controlling access to the channel from the many distributed message sources. The problem is resolved by using a particular multiaccess control scheme where the performance of the network is then mirrored with a family of delay versus channel load curves.

A very good follow-up to Kleinrock's paper is a tutorial called "Satellite Multiple Access Protocols" written by C. Retnadhas. The title very aptly describes the content of this article. Again, the objective is to make efficient use of the satellite channel capacity when there are a large number of users with bursty traffic by enforcing a multiple access technique. For example, we might segment the channel using frequency-division multiple access (FDMA), time-division multiple access (TDMA), or code-division multiple access (CDMA). Alternately, dynamic sharing of the channel can be accomplished by using one of many demand assignment techniques, with or without a reservation system. Retnadhas describes a set of control techniques starting with the pure ALOHA protocol and ending with a contention-based demand assignment protocol. He concludes by noting that a suitable performance/complexity (cost) tradeoff will determine which protocol scheme is finally employed.

The last two papers focus on a particular type of computer

communications network, i.e., a local distribution system where geographically dispersed data sources must be connected to a central facility. J. F. Hayes has written a thorough expository article on this topic entitled "Local Distribution in Computer Communications." He describes the fundamental principles behind three categories of local distribution techniques, i.e., polling, random access, and adaptive techniques. In all cases the performance analysis is based on queueing theory and the fundamental work of Erlang. Here too, average time delay as a function of traffic load is taken as the performance index. Techniques such as hub polling and carrier sense multiple access (CSMA), to name a few, are well documented. The adaptive techniques are the most complex, but perhaps the most interesting also. The paper concludes with a concise literature review including scores of references for the reader who requires further detail.

This section ends with A. G. Fraser's article "The Present Status and Future Trends in Computer/Communication Technology." Although chronologically first, this paper is best appreciated when read last. It initially concentrates on local distribution systems that use demand shared transmission lines. Here the objective is gaining access to a central switch economically and efficiently, knowing that we are constrained by bandwidth and must use a control scheme that will avoid saturation or instability. The ARPA network is used as an example of the trend in packet switching and several levels of protocol are described to provide reliable transmission. Fraser concludes with a set of problems as future prospects in computer/communications technology. His final comment is that the computer, although itself being the source of much complexity in these communication systems, will ultimately be the solution to the complexity problems.

A REVIEW OF COMPUTER-COMMUNICATION NETWORK CLASSIFICATION SCHEMES *

Inder M. Soi and Krishan K. Aggarwal

Definitions of computer terminology and network classification.

COMPUTER networks are used, for the most part, in one of three ways: by people who require computational resources from a distance, by computers interacting with one another, or by people interacting with one another. Due to a remarkable upswing in the complexity of computer networks, arising out of the significant developments in electronic technology, several sophisticated network classification schemes have evolved.

The field of computer communications has witnessed rapid growth and technological innovations in recent years. The term, computer communications, implies a variety of user-to-computer or computer-to-computer interfaces realized by communication links. These range from various forms of teleprocessing (as in today's data processing industry) and time-sharing systems (between collections of terminals and central computers) to the burgeoning computer-to-computer communication networks typified by the Advanced Research Projects Agency Network (ARPANET).

The remarkable growth of computer networks, which has fostered computer-to-computer and terminal-to-computer communications, has opened new opportunities for designers, users, and managers. It has also posed several difficult problems. Knowledge of network characteristics such as topological design alternatives, common carrier communication services, value-added networks, hardware and software networking technology, cost factors, regulatory issues, measurement techniques, and network administration is of paramount importance.

The complexity of computer networks has taken a dramatic upswing following significant developments in

*Reprinted from *IEEE Communications Society Magazine*, March 1981, Vol. 19, No. 2, pp. 24–32.

electronic technology such as medium- and large-scale integrated circuits and microprocessors. Along with this upswing in complexity, several sophisticated network classification schemes have evolved. Abstruse terminology such as packet switching, line switching, circuit switching, value-added networks, remote-access networks, terminal-access networks, mission-oriented networks, message switching, routing algorithms, protocols, network topology, distributed databases, and job migration frequently appears in discussions and papers and tends to confuse communication subnet users. For example, most users will fail to recognize immediately that distributed networks are packet or message switching networks, and that centralized networks inherently use a deterministic routing algorithm.

Classification of computer communication networks often depends on the point of view and background of the persons doing the classification. This paper attempts to clarify the relationship of various data network classification schemes and therefore provide a basis for evaluating a network as a potential resource. The discussion here is as nontechnical as possible and should provide a basis of understanding for persons not previously exposed to data network terminology.

TERMS AND CONCEPTS

A *computer network* is a set of one or more computers, communication links, and terminals interconnected to provide service to a set of users. A *node* is a computer system that is attached to, or part of, a computer network. A *switch node* is a node whose primary function is switching data in the network. A *host* is a node which has primary functions other than switching data for the network. A *homogeneous* network is one consisting of physically or logically identical processors which are capable of executing copies of the same software system. The processors may be physically different, yet appear (by virtue of hardware, firmware, or software) to be logically identical. A *heterogeneous* network is one that is not homogeneous but may contain homogeneous subnetworks.

125

For example, the PDP-10's that are on the ARPA network, and use the tenex operating system, constitute a homogeneous subnetwork of the heterogeneous ARPA network [23]. Some designs permit a single computer to both the node and the host. *Terminals* are interfaces between the user and the computer or network. Transmission *links* connect the hosts, nodes, and terminals together to form a network. A *path* is a series of end-to-end links that establishes a route across a part of the network. The links and nodes, along with the essential control software, make up the communication subnet, or the data network.

Broadly speaking, network users fall into the following three categories.

1) Real-time user—A real-time computer system receives an input, processes it, and returns a result quick enough to affect the function at the input terminal (e.g., aircraft and highway traffic control, and process control, etc).

2) Teleconference user—Users communicate a complete idea or concept among themselves. Due to the context sensitive nature of such messages, these are not easily adaptable for input into a computer. The users are said to be in a teleconference mode of communication.

3) Data-sharing user—A computer that shares data with other computers such as an airline reservation system. The data to be shared may be only a few bits, or it may be an entire database of many millions of bits.

The basic performance criteria to be met by computer networks as interpreted by a designer are reliability, transparency, economy, convenience, and security.

CLASSIFICATION SCHEMES

Table I represents a summary of the different classification schemes for computer networks as based on the point of view and background of the person performing the classification. The six major types of classification schemes are the functional view, the designer's view, the manager's view, the operator's view, the communication view, and a hybrid scheme based on suggestions by Rudin [8] which is based on a scheme combining routing and topological classifications.

Functional View

Three functional forms of computer networks can be distinguished: Remote-Access Networks (RAN), Value-Added Networks (VAN), and Mission-Oriented Networks (MON). The distinction between the first and second forms of networking is reflected in the enhanced technological capabilities, while the difference between the second and third forms is that the third requires significantly more technological capability in order to capitalize on its organizational distinction.

Remote-Access Networks are designed to support interaction between an end user and a given host computer. Generally, the service provided by a RAN can be divided into two categories: terminal access and remote batch. Networks of this type are used by individual organizations and as commercially available utilities such as TYMSHARE [18] and CYBERNET [19]. The underlying technological requirements implicit in providing this limited type are well

TABLE I
COMPUTER NETWORK CLASSIFICATION SCHEMES

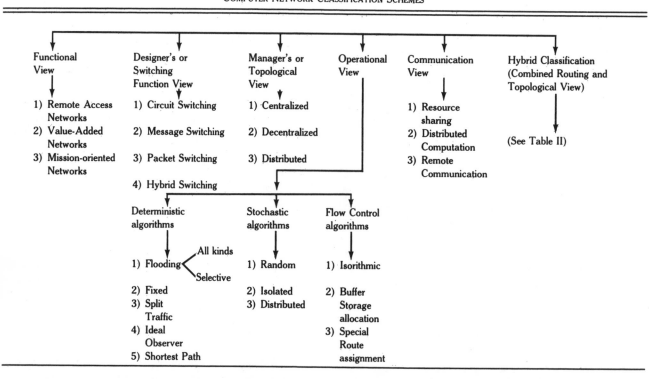

understood. In addition, heuristic approaches for the design of such subnetworks have been developed [17].

In contrast with a RAN which supports communication between a user and an individual host computer, a VAN supports communication directly between host computers, e.g., ARPANET. The scope of potential activity in VAN includes file transfer, querying of remote databases, and geographically dispersed multiprocessing. RAN capabilities can be supported within a VAN by providing limited minicomputer interface between terminals and host computers in the network (e.g., the terminal IMP on ARPANET [7]).

In a VAN, both the subnets providing communication resources and the collection of host computers, are commonly regarded as organizationally independent. As a consequence, interaction among host computers proceeds on the basis of individually arranged agreements. It is intuitively clear that the possibility of sharing can be significantly enhanced if there exists a closer organizational coupling of the host computers and, perhaps, the subnets. The network in which such a coupling exists is known as Mission-Oriented Network (MON). The distinction between a MON and a VAN is organizational and not technological. In short, one can say that a MON is a VAN in which hosts and, perhaps, the subnets are under the control of a single administrative organization. This organizational distinction permits allocation and control of programs and data, and their interaction within the network, thereby maximizing the efficiency of the organizational information processing function.

Designer's View

Designers tend to categorize a network according to its switching function or the technique for interconnecting computers—circuit switching (CS), message switching (MS), packet switching (PS), and hybrid switching (HS).

Circuit switching is analogous to the telephone (voice) network. Call and message routing are set up prior to commencement of message transmission. Once a complete circuit or route is established, the message is ready for transmission. CS has been the primary technology utilized in support of RANS such as CYBERNET. Call setup time plays a critical role in assessing the performance of networks of this type. The TYMNET and the recently defunct DATRAN networks [11], [18] are prominent examples of the circuit switched type. (DATRAN closed for other than technical reasons.) Circuit switching can be either manual or automatic. In the manual mode, the user dials a sequence of digits to obtain access to a particular computer system. If he finds the path unacceptable, or if he desires access to another computer, he terminates the connection and redials to establish a different circuit. In the automatic switching mode, the required path is established on the basis of information in the data stream. Both manual and automatic circuit switching systems are subject to line contention delay when some circuits required for a particular path are busy. Though extensively used, this type of switching is considered inefficient and uneconomical since it requires dedicated links to form a path. However, the prospect of shortened switching delays of all solid-state devices may justify reevaluation of circuit switching as an alternate technique of network design.

In message switching, a message works its way through the network, from link to link, queueing at specific nodal points. Small computers (minicomputers and, increasingly, microcomputers) located at message concentration centers and routing points perform the necessary message coding for entry into the network and combine buffer and route messages to the next destination. These computers are programmable or remote concentrators. The networks introduce buffering or queueing delay, and thus, delay time or response time plays a critical role in their design. The main advantage of MS over CS is increased circuit utilization. The main disadvantage of MS compared to CS is the facilities required to set up MS environment. Analysis of MS subnetworks is difficult because messages can vary in size. Kleinrock [21] has utilized queueing theory to model such systems. AUTODIN [24] is a prominent example of MS type networks.

Packet switching attempts to solve the problems of variable-length messages in MS by dividing a "message" into "packets." A *packet* is a subdivision of a message prefaced with an identifier containing suitable address information and information which will allow the message to be reconstructed. Packets are independent messages which can make their way through the network traveling independently over different channels. ARPANET [11] and MERIT [25] are prominent examples of PS type networks. In contrast to CS, PS attempts to multiply the use of the communications circuits among all attached subscribers. PS subnetworks implement a logical circuit (path) from source to destination with the properties of high availability, low error rate, fluctuating delay, fluctuating bandwidth and per packet overhead. High availability is achieved by means of a routing algorithm with the capability of selecting alternate paths. A low error rate is achieved by the store-and-forward transmission discipline. Large delays associated with MS store-and-forward systems are reduced in PS networks by eliminating secondary storage buffering, and by reducing queueing on the output circuits and permitting "pipelining" along the source-to-destination path. Fluctuating delay and bandwidth are the direct results of the statistical nature of switch and circuit multiplexing. Finally, the advantages of PS are in the low cost of control overhead on each data unit transmitted.

Fig. 1(a) and 1(b) demonstrate CS, MS, and PS. Whether to use CS or PS depends on a number of factors including mean message size, intermessage arrival, and conversation length [15]. A rule of thumb can be stated: If all the messages are long (e.g., file transfers) then CS is probably better, but if most of the messages are short (e.g., database query/ response or interactive traffic), PS is the more effective technology. For a combination of long and short messages, PS seems to have an edge over CS.

Hybrid switching is a recently developed switching approach which combines CS and PS within a single data

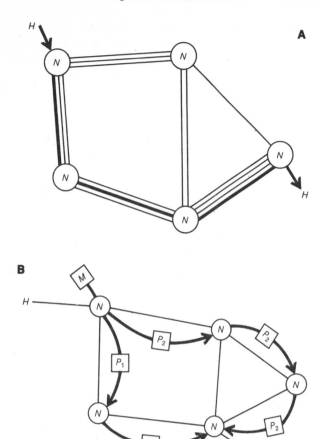

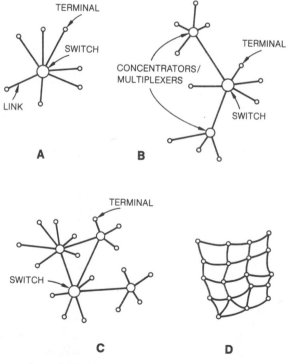

Fig. 1.   (a) Circuit switching. (b) Packet switching: M = message,
P = packet.

reliability of the central node or switch greatly affects the overall reliability of a centralized network. Its failure causes the failure of the complete network whereas an individual link failure will only affect a single device per link. To increase reliability, duplication of the central node is employed.

When several terminals in a geographically dispersed system are close to one another, they are often connected to concentrators or multiplexers (i.e., devices used to obtain more efficient link utilization at the expense of an occasional delay in response time). These in turn are connected to the central computer [(Fig. 2(b)]. A Multiplexer is used if the information transfer rate of all simultaneously active terminals never exceeds the information transfer capacity of the link to the central node. A concentrator is used where the potential input capacity exceeds the link capacity. Both multiplexers and concentrators combine messages from several terminals on to one line. Concentrators, in addition, have a storage capability when the input rates exceed the link capacity. Both these devices fail to save the network in case the central node fails.

The centralized network design model deals with two extremely important problems: the terminal layout problem in which terminals are to be connected in so-called multidrop or multipoint fashion to a specified concentrator, and the centralized network problem in which concentrators themselves are connected to a central processing facility. Combinations are, of course, possible with both terminals and concentrators connected in a tree fashion to a central facility.

network and thus can handle each category of traffic. Networks providing hybrid switching capabilities usually apply dynamic time-division multiplexing techniques to allocate a portion of channel bandwidth to CS applications. The remaining channel bandwidth is then available for PS traffic. Systems providing these capabilities are available in the commercial market although they are still at an early stage of development.

Manager's View

Managers consider those topological aspects of a network which significantly influence the economic consideration of a particular network. Networks are classified according to their topological features, i.e., centralized, decentralized, and distributed. Use of topology as a tool for classification of data communication networks had its origin in graph theory and deals with the properties that include the connection pattern of links and nodes in networks.

A centralized system with all messages flowing inward to some central processing facility is essentially a star configuration [(Fig. 2(a)] with links radiating from a single node. It is the simplest form of network topology and requires a link to be dedicated between the central node and each terminal. The

Fig. 2.   (a) Centralized network. (b) Centralized network with
concentrators/multiplexers. (c) Decentralized network.
(d) Distributed network.

Line costs for a tree-type structure are inherently lower than a star-type configuration for the same centralized system application [11]. Design aspects and techniques for centralized networks have been excellently surveyed by Boorstyn and Frank [2].

A decentralized network [Fig. 2(c)] is an expanded centralized network with some multiplexers or concentrators with switching power independent from that of any other node. Basically, the decentralized network differs from that of the centralized one only in the organization of the switching function. From the graph theory point of view, a decentralized network is described as a mixture of star and mesh components. A decentralized network is obviously more reliable than a centralized one due to the additional computers (nodes) and corresponding connecting links which permit some paths to be duplicated. Most investigators no longer distinguish between decentralized and distributed networks since the latter has at least two disjoint paths between every pair of nodes.

The distributed network consists of a set of mesh subnetworks [Fig. 2(d)] in which each node is connected to at least two other nodes and thus provides a topology which is inherently reliable. Performance of a distributed computer communication network is characterized by parameters of cost, throughput, response time, and reliability. Network design must be concerned with the properties of nodes as well as the network's topological structure. Performance criteria like response time, throughput, and network reliability have to be considered. The evaluation of properties of network nodes is concerned with nodal characteristics such as message handling and buffering, error control, flow control, routing, throughput, and reliability.

The outstanding design problems for large distributed networks are the specification of their routing and topological structure. Boorstyn and Frank have provided an excellent discussion on the design of large distributed networks [2]. A cost-effective structure for a large network is a multilevel hierarchy consisting of a backbone network and a family of local access networks. The backbone network is generally a distributed network, while the local access networks are typically centralized systems. In special cases, the network may consist primarily of either centralized or distributed portions. Due to the inherent reliability in its topology, the distributed network using packet switching probably has the greatest potential for future networks.

Operational View

In this approach, networks are classified according to the method of routing the communications entity (message, packet, etc.). This point of view is adopted by the network operator who tends to refer to networks with respect to their use as deterministic, stochastic, or flow control routing algorithms.

A computer-communication network routing strategy defines a set of rules to determine the path(s) over which messages should flow from one site to another. There are two distinct routing procedures: those actually implemented in the operating network, and those used during the design of the network. A good routing procedure for the design process must be a compromise between three conflicting requirements.

1) It must make full use of available line capacities.
2) It must be computationally efficient and inexpensive to apply.
3) The procedure must be realistic, and similar to the one to be actually implemented in the final operating network.

The evaluation of various routing algorithms is a difficult problem because of the variety of performance criteria.

Generally speaking, a minimum time delay or response time is sought. The following performance criteria may be considered for the evaluation of a routing strategy:

• capability of the algorithm to provide a reasonable response time over a range of traffic intensities;
• complexity of the calculations to be carried out;
• signaling capacity required for transmitting routing information;
• the rate at which the algorithm adapts in the case of adaptive procedures; and
• stability of the algorithm under load fluctuations or in the presence of error bursts.

A number of authors have carried out extensive research in the design and modeling of network algorithms; e.g., Fultz [9] has devised an excellent classification scheme of routing strategies.

These routing algorithms derive routes according to a rule specified in advance. Each rule produces a loop-free routing, so messages can never become trapped in closed paths. These routing strategies do not adapt to traffic changes, rather they are designed to provide satisfactory performance over a range of traffic intensities. There are five basic techniques for the deterministic category: flooding, fixed routing, split traffic, ideal observer, and shortest path.

In flooding, each node receiving a message simply restransmits it over all outgoing links (i.e., all links flooding) or a set of preselected links. After a message has circulated through the network for a specified period, it is returned to the node of origin as confirmation that the flooding cycle has been completed for that message. The flooding technique is simple and messages always arrive. But, as the name implies, the network becomes flooded with multiple copies of messages (i.e., a congestion problem). The technique is, therefore, only appropriate under very low traffic conditions. As the flooding rule becomes more selective the number of multiple messages is reduced but only at the expense of increased complexity, so efficiency considerations rule out flooding as a day-to-day policy for network routing.

Fixed routing assumes a fixed topology and known traffic patterns. It reduces optimal route selection to a multicommodity flow problem with well defined techniques for solution. Routing tables are maintained at every node computer and are fixed for each network configuration. A routing table contains the link address for sending a packet from a node to any location in the network. By referring to the table for a given destination, the node obtains a cross-reference to the corresponding link for transmitting the packet. Being inflexible, fixed routing strategies cannot be used in hostile environments such as combat, but modified algorithms including alternative fixed routes do provide a reasonable degree of survivability under such conditions.

Split traffic routing allows traffic to flow on more than one path between a given source and destination with different probability for each path—the sum of all probabilities, of course, being one. The algorithm uses a routing table that lists all the alternate routes, the probability for each route, and a record of past choices which establishes the current choice. Although the mathematical formulation of such algorithms is not difficult, when put to practice, such algorithms turn out to be less than optimum. Compared to fixed routing, the split traffic routing maintains a better balance of traffic throughout the network, thus achieving smaller average message delays.

Ideal observer routing employs least-time delay algorithms. For each new packet entering the node computes a route that minimizes the travel time to the packet's destination. This computation requires total and continuous knowledge of the system and it assumes that the traffic matrix is given. The routing tables may be updated periodically. This requires a central control that continually estimates network traffic, carries out the routing calculation, and transmits the appropriate routing table changes to the nodes in question. Interestingly enough, multiple routes provide optimum (least time) performance with a specified fraction of entering messages at any node corresponding to a particular source-destination pair routed over each outgoing link. The central control, in this case, instantaneously knows the status of traffic everywhere in the network and can take quick action. This corresponds to a nonrealizable procedure. However, it does provide a bound on other procedures that might be investigated. Therefore, the ideal observer technique is of theoretical interest only.

Shortest path techniques may be used to establish the path containing the fewest number of links. This may not correspond to the least-time path. It requires no updating of routing tables unless a node/link fails. Alternate paths may be assigned in a hierarchical fashion to accommodate such contingencies or to account for traffic variations. A more generalized version weighs each link by a specific factor depending on link cost, length of the link, link propagation delay, estimated traffic on the link, number of errors detected, etc. Thus, the least-time routing can be considered a generalized form of the shortest path algorithm as in TYMNET or SITA. It is apparent that the shortest path class of algorithms requires centralized control.

STOCHASTIC ALGORITHMS

Stochastic routing strategies are probabilistic decision rules as opposed to deterministic ones. Routes are selected in

accordance with the network topology and estimates of the state of the network, based on statistically derived delay information, are transferred from node to node between packets. Routing tables are maintained at each node for this delay information and updating of these routing tables takes place as and when new information becomes available. Use of routing tables is similar to that in split traffic and fixed routing except that the table is not fixed and the number of routes (number of table entries) equals the number of links from the source node, and may be greater. Three specific stochastic algorithms are discussed below: random routing, isolated routing, and distributed routing.

With random routing, each node is assumed to send its received messages forward along a randomly chosen link. The message reaches the destination node after following a "drunkard's walk." It is possible to include a bias in the algorithm to guide the message roughly in the right direction, but substantial randomness is essential to cope with possible link or node failures. Such algorithms are inefficient, but are surprisingly stable in networks having high probabilities of link or node failures.

ARPANET uses an adaptive random routing strategy in which each node carries out least-time estimates and decides, on a decentralized or locally determined basis, which outgoing link to use to minimize the estimated delay to a specified destination.

Isolated Routing

When using isolated routing, the nodes make decisions on their own. There are two kinds of isolated routing: the "backward learning" and "hot potato" approaches.

The underlying assumption of backward learning is that the traffic loads are approximately equal in both directions and, therefore, the elapsed time in one direction provides an estimate of the time in the other direction. The algorithm updates the previous estimate in terms of a newly-measured time by making use of recursion relation [11]. This technique has been found in practice to suffer from a "ping-pong" or looping effect, in which messages sometimes return to a node from which they were previously transmitted. In addition, backward learning has been found to adapt poorly to damaged networks (those in which nodes or links become disabled.

The hot potato technique is credited with stimulating research and development in ongoing packet switching concepts, although the original research was done more than a decade ago by Baran [26]. The procedure requires the intermediate nodes to retransmit a packet as quickly as possible after receiving it. Each node in the network, consulting a ranked list of lines leading to neighboring nodes for every destination, directs packets to the highest ranking free line for a given destination, or, if none are available, to the line having the shortest queue. This is also known as the isolated shortest queue routing procedure.

Distributed algorithms depend on the exchange of observed delay information between the nodes. This approach requires an inordinate amount of measurement information so it is impractical for large networks. Two modified approaches are the minimum delay vector, and the area approach.

A minimum delay vector for a particular node indicates the delay from that node to each of its nearest neighbors rather than to all other nodes, as in the unmodified approach. Each node exchanges this vector with each of its neighbors, where it is updated and passed on to their neighbors. As these vectors pass among the nodes, they eventually provide each node with a matrix of delays to all possible destinations. Exchanges and updates can be repeated periodically or aperiodically.

In the area approach, the network is partitioned into several disjoint areas. Within each area, every node exchanges information with every other node, but similar information is also with adjacent areas considering such areas as single nodes. This approach can be extended to a hierarchy of clusters at many levels. The objective is to reduce the amount of routing information that each node is required to retain.

Another distributed algorithm combining stochastic and topological features uses a network routing center (NRC), to which each network node periodically sends up-to-date traffic load information which is used by NRC to regenerate routing tables. These remain fixed until the next update. The NRC approach suffers from two drawbacks: first, it has a single switching point where the routing strategy can change between any two packets, and second, single paths for each source-destination node pair constrain the system's behavior. The presence of these drawbacks in the network, which is distributed except the existence of NRC, can cause instability in otherwise well-balanced networks.

FLOW CONTROL ALGORITHMS

It is apparent, however, that even with the best possible routing procedure, messages will sometimes encounter congestion and be delayed. This may be due to normal fluctuation in traffic above the level for which the routing algorithm was designed. It may be due to unexpected surges at some points in the network, or it may be due (particularly, in case of adaptive routing) to traffic building up momentarily after routing decisions have been made and routing tables updated. The resulting congestion can cause loss of data or control information, degradation in performance in the form of increased delay, or data arriving out of sequence. The congestion can be taken care of by the use of a hierarchy of protocols that indicate which of several alternate actions is appropriate on the basis of information from message and packet control fields. The hierarchy consists of host-to-host, source-to-destination, and node-to-node protocols, corresponding, respectively, to action at the message, packet, and link levels. The link level protocols for alleviating congestion between nodes are local, while the message or packet level protocols are global, or end-to-end, since they control traffic into the network. One or more of these protocols is utilized in

the following three routing schemes: isarithmic, buffer storage allocation, and special route assignment.

Isarithmic Network

An isarithmic network is one in which the total number of packets in the entire network is constant. Packets not being used to transfer data may travel through the network as "empties" (sometimes called "permits"). This provides a form of flow control since the number of empty packets effectively controls the acceptance of traffic by the network. The impact of this technique is that the supply of packets is a network-wide resource which is not under the direct control of any node. If a node loses packets or invalidly creates new packets, there is no direct way of determining that the size of the packet supply has changed. A possible solution is to have a node take a packet census, setting a bit in each packet it counts, to determine the size of the current packet supply. The National Physical Laboratory of Great Britain [11] has carried out simulation studies of the isarithmic method. One of the studies showed that by optimally choosing the permit pool size at each node, the admission delay experienced by packets in waiting to receive a permit was reduced to negligible values.

Buffer Storage Allocation

The source node requests allocation of message reassembly space in the destination before transmitting the message. The alternate strategy to go ahead and transmit the message would occasionally find the destination node's receiving buffer full. If that node were a source for messages moving the other way, it could not clear its own transmitting buffers. Soon the whole network would lock up. Another alternative, to have the destination notify the source when it could not accept a packet, requires the source to transmit every such packet at least twice. Advance allocation of reassembly buffers, resulting in occasional transmission delays, is more efficient than recovering from discarded packets. Neither of the above two procedures is adequate for real-time or data-sharing users.

An alternate technique to buffer storage allocation is the assignment of special routes based on status information received from adjacent nodes, and on traffic patterns encountered by the node over the past several seconds.

COMMUNICATION VIEWPOINT

Another classification of computer-communication networks, based on categorizing them according to their communicating components and communication characteristics, has been reported by Wecker [27]. By this criteria, networks can be divided into three categories: resource sharing, distributed computation, and remote communications.

In resource sharing networks, resources on one computer system are made available to, or shared with, another system.

Resource sharing activities may include remote file access, intercomputer file transfer, distributed database queries, and remote use of printer output devices. Communication in these networks usually occurs in one of the following modes:

• between a program being executed in one system and an I/O resource or resource manager in another;

• by involving a stream of related sequential data such as in a file transfer; and

• by enhancing short independent messages such as in transaction-oriented database access.

In short, the network provides the communication necessary to make it appear as if the remote resources are locally available.

In distributed computation networks, cooperating programs or processes executed in different computers in the network communicate and exchange information in the performance of an overall larger task. The communication in these networks, consists, in some cases, of short message exchanges, while in others, streams of data are transmitted. Examples of such networks include real-time process control systems, specialized multiple processor systems such as data base computers, and parallel processing structures.

The objective of remote communication networks is to connect users to remote systems in a cost-effective manner. Remote interactive terminals and batch entry/exit stations use remote communication network facilities. They share the communication facilities of the network, usually via concentrators and multiplexers, in the movement of information to and from host computers. Interactive terminals usually communicate via short transaction-like sequences of messages, while batch stations transfer sequential message streams.

In summary, it is I/O devices communicating with programs in resource-sharing networks. It is programs communicating with each other in distributed computation networks; and it is terminals and batch stations communicating with remotely located host programs in remote communication networks.

HYBRID CLASSIFICATION

Rudin [8] has presented a classification scheme which combines routing and topological considerations. All routing algorithms are divided into eight classes (see Table II). No attempt will be made to describe these routing algorithms as the same have already been explained in previous sections of this paper. A comparative evaluation is made below.

Centralized routing techniques give satisfactory performance within their natural constraints such as a high propensity for total system collapse if the central node fails, and inherent inflexibility for adjustments to load variations. Under stable traffic flows, the delay characteristics for centralized routing algorithms show a better performance than distributed routing algorithms. Broadly speaking, routing decisions can be classified into two classes: those that affect the network only locally, implemented at node level, and those with global effect entrusted to a network routing center.

Delta routing is an algorithm which exploits advantages of both classes, but still has the inherent disadvantage of a central control facility. Fixed routing schemes provide optimal routing, but are usable only for analysis due to the assumptions of total reliability and fixed load patterns. Adaptive routing strategies find extensive usage in networks with heterogeneous hosts or a large number of nodes and links due to their requiring knowledge of the state of the current system. Traffic flow measurements are useful tools in an optimum routing policy which follows the pattern of the problems involving the multicommodity flow and priority pricing in management science. Distributed class routing algorithms using cooperative, periodic, or asynchronous updating, possibly with a bias term, appear to have substantial advantages over current network routing procedures and, thus, offer one of the greatest challenges to the large-scale data communication networks.

SUMMARY, CONCLUSIONS, AND FUTURE RESEARCH

As an advancing technology, the evolution of large data networks offers many unusual and challenging problems to

TABLE II

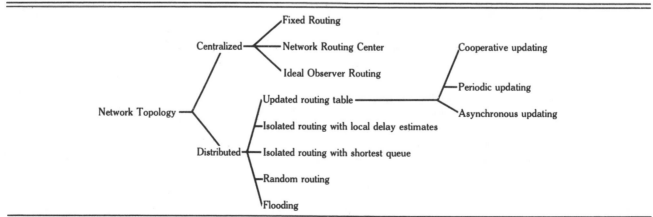

the designer, manager, and operator. The designer must provide new, cost-effective technology to satisfy the users' increasing requirements; the manager must balance available resources with expected demand; and the operator must ensure day-to-day operation consistent with users' expectations.

Large-scale distributed networks still need extensive research before they become a practical and economic reality. There is an urgent need for the development of a unified theory of design for large-scale networks based on the experience gained from prior research. The major problems in computer communication networks, like topological network optimization, for cost, delay, throughput, routing techniques, flow control, queueing problems, the design of efficient protocols, and their effectiveness, still require extensive research.

REFERENCES

[1] S. R. Kimbleton and M. G. Schneider, "Computer communication networks: Approaches, objectives, and performance considerations," *ACM Computing Surveys*, vol. 7, p. 129, Sept. 1975.

[2] R. R. Boorstyn and H. Frank, "Large-scale network topological optimization," *IEEE Trans. Commun.*, vol. COM-25, p. 29, Jan. 1977.

[3] H. Frank and W. Chou, "Topological optimization of computer networks," *Proc. IEEE*, vol. 60, p. 1385, Nov. 1972.

[4] W. Greene and U. W. Pooch, "A review of classification schemes for computer communication networks," *IEEE Trans. Comput.*, vol. 10, p. 12, Nov. 1977.

[5] D. E. Morgan, D. J. Taylor and G. Custeau, "A survey of methods for improving computer network reliability and availability," *IEEE Trans. Comput.*, vol. 10, p. 42, Nov. 1977.

[6] M. Gerla and L. Kleinrock, "On the topological design of distributed computer networks," *IEEE Trans. Commun.*, vol. COM-25, p. 48, Jan. 1977.

[7] R. E. Kahn and W. R. Crowther, "A study of the ARPA computer network design and performance," Bolt, Beranek & Newman Inc., Cambridge, MA, Rep. 2151, Aug. 1971.

[8] H. Rudin, "On routing and delta routing: A taxonomy of techniques for packet switched networks," IBM Corp. Res. Div., Zurich, Switzerland, Rep. RZ-701, June 1975.

[9] G. L. Fultz and L. Kleinrock, "Adaptive routing techniques for store-and-forward computer communication networks," NTIS Rep. AD-727-989, July 1972.

[10] J. Martin, *Systems Analysis for Data Transmission*. Englewood Cliffs, NJ: Prentice-Hall, 1970.

[11] M. Schwartz, *Computer-Communication Network Design and Analysis*. Englewood Cliffs, NJ: Prentice-Hall, 1977.

[12] W. Chou and H. Frank, "Routing strategies for computer network design," in *Proc. Symp. on Commun. Networks Teletraffic*, Polytech. Inst. Brooklyn, Brooklyn, NY, vol. 22, Apr. 1972.

[13] L. Kleinrock, "Scheduling, queueing and delays in time-shared systems and computer networks," in *Computer-Communication Networks*, N. Abramson and F. Kuo, Eds., Englewood Cliffs, NJ: Prentice-Hall, 1973.

[14] D. W. Davies, "The control of congestion in packet switching networks," *IEEE Trans. Commun.*, vol. 20, p. 546, June 1972.

[15] G. Falk and J. M. McQuillan, "Alternatives for data network architectures," *IEEE Trans. Comput.*, vol. 10, p. 22, Nov. 1977.

[16] B. W. Boehm and R. L. Mobley, "Adaptive routing techniques for distributed communications systems," The Rand Corp., Santa Monica, CA, Rep. RM-4781-RR, Feb. 1966.

[17] W. Chou, "Planning and design of data communication networks," in *Proc. AFIPS 1974 Nat. Comput. Conf.*, vol. 43, 1974, p. 553.

[18] L. Tymes, "TYMNET—A terminal-oriented communication network," in *Proc. AFIPS 1971 SJCC*, vol. 38, 1971, p. 211.

[19] W. J. Luther, "The conceptual basis of CYBERNET," in *Comput. Networks*, R. Rustin, Ed. Englewood Cliffs, NJ: Prentice-Hall 1973.

[20] *Large-Scale Networks: Theory and Design*, F. Boesch, Ed. New York: IEEE Press, 1976.

[21] L. Kleinrock, *Communication Nets: Stochastic Message Flow and Delay*. New York: McGraw-Hill, 1964.

[22] D. W. Davies and D. L. Barber, *Communication Networks for Computers*. New York: Wiley, 1973.

[23] H. R. Thomas and D. A. Henderson, "McRoss—A multicomputer programming system," in *Proc. AFIPS 1972 SJCC*, vol. 40, pp. 281-293.

[24] L. M. Paoletti, *"AUDODIN"—Computer Communication Networks*. Leyden, The Netherlands: Noordhoff International Publications, 1975, p. 345.

[25] E. M. Aupperle, "The merit network re-examined," MERIT Computer Network, Univ. Michigan, Ann Arbor, Rep. MCN-0273-TP-13, Feb. 1973.

[26] P. Baran, "On distributed communications: An introduction to distributed communication networks," The Rand Corp., Santa Monica, CA, Rep. RM-3420-PR, Aug. 1964.

[27] S. Wecker, "Computer network architecture," *IEEE Trans. Comput.*, vol. 9, Sept. 1979.

Inder M. Soi received the B.Sc. (Eng.) degree in electrical engineering from Punjabi University in 1966 and the M.Sc. (Eng.) degree in control systems in 1978 from the Regional Engineering College, Kurukshetra, India. He is presently working toward the Ph.D. degree in computer communication networks topological design. His fields of research are computer networks reliability, software reliability, control systems, digital circuits, and system engineering. He has published/ presented about 15 papers in India and abroad. He is a student member of the following IEEE societies: Computer, Reliability, Communications, Systems, Man, and Cybernetics, and Control Systems.

Krishan K. Aggarwal was born in 1948. He received the degree in electronics and communication engineering from Punjab Engineering College, Chandigarh, India, and was graduated with first position. After serving in industry for a period of time, he joined the teaching profession at Regional Engineering College, Kurukshetra, India. There, he also obtained the M.Sc. degree in applied electronics and automatic control and the Ph.D. degree based on his thesis entitled "Reliability Evaluation and Optimization of General System."

He is presently Professor and Head of the Department of Electronics and Communication Engineering, Regional Engineering College, Kurukshetra. He is a member of various academic bodies of several universities in India. He is a member of the Institution of Engineers (India), and he also received certification as a Chartered Engineer (India). He is the coauthor of *Waveshaping and Digital Circuit*. His main research interests are digital electronics, system engineering, and reliability engineering. In these areas, he has published about 60 papers in India and abroad. He has been selected for the IEEE Reliability Society Guest Editor Award to be presented to him at the Annual Reliability and Maintainability Symposium scheduled for 1982 in Los Angeles, CA.

ON RESOURCE SHARING IN A DISTRIBUTED COMMUNICATION ENVIRONMENT *

Leonard Kleinrock

The efficiency of resource sharing provides the cost-effectiveness of packet switching in many of our computer-communications systems.

A revolution is in the making! We are witnessing a growth rate in technological change which is overwhelming. Thanks to enormous advances in data communications and in integrated chip technology, we are in the midst of a computer communication explosion which has already made significant changes in the field of data processing. The early phase of the revolution has passed— we have developed cost-effective data communication systems. Indeed in the last five years we have witnessed the rise of *computer networks* whose function it is to span intercontinental distances and provide communication among computers across nations and across the world. There now exists a large number of national networks which are in the process of interconnecting to each other in such a world network.

These networks have hastened the next phase of the revolution, namely, the widespread acceptance and application of teleprocessing and networking by the business sector of our economy. As this second phase proceeds, we will see a stress placed on our computer networks in two areas. First, in the need for long-haul, wide-band inexpensive communications deep in the backbone of our networks; one answer to this need is the introduction of sophisticated packet **satellite** radio data communication systems. The other environment in which we will see stress is at the periphery of our networks where local access is the major problem. The early generation computer networks did not properly solve the local interconnection problem, namely, how to efficiently provide access from the user at the terminal to the network itself. A potential solution to this problem is the use of **ground radio** packet communications. In this article we describe some of the recent technological advances which have provided solutions to these two problems. Indeed the systems issues involved for both radio communications problems are very similar, although the technological implementations are quite different as we shall see. To begin with, we discuss the general principles of resource sharing which provide the key to the cost-effectiveness of radio packet switching.

RESOURCE SHARING

A privately owned automobile is usually a waste of money! Perhaps 90 percent of the time it is idly parked and not in use. However, its "convenience" is so seductive that few can resist the temptation to own one. When the price of such a poorly utilized device is astronomically high, we do refuse the temptation (how many of us own private jet aircraft?). On the other hand, when the cost is

Cost-effective computer networks have hastened the widespread acceptance and application of teleprocessing and networking by the business sector of our economy.

The author is with the Department of Computer Science, University of California, Los Angeles, CA 90024.

*Reprinted from *IEEE Communications Society Magazine*, January 1979, Vol. 17, No. 1, pp. 26–34.

extremely low, we are obliged to own such resources (we all own idle pencils).

An information processing system consists of many poorly utilized resources. (A resource is simply a device which can perform work for us at a finite rate.) For example, in an information processing system, there is the CPU, the main memory, the disk, the data communication channels, the terminals, the printer, etc. One of the major system advances of the early 1960's was the development of multiaccess time-sharing systems in which computer system resources were made available to a large population of users, *each* of whom had relatively *small* demands (i.e., the ratio of their peak demands to their average demands was very high) but who *collectively* presented a *total* demand profile which was relatively smooth and of medium to high utilization. This was an example of the advantages to be gained through the smoothing effect of a large population (i.e., the "law of large numbers") [1]. The need for resource sharing is present in many many systems (e.g., the shared use of public jet aircraft among a large population of users).

In computer communication systems [2] we have a great need for sharing expensive resources among a collection of high peak-to-average (i.e., "bursty") users. In Fig. 1, we display the structure of a computer network in which we can identify three kinds of resources:

1) the *terminals* directly available to the user and the *communications resources* required to connect those terminals to their HOST computers or directly into the network (via TIPS in the ARPANET, for example)—this is an expensive portion of the system and it is generally difficult to employ extensive resource sharing here due to the relative sparseness of the data sources;

2) the *HOST machines* themselves which provide the information processing services—here multiaccess time-sharing provides the mechanism for efficient resource sharing;

3) the *communications subnetwork*, consisting of communication trunks and software switches, whose function it is to provide the data communication service for the exchange of data and control among the other devices.

The HOST machines in 2) above contain hardware and software resources (in the form of application programs and data files) whose sharing comes under the topic of time-sharing; we dwell no further on these resources. Rather, we shall focus attention on those portions of the computer communications system where packet communications has had an important impact. Perhaps the most visible component is that of the communications subnetwork listed in item 3) above. Here packet communications first demonstrated its enormous efficiencies in the form of the ARPANET in the early 1970's (the decade of computer networks) [2]. The communication resources to be shared in this case are *storage capacity* at the nodal switches (the IMP's in the ARPANET), *processing capacity* in the nodal switches, and *communications capacity* of the trunks connecting these switches. Packet switching in this environment has proven to be a major technological breakthrough in providing cost-effective data communications among information processing systems.

As stated earlier, deep in the backbone of such packet switched networks there is a need for long-haul, high-capacity inexpensive communications, and it is here where we see the second application of packet communi-

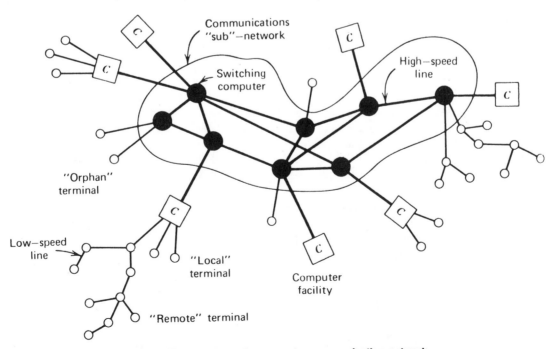

Fig. 1. The structure of a computer communication network.

ARPA — Advanced Research Projects Agency of the Department of Defense.

ARPANET — The ARPA computer network which interconnects university-computing locations and other research establishments.

Packet — A group of binary digits including data and call control signals which is switched as a composite whole. The data, call control signals, and possibly error control information are arranged in a specified format. (CCITT definition).

Packet Switching — The transmission of data by means of addressed packets whereby a transmission channel is occupied for the duration of transmission of the packet only. The channel is then available for use by packets being transferred between different data terminal equipment. (CCITT definition).

IMP — Interface Message Processor. A device at each node of the ARPANET which performs message switching and interconnects the research computer centers or "hosts" with the high bandwidth leased lines.

TIP — Terminal IMP. An IMP with multiplexing and demultiplexing equipment that collects characters from terminals, packages them in the form suitable for processing by the IMP, and sorts out characters destined for particular terminals.

Contention — The condition where two or more users may want access to a single resource with the possibility that their demands could come at the same time.

cations for resource sharing in the form of *satellite packet switching*. The third application may be found in the local access problem stated in item 1) above which also lends itself to the use of packet switching to provide efficient communications resource sharing; this takes the form of the use of a multiaccess broadcast channel in a local environment, commonly known as *ground radio packet switching*. The common element running through all these systems is the application of the smoothing effect of a large population to provide efficient resource sharing, an exquisite example of which is provided by packet communications.

Let us consider two important examples of the effectiveness of resource sharing. Both cases involve the sharing of communication lines. In the first case we consider a voice communication system in which we provide m trunks to serve a population of users attempting to place telephone calls. We assume that a user call is blocked (and hence lost) if all trunks are occupied when he attempts to place his call. The common measure of load on such a system is expressed in *Erlangs* (one Erlang represents the full-time use of a single trunk). Clearly, we cannot handle more than m Erlangs, but if we even approach a load of m Erlangs, then due to statistical fluctuations in customer behavior, we know that many calls will be blocked and this is a situation we wish to avoid. The game, therefore, is to design enough trunks into a system to satisfy a given load so that the probability of blocking is small enough to satisfy the users' needs in an economical fashion. The analysis for this classic problem

was solved 70 years ago by A. K. Erlang [1] and that solution has the amazing characteristic shown in Table I. Here we see the way which the percentage of blocked calls varies with the number of trunks (m) and the load level. Now for the resource sharing. Suppose we had a population presenting a load of 2/3 Erlangs to a single trunk; we then see that the percentage of blocked calls is an atrocious 40 percent. Obviously, if we had four such populations, each accessing their own single trunk, then each group would experience the same 40 percent blocking. However, if these four groups would simply pool their trunks yielding a total population load of 8/3 Erlangs sharing the set of 4 trunks, then we see that the percentage blocked has now drastically reduced to 17 percent. By pooling 16 times the load (32/3 Erlangs) onto 16 trunks, the percentage blocked drops to 3 percent and if we go further down the table to 64 times the original load (128/3 Erlangs) sharing a pool of 64 trunks, we reduce the percentage blocked to a mere 0.05 percent. This represents a gain of almost three orders of magnitude! Thus, in effect, by the creation of a larger population sharing pooled resources, we have gained enormously in system performance.

By the creation of a larger population sharing pooled resources, we have gained enormously in system performance.

Our second case involves a data communication system in which random blocks of data (i.e., messages) arrive (at a rate of λ msg/s). Let us assume that we have available a single data communication channel of C bits/s to satisfy these demands. When an arriving demand finds the channel busy transmitting some other message, this arriving demand will queue up and wait its turn for service (rather than being blocked, and therefore lost, as in the first case). As before, we can describe the load on the system in terms of a number ρ which is the fraction of time the channel is transmitting data (in this case of a single channel, ρ is exactly equal to the load in Erlangs); as ρ approaches one, the average delay increases without bound. We wish to provide a large enough channel capacity to keep the delay within acceptable bounds. For this case (where a given population

TABLE I
Resource Sharing of Telephone Trunks

Load (Erlangs)	Number of Trunks (m)	Percentage Blocked
2/3	1	40%
8/3	4	17%
32/3	16	3%
128/3	64	0.05%

presents a load ρ to a channel of capacity C) let us denote the resulting average delay by $T(\lambda,C)$. If we had two populations each presenting a load ρ to their own channel of capacity C, then clearly each group would suffer an average delay of $T(\lambda,C)$. On the other hand, let us consider merging these two data streams, thereby resulting in a total rate of 2λ msg/s (with each message having the same average length as before) and merging the two capacities into a single channel of capacity $2C$; the resultant delay may be denoted by $T(2\lambda,2C)$. The load ρ in this merged case would be the same as it was before the merging. However, the startling fact is that average delay in the merged case (the case of resource sharing) is *half* that of the former situation. That is,

$$T(2\lambda,2C) = \tfrac{1}{2}\, T(\lambda,C).$$

In fact if we merged N such groups, then the overall delay will drop by a (sharing) factor of N [3]. Again we see the remarkable advantages of resource sharing. The principle here is that serving a large population with a large shared resource is extremely efficient in terms of performance.

It is this efficiency of resource sharing that provides the cost-effectiveness of packet switching in so many of our computer communication systems. This is accomplished by assigning network resources (channel capacity, storage, logical links, etc.) on a demand basis; these resources are designed to be shared among a large population of users and are therefore endowed with a large capacity for serving these users. It is precisely when large populations share large-capacity resources that we enjoy the performance efficiencies of resource sharing.

MULTIACCESS BROADCAST CHANNELS

Let us now characterize the satellite and ground radio switching systems which we introduced earlier. Such channels may be described as *multiaccess broadcast channels in a distributed environment*. Indeed, the object is to properly share this precious communications resource among a collection of user terminals where the resource is to be used to provide communications among the users. From the previous section we recognize that if the users are bursty, then we should not partition the channel into small pieces (each such piece assigned to a fraction of the user population), but rather, we should look for ways to share the full channel among all users on a demand basis. Immediately we can separate two cases: the first case is that in which all users are within radio range and line-of-sight of each other (we speak of this as a *one-hop* system); the second case is where not all users are within range and sight of each other (in which case we have a *multihop* environment). The adjective *multiaccess* refers to the fact that the many users are trying to share the channel simultaneously in some cooperative fashion. The adjective *broadcast* refers to the fact that

each channel can hear the transmission from many or all other terminals.

The important characteristic is described by the adjective *distributed*, which refers to the fact that our terminals are geographically distributed in a way which makes controlling their behavior an issue of importance. The key parameter describing this notion of distributed sources is usually taken to be the ratio a of the propagation delay (the time it takes electromagnetic energy moving at the speed of light to pass between two separated terminals) to the transmission time of a packet. For example, consider 1000 bit packets transmitting over a channel operating at a speed of 100 kbits/s. The transmission time of a packet is then 10 ms. If the maximum distance between the source and destination is 10 mi then the (speed of light) packet propagation delay is on the order of 54 μs. (This is a typical example for a ground radio packet-switching system.) Thus the propagation delay constitutes only a very small fraction ($a = 0.005$) of the transmission time of a packet. On the other hand, in a satellite environment, this ratio is more often on the order from 10 to 30; for example, a geostationary satellite introduces a propagation delay on the order of 250–270 ms, and for the 10 ms packet transmission time mentioned above, we would then have a ratio of propagation delay to packet transmission time of roughly $a = 25$.

Now how do we pull off the "resource sharing"? An ever-increasing number of access schemes have recently been described in the published literature [4] which more or less succeed at this; these we describe shortly. Before doing so, however, let us discuss the *price* one must pay for sharing a communication channel in such a distributed environment.

THE UNAVOIDABLE PRICE

As with most contention systems, two factors contribute to a degradation in performance: first, there are the usual queueing effects due to the random nature of the message generation process; second, there is the cost due to the fact that our message sources are geographically distributed. If all the terminals were collocated (i.e., coordination among them was free and instantaneous) then we could form a common queue of the generated message packets and achieve the optimum delay-throughput profile, namely, that of the $M/D/1$ queueing system [1] described later. Unfortunately, coordination is not free and we must expend some effort in organizing our many terminals which are distributed and which independently generate messages. The total capacity we have available is fixed and we are faced with controlling access to this channel from these distributed message sources in which the control information must pass over the same channel which is being controlled (or over a subchannel which is derived from the data channel).

We have a spectrum of choices for introducing this

control, ranging from no control at all to dynamic control, and finally to extremely tight static control. For example, we could allow the terminals to access the channel using PURE (i.e., unslotted) ALOHA in which a terminal transmits a packet as soon as it is generated hoping that it will not collide with any other packet transmission; if there is a collision, then all packets involved in that collision are "destroyed" and must be retransmitted later at some randomly chosen time. This *uncontrolled* scheme is extremely simple, involves no control function or hardware, but extracts a price from the system in the form of wasted channel capacity due to collisions. At the other extreme, we could have a very tight *fixed control* as for example in FDMA or TDMA (see next section) where each terminal is permanently assigned a subchannel derived from the original channel. Such a fixed control scheme certainly avoids any collisions, but is inefficient for two reasons: first, because terminals tend to be bursty sources and therefore much of their permanently assigned capacity will be wasted due to their high peak-to-average ratio; and second, the response time will be far worse in this channelized case due to the scaling effect which is especially apparent in FDMA (see the second section). A *dynamic control* scheme such as reservation-TDMA, or Roberts' reservation scheme [5] makes use of a reservation subchannel through which terminals place requests for reserved space on the data channel; this system permits dynamic allocation of channel capacity according to a terminal's demand, but requires overhead in order to set up these reservations.

Thus we see that the issue of allocating capacity in a distributed environment is a serious one. In one form or another nature will extract her price! This price will appear in the form of collisions due to poor or no control, wasted capacity due to rigid fixed control, or overhead due to dynamic control. These comments are summarized in Table II below.

In general, as the number of terminals grows, and as the geographical separation grows, then also grows the price we pay for distribution.

TABLE II
The Price for Distributed Sources

	Collisions	Idle Capacity	Overhead
No Control (e.g., ALOHA)	Yes	No	No
Static Control (e.g., FDMA)	No	Yes	No
Dynamic Control (e.g., Reservation Systems)	No	No	Yes

A FAMILY OF MULTIACCESS METHODS [4]

Multiaccess methods for distributed computer communication systems have recently been evaluated. In this section we describe a variety of these suitable for one-hop systems. It is perhaps best to think of all terminals as transmitting fixed-length packets to a central station which is the destination for these transmissions (this is not a necessary assumption since point-to-point communication also fits this model).

We now consider nine random multiaccess broadcast schemes, and for each we give a reference and an extremely concise definition:

PURE (UNSLOTTED) ALOHA [6]: A newly generated packet will be transmitted by its terminal at the instant of its generation; collided packets destroy each other and must be retransmitted.

SLOTTED ALOHA [2], [7]: The same as PURE ALOHA except that new packet transmissions must begin at the next slot point, where time is slotted into lengths equal to a packet transmission time.

CSMA (Carrier Sense Multiple Access) [2], [8]: The same as PURE ALOHA except that a terminal senses (listens to) the channel and can hear the carrier of any other terminal's transmission; if such a carrier is detected, then the terminal refrains from transmitting and follows one of many defined protocols for later attempts.

POLLING [9]: A central controller sends a "polling message" to each terminal in turn; when a terminal is polled, it empties all of its data before indicating its empty buffer condition whereupon the next terminal is polled in sequence.

FDMA (Frequency-Division Multiple Access) [9]: The bandwidth of the channel is divided into M equal subchannels, each reserved for one of the M terminals.

TDMA (Time-Division Multiple Access) [9]: Time is slotted and a periodic sequence of the M integers is defined such that when a terminal's number is assigned to a slot, then that terminal (and only that terminal) may transmit in that slot; typically each terminal is given one out of every M slots.

MSAP (Mini-Slotted Alternating Priority) [10]: A carrier-sense version of polling whereby a polling sequence is defined and when a terminal's buffer is empty, it simply refrains from transmitting; after a time interval equal to the propagation delay, the next terminal in sequence senses the channel idle and proceeds with its transmission, etc. (This is also known as hub go-ahead polling.)

RANDOM URN SCHEME [11]: An optimal distributed control adaptive scheme in which a fraction N/M of the terminals is given permission to transmit (and each will do so if it has anything to transmit). M is the total number of terminals and N is the number (assumed to be known) that wish to transmit.

The $M/D/1$ Queue

The shoeshine boy above illustrates the simplest kind of waiting-line situation, the $M/D/1$ queue. This queueing system has Poisson-distributed arrivals at a rate λ customers per second with a constant service time of x seconds. With the Poisson distribution, the probability of k arrivals in a time interval t seconds is given by

$$P(k,t) = e^{-\lambda t}(\lambda t)^k/k!$$

for $k=0,1,2,\cdots$. The average response time T (waiting time plus service time) is [1]:

$$T = \frac{\lambda x^2/2}{1 - \lambda x} + x.$$

Therefore, for customers arriving for a shine at an average rate of one every 10 min, and a service time of 6 min a shine, on the average a customer can expect to spend a total of 10.5 min at the shoeshine stand. In the packet network application, λ is messages per second and x is equal to a packet transmission time.

$M/D/1$ [1]: The classic first-come-first-serve single-server queueing system with Poisson-distributed arrivals and constant service time equal to a packet transmission time. This is an ideal system which neglects the fact that the terminals are geographically separated.

In Fig. 2, we plot the mean response time of the system (the time from when the packet wants to transmit until it is correctly received at the destination) as a function of the channel load ρ, for a number of these schemes. This figure is for the case of $M = 100$ terminals and shows the relative performance of the various access schemes. We note, for example, that PURE ALOHA gives the best per-

formance at extremely small loads whereas the MSAP scheme seems to give the best performance at high loads (excluding the ideal scheme $M/D/1$). Indeed a well-designed static control system such as TDMA will also perform very well at high loads. What is important is to find a scheme which adapts its behavior between that of an ALOHA-like scheme at light loads to a static control scheme at heavy loads. Such schemes are beginning to appear in the literature and an example of one is the URN scheme described above. Another example is a scheme known as Scheduled Retransmission Upon Collision (SRUC) which was recently described in [12].

The performance profiles shown in Fig. 2 represent

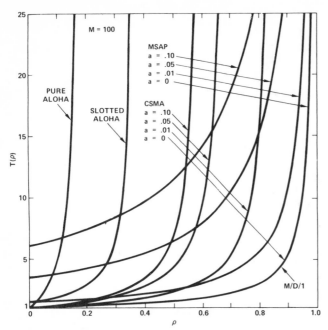

Fig. 2. Delay-throughput profile for multiaccess broadcast schemes (100 terminals). The parameter *a* is the ratio of propagation delay to packet transmission time.

some of the better known access schemes currently available. Many more are being studied and will soon be available in the literature. Again, the idea is to create access schemes which perform well in this multiaccess broadcast distributed environment. To perform well means to find an efficient way to share the common channel capacity.

APPLICATIONS AND THE FUTURE

Two applications we have mentioned are: wideband satellite systems with an enormous propagation delay and ground radio systems with a tiny propagation delay. It is worthwhile observing that the Advanced Research Projects Agency (ARPA) has been conducting experiments for both of these systems. The first is an experimental satellite network which currently connects three countries across the Atlantic [13]; the measurement and implementation results from this system have been quite encouraging and indicate that satellite packet switching is a cost-effective and viable technology. Also, ARPA is conducting an experiment in the Palo Alto, CA area for a ground radio packet-switching environment including mobile terminals [14]. Here too the early measurements indicate that the system is feasible and effective.

It is fair to say that radio packet switching is a new technology about to explode in the applications area. The satellite application deep in the backbone of our computer networks is clear and needs no further justification. Access for local terminals has been a long outstanding problem to teleprocessing system designers in that the

cost of the local access portion of the network has been far too high relative to the rest of the communication system. Radio packet switching promises to reduce that cost significantly, and we can expect to see such systems available in the near future; an interesting example is Xerox's proposed XTEN network. Indeed the ground radio applications include such things as communications among moving vehicles (taxicabs, police cars, ambulances, private fleets), communications among aircraft, and indeed communications among any mobile units or any widely distributed units in a sparse environment. We may even see the use of radio packet broadcasting on a tiny scale down at the integrated chip level if cost-effective lasers can be implemented on a chip; laser packet switching among logic elements on a chip may greatly simplify the interconnection and/or prototype problem in chip design.

An exciting application of these radio packet access schemes has been under development recently and we are already beginning to see products and services based upon this new development. The application is to use packet radio access schemes *not* in a *radio* environment, but rather on a coaxial cable- or other *wire*-communication media. Indeed the entire technology of loop and ring structures has recently taken advantage of these access schemes. For example, consider the case of a data communication bus to which are attached a number of devices (e.g., a CPU, a disk, a drum, terminals, minicomputers, etc.). Until recently, contention for access to this bus had usually been resolved by a central controller. Clearly, any of the access schemes which we described in the previous section also lend themselves for application to this wire-communication bus. Indeed we are already seeing products based upon this idea in which demand access and random access are used to govern the use of a communication bus; a prime example of this is the ETHERNET [15] developed by Xerox. Here, a 1 km coaxial cable is being used to connect up to 256 devices which transmit data at 3 Mbits/s, using a variation of CSMA. The variation is simply that a device cannot only listen *before* it transmits, but it can also listen *while* it is transmitting; this permits it to detect collisions and then to abort its own transmission in the event of a collision, thereby saving considerable wasted time on the channel. The early indications are that such a scheme works extremely well and we can look forward to many more applications of packet radio access schemes to wire communications. For example, there is no reason why all future aircraft and naval vessels should not be wired up in such a fashion. Furthermore, all office buildings could have a common pipe running throughout the building, attached to which are all terminals requiring access to each other and to a centralized computer located perhaps in the basement. Indeed, not only in-building but in-plant communication should be run this way among a number of buildings at a given site. The applications here are unlimited and in fact, one may expect to see the applications to wire-based communications appearing on the market before the radio schemes are available.

The key to the success in all these developments is simply that large populations sharing large resources provide enormous efficiencies in performance and are to be incorporated whenever possible. The analytic and design problems which remain in studying single-hop and multihop schemes continue to occupy the analysts and designers in computer communications. The applications have not, should not, and will not wait for analytic results as long as a cost and performance savings can be demonstrated. We can expect much future work in this area, both in analysis as well as in development.

REFERENCES

[1] L. Kleinrock, *Queueing Systems, Vol. 1: Theory.* New York: Wiley, 1975.

[2] ——, *Queueing Systems, Vol. 2: Computer Applications.* New York: Wiley, 1976.

[3] ——, "Resource allocation in computer systems and computer-communication networks," *Proc. IFIP Cong.-74*, pp. 11-18, 1974.

[4] ——, "Performance of distributed multi-access computer-communication systems," *Proc. IFIP Cong.-77*, pp. 547-552, 1977.

[5] L. G. Roberts, "Dynamic allocation of satellite capacity through packet reservation," *AFIPS Conf. Proc. 1973 Nat. Comput. Conf.*, vol. 42, pp. 711-716, 1973.

[6] N. Abramson, "Packet switching with satellites," *AFIPS Conf. Proc. 1973 Nat. Comput. Conf.*, vol. 42, pp. 695-702, 1973.

[7] L. Kleinrock and S. Lam, "Packet switching in a slotted satellite channel," *AFIPS Conf. Proc. 1973 Nat. Comput. Conf.*, vol. 42, pp. 703-710, 1973.

[8] L. Kleinrock and F. Tobagi, "Random access techniques for data transmission over packet-switched radio channels," *AFIPS Conf. Proc. 1974 Nat. Comput. Conf.*, vol. 44, pp. 187-201, 1975.

[9] J. Martin, *Teleprocessing Network Organization.* Englewood Cliffs, NJ: Prentice-Hall, 1970.

[10] M. Scholl, "Multiplexing techniques for data transmission over packet-switched radio systems," Ph.D. dissertation, Comput. Sci. Dep., Univ. of California, Los Angeles, Eng. Rep. UCLA-ENG 76123, Dec. 1976.

[11] L. Kleinrock and Y. Yemini, "An optimal adaptive scheme for multiple access broadcast communication." *ICC'78 Conf. Rec.*, vol. 1, pp. 7.2.1-7.2.5, June 4-7, 1978.

[12] F. Borgonovo and L. Fratta, "SRUC: A technique for packet transmission on multiple access channels," *Proc. of 4th Int. Conf. on Comput. Commun.*, pp. 601-607, 1978.

[13] I. M. Jacobs, R. Binder, and E. V. Hoversten, "General purpose packet satellite networks," *Proc. IEEE*, vol. 66, pp. 1448-1467, Nov. 1978.

[14] R. E. Kahn, "The organization of computer resources into a packet radio network," *IEEE Trans. Commun.*, vol. COM-25, pp. 169-178, Jan. 1977.

[15] R. M. Metcalfe and D. R. Boggs, "Ethernet: Distributed packet switching for local computer networks," vol. 19, no. 7, pp. 395-404, July 1976.

Leonard Kleinrock received the B.E.E. degree from City College of New York, New York, NY in 1957 and the M.S.E.E. and Ph.D.E.E. degrees from Massachusetts Institute of Technology, Cambridge, in 1959 and 1963, respectively.

He was an Assistant Engineer at the Photobell Co. in New York from 1951 to 1957 and a Staff Associate at the MIT Lincoln Laboratory from 1957-1963. In 1963 he joined the faculty of the School of Engineering and Applied Science, University of California, Los Angeles, where he is now Professor of Computer Science. His research spans the fields of computer networks, computer systems modeling and analysis, queueing theory and resource sharing, and allocation in general. At UCLA, he directs a large group in advanced teleprocessing systems and computer networks. He is the author of three major books in the field of computer networks: *Communication Nets: Stochastic Message Flow and Delay* (New York: McGraw Hill, 1964; also New York: Dover, 1972); *Queueing Systems, Vol. I: Theory* (New York: Wiley-Interscience, 1975); and *Queueing Systems, Vol. II: Computer Applications* (New York: Wiley-Interscience, 1976). He has published over 85 articles and contributed to several books. He serves as consultant for many domestic and foreign corporations and governments and he is a referee for numerous scholarly publications and a book reviewer for several publishers.

Dr. Kleinrock was awarded a Guggenheim Fellowship for 1971-1972 and is an IEEE Fellow "for contributions in computer-communication networks, queueing theory, time-shared systems, and engineering education." He was the co-winner of the 1976 Lanchester Prize for his book *Queueing Systems, Vol. II: Computer Applications* (J. Wiley, 1976).

SATELLITE MULTIPLE ACCESS PROTOCOLS*

C. Retnadhas

Protocol for accessing satellites efficiently: a tutorial.

SATELLITES provide a convenient medium for data communication between widespread geographic areas. Compared to a terrestrial data network, the satellite system has wide bandwidth, high accuracy of transmission, and high availability of transmission medium. The main disadvantages of the satellite system are the inherently long transmission delays (270 ms one way, the effect of local weather conditions and interferences, and the high cost of the system. Technological advances can reduce the cost and the effect of weather conditions on the transmitted signal. The effect of long transmission delay can be minimized by using effective transmission protocols. Because of the above advantages, satellite technology has aroused a great deal of interest in recent years [7],[14].

This paper presents a tutorial on the various protocols used in satellite data transmission. The most important characteristic of the satellite system is the ability of the earth stations, located at geographically dispersed areas, to access the satellite to transmit and to receive data. The area covered by a geostationary satellite is a function of the satellite's receiving and transmitting antenna(s). For a large number of users with bursty traffic, a highly efficient way of using the channel capacity is to use multiple access techniques. In a multiple accessed channel, two or more users may nominally share the channel. The satellite system can provide broadcast capability at any given time to all earth stations within its transmission coverage area. The combination of multiple access and broadcast capability makes it possible to configure the earth stations into a fully connected "one-hop" network.

The author is with the Department of Quantitative and Information Science, Western Illinois University, Macomb, IL 61455.

*Reprinted from *IEEE Communications Society Magazine*, September 1980, Vol. 18, No. 5, pp. 16–20.

CHANNEL DERIVATION

There are three ways to obtain channels in a satellite system [8]. In the first method, the channels are obtained by using the built-in satellite channelization due to the use of multiple transponders operating in different frequency bands. Each one of the independent transponders in the satellite is designed to accept transmission at a selected frequency band, i.e., the uplink frequency. The satellite carries out a frequency translation to a well-defined frequency band, i.e., the downlink frequency. This scheme thus divides the total bandwidth of the satellite into well-defined channels. The advantages provided by this scheme are reduced interference problems and improved reliability in that the possibility of losing all the channels due to satellite failure is small.

The second method uses the basic multiple access techniques of frequency-division multiple access (FDMA), time-division multiple access (TDMA), and code-division multiple access (CDMA).

A number of multiple access protocols are presented. In the final analysis, it is cost which will dictate which protocol is suitable for a particular application.

A simple form of obtaining an FDMA channel is to divide the bandwidth of a transponder into separate nonoverlapping subchannels, with each user assigned a separate subchannel. In FDMA, each user has access to a dedicated portion of the channel at all times. The main advantages of FDMA are that it is simple to implement in that no real-time coordination among transmission of data is needed and can be used to transmit either analog or digital signals. For bursty traffic, the channel utilization is poor. This scheme is cost effective for applications that involve point-to-point trunking.

In TDMA, each user is scheduled to transmit in short nonoverlapping intervals. Therefore, a TDMA scheme requires some form of frame structure and a global timing mechanism to achieve nonoverlapping transmission. For this reason, a TDMA system is more complex to implement than an FDMA. However, an important advantage is the connectivity. This is obtained because all receivers listen on the same channel, while all sources in a TDMA system transmit on the same common channel at different times.

The third method uses dynamic sharing of a channel using demand assignment techniques. This method may be used for circuit-switched voice traffic or packet-switched data traffic [8]. In this paper, we confine ourselves to packet-switched data traffic only. The packet-switched data traffic system can be divided into random access, implicit reservation, and explicit reservation systems. In the following discussion, we will assume that the whole of a transponder bandwidth is devoted to multiaccess operation, the up channel at one frequency operating in multiaccess mode and the down channel at another frequency operating in broadcast mode. The earth stations which are visible to the satellite antenna transmit packets at the full available bandwidth. The satellite after frequency translation retransmits packets at the full available bandwidth. The downward packets are received at all earth stations within the satellite's coverage. The earth stations identify packets destined to them by looking at the packet header address. All packets addressed to other stations are ignored and those addressed to the station are passed on to it.

RANDOM ACCESS SYSTEM

One of the protocols used for transmitting packets in a random access satellite system is the ALOHA protocol. In this protocol, each one of the earth stations transmit packets as soon as each one of them has a packet to transmit without regard for other stations. Due to the lack of coordination among the distributed ground stations, packets from different stations may reach the satellite at the same time and collide, thereby destroying the information content. Therefore, a subsequent retransmission of the packet is required. Because of the broadcast capability posed by the down channel, the transmitting station will be able to detect any collision. No acknowledgment is necessary in the satellite system in the event of collision. The collided packets are retransmitted after a further random delay in order to avoid the risk of repeated collisions. The maximum channel capacity that is usable is about 18 percent in the ALOHA protocol.

A substantial increase in usable channel capacity can be obtained by using the S-ALOHA (slotted-ALOHA) protocol. In the S-ALOHA protocol, the satellite channel is slotted into segments whose duration is exactly equal to the transmission time of a single packet (assuming fixed size packets). If the earth stations are synchronized to start the transmission of packets at the beginning of a slot, the channel utilization efficiency increases. In the ALOHA protocol, when a collision takes place, the packets may overlap fully or partially. By using the S-ALOHA protocol, the partial overlap is eliminated. Under certain assumptions about the

message traffic generated by the earth stations, the channel utilization efficiency is about 36 percent [1],[9]. This increase in channel utilization efficiency is obtained at the cost of increased complexity in control compared to the ALOHA system.

One of the drawbacks of the random access system is the problem of instability. When large numbers of stations are active, excessive traffic leads to more collision. After collision, the channel traffic consists of both the newly generated packets and the retransmitted packets. As the number of newly generated packets increase, the chance of collision increases. This, in turn, increases the number of retransmissions which, in turn, increases the chance of a collision, and a runaway effect occurs; thus, the channel becomes unstable. In the absence of a control mechanism [5],[10],[13], the collision retransmission may produce a congested condition with the system throughput becoming zero. The purpose of the control is to prevent the channel from reaching the unstable condition, while optimizing channel efficiency and performance during normal operating conditions [13].

The low bandwidth utilization of the ALOHA and the S-ALOHA systems have led to many proposals for increasing utilization by means of slot reservation schemes. The object of slot reservation schemes is to reserve a particular time slot for a given station. This ensures that no collision will take place. In general, it may be possible to achieve potentially high channel efficiency using some form of a reservation technique. This increase in channel utilization efficiency is obtained at some overhead cost, either in terms of allocation of part of the bandwidth for reservation purposes and/or increased complexity of the control mechanisms in transmitting stations. All reservations methods use some form of framing approach, and the reservation scheme can be either implicit or explicit.

IMPLICIT RESERVATION

The implicit reservation protocol uses a frame concept to the S-ALOHA channel to permit implicit reservation. A frame may consist of more than one slot. The total number of slots can be grouped into a set of reserved slots and a set of slots which can be accessed using the S-ALOHA contention protocol. Efficient channel utilization is obtained by allowing stations with high traffic rate access to one or more slots from the reserved set of slots in each frame.

The reservation-ALOHA utilizes this principle with implicit reservation-by-use allocation. Reservation-ALOHA uses distributed control, and each earth station executes an identical allocation algorithm based on the global information available from the channel. Whenever a station successfully transmits in a slot, all the stations internally assign that slot in subsequent frames for exclusive use by the successful station. Thus, each station maintains a history of usage of each channel slot for one frame duration. This slot is reserved to this station until the station is finished using it. The stations use the S-ALOHA contention method to access the unassigned slots in each frame. In this scheme, there is no way to prevent a station from successfully capturing most or all of the slots in a frame for an indefinite time period.

EXPLICIT RESERVATION

These reservation schemes try to make better use of the channel bandwidth by explicitly reserving future channel time for transmitting one or more messages for a specific station. To obtain good performance, the ground stations should cooperate with one another to maintain synchronism. Only by conforming to the reservation discipline can the earth stations ensure that packet collisions will either be eliminated or reduced drastically. In the explicit reservation scheme, the earth stations use part of the channel bandwidth for sending reservations for future time slots. This, to some extent, reduces the total bandwidth available for data transmission. By keeping the bandwidth required for reservation proportionately small compared to that available for data transmission, high channel utilization efficiency can be obtained. Compared to nonreservation schemes, more complex control mechanisms are needed in the earth stations. The reservations may be sent in separate time slots which are distinct from the time slots used for data transmission or they may be combined with data transmission (piggybacked) or both. The control technique used to allocate the reserved time slot may be central control, distributed control, or a combination of both.

RESERVATION ALOHA

This scheme makes use of separate time slots for reservation, with the control function distributed in all the stations. The satellite channel is divided into time slots of fixed size [11]. Every $M + 1$th slot is subdivided into V small slots as shown in Fig. 1.

The V small slots are used by all the active stations to send reservations for future time slots and acknowledgments. These V small slots are accessed using the ALOHA contention technique. The M large slots carry reserved data packets.

Whenever a station receives data packets to transmit, it randomly selects one of the V slots and transmits its reservation. This reservation is heard by all the stations. The distributed control in each of the earth stations adds the broadcasted reservation to the existing reservation count. Effectively, all the waiting packets for which a reservation has

been made join one "queue in the sky," the length of which is known at all times to all ground stations. The number of reserved data slots that can be reserved in one request range from one to eight. The requesting station has now successfully reserved a sequence of future time slots for data transmission. Once a reservation is made, each one of the stations knows which future slots belong to them, and no other station need concern itself with the details of reservations made by other stations.

Fig. 2 shows an example taken from [11] which illustrates how this reservation scheme functions. Let us assume that the total roundtrip delay for signal travel is 10 slot time and there are five data slots (M) and six small slots (V). If a station transmits a reservation for three future data slots so as to fall in a small slot (ALOHA) at $t = 5$, then all stations will receive this reservation request at $t = 10$ (the roundtrip delay). If no collision has taken place, then the future data slots that can be used for data transmission are easily calculated, provided the current queue length is known. Assuming the current queue length to be 13, then the station which requested the reservation has to wait until 13 data slots have passed by before it can transmit data. In our example, the slots are at time $t = 21, 22,$ and 24, 23 being the ALOHA slot. Because it takes 5 slot time for the data packets to reach the satellite, the ground station starts transmission at $t = 16, 17,$ and 19.

The performance of the system is a function of the value M, the number of data slots available between each reservation slot. Assuming that there are N ground stations, and if each one of them is allowed to reserve up to eight slots, the maximum allowed, then some reservations may carry over beyond the next reservation slot if $8N$ is greater than M. The system becomes overloaded if each station is allowed to reserve eight further slots. This increases the queue length of future reservations, thereby increasing packet delay. This situation can be avoided if each ground station is constrained to a limit of eight future reserved slots at any time [4]. Another factor which may degrade the performance of the system is excessive contention for reservation slots. The number of V slots must be related to the number of earth stations and to the likely traffic activity to be expected.

In this scheme, the channel may be in any one of the two states called the ALOHA and RESERVED states. On start up and when it so happens that no reservations are

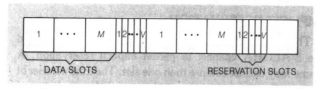

Fig. 1.   Satellite channel for reservation ALOHA.

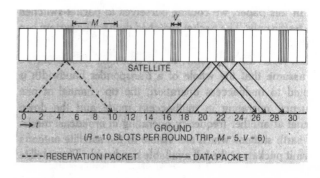

Fig. 2.   Satellite channel for reservation.

outstanding, the channel is in ALOHA state. In ALOHA state, the channel consists of only slots of type V. It is possible to send acknowledgments, reservations, and even data which will fit into the small slots. In this state, a reservation request may be transmitted in any of the small slots, with no requirement to wait for up to M data slots to pass by. Once a successful reservation has been established, the channel enters the reservation state and any further reservation can be made in the small slot. Once again the channel enters the ALOHA mode if the number of reservations goes to zero.

R-TDMA

This explicit reservation protocol is a modified version of the contention and fixed assignment reservation method used in [2]. This scheme uses a fixed-assignment technique for making reservations and allows the total available channel capacity to be shared among all stations that are busy [14].

Fig. 3 shows the R-TDMA channel. One routing frame on the channel is divided into a number of reservation frames. The reservation frame consists of a set of reservation slots and a number of fixed length data slots. These data slots are grouped together to form a data frame. A reservation frame may have one or more data frames. Each station is assigned a fixed slot in each reservation frame. Each of the stations is assigned a fixed slot in each one of the data frames. Therefore, each data frame has as many slots as there are stations.

Each earth station keeps a reservation table to track the data slot allocation. To make reservation for data slots, the earth station transmits its "new reservation" count in its reservation slot. The stations which do not have data to send place a value of zero in their fixed reservation slots. The new reservation count represents the number of data packets that arrived after the last reservation took place. All the earth stations receive the reservation packet and adjust their reservation table values by adding the new reservation counts at a globally agreed upon time.

The allocation of data slots now becomes straightforward. Those stations whose reservation table entries are not zero transmit their data packet in their fixed slots. The data slots which belong to station with no data packets are assigned in a round-robin manner among those stations with outstanding data packets. The sender for each slot is determined just prior to the slot transmission time. In this scheme, synchronization is acquired and maintained by having each station send its own reservation table entry in its reservation slot.

CONFLICT-FREE MULTIACCESS (CFMA)

This scheme [6] eliminates all conflict on the satellite multiaccess channel. The channel is divided into frames. Each frame is subdivided into an R-vector, an A-vector, and an I-vector. Fig. 4 shows the frame structure and three vectors. The R-vector is used to request future reservations and is divided into a number of reservation slots. The number of reservation slots in the R-vector is equal to the number of earth stations. Each one of the earth stations is assigned a reservation slot in the R-frame. This avoids contention for the reservation slot. The A-vector is divided into a number of mini-slots which are used to send acknowledgment for previously received packets. An I-vector in a frame is divided into data slots. In this scheme, the maximum number of slots a station may request is equal to the number of slots in the I-frame. Assuming that there are m data slots in an I-frame, the allocation of data slots is based on assigning a priority order for each of the m slots. For example, if the number of stations equals the number of data slots ($N = m$) in an I-vector, then the priority order for each data slot is different. Every station has one data slot for which it has first priority, another for which it has second priority, and so on down to the least priority. If a station does not use its first priority data slot, then a station with second priority to that slot gets a chance to use that data slot. If all stations are busy, then each of the stations will be allocated its first priority data slot and no station will be allocated more than one slot in the above example. The overhead involved in this system does not seem to be high in terms of channel bandwidth.

CONTENTION-BASED DEMAND ASSIGNMENT PROTOCOL (CPODA)

This protocol is designed to handle packetized data and voice traffic [7]. It can handle traffic with multiple priority and delay class distinctions, variable message lengths, and arbitrary load distribution among the stations.

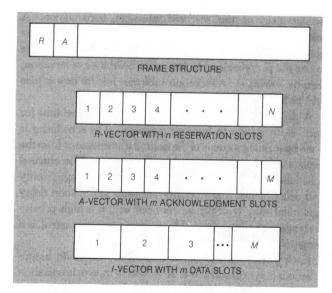

FRAME STRUCTURE

R-VECTOR WITH n RESERVATION SLOTS

A-VECTOR WITH m ACKNOWLEDGMENT SLOTS

I-VECTOR WITH m DATA SLOTS

Fig. 4. Frame structure for CFMA channel.

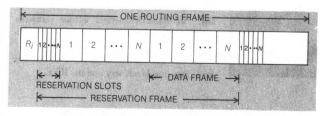

ONE ROUTING FRAME

RESERVATION SLOTS

DATA FRAME

RESERVATION FRAME

Fig. 3. R-TDMA channel.

The channel is divided into fixed size frames, and each frame consists of reservation and information subframes. The reservation subframe is divided into fixed size reservation slots. In this scheme, the reservation subframe is allowed to grow or shrink according to the amount of traffic. Therefore, when the number of reservations for the information frame is zero, the reservation subframe expands to occupy the whole frame. On the other hand, when the system is fully loaded, the reservation subframe contracts to the minimum number of slots required to allow reservations by high priority traffic or previously idle stations.

There are two ways in which reservation for information subframes can be done. The first way is to send a reservation in the slots in the reservation subframe. The stations use contention to gain access to the reservation slot. The second way is to send the reservation by piggybacking them in the header field of the reserved message transmission. A maximum of only two new reservations is allowed in each one of the messages. This allows a station transmitting messages to use the piggybacking technique to build their reservation, thereby leaving the reservation subframe free for new entries and/or higher priority traffic.

A distributed control is used to schedule channel time for each earth station to transmit messages. The scheduling is done by forming a queue of the desired transmissions from the explicit reservation requested by the stations. The channel scheduling in this scheme is some function of message priority and delay. Thus, a low priority message with a short delay constraint may typically be serviced before a high priority message with a long delay. The ordering, to some extent, is a weighted function of priority and delay.

Each station carries out a consistency check to assure scheduling synchronization. A station is in synchronization when its scheduling decision agrees with the actual transmission in the channel. A station can be in one of three states as shown below.

A station in the in-sync state is in synchronism with the actual transmission taking place in the channel. Hence, it can continue sending messages at the scheduled time. Whenever a station detects a number of inconsistent scheduling within a specified time period, it moves to the out-of-sync state. In this state, the station is not allowed to send any message; instead it carries out channel scheduling and closely monitors the channel. If the station, in the monitoring channel, finds itself in synchronism again within a fixed period of time, it moves back to the in-sync state and participates in message transmission. Otherwise, it moves to the initial acquisition state. In this state, the station listens to the new reservations on the channel and builds up its channel scheduling information. The station does not transmit any message. Once this station has constructed a reservation list compatible with other stations, it can move to the out-of-sync state.

CONCLUSIONS

A number of multiple access protocols have been presented, some of which are undergoing testing for satellite

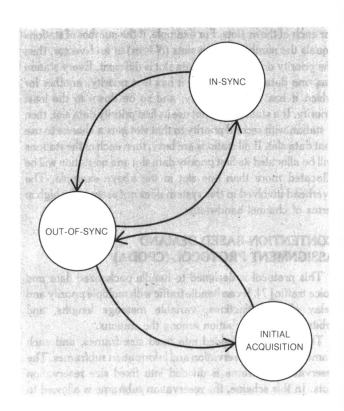

communication. These reservation methods provide a means to increase channel utilization compared to nonreservation schemes. In all the schemes, one must trade off complexity of implementation with suitable performance. Therefore, in the final analysis, it is cost which will dictate which of the protocol schemes is suitable for a particular application.

REFERENCES

[1] N. Abramson, "Packet switching with satellites," in *Proc. AFIPS Conf.*, vol. 42, June 1973.

[2] R. Binder, "A dynamic packet-switching system for satellite broadcast channel," in *Proc. ICC '75*, San Francisco, CA, June 1975.

[3] W. R. Crowther *et al.*, "A system for broadcast communication: Reservation-ALOHA," in *Proc. 6th Hawaii Int. Conf. Syst. Sci.*, Jan. 1973.

[4] D. W. Davis *et al.*, *Computer Networks and Their Protocols.* New York: Wiley, 1979.

[5] M. Gerla and L. Kleinrock, "Closed loop stability control for S-ALOHA satellite communication," presented at the 5th Data Commun. Symp., Sept. 1977.

[6] H. R. Hwa, "A framed ALOHA system," in *Proc. PACNET Symp.*, Sendai, Japan, Aug. 1975.

[7] I. Jacobs *et al.*, "CPODA—A demand asssignment protocol for SATNET," in *Proc. 5th Data Commun. Symp.*, 1977.

[8] I. M. Jacobs *et al.*, "General purpose packet satellite network," *Proc. IEEE* Nov. 1978.

[9] L. Kleinrock and S. S. Lam, "Packet-switching in a slotted satellite channel," in *Proc. AFIPS Conf.*, vol. 42, June 1973.

[10] S. Lam and L. Kleinrock, "Packet switching in a multi-access broadcast channel: Dynamic control procedures," *IEEE Trans. Commun.*, vol. COM-23, Sept. 1975.

[11] L. G. Roberts, "Dynamic allocation of satellite capacity through packet reservation," in *Proc. AFIPS Conf.*, vol. 42, June 1973.

[12] "Satellite carrier posed for increasing demands," *Commun. News*, Mar. 1979.

[13] F. A. Tobagi *et al.*, "Modeling and measurement techniques in packet communication networks," *Proc. IEEE*, vol. 66, Nov. 1978.

[14] R. Weissler *et al.*, "Synchronization and multiple access protocols in the initial satellite IMP," in *Proc. COMPCON*, Fall 1978.

C. Retnadhas received the Ph.D. degree from Iowa State University. He is a faculty member in computer science at Western Illinois University. His research interests include computer architecture, computer communication networks, and distributed processing.

Dr. Retnadhas is a member of the Association for Computing Machinery and the IEEE.

LOCAL DISTRIBUTION IN COMPUTER COMMUNICATIONS *

Jeremiah F. Hayes

A sophistication of early queueing theory solves bursty transmission problem.

A significant part of the field of Computer Communications is concerned with providing transmission facilities for data sources which may be characterized as bursty, i.e., short spurts interspersed with relatively long idle periods. It has been estimated, for example, that terminals in interactive data networks are active from 1 to 5 percent of the time [1],[2]. This burstiness allows one to share channels among a number of sources. In this paper, we shall consider a particular context in which transmission facilities must be provided—local distribution. In local distribution a number of geographically dispersed sources are to be connected to a central facility. The importance of local distribution systems lies in the fact that they are the most common class of computer communication networks. Further, as part of large systems, local distribution consumes a significant portion of the total cost. In this paper, we shall describe the basic approaches to local distribution. Our discussion encompasses certain adaptive techniques which have been discovered recently.

The focus of our discussion is on the fundamental principles of the techniques used in local distribution without dwelling on details of implementation. We distinguish three main categories: polling, random access, and adaptive techniques. A fourth category is techniques which are suited to a particular topology—the ring or loop structure. As well as describing the techniques we shall summarize the results of studies of performance. These results quantify the effect of various system parameters on performance. An extensive review of the literature is given in a final section of the paper.

To a large extent the analyses of performance that we shall discuss are based on queueing theory. Queueing theory began with the work of a Danish mathematician, A. K. Erlang (1878-1929) [3], on telephone switching systems. His first paper on the subject was published in 1909 [4]. Amazingly, the formulas derived by Erlang in a 1917 paper [5] are in constant use in engineering the modern telephone office. Furthermore, although queueing theory finds wide use as one of the basic components of operations research, telecommunications remains as its most successful application. As one might expect, voice traffic was the primary concern in telecommunications applications. However, recently there has been an upsurge of interest in data traffic and a corresponding reapplication of queueing theory in connection with computer communications [6],[7].

Although Erlang's work was concerned with voice traffic, certain of his basic concepts are appropriate to data networks. In the generic queueing model customers randomly arrive at a facility with service requirements that may be random in nature. The theory attempts to find probabilistic descriptions of such quantities as the size of the waiting lines, the delay of a customer and availability of a serving facility. In the voice telephone network, demands for service take the form of telephones going off hook or call attempts. Erlang found that given a sufficiently large population, the random rate of such calls can be described by a Poisson process.[1] The service time of a customer is the duration (holding time) of a call and was found to have an exponential distribution which is closely related to the Poisson distribution. In computer communications applications, the generation of data messages at a terminal is the analog of customer arrival. The service time is the time required to transmit the data message. In many cases of interest, the arrival process is approximated by a Poisson process. The duration of messages is commonly taken to be constant or to be exponentially distributed.

*Reprinted from *IEEE Communications Society Magazine*, March 1981, Vol. 19, No. 2, pp. 6-14.

[1] A definition of the Poisson process will be given in the sequel.

POLLING SYSTEMS

A ubiquitous example of a local distribution system is shown in Fig. 1 where we depict part of the communications facilities in the Bonanza of Bargains Shopping Mall and Family Entertainment Center. A number of terminals situated throughout the B of B are bridged across a common line and connected to a common computer. The common line may be wire or coaxial cable.[2] The terminals are engaged in such commercially useful activities as credit checking and inventory control. However, even in the best of seasons the traffic produced by an individual terminal is bursty and a number of terminals can share the same line. Located at the computer is a controller, one of whose functions is effecting this sharing efficiently and equitably.

A common technique for parcelling out bits per second among users sharing a common line is *roll-call polling*[3]. It is assumed that the common line is such that the controller can broadcast to all terminals simultaneously. Each terminal has an address which is transmitted in sequence by the controller over the common line. After broadcasting a terminal's address the controller pauses for a message from the terminal. If a terminal has a message the polling cycle is interrupted while the message is transmitted.

The ability to achieve economies by sharing transmission facilities is limited by performance requirements usually expressed in the delay experienced by a user in obtaining service. If there are too many terminals on the line, for example, the time required to cycle through all terminals is too large and user dissatisfaction ensues. The parameters of the mathematical models of performance are: the number of terminals, the volume of traffic generated by each terminal, the line speed in bits/s and the line required by the polling

protocol. As we shall see in connection with the analysis of polling models a significant factor in performance is overhead, i.e., the time required to poll all terminals even when there are no messages. In the case of terminals equipped with voiceband modems, for example, this may involve equalizer training as well as the phase and timing recovery associated with the transmission of each polling message.

A close relative to roll-call polling is *hub polling* [8],[9]. Again, we have the geographically dispersed terminals of Fig. 1. The controller begins a polling cycle by broadcasting the address of the most distant terminal thereby granting to this terminal exclusive access to the line. After this terminal has transmitted any messages that it might have, it transmits an "end of message" symbol which acts to grant access to the next most distant terminal. Upon receiving this symbol, the next most distant terminal repeats the process, passing on access to the third most distant terminal when its messages have been transmitted. The process continues until all terminals have been given an opportunity to transmit messages whereupon the controller initiates a new cycle. This model contains the same parameters as roll-call polling. The salient difference between the two is the time required to grant access to a terminal. In roll-call polling, the time required to transmit a message and receive a reply is typically much larger than the time required to transmit a symbol from one terminal to another. However, hub polling requires that the line be such that terminals reliably receive transmissions from other terminals.

The hub-polling technique has been implemented in the ring topology depicted in Fig. 2 [10]. Flow around the ring is clockwise as shown. The central processor grants first access to the first terminal downstream. As in the previous implementation, access is passed from a terminal to its nearest neighbor downstream. An end of message character is appended to the data from a terminal. We have the same set of parameters as in the previous cases. The time required

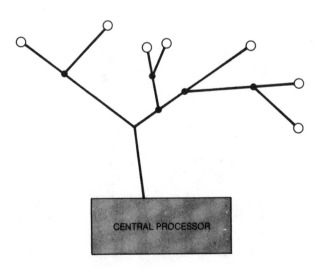

Fig. 1.   Geographically dispersed users tree topology.

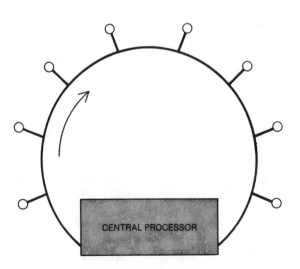

Fig. 2.   Ring topology.

[2]The required properties of this common line are related to the particular local distribution technique employed and will be discussed in due course.
[3]For implementation of polling systems, see [8] and [9].

to pass access from one terminal to another is the time required to transmit an end of message character.

Before going on to consider other local distribution techniques we pause to consider the performance of polling systems as related to network parameters and to traffic. A useful measure of performance is the cycle time which is the time required to grant access to all terminals in the system at least once and to transmit messages from the terminals. We may view the cycle time as having two components—fixed and random. The overhead or fixed component is the time required to grant access to all terminals. In roll-call polling, for example, it is the time required to broadcast all of the terminal addresses and to listen for replies. In the hub-polling technique, overhead is the total time in a cycle that is required to pass access from one terminal to another.

The random component of a polling cycle is due to the random nature of the message generation process. The number of messages transmitted in a cycle varies from one cycle to the next. The analysis of polling systems is complicated by the fact that there are correlations between the number of messages encountered in successive cycles and in the number of messages in adjacent terminals.

The most studied model assumes Poisson arrival[4] at a terminal having storage facilities which may be regarded as being infinite, i.e., compared to the arrival rate of messages the terminal buffer is so large that the probability of overflow is negligible. The time required to poll n terminals may be written

$$T_c = \sum_{i=1}^{n} t_i + W \qquad (1)$$

where W is the overhead in a cycle, assumed to be constant, and t_i is the time spent at terminal i to read out messages. Even though there are dependencies between buffer contents, the average cycle time is easily found since the expected value of a sum such as shown in (1) is the sum of the expected values. Assuming that the arrival rates and message transmission times are the same for all terminals we find that $\overline{T}_c$ the average duration of a cycle is given by

$$\overline{T}_c = W/(1 - S) \qquad (2)$$

where $S = n \, \overline{m} \, \lambda$, $\overline{m}$ is the average duration of a message and λ is the average arrival rate at each terminal. Equation (2) has a characteristic queueing theory form. The numerator represents overhead, the amount of time during a cycle for which a message is not being transmitted. All of the traffic dependency is contained in the quantity S in the denominator. This load S is the average work presented to the system normalized to the capacity of the channel. In voice networks, a similar quantity has been given the unit of Erlangs. There is a point of instability when $S = 1$ since the average amount of

work that is arriving is just equal to the capacity of the system and there is no allowance for overhead. We note that when $W = 0$, the average cycle time is zero. This is consistent if we consider that an infinite number of cycles occur in zero time when the terminals have no messages to transmit. Equation (2) indicates the effect of overhead on performance. Suppose, for example, that the total traffic load into the system is kept constant (i.e., S constant) while the number of terminals is doubled. If overhead is incurred on a per terminal basis, the cycle time is doubled with no increase in traffic.

A more tangible measure of performance for the user is message delay which we define to be the time elapsing between the generation of a message and its transmission over the common line. This delay consists of several components. A message generated at a terminal must wait until it is the terminal's turn to be polled. If the terminal can store more than one message at a time, a queue is formed at each terminal which implies further delay. Finally, a certain amount of time is required simply to transmit the message. There have been a number of analyses of the performance of polling systems.[5] In connection with cycle time, we considered the case of infinite buffers and Poisson message generation. Results on the average delay for this case with constant length messages are shown on Fig. 3. The average delay normalized to the time required to transmit a message is shown as a function of the total load into the system, $S = \overline{m} \lambda n$. The parameters are n, the total number of terminals, and $W/n\overline{m}$ is the overhead per terminal normalized to the message transmission time. The curves show the characteristic rapid increase in delay as the load approaches one. A strong dependence on overhead is also evident. For loadings

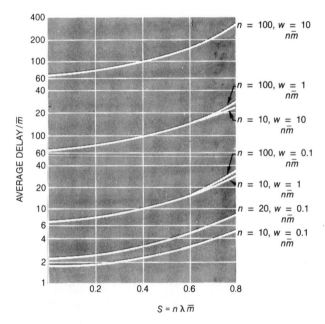

Fig. 3.   Polling delay versus load.

---

[4]For the Poisson distribution the probability of k message arrivals in T seconds is $P_k = (\lambda T)^k \exp(-\lambda T)/k!$ $k = 0,1,2\cdots$ where λ is the average arrival rate. The interarrival time is exponentially distributed, i.e., P_r [interarrival time $\leq \tau$] = $1 - \exp(-\lambda \tau)$; $\tau \geq 0$.

[5]See the review of the literature at the end of the paper.

less than 0.5, which is the region where the system will be operated, overhead dominates. These points can be illustrated by an example. Suppose that ten terminals share a common 2400 bit/s line. Each terminal generates messages at an average rate of 28.8 messages per busy hour (0.008 messages/s). The messages are each 1200 bits long. Finally, assume that in order to poll each terminal and listen for a response 50 ms are required. The load in the system is $S = .004$. From Fig. 3, the average delay is approximately 0.9 seconds. Now suppose that as a convenience to users the number of terminals sharing the line is multiplied by ten without increasing the load. We see from Fig. 3 that the average message delay is multiplied by ten. Of the two polling techniques hub polling tends to have lower overhead than roll-call polling. We may view hub polling as a more distributed form of control of access to the system.

LOOP NETWORKS

The ring or loop topology shown in Fig. 2 finds extensive application in distributed processing where computers and peripherals in close proximity (within a kilometer) are tied together. In this application, it is necessarily true that the traffic is bursty. However, the loop structure lends itself to interesting multiplexing techniques which may be appropriate to bursty sources. The most obvious technique is a form of time division multiplexing which in this context is commonly called Time Division Multiple Access (TDMA). Assuming synchronous transmission, the flow on the line is partitioned into segments each of which is dedicated to a particular terminal. A terminal simply inserts messages into segments assigned to it. The shortcoming of this system in the case of many lightly loaded terminals is that very often terminals have nothing to send and segments are wasted. At the same time, empty segments may be passing by terminals which do have messages. The same drawback applies to Frequency Division Multiple Access (FDMA) in which each terminal is allocated a fixed bandwidth. A recent study has shown that FDMA is inferior to TDMA from the point-of-view of performance [11].

An alternate technique to TDMA in a loop context is Demand Assignment (DA). The flow is the same as in TDMA except that the blocks are not assigned to any terminal. When an empty block passes by a terminal which has a message to transmit, the block is seized by the terminal and the message along with addressing information is inserted. There is an increase in the utilization of the line over TDMA at the cost of an increase in the complexity of the terminals. On Fig. 4, the average delay is shown as a function of the load with the number of terminals in the system as a parameter. The results illustrate the inefficiency of TDMA for lightly loaded systems where the dominant factor is the time required to transmit a single message. At light loading, demand multiplexing is superior to TDMA by a factor equal to the number of terminals sharing the line. As the load increases, the difference between the two systems decreases

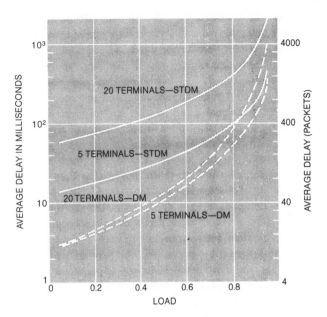

Fig. 4. Average delay versus load in STDM and DM [34].

since in demand multiplexing different terminals will tend to have messages at the same time. Once again, the lesson that we carry away from this study of loop systems is that in lightly loaded systems, a distributed control of access to the channel is more sufficient.

The TDMA technique is also appropriate to the tree topology of Fig. 1. It is necessary to establish synchronization among the terminals. Each terminal is assigned a periodically recurring time slot. However, in the tree topology as well as in the loop topology the TDMA technique is not efficient for bursty sources.

RANDOM ACCESS (ALOHA)

Random access techniques hitherto associated with radio and satellite systems have recently been applied to local area networks [12]. The origin of these methods is the ALOHA protocol which is the ultimate in distributed control. Again, we assume that n terminals are sharing the same channel as depicted in Fig. 1. As soon as a terminal generates a new message it is transmitted on the common line. Along with the message, the terminal transmits address bits and parity check bits. If a message is correctly received by the central controller a positive acknowledgment is returned to the terminal on the return channel. Since there is no coordination among the terminals it may happen that messages from different terminals interfere with one another. If two or more messages collide, the resulting errors will be detected by the controller which returns a negative acknowledgment or no acknowledgment. An alternative implementation is to have the terminal itself detect collisions simply by listening to the channel. After a suitable timeout interval a terminal involved in a collision retransmits the message. In order to avoid repeated collisions the retransmission intervals are chosen randomly. The key element of the ALOHA protocol and its descendants is the retransmission traffic on the common line. If the rate of newly generated traffic is increased, the rate of conflicts among terminals increases to the point where retransmitted messages dominate and there is saturation. This effect is expressed succinctly in the formula

$$S = G \exp(-2G). \qquad (3)$$

where S is the normalized load into the system generated at all terminals and G is the total traffic on the line including all retransmissions. In the derivation of (3), it is assumed that all messages are the same length and that they are generated at a Poisson rate. The plot of (3) on Fig. 5 shows that the channel saturates at 18 percent of its capacity inasmuch as the input cannot be increased beyond this point. Thus, it appears that simplicity of control is achieved at the expense of channel capacity.

The basic ALOHA technique can be improved by rudimentary coordination among the terminals. Suppose that a sequence of synchronization pulses is broadcast to all terminals. Again, let us assume constant length messages or packets. A so-called slot or space between synch pulses is equal to the time required to transmit a message. Messages, either newly generated or retransmitted, can only be transmitted at a pulse time. This simple device reduces the rate of collisions by half since only messages generated in the same interval interfere with one another. In pure ALOHA, the "collision window" is two message intervals. The equation governing the behavior of slotted ALOHA is

$$S = G \exp(-G) \qquad (4)$$

We see from the plot of (4) on Fig. 5 that the channel saturates at approximately 36 percent of capacity.

An extension of the ALOHA technique that is particularly appropriate for local distribution is Carrier Sense Multiple Access (CSMA). Before transmitting a message a terminal listens on the common channel for the carrier of another terminal which is in the process of transmitting. If the channel is free the terminal transmits; if not, transmission is deferred.

Variations on the basic technique involve the retransmission strategy. We illustrate retransmission strategies by means of the P-persistent CSMA strategy. If the channel is busy then the terminal transmits at the end of the current transmission with probability P. With probability $1-P$, transmission is delayed by τ seconds which is the maximum propagation time between any pair of terminals. Due to propagation delay there may be more than one terminal transmitting at the same time in which case messages are retransmitted after random timeout intervals. The value of P is chosen so as to balance the probability of retransmission with channel utilization. The characteristic equations for CSMA are plotted on Fig. 5. The form is similar to pure and slotted ALOHA. The ability to sense carrier from other terminals leads to considerable improvement in throughput. As indicated, decreasing P leads to improved throughput which is obtained at the expense of increased delay. The curves shown in Fig. 5 are for a propagation delay 0.01 normalized to message transmission time. As this normalized delay is increased the performance of CSMA degrades.

There have been a number of analyses of random access protocols focusing on message delay as a function of throughput. On Fig. 6, we summarize the results of this work

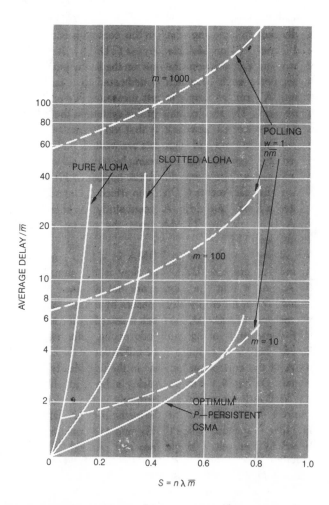

Fig. 6.   Delay in random access systems [47].

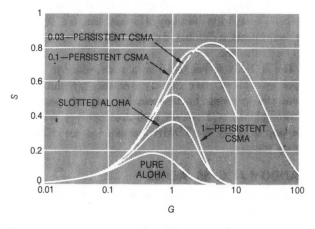

Fig. 5.   Input load as a function of channel traffic for several random access techniques [47].

in the form of normalized message delay as a function of load. For lightly loaded systems, pure and slotted ALOHA perform well. However, as the load increases the increasing rate of retransmission rapidly degrades performance. Since the carrier sense protocol keeps the channel clear by avoiding retransmission, it has a graceful degradation. Also shown on Fig. 6 is delay for roll-call polling. As we have seen earlier, there is a severe penalty for overhead required when the number of terminals is increased. The curves also show that the performance of the polling protocol degrades more gracefully than that of the random access protocols. This is where the beneficial effect of the controller is seen. By scheduling transmission, the avalanche effect of retransmissions in the random access protocols is prevented.

The curves for the random access techniques in Fig. 5 show the same basic form in which a level of input traffic S can lead to two possible levels of line traffic G. It can be shown that this characteristic may lead to an unstable state resulting in saturation of the channel and a drop in throughput. By choosing system retransmission parameters properly, unstable states can be prevented. An example of this is decreasing the parameter P in the P-persistent CSMA protocols. In ALOHA, the range of the retransmission interval can be increased. In both cases, there is a penalty in increased delay.

ADAPTIVE TECHNIQUES

The deleterious effect of overhead on the performance of polling systems is abundantly clear from the foregoing results. In order to ameliorate this effect an adaptive technique has been devised recently. The essence of the technique, which has been designated probing, is to poll terminals in groups rather than one at a time. In order to implement the technique, it is assumed that the central controller can broadcast to all terminals in a group simultaneously. If a member of a group of terminals being probed has a message to transmit, it responds in the affirmative by putting a noise signal on the common line. Upon receiving a positive response to a probe, the controller splits the group into two subgroups and probes each subgroup in turn. The process continues until individual terminals having messages are isolated whereupon messages are transmitted. The probing protocol is illustrated on Fig. 7 for a group of eight terminals of which terminal 6 has a message. The algorithm is essentially a tree search which the controller begins by asking, in effect, "Does anyone have a message?" Branches with affirmative responses are split into subbranches.

If only one terminal in a group of 2^k has a message, the probing process requires the controller to transmit at most $2k + 1$ inquiries. In contrast, conventional polling requires 2^k inquiries. The comparison may not be so favorable when more than one terminal has a message. For example, if all terminals have messages, $2^{k+1} - 1$ inquiries are required for probing. This consideration leads to adaptivity where the size of the initial group to be probed is chosen according to the probability of an individual terminal having a message. Thus,

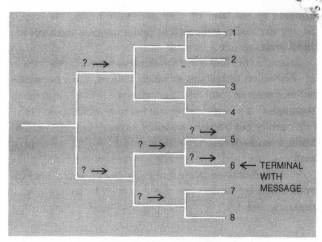

Fig. 7. Probing illustration.

in Fig. 7, for example, one may begin a cycle by probing two groups of four rather than one group of eight. The criterion for choosing the sizes of the groups is the amount of information gained from an inquiry. If the initial group is too large the answer to an inquiry is almost certainly, "Yes there is a message." However, if the group is split into too many subgroups, the answer to an inquiry is too often, "No." If the arrival of messages to terminals is Poisson, the probability of a terminal having a message can be calculated by the controller given the duration of the previous probing cycle. Given this probability, the optimum group size can be found. Notice that if the probability is high enough, the best strategy may be to poll every terminal.

The results of simulation for the adaptive technique are shown on Fig. 8 where the average time to probe all terminals in a 32 terminal network is shown as a function of message arrival rate. In Fig. 8, the cycle time and the message length are normalized to the amount of time required to make an inquiry. The comparison made with conventional polling shows a considerable improvement in performance for light loading. Moreover, due to the adaptivity there is no penalty for heavy loading.

Although the probing concept was devised in connection with polling systems it is also appropriate in a random access context. Suppose that in response to a probe a terminal transmits any messages that it might be harboring. Conflicts between terminals in the same group are detected by the controller and the group is divided in an effort to isolate individual terminals. Each subgroup is given access to the line in turn. Optimal initial group sizes can be chosen by means of very much the same criterion as in polling systems. Probing too large a group results in almost certain conflict. The opposite extreme gives too many probes of empty groups of terminals. Again, the optimum group size can be chosen adaptively as the process unfolds. The probability of a terminal having a message is a function of the previous cycle and the average message generation rate at a terminal. This probability determines optimum initial group size.

Control of the adaptive process need not be as cen-

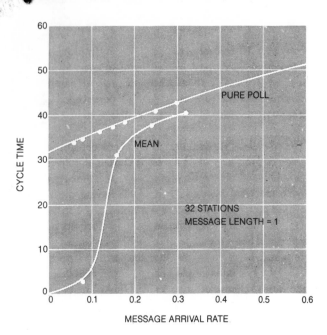

Fig. 8.   Average cycle time versus message arrival probing technique [49].

8 conflict. The conflict is resolved in subsequent slots. After this conflict resolution process has begun, any newly arrived messages are held over until the next cycle. Again, the algorithm can be made adaptive by adjusting the size of the initial groups to be given access to the channel according to the probability of a terminal having a message.

Upper and lower bounds on average delay as a function of the input load are shown in Fig. 10. Notice that the system saturates when the load is 43 percent of capacity. This contrasts with the case of slotted ALOHA where this maximum is 36 percent of capacity. Recent improvements of the technique have pushed this figure to over 50 percent. There are no unstable states where the system is saturated by retransmissions and conflicts. If conflicts persist each terminal is assigned an individual slot and the system reverts to TDMA.

The so-called "random urn" is another technique in which the size of groups granted access is chosen adaptively. The assumption underlying this protocol is that at the beginning of a cycle the total number of terminals having messages is known to all terminals. Access is granted to groups of size k where k is chosen so as to maximize the probability that only one terminal has a message. If, as in heavily loaded systems, all terminals have messages, then the optimum group size is one and the system is simple TDMA. Under light loading, the random urn scheme behaves as ALOHA. The key issue in implementing this scheme is determining the number of terminals having messages. One possibility is a reservation interval at the beginning of a cycle. In this interval, terminals having messages indicate as such. From this, terminals can estimate the number of other terminals having messages. Simulation studies indicate that performance is insensitive to small errors in this estimate.

Related to random access multiplexing are a large number of reservation techniques in which sources, upon becoming active, reserve part of the channel. The reservation techniques are appropriate to sources which are active

tralized as in the foregoing. Suppose that as in slotted ALOHA synchronizing pulses are broadcast to all terminals. Suppose further that the slots between synch pulses are subdivided into two equal subslots. In the tree search protocol, the first subslot is devoted to an upper branch and the second to a lower. Consider the example in Fig. 9(a) and (b) depicting an eight terminal system of which 5, 7, and 8 have messages. The first subslot is empty since it is dedicated to terminals 1-4. In the second subslot, terminals 5, 7, and

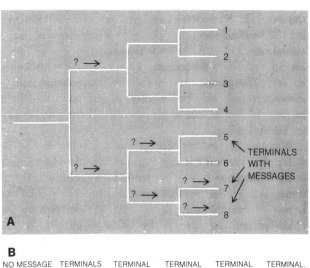

Fig. 9.   (a) Tree search illustration. (b) Tree search illustration.

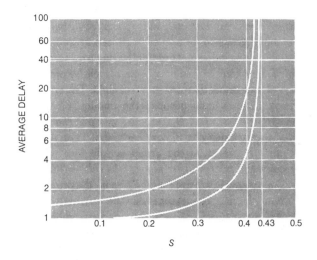

Fig. 10.   Average delay versus load tree search [50].

infrequently but transmit a steady stream while active. The traffic from such sources is not bursty. However, the request methods are and consequently may be treated by the techniques discussed in the foregoing. For example, reservations could be made using the ALOHA technique over a separate channel.

REVIEW OF LITERATURE

There is an analogy between polling systems and machine patrolling in which a repairman examines n machines in a fixed sequence. If a machine is broken he pauses to make repairs. This is the analog in polling systems to message transmission. The overhead that is incurred is the time required to walk between machines. This walktime corresponds to the time required to poll a terminal. The earliest work on this problem was done for the British cotton industry by Mack et al. [13]. Based on this work, Kaye [14] derived the probability distribution of message delay for the case where terminals store a single fixed length message. This result is the one shining example of a simple expression for probability distributions in polling models. Some idea of the delicacy of the model may be gained from Mack's analysis of very much the same situation but with a variable repair time [15] (corresponding to variable length messages in polling systems). In order to find a solution it is necessary to solve a set of 2^{n-1} linear equations. For a treatment of work on related problems, see Cox and Smith [16].

A great deal of work has been devoted to the case of the infinite buffer. The earliest work in this area involved just two queues with zero overhead [17],[18]. Later, this was generalized to two queues with nonzero overhead [19],[20]. In terms of the models that we are concerned with, the first papers of interest are those of Cooper and Murray [21] and Cooper [22]. The number of buffers is arbitrary and both the gated and exhaustive services models are considered. The drawback is that the analysis assumes zero overhead. The characteristic functions of the waiting times are found. Also found is a set of $n (n + 1)$ linear equations whose solution yields the mean waiting time at each buffer when the message arrival time is different for each. The assumption of zero overhead here may yield useful lower bounds on performance.

For a long time, the only work on an arbitrary number of queues with nonzero overhead was by Liebowitz [23] who suggested the independence assumption. In 1972, both Hashida [24] and Eisenberg [25] separately published results on multiple queues with nonzero overhead. Both used imbedded Markov chain approaches. (Some of Hashida's results are plotted on Fig. 3.) Computer communications stimulated the next significant step in polling models. Konheim and Meister [26] studied a discrete time version of model. Transmission time over the channel is divided into fixed size discrete units called slots. Messages are described in terms of data units which fit into these slots. (An 8-bit byte is a good example of a data unit.) The analysis is carried out by imbedding a Markov chain at points separated by slots. In most of this work, the emphasis was upon symmetric traffic.

Recently, Konheim and Meister's work was extended to the case of asymmetric traffic [27]. Interestingly, it was found that in the case of asymmetric traffic, the order in which terminals are polled affects performance.

A significant remaining problem involves nonexhaustive service where, at most, a fixed number of messages are transmitted from a particular buffer. If there are more than the fixed number of messages at the buffer they are held over until the next cycle. If there are less than this fixed number the next terminal is polled immediately after the buffer is emptied. At the present writing no exact analysis is available. There have been several analyses of systems of this kind based upon approximations [28]-[30]. The latest of these is by Kuehn who obtains results when at most one message is removed at a time. Kuehn evaluates his results by comparing them to earlier results by Hashida and Ohara and to simulation.

Pioneering work on loop systems was carried out by Farmer and Newhall [10] who proposed the hub-polling technique discussed above. The demand multiplexing approach in loop systems is due to Pierce [31],[32]. A version of demand multiplexing was used by Fraser in the implementation of the Spider network [33]. There have been several analyses of demand multiplexing [34]-[37]. The curves shown on Fig. 4 were taken from [34]. A nice summary of later work on the implementation and the analysis of performance of loop networks is contained in [38].

Recently, two thorough survey papers emphasizing random access techniques have appeared [39],[40]. These allow us to be more terse in our survey. The first publication in the area is due to Abramson [41],[42] who derived (3) under the simplifying assumption of Poisson retransmitted traffic. A great deal of subsequent work has shown (3) to be an accurate description of ALOHA. The slotted ALOHA technique was proposed by Roberts [43] who derived (4). An analysis of message delay as effected by retransmission strategy for the pure ALOHA technique is contained in [44]. Also given in [44] is a comparison of random access and polling. Instability in random access systems was brought to light by Carleial and Hellman [45] and by Kleinrock and Lam [46]. The results on carrier sense multiple access given in Figs. 5 and 6 are drawn from work by Tobagi and Kleinrock [47]. For several extensions of the basic ALOHA concept and for work on reservation systems, the reader is referred to the survey papers mentioned above. The reader is also referred to an insightful tutorial paper on this material [48].

The probing technique discussed in connection with adaptive systems is due to Hayes [49]. The distributed adaptive protocol described above was devised by Capetanakis [50] who also found the increase in the capacity given by adaptive techniques. More recent work in this area is contained in [51]-[55]. Kleinrock and Yemini devised the random urn scheme [56].

CONCLUSION

We have reviewed the basic techniques of implementing local distribution for bursty data sources. A couple of generalizations emerge from this study. It seems that under conditions of light loading distributed control is best. However, as the loading increases distributed control leads to difficulties and centralized control gives the better performance. This is entirely in conformity with everyday experience with automobile traffic. At 4 A.M., stop signs minimize delay. However, along heavily traveled routes at rush hour, stop signs would cause collisions (in the usual sense of the word) and the centralized control of traffic lights is required.

ACKNOWLEDGMENT

The author expresses his thanks to Pauline Fox for her efforts in preparing the manuscript.

REFERENCES

[1] P. Jackson and C. Stubbs, "A study of multiaccess computer communications," in *AFIPS Conf. Proc.*, vol. 34, p. 491.

[2] E. Fuchs and P. E. Jackson, "Estimates of distributions of random variables for certain computer communications traffic models," *CACM*, vol. 13, no. 12, pp. 752-757, 1970.

[3] E. Brockmeyer, H. L. Halstrøm, and A. Jensen, "The life and works of A. K. Erlang," *Trans. Danish Academy Tech. Sci.*, ATS no. 2, 1948.

[4] A. K. Erlang, "The theory of probabilities and telephone conversations," *Nyt Tidsskrift Matematik*, B.V20, pp. 33-39, 1909.

[5] ——, "Solution of some problems in the theory of probabilities of significance in automatic telephone exchanges," *Electroteknikeren*, vol. 13, pp. 5-13, 1917; in English: *PO Elect. Eng. J.*, vol. 10, pp. 189-197, 1917-1918.

[6] L. Kleinrock, *Queueing Systems, Vol. 1: Theory* and *Vol. 2: Computer Applications*. New York: Wiley, 1975.

[7] H. Kobayashi and A. G. Konheim, "Queueing models for computer communications system analysis," *IEEE Trans. Commun.*, vol. COM-25, pp. 2-29, Jan. 1977.

[8] M. Schwartz, *Computer Communication Network Design and Analysis*. Englewood Cliffs, NJ: Prentice-Hall, 1977.

[9] J. Martin, *Teleproccessing Network Organization*. Englewood Cliffs, NJ: Prentice Hall, 1970.

[10] W. D. Farmer and E. E. Newhall, "An experimental distributed switching system to handle bursty computer traffic," in *Proc. ACM Symp. Problems Optimization Data Commun. Syst.*, pp. 1-34, Pine Mountain, GA, Oct. 1969.

[11] I. Rubin, "Message delays in FDMA and TDMA communications channels," *IEEE Trans. Commun.*, vol. COM-27, pp. 769-777, May 1979.

[12] R. M. Metcalfe and D. R. Boggs, "Ethernet: Distributed packet switching for local computer networks," *Commun. ACM*, vol. 19, pp. 395-404, July 1976.

[13] C. Mack, T. Murphy, and N. L. Webb, "The efficiency of *N* Machines unidirectionally patrolled by one operative when walking and repair times are constant," *J. Royal Stat. Soc. Ser. B.*, vol. 19, pp. 166-172, 1957.

[14] A. R. Kaye, "Analysis of a distributed control loop for data transmission," in *Proc. Symp. Comput. Commun. Network Teletraffic*, Polytech. Inst. Brooklyn, Brooklyn, NY, Apr. 4-6, 1972.

[15] C. Mack, "The efficiency of *N* machines unidirectionally patrolled by one operative when walking time is constant and repair times are variable," *J. Royal Stat. Soc. Ser. B.*, vol. 19, pp. 173-178, 1957.

[16] D. R. Cox and W. L. Smith, *Queues*. London: Methuen, 1958.

[17] B. Avi-Itzhak, W. L. Maxwell, and L. W. Miller, "Queues with alternating priorities," *J. Oper. Res. Soc. America*, vol. 13, no. 2, pp. 306-318, 1965.

[18] L. Takacs, "Two queues attended by a single server," *Opns. Res.*, vol. 16, pp. 639-650, 1968.

[19] J. S. Sykes, "Simplified analysis of an alternating priority queueing model with setup time," *Opns. Res.*, vol. 18, pp. 399-413, 1970.

[20] M. Eisenberg, "Two queues with changeover times," *Opns. Res.*, vol. 19, pp. 386-401, 1971.

[21] R. B. Cooper and G. Murray, "Queueing served in cyclic order," *Bell Syst. Tech. J.*, vol. 48, pp. 675-689, Mar. 1969.

[22] ——, "Queues served in cyclic order: Waiting times," *Bell Syst. Tech. J.*, vol. 49, no. 3, pp. 399-413, Mar. 1970.

[23] M. A. Liebowitz, "An approximate method for treating a class of multiqueue problems," *IBM J.*, vol. 5, pp. 204-209, July 1961.

[24] O. Hashida, "Analysis of multiqueue," *Rev. Elect. Commun. Lab.*, NTT vol. 20, Nos. 3 and 4, pp. 189-199, Mar. and Apr. 1972.

[25] M. Eisenberg, "Queues with periodic service and changeover times," *Opns. Res.*, vol. 20, pp. 440-451, 1972.

[26] A. G. Konheim and B. Meister, "Waiting lines and times in a system with polling," *J. ACM*, vol. 21, pp. 470-490, July 1974.

[27] G. B. Swartz, "Polling in a loop system," *J. ACM*, vol. 27, pp. 42-59, Jan. 1980.

[28] O. Hashida and K. Ohara, "Line accommodation capacity of a communication control unit," *Rev. Elect. Commun. Lab.*, NTT vol. 20, pp. 231-239, 1972.

[29] S. Halfin, "An approximate method for calculating delays for a family of cyclic type queues," *Bell Syst. Tech. J.*, vol. 54, pp. 1733-1754, Dec. 1975.

[30] P. J. Kuehn, "Multiqueue systems with nonexhaustive cyclic service," *Bell Syst. Tech. J.*, vol. 58, pp. 671-699, Mar. 1979.

[31] J. R. Pierce, "How far can data loops go?," *IEEE Trans. Commun.*, vol. COM-20, pp. 527-530, June 1972.

[32] ——, "A network for the block switching of data," *Bell Syst. Tech. J.*, vol. 51, pp. 1133-1145, July/Aug. 1972.

[33] A. G. Fraser, "Spider—A data communications experiment," Computing Sci. Tech. Rep. 23, Bell Laboratories, Murray Hill, NJ, 1974.

[34] J. F. Hayes, "Performance models of an experimental computer communications network," *Bell Syst. Tech. J.*, vol. 53, pp. 225-259, Feb. 1974.

[35] J. F. Hayes and D. N. Sherman, "Traffic analysis of a ring switched data transmission system," *Bell Syst. Tech. J.*, vol. 50, pp. 2947-2978, Nov. 1971.

[36] A. G. Konheim and B. Meister, "Service in a loop system," *J. ACM*, vol. 19, pp. 92-108, Jan. 1972.

[37] J. D. Spragins, "Loop transmission systems—Mean value analysis," *IEEE Trans. Commun.*, vol. COM-20, Part II, pp. 592-602, June 1972.

[38] B. K. Penney and A. A. Baghdadi, "Survey of computer communications loop networks," *Comput. Commun.*: Part 1, vol. 2, no. 4, pp. 165-180; Part 2, vol. 2, no. 5, pp. 224-241.

[39] F. A. Tobagi, "Multiaccess protocols in packet communications systems," *IEEE Trans. Commun.*, vol. COM-28, pp. 468-489, Apr. 1980.

[40] S. S. Lam, "Multiple access protocols," TR-88, Dep. of Comput. Sci., Univ. of Texas at Austin, to appear in *Computer Communication: Start of the Art and Direction for the Future*, W. Chou, Ed. Englewood Cliffs, NJ: Prentice-Hall.

[41] N. Abramson, "The ALOHA system—Another alternative for computer communications," in *1970 Fall Joint Comput. Conf. AFIPS Conf. Proc.*, vol. 37, pp. 281-285.

[42] ——, "The ALOHA system," *Comput. Commun. Networks*, N. Abramson and F. Kuo, Eds. Englewood Cliffs, NJ: Prentice-Hall.

[43] L. G. Roberts, "ALOHA packet system with and without slots and capture," *Computer Commun. Rev.*, vol. 5, pp. 28-42, Apr. 1975.

[44] J. F. Hayes and D. N. Sherman, "A study of data multiplexing techniques and delay performance," *Bell Syst. Tech. J.*, vol. 51 pp. 1985-2011, Nov. 1972.

[45] A. B. Carleial and M. E. Hellman, "Bistable behaviour of ALOHA-type systems," *IEEE Trans. Commun.*, vol. COM-23, pp. 401-410, Apr. 1975.

[46] S. S. Lam and L. Kleinrock, "Packet switching in a multiaccess broadcast channel: Performance evaluation," vol. COM-23, pp. 410-423, Apr. 1975.

[47] F. A. Tobagi and K. Kleinrock, "Packet switching in radio channels," *IEEE Trans. Commun.*: Part I: Carrier Sense Multiple Access Modes and Their Throughput Delay Characteristics, vol. COM-23, pp. 1400-1416, Dec. 1975; Part III: Polling and (Dynamic) Split Channel Reservation Multiple Access, vol. 2, COM-24, pp. 832-845, Aug. 1976.

[48] L. Kleinrock, "On resource sharing in a distributed communications environment," *IEEE Commun. Mag.*, vol. 17, pp. 27-34, Jan. 1979.

[49] J. F. Hayes, "An adaptive technique for local distribution," *IEEE Trans. Commun.*, vol. COM-26, pp. 1178-1186, Aug. 1978.

[50] J. Capetanakis, "Tree algorithms for packet broadcast channels," *IEEE Trans. Inform. Theory*, vol. IT-25, pp. 505-515, Sept. 1979.

[51] A. Grami and J. F. Hayes, "Delay performance of adaptive local distribution," in *Proc. ICC '80*, Seattle, WA, pp. 39.4.1-39.4.5, June 1980.

[52] N. Pippenger, "Bounds on the performance of protocols for a multiple access broadcast channel," Report RC 7742, Math Science Dep., IBM Thomas J. Watson Research Center, Yorktown Heights, NY, June 1979.

[53] P. A. Humblet and J. Mosely, "Efficient accessing of a multiaccess channel," presented at the IEEE Conf. Decision Contr., Albuquerque, NM, Dec. 1980.

[54] C. Meubus and M. Kaplan, "Protocols for multiaccess packet satellite communication," in *Proc. NTC '79*, Washington, DC, Dec. 1979, pp. 11.4.1-11.4.5.

[55] E. P. Gundjohnsen *et al.*, "On adaptive polling technique for computer communication networks," in *Proc. ICC '80*, Seattle, WA, June 1980, pp. 13.3.1-13.3.5.

[56] L. Kleinrock and Y. Yemini, "An optimal adaptive scheme for multiple access broadcast communication," presented at the ICC '78, Toronto, Ont., Canada, June 1978.

Jeremiah F. Hayes received the B.E.E. degree from Manhattan College, New York, NY, in 1956. He received the M.S. degree in mathematics from New York University, New York, NY, in 1961, and the Ph.D. degree in electrical engineering from the University of California, Berkeley, CA, in 1966.

From 1956 to 1960 he was a Member of the Technical Staff at Bell Laboratories, Murray Hill, NJ. He worked at the Columbia University Electronics Research Laboratories, New York, NY, from 1960 to 1962. In the interval 1966 to 1969 he was a member of the faculty at Purdue University, Lafayette, IN. During the summer of 1967 he was employed by the Jet Propulsion Laboratory, Pasadena, CA. From 1969 to 1978 he was a member of the Technical Staff at Bell Laboratories, Holmdel, NJ. Since September 1978 he has been a member of the Electrical Engineering Department at McGill University, Montreal, P.Q., Canada.

Professor Hayes is a Senior Member of the IEEE. He is currently the Editor for Computer Communication of the IEEE TRANSACTIONS ON COMMUNICATIONS. His research interest is primarily in the area of computer communications.

THE PRESENT STATUS AND FUTURE TRENDS IN COMPUTER/COMMUNICATION TECHNOLOGY *

A. G. Fraser

Abstract   This paper[1] reviews some aspects of computer communication network design in the light of recent research. The paper first concentrates on local distribution systems that use demand shared transmission lines. Then there is a discussion of communication protocol and the impact that it has on network performance. Finally there is some discussion of issues that face the designer of a general-purpose data network.

I. INTRODUCTION

It is not difficult to observe the rapidly growing number of private and mutually incompatible data networks [1]. There are networks for credit-checking, for banking, for hotel reservations, for time-sharing services, for local government, and for monitoring continuously running equipment of many types. We may come to regret this diversity. In the modern world no industry is independent of all others and, as more processes become automated, the need for communication between subsystems will become more pressing. Already there are connections between the various airline reservation systems for such purposes as multi-carrier flight reservations and baggage claim. Some airlines have links to hotel-booking and car-rental systems. An increasing number of large corporations and banks have networks of their own and it seems very likely that these will become interconnected at least for the electronic transfer of payments. The trend can only continue and it poses a serious question. Is it practical to build a single network that can handle the great variety of terminal types, computers, and transmission systems that now characterize computer communications? And, if practical, what can current research tell us about the form that it can take?

Some initial steps have already been taken. While private companies have been building their own networks, the common carriers have been installing long-haul digital transmission facilities [2], [3]. These synchronous systems, by using regenerative repeaters, give much lower error rates than the circuits presently used for voice communication. Of course digital transmission is already used in the voice network but it carries data much less efficiently than the new facilities. When data are sent over the voice network they are transmitted by modulating an analog signal in a modem and, if digital transmission is subsequently used, the analog signal is then digitized. The result is that a 9.6-kbits/s data stream gets converted into an analog signal with nominal 4-kHz bandwidth and that, when digitized, becomes a 64-kbits/s bit stream. Obviously, direct transmission of binary data over a digital circuit is more efficient than this.

However, there is some indication that making best use of available bandwidth is not the major problem in making a computer network. The cost of connecting a terminal to its nearest switching machine is a much more pressing issue. The equipment used to make that connection, called the local distribution system, involves costs that are multiplied by the number of subscribing terminals. Furthermore, an increasing part of that cost is for manpower that is difficult to replace by automation. Equally important and difficult are the problems of control that arise in computer communication systems. The principle problem is not that some required functions are necessarily expensive to perform but that we understand so little of the task that it is difficult to be

[1]The text of a talk given at a meeting of the Japanese Electronic Industries Development Association, Tokyo, Japan in October 1975.
The author is with Bell Laboratories, Murray Hill, NJ 07974.

*Reprinted from *Communications Society*, September 1976, Vol. 14, No. 5, pp. 10–19 and 27.

certain that we are doing the right thing. By providing the wrong facilities we may invest in a network that eventually turns out to be as much a liability as it is an asset.

For these reasons this review of computer communications focuses first on research that has possible application in a local distribution system, and then on the role of communication protocols in packet-switched systems. The paper ends with some speculation on the prospects for new technological advances towards achieving a shared data network.

II. LOCAL DISTRIBUTION SYSTEMS

One way in which one can reduce local distribution costs is to minimize the number of transmitters and receivers required to connect N terminals to a switching center. Polling is one example of such a technique. Consider, if all terminals had a separate transmission line to the switching center there would be one transmitter and one receiver at each end of each line as shown in Fig. 1(a). A total of 2N transmitters and 2N receivers would be required. In a polling system the terminals each have one transmitter and one receiver but they use a common line and the switch has only one transmitter and one receiver for that line [Fig. 1(b)]. The total is therefore N+1 transmitters and N+1 receivers.

There are, of course, many variations of the polling technique, but they can all be typified as follows. Each

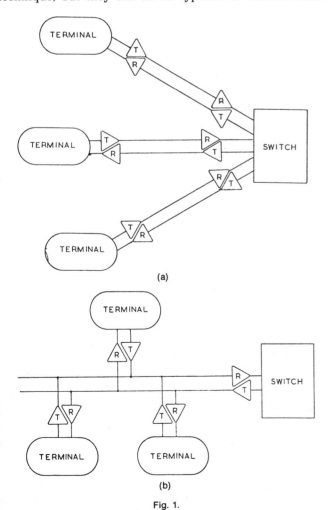

Fig. 1.

terminal contains a buffer in which to queue data for transmission to the central switch. At regular intervals the switch polls each terminal by transmitting the terminal's address. The terminal is expected to respond by transmitting any data that it has ready for transmission. In this way each terminal gets a chance to use the transmission line, and the frequency of use is determined by the switch. A problem with the scheme is that much transmission line time is taken up by polling terminals that turn out not to have any data ready for transmission.

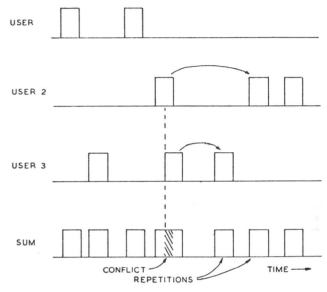

Fig. 2. Saturation load = 18 percent.

The ALOHA Technique

In 1969 Abramson [4] described an alternative to polling that still uses a single shared transmission path. (Abramson was in fact concerned with the use of a radio channel rather than a transmission line but the logic of what he did could be applied to either medium.) The technique is used in the ALOHA system at the University of Hawaii. Like a polling system, every terminal keeps a buffer with data queued for transmission. But unlike a polling system, the central switch does not attempt to coordinate the terminals (see Fig. 2). As a result several terminals may transmit at once. Of course, if there is not much traffic, the chance of two terminals transmitting at once will not be great, but as traffic levels increase, so does the chance of a collision.

Abramson gave a simple analysis of the technique and showed that the channel becomes saturated when the offered load is about 18 percent of that which the channel could carry if there were no sharing. When the load is very light the channel is idle most of the time and collisions are infrequent. As load increases the channel becomes more busy and collisions occur more frequently. Collisions cause retransmissions and the retransmitted packets cause a further increase in channel traffic with a consequent increase in the collision rate. The situation can be improved somewhat by arranging that all terminals synchronize to a common clock [5] (Fig. 3). Time is divided into time slots equal in length to the time to transmit one packet. Packets are transmitted in these time slots. The result is to raise the channel saturation point from an 18 percent load to a 36 percent load.

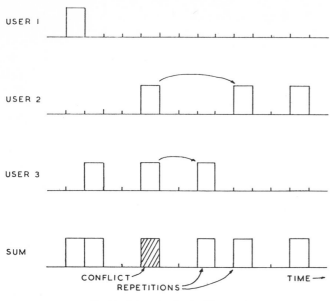

Fig. 3.   Saturation load = 36 percent.

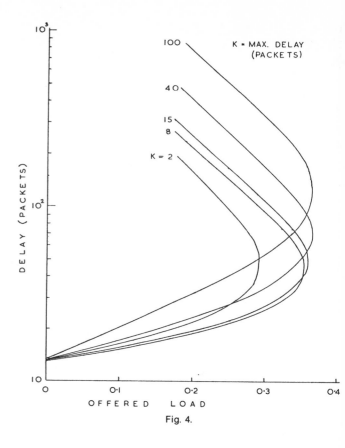

Fig. 4.

Each packet carries a checksum. When the switch receives a packet, the checksum is verified and, if good, an acknowledgment is transmitted back to the terminal that sent the packet. Packets with bad checksums are assumed to result from collisions when two terminals transmit simultaneously. They are ignored. When a terminal has a packet to send it transmits it immediately without regard to the data that are already being broadcast by other terminals. If an acknowledgment is not received, the terminal waits a randomly chosen length of time and then transmits again. The system assumes that if two terminals should transmit at the same time both transmissions will be received erroneously and so both terminals will have to retransmit. By requiring a random delay before retransmitting, the terminals reduce the chance of further conflict.

The expected delay for a message passing through an ALOHA channel depends upon how long it takes before a packet involved in a collision gets retransmitted. That delay has two components. A constant part of the delay is the time taken before the transmitting terminal knows that its transmission failed. The variable part is the random delay introduced by a terminal in order to minimize the chance of subsequent collision. To minimize the constant part of the delay one need not wait, as Abramson suggested, for a returning acknowledgment but the sending terminal could look at the transmitted signal and determine directly when conflict has occurred. To minimize the variable part of the delay is to increase the chance of a subsequent collision and so reduce the load at which the channel becomes saturated.

Suppose that the retransmission delay is chosen randomly to be between 1 and K packet transmission times. The relationship between delay, offered load, and K has been estimated by Kleinrock and Lam [6], and Fig. 4 shows their results for a 50 kbits/s satellite channel carrying 1125-bit packets. (It takes a radio signal about 0.27 seconds to travel from earth to a satellite and back, so that a transmitter cannot detect that a clash has occurred until that period has elapsed.) One can see that the optimum K, in this case at least, lies between 8 and 15 packet times. But we can also see that there are two values for delay at a given level of offered load. The system is apparently bistable.

The problem is that, although the average load may be less than 36 percent of the line's capacity there will be fluctuations when the instantaneous load exceeds that value. Temporarily, there will be heavy traffic on the channel. Clashes will be more frequent and so the traffic on the channel will become heavier. And so the vicious cycle proceeds, bringing the system to a standstill. For a given statistical distribution of message arrivals there is a relationship between the number of users and the value of K which will cause the system to become unstable. For example, if the system whose behavior is shown in Fig. 4 is operated with $K = 10$, and if its users each contribute a load of about 0.11 percent, the system will be stable providing that there are no more than 110 users. Actually, if the number of users is increased slightly above this number, the channel will operate satisfactorily for a predictable length of time before a temporary fluctuation in the load causes it to seize up. For example, with 220 users the channel will fail approximately once in every two days.

An ALOHA system uses essentially the same transmission line layout as a polling system. In the ALOHA system bandwidth is wasted as the result of clashes between competing terminals. In a polling system bandwidth is wasted by polling terminals that have nothing to send. A compromise is possible. Suppose that we add one more transmission path, a "reservation channel," to a polling system. It is provided to allow the terminals to tell the central machine when they have something to transmit, which they do by transmitting their address using an ALOHA technique. The central machine now only polls those terminals that have something to transmit. Polling with a reservation channel is more efficient than a pure ALOHA system because it is only the reservation channel that must operate with less than 36 percent efficiency and that channel, having only addresses to carry, need not have a large bandwidth. It is more efficient than a pure polling scheme if fewer than 1/6 of the terminals have anything to transmit when polled. A more sophisticated transmission system based upon the reservation-channel technique has been described by Roberts [7] in connection with satellite communications.

Data Loops

The ALOHA and polling techniques lose performance because of the need to allow for signal distortion introduced by a transmission line. In general, when a transmitter sends a signal to a receiver the receiver must be adjusted, or must automatically adapt, to the distortions introduced by the line. When the receiver must listen to a series of different transmitters time must be allowed between each transmission for the receiver to adapt to the new transmitter's signals. Thus there is always a necessary delay at the start of packet transmission in ALOHA and polling systems.

In 1969 Newhall and Farmer [8] described a technique in which these delays do not occur and yet their system uses the same number of receivers and transmitters as a polling system. Newhall and Farmer connected their terminals into a loop so that the output of one terminal fed the input of the next and the output of the last fed the input of the first (Fig. 5). (Actually they were not so much concerned with local distribution from a switching center as with providing a symmetric means of communication between all pairs of terminals on the loop.) Under this arrangement data can be passed in either direction between any two terminals provided that intervening terminals pass the information around the loop without alteration. For this purpose it is convenient to install special hardware, a "node," at the point where a terminal connects to the transmission lines. The node serves to regulate a terminal's direct access to the loop and to bypass the terminal when other conversations are in progress.

In the scheme developed by Newhall and Farmer, terminals transmitted messages to one another. Transmission on the loop itself was by means of a bipolar signal in which a single binary digit was represented by a pair of pulses with opposite polarities. Pulse strings that violated this format were used to denote start of message (SOM) and end of message (EOM). The message format is shown in Fig. 6. The messages are variable length and have the source and destination addresses in the first 12 bits. A terminal expecting to receive data scans the line for a passing message having the terminal's address in the destination address field. A terminal wishing to transmit must await its turn. Following the EOM symbol is a single bit, the "token." When a terminal completes sending a message it affixes a "one" token and passes it on to the next node on the loop. If that node has no data to transmit the token is passed on unaltered, but if there are data to transmit, the node changes the token to a "zero" and then proceeds to send its message.

Several schemes have since been used for controlling the sharing of loops (9)-(11), (13). One that we have used in my laboratory treats the loop like a conveyor belt with fixed size slots for packets of data (see Fig. 7). The network in which the loop is used is called "Spider" [10]. It connects together a number of small computers. Data are transmitted and received in packets each bearing a seven-bit address. Each packet occupies one time slot on the conveyor belt and empty slots have zero in the address position. When a computer has data to transmit, its node inserts its own address in the packet address field and places the packet in the next empty slot on the transmission line. The packet is then carried to the switch. The packet address then tells the

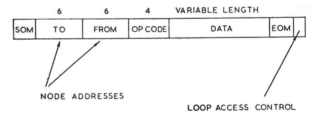

Fig. 6. Newhall/Farmer packet structure.

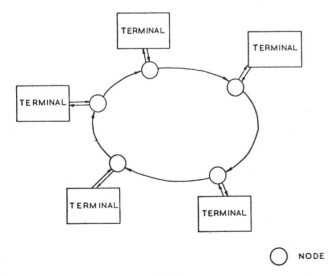

Fig. 5.

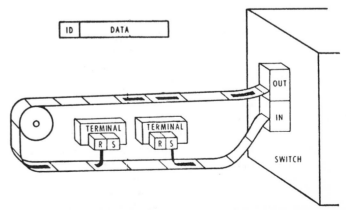

Fig. 7. T1 carrier as a data packet conveyor belt (10^6 bits/s effective dat rate).

switch where the packet came from, and by referring to a route table, established by a previous call setup procedure, the switch decides to which computer the packet must be sent. The address of the destination computer is then written in the packet and it is sent out over the loop. The receiving computer's node recognizes its address in the packet, reads the packet from the line, and leaves the slot empty.

Loops do not have to be used in a demand-shared manner. The IBM 2790 [11] uses synchronous time-division multiplexing on a loop that connects several terminals to a central machine. Each terminal is assigned a time slot and must hold its data until the proper time slot arrives. If the terminal has no data when its time to transmit comes round, the time slot goes unused. In general that scheme requires less complex control circuitry in each terminal, but uses the transmission line bandwidth less efficiently than the Farmer/Newhall and Spider schemes.

Hayes has published a queueing analysis [12] in which synchronous and asynchronous data loops are compared. He used the Spider system as a model and compared it with a synchronous system that uses the same packet size and line speed. He assumes that terminals generate messages at random and that these are queued up in the terminal until such time as they can be transmitted. Typical results are shown in Fig. 8. The delay, in packet times, is the time between a message being generated at a terminal and the last bit of the message being transmitted on the loop. The load is the total rate of information generated by all terminals expressed as a fraction of the capacity of the transmission line. Clearly, when the load is very small the line is usually available for asynchronous transmission and such transmission can proceed at the raw speed of the line. In the synchronous case transmission always proceeds at $1/N$ of the raw line speed where N is the number of terminals sharing the line. At high load levels both systems give long delays because the frequent bursts of message arrivals must be evened out over a long period and the queues are therefore long.

Loops are attractive because they give better line utilization than the ALOHA and polling systems. They have one major problem: they are vulnerable to mischief and malfunction in ways that the other systems are not. The problem is that one terminal node carries the traffic for many other terminals. There is a risk that a malfunction at one user site might deny service at many other sites. To overcome this last problem various schemes have been suggested to provide automatic error detection and loop reconfiguration. They all suffer from essentially the same problem. Since many users are inconvenienced when one part of a loop fails the loop must be reconfigured quickly to exclude the faulty component. That means quick and automatic diagnosis of the problem, and that is not easy.

III. PACKET SWITCHES

Pierce [13] has suggested that a network be made entirely out of interconnected loops (Fig. 9). Between one loop and another, he would install a switching machine much like a packet switch. However, his would be a much simpler device than the packet switches that are in use in some

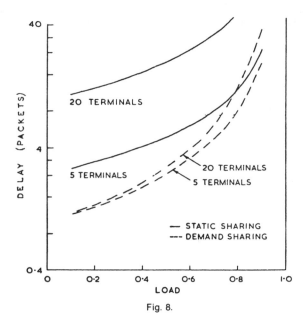

Fig. 8.

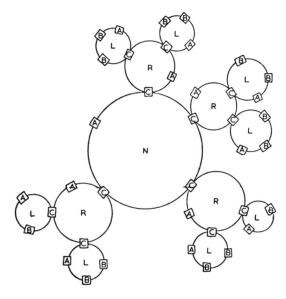

A	LOOP TIMING	L	LOCAL LOOP
B	TERMINAL ACCESS NODE	R	REGIONAL LOOP
C	LOOP INTERCONNECTION	N	NATIONAL LOOP

Fig. 9. Pierce loop system.

networks today. It would be concerned solely with switching, whereas current packet switches tend to include extra facilities such as automatic error control and flow control.

I shall use the ARPA network [14]-[17] as an example of the current trend in packet switching.

The ARPA network provides error control by automatically retransmitting packets that fail to be transmitted successfully. It provides flow control by limiting the size and number of messages that can be in the network at any one time. The mechanics of this rely upon a hierarchy of communication protocols [14], [15].

Level-0 Protocol: Suppose that computer *A* is sending messages to computer B (Fig. 10). A is connected to switch Sa and B to switch Sb. Between Sa and Sb are other switches S1, S2, etc. A message generated by A is passed to Sa where it is split into packets. The packets are routed through S1, S2, etc., until they reach Sb. (I shall discuss packet routing later.) At Sb the packets are reassembled into a message and the message is then delivered to B. The transfer of a packet from one switch to another proceeds as follows. Switch S1, say, has the packet in its memory. It gives the packet a sequence number and checksum and sends it to S2. S2 is expected to acknowledge receipt of the packet and S1 will then discard its copy. If S1 does not get an acknowledgment in a certain time, it retransmits the packet. There are several reasons why an acknowledgment may not reach S1. The packet may have been corrupted during transmission, or the acknowledgment itself may have been corrupted. Alternatively, S2 may not have enough storage to hold the packet or, because it is too congested, may have decided to ignore the packet temporarily. In any event, once S2 has acknowledged the packet, it is S2's responsibility to make sure that the packet does not get lost.

Level-1 Protocol: What I have just described, very briefly, is the lowest level communication protocol of the ARPA network. The next level of protocol is between Sa and Sb. It operates in terms of messages. Each message arriving at Sa from A is transmitted to Sb where it is delivered to B and at the same time an acknowledgment message is sent back to Sa. Flow control at this level is done in two ways. First, Sa cannot send more than four messages at a time to Sb. After sending the fourth such message Sa must wait for an acknowledgment from Sb before sending any more. Second, Sa must request message storage space in Sb before sending a long message to Sb. (A long message is one that occupies more than one packet.) Sb will respond to a request for space by returning an allocation message to Sa whereupon Sa can transmit its long message. (I shall discuss this again later.) When Sa receives a message from A, a software checksum is appended and is carried with the message for the rest of its journey to Sb. The problem is that the checksum used in the lowest level protocol between switches does not check that the switches themselves are working properly; it only checks the transmission lines.

Level-2 Protocol: The next level of protocol is between A and B where virtual circuits are established. Messages simply become the vehicles for carrying data and control information on these circuits. Each computer can support 256 different virtual circuits at one time, and each is identified by an 8-bit "socket" number. When a connection is established between A and B, a socket belonging to A is connected to a socket belonging to B. To establish such a connection, A and B must exchange messages called "requests for connection." (Actually, connections are usually made in pairs because a single connection is only a simplex path.) Once the connection is established, data can be transmitted between the computers but only in accordance with a flow control mechanism that operates independently on each connection. If A is to send data to B, B must first send an allocation to A and that allocation is for a certain number of messages and a certain amount of data. A must not send more of either than it has been specifically allocated.

So far I have described the lower three levels of protocol. There are others [16], [17] but typically the hierarchy does not get more than about five deep.

Data formats used by the ARPA network are illustrated in Fig. 11. At each protocol level, we find that there are both data and control to be transmitted. The level-0 protocol uses 48 bits for frame alignment and 56 bits for control in every packet. That leaves 1088 bits for data and higher level protocols. The level-1 protocol takes a further 80 bits in each data packet, but also generates control packets that carry no data at all. The level-2 protocol uses 40 bits for control purposes in each data message, and it too generates messages that carry no data. And so it goes for each level of protocol. Measurements made during one week in 1974 [18] show that, on average, 6.7 percent of the line capacity was being used and, of that, only 0.6 percent was user data. Figures taken with so little traffic can be misleading unless used with care. If one assumes that the profile of user behavior remains constant, but that usage builds up to give a traffic intensity of 75 percent, one gets the following forecast of how the lines

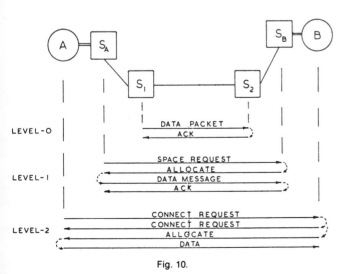

Fig. 10.

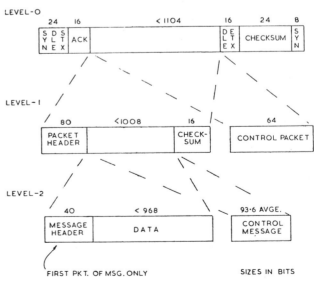

Fig. 11.

would be used in a fully loaded network (see Fig. 12). (Anything higher than 75 percent traffic intensity would result in undesirably long queues.) About 26 percent of line capacity would be used by the level-0 protocol, 18 percent by the level-1 protocol, and 9 percent by the level-2 protocol. There is a small amount (4 percent) of background traffic for network management, and the remaining 18 percent would be user data. If the users were to change their pattern of behavior to make the best possible use of the network line, they could use up to about 59 percent of the available line capacity. If they were all to operate terminals that generate, on the average, 12 characters per message (typical for users of time-sharing systems (19)), the figure would fall to 14 percent.

There is no doubt that some of this overhead is required, but it is plain that we have a lot to learn about protocol design and how to efficiently control computer communications.

The Importance of Delay

Each one of the protocols described above involves a handshake procedure. The level-0 protocol requires that acknowledgments be transmitted for each packet received. At level-1 there are acknowledgments and space allocations. At level-2 space allocation messages must be sent from receiver to sender before data can flow. Handshakes are apparently typical of the control procedures used at all protocol levels. Consider, for example, a file transfer in the ARPA network. I estimate that it involves at least 28 end-to-end handshakes. At level-1 there is an end-to-end handshake for each message transmitted. At level-2 there is a protocol for creating and taking down a virtual circuit, and each of these actions requires an exchange of messages. At higher protocol levels there are handshakes when the user logs in, provides his password, establishes the parameters that are to govern the data transfer, and commands the transfer to take place. Therefore, although a user may think of a file transfer as a simple unidirectional data flow, it is in reality a quite extensive exchange of messages. The time required for a file transfer depends upon the speed with which the required handshakes can be completed.

In my experience, a network user estimates the performance of the system by dividing the time that it takes to transmit a file by the size of the file. A network with very high-speed lines will be regarded as slow if queueing delays within the network are long and handshaking is therefore slow. For this reason I believe that, as users become more sophisticated in their use of computer networks, the demand will grow for a network that has low queueing delays.

Static and Dynamic Routing

Many of the design features of a packet-switched network stem from one decision: whether the packets of one conversation are allowed to get out of sequence. In the telephone system, the user dials up a connection before he starts to talk. Part of the procedure for setting up the connection is to choose a route passing from the person making the call, through several switching machines, to the person being called. All information transmitted as part of the conversation then follows that route. The same procedure could be used in a packet-switched system and I shall call it a "fixed route" strategy. Another possibility is to choose a route separately for each packet that enters the network. Suppose that terminal A transmits a message consisting of several packets. The first packet enters switching machine Sa. That machine examines tables telling how busy the other parts of the network are and which transmission lines are inoperative. From that information it chooses the next switch which the packet should visit. The next switch does the same until eventually the packet reaches switch Sb. The second packet of the message from A undergoes the same procedure and eventually reaches Sb, but there is no guarantee that the second packet will follow the same route as the first. I call this a "dynamic route" strategy.

A number of network designers seem to have followed the ARPA lead and have used a dynamic route strategy. Thus we can now examine how successful that strategy has been. It has problems. Most of the difficulties arise when the packets transmitted from A arrive in a different sequence at B. That is a distinct possibility with a dynamic route strategy since there is nothing to prevent the first packet from being assigned a slow route while the second goes by a faster route.

It is impractical, in the space available, to show you how complex life can get when packets get out of sequence, so we must be content with one simple example (see [20] for other examples). The ARPA network promises its users that the data bytes of a message will leave the network in the sequence that they entered the network. Suppose that packet 2 reaches B's switch before packet 1. The switch must store 2 until 1 arrives. In general B's switch must be prepared to collect all the packets of one message before delivering any to B. That requires storage in the switch. Suppose that the storage to be used for that purpose is shared among the many users connected to the switch. In that way we take advantage of the bursty nature of data flow and need not provide as much storage as would be required to hold all messages for all terminals. But now suppose that, on a particularly busy occasion, the 2 packets of many messages arrive at the switch. Perhaps they will

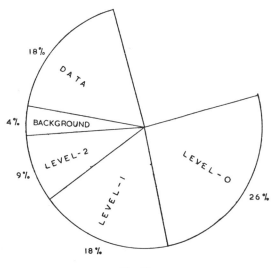

Fig. 12.

use up so much space that there is no room in which to put any packet 1. Thus the switch must turn away the very packets that will allow it to complete the delivery of data which are now blocking up its store. No further traffic can now pass through this switch. To overcome this problem the ARPA network has introduced an extra handshake procedure into the level-1 protocol. Before a multipacket message can be sent from A to B, A's switch must ask for, and receive, an allocation of the necessary storage space in B's switch. To do that adds an extra round-trip delay to the transmission delay for a message. So, to reduce the average additional delay, another bit of control procedure is added. After B's switch receives one long message from A it automatically allocates storage for another. Now A must tell B if it does not need the extra storage.

When part, or all, of a network gets into a state where no further traffic can flow it is said to be in a "lock up" condition. There is a distinct chance of such a condition arising in many packet-switched networks, even when they preserve the sequence of data for one message. It can arise if the network attempts to control transmission errors separately on each link of the path between source and destination terminals. For example, suppose that switch S1 sends data to S2 but keeps a copy of those data until S2 acknowledges their receipt. Unless precautions are taken, there may come a time when S1 has its memory full of packets for switch S2, while S2 has its memory full of packets for S1. Neither switch has room for more data so each must discard all further inputs. No further progress is made. That, of course, is a simple condition to anticipate and prevent, but more complex situations, involving many switches, can arise and are very difficult to handle. So far as I know, it is not possible to say for sure that any packet-switched network that uses a dynamic route strategy or does switch by switch error control is free from the possibility of lock-up.

For these reasons, I anticipate that a dynamic route strategy will not be a popular choice for future large-scale data networks. For years the common carriers have faced the problems of making a reliable switching machine and have developed the technique of switching in a new circuit when another fails. Those techniques can be applied to data communications and the control problems associated with dynamic routing can be avoided.

IV. FUTURE PROSPECTS

Turning now to the future, let us consider what advances in technology might be looked for in the next few years. I shall consider those aspects of computer/communications technology that will improve the prospects for achieving a single network to serve a wide variety of terminal types and usage patterns. Solutions to the following problems are required.

1) How to minimize the cost of local distribution and yet accommodate, in one network, a wide variety of terminal types and speeds.

2) How to control communications so that there can be full connectivity between devices that talk at different speeds and with different protocols.

3) How to design efficient protocols and how to prove that they work.

4) How to simultaneously obtain low delay and reasonable line utilization.

5) How to control the flow of information in a network so that there is a low probability of data loss due to transmission error or queue overflow.

6) How to reconcile the need for heavy investment, characteristic of communication networks, with the continuing high rate of change in computing technology.

Protocol design problems may turn out to be the most difficult. Our present ability to handle communication protocols reminds me of our abilities with programming language in the 1950's. At that time there was no convenient formalism in which syntactic constructs could be described. Without such a means of expression we were in a poor position to make proofs about the languages we invented and there was little that could be done to automate the process of compiler construction. The breakthrough came in about 1958 with the official description of Algol 60 and its use of the Backus-Naur Form (BNF) to describe the syntax of the language. Today we anxiously await a similar breakthrough for the description and manipulation of communication protocols.

For the experimentalist seeking to make progress on the other problems mentioned above, there are at least two promising possibilities. One is clearly the application of LSI, particularly microprocessors, to switching and control. (I shall say more about that later.) The other is to reevaluate ideas originally conceived in connection with digitized voice communication. Until recently there has been little interchange of ideas between communications engineers and those who specialize in computing science. As a result techniques known in one field have been slow to be applied to the other and there is little exchange of information about problems which are essentially common to the two fields. For example, a great deal is known about the construction of switching machines for digitized voice, yet apparently very little of that knowledge has been applied to data networks. Existing machines switch digital signals in time and space; addressed packets just represent one more dimension in which these machines might work. The computing literature has, in recent years, contained much discussion on the problems of finding a good model for interprocess communication. Even now few operating systems provide good facilities of that type. But signaling systems are the basis of all communication networks and they fulfill essentially the same role. Actually neither industry seems to have the subject under control and each is approaching the problem in a different way.

Let us now speculate a little on how some of the problems and opportunities just mentioned might impact computer communications systems in the next few years.

In order to understand how delay might be minimized, we can start with analytical results obtained by Chu[21]. Fig. 13 shows the relationship between delay and traffic intensity in a single packet switch. There are several curves, one for packets of fixed size and others for variable-size packets with different statistical distributions. (k is the coefficient of variation of packet size.) The main thing to notice about these results is that delay measured in packet transmission times is a constant for a given traffic intensity. Delay increases linearly with packet size and therefore so does the

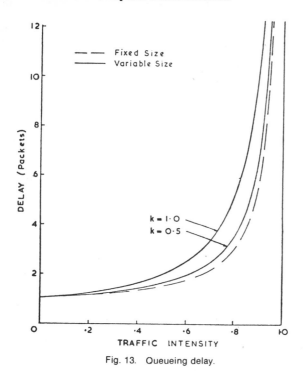

Fig. 13. Queueing delay.

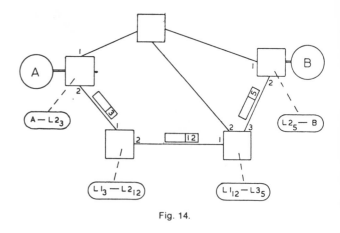

Fig. 14.

amount of queue storage space that must be provided in a switch. In order to obtain minimum delay one should use quite small packets. But small packets could result in low line utilization. In order to estimate how packet size affects line utilization, one can refer to a paper by Hayes and me [22] where it is shown that a good compromise is obtained by choosing a packet in which the data occupy about eight times as much space as the header.

Low delay therefore requires small packets which, in turn, require small packet addresses. To obtain a small packet address in a large network, one cannot afford to put the complete address of a destination in each packet. Instead, one should use an abbreviated addressing scheme (Fig. 14). When a conversation is first established, its route can be chosen, and at the same time the packet addresses to be used on each leg of the journey can be chosen. The address used on a given transmission line need only be sufficient to distinguish between the conversations being carried by that line. (Assigning a packet address to a conversation in a packet-switched system can be like assigning a time slot in a synchronously multiplexed system.) The packet address might be as short as one byte.

A small packet may give rise to minimum delay but it poses a problem in switch design. Most current computer networks use switches constructed out of general-purpose minicomputers. The more high-performance of these transfer packets into and out of memory by means of direct memory access channels. The processor is only called in to process the packets when they have been completely assembled in core. Thus the throughput of a switch is most accurately stated in terms of packets, not bytes, handled per second. If the packets are small, the machine will be able to handle fewer bytes per second.

Small packets will be one reason for exploring alternative mechanisms for packet switching. Other reasons will be to

get more throughput for a given amount of switching hardware, to get better modularity so that small switches can be cheap and yet can grow gracefully as demand dictates, and to provide greater versatility in networks that employ a mixture of packet addressing and time-division techniques. Packet switches of the future may look less like general-purpose computing systems and a lot more like the synchronous machines used for switching digitized voice signals.

Incompatibilities between terminals, and variety in protocols will require the network to perform some processing on transmitted data. The ARPA terminal interface processor (TIP) is an example of what can be done. Terminals whose data are routed through that processor appear to other machines on the network as if they conformed to some standard definition of a terminal. In particular, the TIP does flow control so that a computer talking to a terminal does not have to acquire explicit knowledge of the terminal's speed. Protocol translations of this type are an appropriate application for microprocessors, and a suitable way of providing such services is to give the user the option of routing his conversation through such machines when he "dials" his call. The technique can be extended to other application areas. For example, there is emerging a standard format for credit cards. The data generated by a credit card reader could be interpreted by a service processor, and thereafter routed to the appropriate credit checking agency.

There is apparently a big role for microprocessors in data communications and it comes about because of the complexity that seems inevitably to accompany data communication protocols. By way of illustration, I shall conclude by describing how microprocessors are used in the Spider network [10].

Computers connect to Spider through an "intelligent" terminal interface unit (TIU) (Fig. 15). The transmission line is connected to one side of a TIU and the computer to the other. The computer interface with the TIU is asynchronous: one byte is transferred for each handshake on a pair of control leads. The TIU transmits packets into the network and it is the TIU's job to generate the necessary headers and checksums. Within the network the packets are routed without regard to their content; only the packet address is examined. The destination TIU has the task of checking packet sequence and checksums. If all is well, the destination TIU sends an acknowledgment packet to the transmitting TIU.

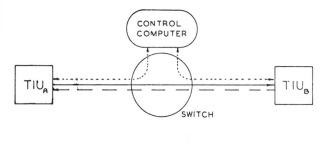

CONTROL SIGNALS
ACKNOWLEDGEMENTS
DATA PACKETS

Fig. 15.

The users see nothing of this and have, in effect, a network with automatic error and flow control.

The crucial element in a TIU is a small computer, called "Fly," which we made for the purpose. It is an 8-bit computer having 16-bit instructions. There are 256 words in its read-only program store and 16 words in its data store. In addition the TIU contains two data buffers each big enough to hold one packet. Since the transmission line is quite fast, 1.544 Mbits/s, Fly must also be fast. It executes one instruction in 200 ns.

It is of course not reasonable to attempt to put all the code required for a practical error control system in such a small machine. Fortunately, most of the code is required for events that occur only very infrequently. We have placed that code in the central switching machine and have arranged that a TIU can call upon the central machine for help when errors occur. The ultimate mechanism provided for this purpose is a means by which the central machine can read any word of Fly's data store.

V. CONCLUSION

At the start of this paper we noted the wide variety of computer and terminal types which a network might have to support. It seems that there will also be variety in local distribution systems. No one distribution system is best for all circumstances.

With such variety, control and signaling become a central design issue for a data network. As yet there is no sign that these matters are even adequately understood. We must hope that a better understanding of protocol design comes soon.

The computer, besides being the source of much of this complexity, will be its solution. We are fortunate that the large-scale integration of digital circuits has already advanced to the point where we can consider putting small processors in communications equipment and terminals. I expect that microprocessors will play a big part in future computer communications systems.

REFERENCES

[1] D. Bernard, "Intercomputer networks: An overview and a bibliography," rep. NITS AD-769-232, May 1973.
[2] Various papers, *Bell Syst. Tech. J.*, vol. 54, May-June 1975.
[3] D. J. Horton and P. G. Bowle, "An overview of DATAROUTE: System and performance," and other papers in *Proc. Int. Conf. Commun.*, IEEE catalogue 75 CHO 859-9-CSCB, June 1974.
[4] N. Abramson, "The ALOHA Sytem—Another alternative for computer communications," in *AFIPS Conf. Proc.*, vol. 37. Montvale, NJ: AFIPS Press, 1970, p. 281.
[5] L. Kleinrock and S. S. Lam, "Packet switching in a slotted ALOHA channel," in *AFIPS Conf. Proc.*, vol. 42. Montvale, NJ: AFIPS Press, 1973, p. 703.
[6] ——, "Packet switching in a multiaccess broadcast channel: Performance evaluation," *IEEE Trans. Commun.*, vol. COM-23, p. 410, Apr. 1975.
[7] L. G. Roberts, "Dynamic allocation of satellite capacity through packet reservation," in *AFIPS Conf. Proc.*, vol. 42. Montvale, NJ: AFIPS Press, 1973, p. 711.
[8] W. D. Farmer and E. E. Newhall, "An experimental distributed switching system to handle bursty computer traffic," in *Proc. ACM Symp. Problems in the Optimization of Data Communications Systems*, Pine Mountain, GA, Oct. 1969.
[9] D. J. Farber and K. Larson, "The structure of a distributed computer system—The communications system," in *Proc. Symp. Computer Networks and Teletraffic.* New York: Polytechnic Inst. of Brooklyn Press, 1972, p. 21.
[10] A. G. Fraser, "Spider—An experimental data communications system," in *Proc. Int. Conf. Commun.*, IEEE catalogue 74 CHO 859-9-CSCB, June 1974.
[11] E. H. Steward, "A loop transmission system," IBM Corp., Research Triangle Park, NC, 1970.
[12] J. F. Hayes, "Performance models of an experimental computer communications network," *Bell Syst. Tech. J.*, vol. 53, Feb. 1974.
[13] J. R. Pierce, "Network for block switching of data," *Bell Syst. Tech. J.*, vol. 51, July-Aug. 1972.
[14] F. E. Heart *et al.*, "The interface message processor for the ARPA computer network," in *AFIPS Conf. Proc.*, vol. 36. Montvale, NJ: AFIPS Press, 1970, p. 551.
[15] C. S. Carr, S. D. Crocker, and V. G. Cerf, "HOST-HOST communication protocol in the ARPA network," in *AFIPS Conf. Proc.*, vol. 36. Montvale, NJ: AFIPS Press, 1970, p. 589.
[16] J. Malman, "Terminal interface message processor, user's guide," NTIS AD-782 172, June 1974.
[17] A. Bjushan, "The file transfer protocol," Advanced Research Projects Agency, ARPA Network Working Group, rep. RFC 354 NIC 10596, July 1972.
[18] L. Kleinrock, W. E. Naylor, and H. Opderbeck, "A study of line overhead in the ARPANET," in *Proc. Nat. Telecommun. Conf.*, 1975.
[19] P. E. Jackson and C. D. Stubbs, "A study of multiaccess computer communications," in *AFIPS Conf. Proc.*, vol. 34. Montvale, NJ: AFIPS Press, p. 491, May 1969.
[20] H. Opderbeck and L. Kleinrock, "The influence of control procedures on the performance of packet-switched networks," in *Proc. Nat. Telecommun. Conf.*, San Diego, CA, 1974.
[21] W. W. Chu, "A study of asynchronous time division multiplexing for time-shared computer systems," in *AFIPS Conf. Proc.*, vol. 35. Montvale, NJ: AFIPS Press, 1969, p. 669.
[22] J. F. Hayes and A. G. Fraser, "Optimum packet size for data communications," *Proc. IEEE*, p. 1397, Oct. 1974.

A. G. Fraser has specialized in computer communications since joining Bell Laboratories in 1969. His work includes the Spider network of computers and a network-oriented file store. Prior to joining Bell Labs, he was at Cambridge University where he wrote the file system for the Atlas 2 computer. Dr. Fraser has a B.Sc. degree in aeronautical engineering from Bristol University and a Ph.D. degree in computing from Cambridge University. He is a member of IEEE, ACM, and the British Computer Society.

Section 4

Transcription

EDITORS' COMMENTS

The six tutorial papers in this section cover analog and digital transmission systems. They summarize a vast body of knowledge central to problems arising in the design and planning of transmission systems. They examine the various factors that affect a modern transmission system, specify their relative significance, and introduce the reader to basic terminologies used in the professional journals.

The first paper, "Communication Technology: 25 Years in Retrospect, Part III, Guided Transmission Systems, 1952-1977," by F. T. Andrews, Jr. traces the twenty-five year history, growth, and development of guided transmission systems from 1952 to 1977. It discusses the main challenges in the early 1950s, i.e., overseas communications, direct distance dialing, data communications for the computer industry, and digitized voice for military applications. The paper includes the impact of solid-state technology and the possibility for integrating digital subscriber transmission with digital local switching. The need for good quality overseas communications that was also economical led to the development of time assigned speech interpolation (TASI) and six generations of ocean cable systems.

The challenges imposed by direct distance dialing resulted in a family of operating systems for transmission maintenance. The need for data transmission motivated improvements in the communications network. Impairments such as nonlinear distortion, impulse noise, phase jitter, and frequency offset (which affected data but not voice) were corrected in existing systems and avoided in new systems. Andrews reviews the extent to which the communications network has penetrated all areas of the civilized world and concludes by discussing the principal challenges for guided transmission in the future.

In the second paper, "Digital Communications—The Silent (R)evolution?," M. R. Aaron traces the history and growth of digital transmission and gives a brief tutorial on the process of digitizing analog signals. He expands on the advantages of digital communications and explains the differences between time division multiplexing (TDM) and frequency division multiplexing (FDM) systems. He goes beyond these to indicate the status of digital communications, showing the significant strides made since the 1950s. The main difficulty in the (r)evolution is the transition from analog to digital and how to manage this change. An excellent view of a nationwide communications network together with the application factors that control digital penetration of the network are enunciated. The paper also discusses the interdependence of digital transmission and switching. Summaries of the statistics for the worldwide use of digital communications by various telecommunications administrations are also given. Finally, Aaron provides a brief glimpse of the future—the ultimate dominance of digital transmission and its impact on new services.

The third paper, entitled, "Digital Transmission Building Blocks," by S. D. Personick, traces the history of T-Carrier systems used in the telephone network and provides an overview of a variety of hardware systems that are developed to provide for the rapidly expanding digital telecommunications network. The architecture of several systems and the specific hardware used to provide digital transmission between telephone buildings, and to interconnect a customer to a telephone office, are considered. Finally, maintenance and protection of these systems, in addition to their cost and the quality of service, are covered by Personick.

A particular type of hardware described in the previous paper is termed the hybrid. It connects a two-wire to a four-wire circuit in a transmission path. Imperfect hybrids give rise to echoes. These echoes degrade the performance of transmission systems, especially satellite circuits. In the fourth paper, "Echo Cancellation in the Telephone Network," S. B. Weinstein explains in detail why echoes are generated and how they are eliminated using echo cancelers. He explains the theory of echo cancelers, and discusses the effects of double talking, phase roll, and nonlinear distortion on the convergence of various echo cancelers. He presents several structures for echo cancelers including structures that can support full-duplex simultaneous two-way data transmission.

The fifth and sixth papers, both authored by S. D. Personick, deal with optical fiber communications. The fifth paper, "Optical Fibers, A New Transmission Medium," is theoretical in nature and analyzes signals, noise, and distortion associated with digital fiber systems. It outlines a mathematical model of the fiber transmission system and shows the differences and similarities between the theories for fiber and metallic cable systems.

The sixth paper, "Fiber Optic Communication, A Technology Coming of Age," deals with the technology of optical communications. It reviews the progress and summarizes the characteristics of commercial and military prototype systems in service. These systems carry voice, data, and video signals, and represent the current advancement in wideband transmission systems.

COMMUNICATIONS TECHNOLOGY: 25 YEARS IN RETROSPECT, PART III, GUIDED TRANSMISSION SYSTEMS, 1952–1977 *

F. T. Andrews, Jr., Fellow, IEEE

Progress made since 1952 within the technical scope of ComSoc's Transmission Systems Committee includes some very impressive accomplishments. This particular Technical Committee of the IEEE Communications Society focuses its attention on communications systems in which the electromagnetic transmission of information in or along conductors is the dominant factor. Twenty-five years ago, this was "wire" transmission, but has since broadened to include waveguide and optical fibers. In reviewing progress in this area during the past 25 years, it is useful to first consider the state of the art in the area of wire transmission in 1952.

At that time, telecommunications was almost completely dominated by wire transmission. Only 11 percent of toll circuit mileage in the Bell System was on point-to-point radio. Continental distances had been spanned first by voice frequency transmission on open wire, and later by carrier frequency transmission on open wire, multipair cable, and coaxial cable. By 1952, the basic principles of carrier transmission had been well established and carrier systems had taken over long distance communications. The division of circuit miles among these media in the Bell System was roughly 17 percent open wire, 41 percent multipair cable, and 31 percent coaxial cable.

Both double sideband and single sideband frequency division carrier systems had been successfully applied to open wire lines and, by 1952, the use of these techniques on multipair cable was also well established. A basic problem which had to be overcome in the use of multipair cable was the control of far-end crosstalk between paralleling systems. The first approach used in North America was the careful balancing out of crosstalk coupling inbalances as in the K carrier system in 1937. This approach was later abandoned in favor of the use of compandors when the N carrier family was introduced by U.S. manufacturers about 1950. In Europe, carrier systems for multipair cable depended on special cable construction, the star quad, together with balancing to overcome the intersystem crosstalk problem.

The coaxial cable had long been recognized as a superior transmission medium for very long distance transmission. Not only was it possible to avoid crosstalk between systems because of the natural shielding of the coaxial structure, but it was also possible to achieve well-equalized wideband channels to support the transmission of video signals. Systems capable of transmitting 600 channels on a coaxial tube had been introduced in 1941.

In 1952, the dominant telephone set in North America was the 300-type. This telephone represented a significant step forward over earlier telephone sets in that the transmission was relatively flat over the band, from 300 to 3000 Hz, and the transmission of speech was quite natural, free of distortion, and easily understood. Furthermore, the improved 500-type set had just been introduced in 1949 to improve the grade of service through greater sensitivity and automatic equalization.

Local loop transmission was largely on multipair cables. Open wire on cross arms was relegated to long rural routes where low attenuation was required. A composite of polyethylene plastic and aluminum had begun to be used as an alternative to lead cable sheath, but the conductor insulation was still paper or paper pulp.

CHALLENGES OF THE 1950's

What, then, were the challenges remaining for the telecommunications engineer concerned with wire transmission in 1952? A few of the most significant forces which shaped the future are the following.

1) Overseas communications.
2) Direct distance dialing.
3) Data communications.

The author is Director of the Loop Transmission Systems Laboratory, Bell Laboratories, Whippany, NJ 07981.

*Reprinted from *IEEE Communications Society Magazine,* January 1978, Vol. 16, No. 1, pp. 4–10.

It is difficult to remember that, in 1952, overseas voice communications was entirely based on high frequency radio, and there were only 1.4 million calls per year between the United States and overseas points. While a great deal of excellent work had been done in understanding and improving radio transmission over long distances, it remained an uncertain and imperfect medium. The numbers of circuits available were far from adequate to support the ever-increasing community of interest between North America and Europe, which had developed a very strong need for communications in the post-World War II era. Even when circuits were available, the quality of transmission was hardly conducive to good communications.

While the advent of direct distance dialing is often thought of in terms of a switching achievement, it had very important transmission requirements as well. Since operators were no longer involved in establishing connections, there was no check on the quality of transmission before the connection was turned over to the customer. The switching machines, in effect, manufactured a connection as an assembly of individual trunks in response to the customer's dialing. Quality control of the overall connection required quality control of each of the separate parts. This requirement for tighter control, together with demand for rapid growth, opened up a whole new era of automated transmission maintenance systems.

In 1952, the dominant method of data communications was with telegraph and teletype systems limited to just a few hundred bits per second. The burgeoning data processing industry generated pressure for much more rapid transmission of digital information. Also, the military had requirements for transmitting speech in digital form so that encryption could be applied. Transmitting signals at speeds of 1000 to 1500 bits/s over a network designed for speech transmission was not a trivial problem.

Over the next 25 years, telecommunications engineers concerned with transmission, together with their colleagues in the switching and data communications fields, were able to meet these challenges.

THE IMPACT OF SOLID-STATE TECHNOLOGY

The innovation with the most fundamental impact on wire transmission was the introduction of solid-state technology. The transistor had been invented in 1947, but had yet to find practical application in communications in 1952. In fact, during the early 1950's, the development of open wire and multipair cable and coaxial carrier systems continued to be based on well-established electron tube technology. The last major dry land transmission systems introduced by the Bell System which were based on electron tube technology were the L3 coaxial cable system and the A4 channel bank in 1954. The switch to solid-state technology took place throughout the telecommunications industry in about the same time frame, with ocean cable systems being the principal exception.

The first major attempt to use transistors in a wire transmission system was for subscriber loop application. Solid-state technology seemed to offer the first real hope of attacking the very high cost of long rural loops. Previous attempts with the electron tubes (for example, the M1 power

line carrier system) were fraught with reliability and maintenance problems.

This first solid-state transmission application was the P carrier system which, in addition to using transistors, pioneered in the use of silicon-aluminum diodes, ferrite cores, and printed circuit cards. Unfortunately, this four-channel system, designed for use on open wire, was still too expensive to find wide application. However, within ten years of the introduction of the P carrier in 1954, a wide range of reliable, low-cost components had become available, which stimulated successful development activity on subscriber carrier throughout the industry.

The Rural Electrification Administration developed a frequency plan to help coordinate the use of local telephone cables by these systems. This plan prescribed 8 kHz to 56 kHz in the direction towards the central office and above 64 kHz towards the subscriber. Today, a large number of systems are available which conform to this plan, in some instances with minor variations to gain an eighth channel. Subscriber carrier systems are well established as an economic alternative to the use of individual cable pairs to serve customers on the long, skinny rural routes which have been most troublesome to telephone companies.

The systems in widest use have been based on the use of discrete transistors and operational amplifiers. Today, custom integrated circuits are finding their application in subscriber carrier systems to perform complete functions like companding and modulation. The first use of custom IC's was in single channel carrier systems, simple systems which derive a second subscriber line from an existing subscriber line for temporary relief of cable congestion in urban/suburban situations. However, extension of custom IC's into the multichannel carrier field is expected to occur rapidly over the next several years.

Soon after solid-state technology was being tried in the area of subscriber carrier, it was also being designed into new versions of the N carrier family of systems for exchange trunk transmission, and with more immediate success. In 1963, N2 carrier was introduced by Western Electric to replace the N1 system of the 1940's. The design was based on

Computers have revolutionized transmission maintenance of the whole telecommunications network. Shown in the photograph on left are the "tub" files which have served for many years to contain maintenance records of each customer line served by a repair service bureau. On the right is a computer terminal at which more complete information about customers' lines can be retrieved from a mechanized file, and basic maintenance functions can be performed, including informed interaction with the customer.

The great simplification possible through the use of integrated circuit technology in analog carrier systems is illustrated by the contrast between the 1963 vintage N2 compandor unit on the left and the components performing the same basic function in today's systems on the right. The small black rectangle indicates the approximate area occupied by the active semi-conductors involved.

the discrete transistors available at the time and achieved important space and power savings.

Shortly after N2 came N3, which used single sideband modulation to get 24 instead of 12 channels in the same 36 to 268 kHz frequency band occupied by N2. This system was a replacement for the earlier ON electron tube system and was attractive over longer distances where line costs, rather than terminal costs, dominated.

While N2 and N3 did incorporate some system refinements, the real advances were in circuit design made possible by solid-state technology. The basic system principles had been well-established and proven in the earlier electron tube versions.

A similar kind of evolution took place with regard to the terminals for broadband systems, both coaxial cable and microwave radio. For example, the A4 channel bank, introduced in 1954 by Western Electric, was still based on electron tube technology. However, by 1961, it had been superseded by the A5 channel bank which used solid-state technology to achieve nearly 4-to-1 size reduction, reduced power consumption, and improved net loss stability. The A6 channel bank, introduced in 1973, further exploited the possibilities of solid-state technology and introduced the monolithic crystal filter in place of the quartz crystal filters that had been used theretofore.

A similar kind of transition was taking place in the LMX equipment used to stack the 60 to 108 kHz 12-channel groups from the A channel banks for transmission on broadband systems. The LMX 2 system, introduced in 1962 by Western Electric, used solid-state technology and automatic regulation. This regulation greatly improved the net loss stability, making it possible to meet the strict requirements imposed by direct distance dialing without regular adjustments by the telephone craft people. Edge pilots were introduced a few years later, which left most of the band clear for the transmission of wideband data signals within a group band. More recently, the stacking arrangements have been modified and extended for more efficient multiplexing of

channels on the larger capacity coaxial cable and microwave radio systems which have now become available. The latest generation of coaxial cable systems provides 13 200 channels on a pair of coaxial tubes using sophisticated solid-state repeaters at 1-mi intervals.

THE BEGINNING OF A DIGITAL ERA

Paralleling the improvements in analog carrier systems, solid-state technology made practical a whole new era of digital systems. While the concepts of PCM transmission had been available for some years, it was not until the middle 1950's that it became apparent that PCM systems based on solid-state technology were a practical answer to reducing the cost of exchange trunks. Studies showed that multiplexing functions could be performed more economically using time division digital techniques than with frequency division analog techniques. The output of such multiplexers were digital bit streams which required wider bandwidths than analog carrier signals, but which were amenable to complete regeneration by close-spaced repeaters. It was the transistor which made practical the regenerative repeaters to combat the cumulative effects of imperfect transmission media.

Serious development of a PCM system for exchange trunks began in 1956, and six years later, the T carrier, a combination of the T1 repeatered line and D1 channel bank, was introduced. The system provided 24 channels over a 1.544 Mbit digital line with repeaters spaced at about 6000 ft on 22 gauge cable pairs. A very important feature of this system was the rugged bipolar digital transmission format which enabled it to work on existing cable pairs in the telephone plant with a minimum of conditioning and grooming.

Because of its low cost and suitability to the exchange plant cable environment, the use of T carrier is growing at an astounding rate in North America, about 33 percent per year. It became clear throughout the world that digital transmission would have an important role in the evolution of the telephone network, and a great deal of research and development work began.

The 7-bit PCM encoding and the specific companding law used in the D1 bank, while acceptable for exchange trunks,

The use of digital transmission systems is growing at the rate of about 33 percent per year in North America. Two generations of PCM channel banks are shown in the photographs. On the left is Western Electric's 24-channel D3 bank introduced in 1972, and on the right the 48-channel D4 bank introduced in 1977. Current PCM systems make extensive use of integrated circuits to achieve smaller size, better partitioning of functions, and lower costs.

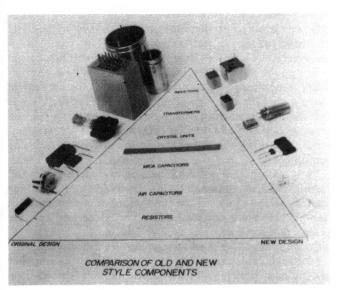

COMPARISON OF OLD AND NEW
STYLE COMPONENTS

New components had a major effect on the design of transmission terminals during the past 25 years. Group and supergroup terminals introduced in the early 1960's achieved a 10 to 1 size reduction, along with the greatly reduced power consumption possible with transistor technology.

Digital transmission systems completely regenerate the information-bearing pulse patterns to avoid the cumulative effects of noise and pulse distortion, and are therefore well suited to the exchange plant environment. The repeater shown in the photograph is the Western Electric 239-type and is representative of the 24-channel, two-way repeaters available today which use the original T1 bipolar pulse format. Very recently, other formats have been introduced to double the rate of transmission.

were not judged satisfactory for more universal use of PCM in the telecommunications network. International agreement was reached on 8 kHz sampling and 8-bit encoding, but two different standards evolved for the basic companding law. In North America, a companding law known as mu 255* was incorporated in D2 and subsequent channel banks developed by the industry. In Europe, the choice was A law companding which differed from mu 255 primarily at low levels. Within the CCITT, it was agreed to accept dual standards for an interim period with the understanding that conversions could be made at interfaces at international boundaries as necessary.

There are also differences in the number of channels used in North America and Europe, 24 versus 30, and in the methods of signaling and framing. This requires different system designs for different areas of the world, depending on whether the North American or European recommendation is being followed. The different numbers of channels leads to a different hierarchy for combining the basic PCM multiplexes on higher speed lines. This further complicates life for manufacturers seeking a world market for their PCM systems.

Since the first widely used PCM systems in 1962, research and development have reduced costs and improved performance through the application of more advanced solid-state technology. Within the past year, fourth generation PCM systems have been introduced which make extensive use of custom integrated circuits to perform on a single chip, functions which required entire plug-ins in the original first generation systems. It is expected that this kind of evolution will continue as device technology permits a larger and larger scale of integration.

Another factor which will affect the growth of digital transmission is the introduction of digital switching. All signals entering the digital domain of a digital switch must

be converted to digital format. Whether this is done immediately adjacent to the switch, or at some remote location, has little effect on the cost of these conversion functions. The net result is a great impetus towards the use of digital transmission for interexchange trunks. This effect has already been well demonstrated in the installation of No. 4 ESS machines for toll switching, and a similar effect can be expected as local digital switches, now offered by several manufacturers, are introduced into the exchange plant.

The introduction of local digital switching offers the possibility of very low cost digital subscriber carrier systems. Since all signals entering the realm of the digital switch must be digitized and time-multiplexed anyway, the added cost of gaining pairs through multiplexing will be that of supporting these functions in the vicinity of the customer, rather than within the central office, plus the cost of digital line transmission to the central office.

It has already been clearly established that digital transmission at T carrier rates is quite practical in subscriber cables. During the early 1970's, several carrier/concentrator systems were introduced for subscriber line use. These employ digital lines at the 1.544 Mbit rate with engineering rules appropriate to subscriber cables. Delta modulation has been used, as well as PCM, to meet the requirements of subscriber line transmission with a somewhat lower bit rate and larger number of channels. By the end of 1977, several thousand 1.544 Mbit lines will be in use to provide service in subscriber cables. Hence, the stage is set for the integration of digital subscriber transmission with digital local switching. In fact, such systems are now promised by a number of manufacturers to work with their local digital switches.

OCEAN CABLE TECHNOLOGY

Today, overseas communications is so satisfactory that it is difficult to believe that the world was still dependent on high-frequency radio in 1952. The first trans-Atlantic cable was completed in 1956, and provided 48 circuits on two 0.62-in coaxial cables. The first system had a 160 kHz top

*See A. Gersho, "Quantization," *IEEE Communications Society Magazine*, pp. 16-29, Sept. 1977.

frequency and spanned an underwater distance of 2000 mi between North America and Europe with 52 repeaters. Each circuit cost about $1M, but the improvement in communications was well worth it.

The high cost of the undersea cable line facilities provided the incentive for the development of a system for more efficient transmission of speech, known as TASI (Time Assignment Speech Interpolation). This system took advantage of the natural pauses in speech to interpolate twice as many active conversations onto the expensive trans-Atlantic circuits. TASI was introduced in 1959 and continues to be used on overseas cable circuits. Digital technology now makes it possible to produce lower cost implementations of this same basic principle, and its use can be expected to expand with the growth of a digital network.

The availability of good overseas transmission generated rapidly rising demand for circuits, resulting in a proliferation of ocean cable installations throughout the world, spanning all the major water barriers. By 1972, calls between the U.S. and overseas points had increased to 125 million annually, 100-fold greater than in 1952. A second generation ocean cable system introduced in 1963 provided 140 channels on a single 1.25-in diameter coaxial cable. The top frequency of this system was 1.1 MHz and the repeater spacing was 20 mi. The year 1968 marked the installation of the first transistorized ocean cable system. This system expanded the capacity to 845 channels and extended the upper frequency to 5.9 MHz. That the use of transistors for ocean cable lagged behind their use for other transmission systems by about ten years is due to the very stringent requirements on both reliability and performance that ocean cable repeaters must meet. This challenge has been successfully met, and ocean cable designs have continued to evolve on the basis of solid-state technology. The sixth trans-Atlantic cable was introduced in 1975 with a capacity of 4000 voice circuits. This evolution continues to make ocean cables a most competitive alternative to satellite systems for international communications.

The second generation submarine cable systems, which fostered the rapid expansion of overseas links in the early 1960's, used rigid repeaters at 20 mi intervals. The repeater used in the Hawaii-Japan cable in 1964 weighed 500 lb and contained 5000 precise, highly reliable components including directional filters that permitted the transmission of 150 channels in both directions over a single cable.

MULTIPAIR CABLE

The development of ocean cable systems required obvious advances in coaxial cable structures for the undersea environment. Not to be overlooked is the equally significant research and development on large multipair cables.

Polyethylene came into wide use as conductor insulation in place of paper and pulp in the late 1950's. Solid plastic insulation now dominates in cables up to 900 pairs. The lower dielectric coefficient of paper makes it the preferred choice for larger cables because the overall diameter of the sheath can be smaller.

Communications cables form a web above or just beneath the surface of the earth wherever people require telephone service. Where there are people, there is also a variety of activities likely to cause damage to cable sheaths. Paper insulated cables fail at the point of damage to the sheath or splice case as the result of water entry. Air under pressure has long been used to prevent this water entry. In recent years, more and more sophisticated methods for planning the combinations of air dryers and distribution pipes and systems for monitoring and analyzing the air pressure have become available from the industry. Today, a computerized central terminal can reach out to 150 remote terminals, more than 20 000 pressure and air flow sensors, and serve 15 000 mi of pressurized cable.

Since plastic insulation is not affected by water, the effect of water entry is not catastrophic, but troublesome nonetheless. Water will flow through the interstices, collect in low spots, and corrode conductors and shields at any pinholes which may exist. In the late 1960's, the concept of completely filling the core of multipair cables with a viscous petroleum jelly mixture was introduced. The use of foamed insulation with a tough outer skin, begun in 1974, makes it possible to offset the higher dielectric constant of the filling compound.

This combination of filling the core to exclude water and foaming the insulation to minimize insulation thickness makes the use of aluminum conductors in telecommunications cables more practical. However, at present relative prices of aluminum and copper, the cost advantage of aluminum is not great and usage has been modest.

Multipair cables are normally provided in a small number of conductor sizes (e.g., 26, 24, 22, and 19 AWG), all with the same nominal mutual capacitance per unit length. This allows pairs of different gauge to be connected in tandem without serious impedance discontinuities. Multipair cables installed specifically for carrier transmission can depart from these standards to achieve optimum tradeoffs between cable costs and repeater spacing. This was the objective of the Bell System LOCAP cable introduced in 1972 for T2 carrier short haul toll trunks and the MAT cable introduced in 1977 primarily for T1 carrier exchange trunks.

OPERATIONS SYSTEMS

Earlier in this paper, the challenges imposed by direct distance dialing with regard to the maintenance of transmission facilities were mentioned. The facilities for such maintenance have evolved from very crude beginnings in the early 1950's, under the combined pressure for better quality control and for increased productivity of craft personnel in the face of a telecommunications network growing between 10 and 20 percent per year. The first step toward more automation was the automatic measurement of the loss of trunks on a periodic basis. These early automated trunk testing systems were helpful, but tended to produce more data than the maintenance personnel could effectively translate into corrective action.

In 1966, the Bell System introduced an automatic trunk measuring system, ATMS, which measured noise as well as loss under programmed control via punched tape or cards. Most important, it made possible the statistical analysis of data and the presentation of test results in a way useful for corrective action. This approach of software-controlled testing has now been expanded to cover very large networks of trunks, all from one centralized test location. A minicomputer-based system introduced in 1973 gave maintenance personnel a far more comprehensive testing and analysis capability than had ever been visualized before.

Also, 1973 saw the introduction of the first fundamental improvement in subscriber line testing since the development of the test desk with its versatile ballistic galvanometer many years before. A new programmed test system made a number of dc measurements of a subscriber loop automatically when initiated by a repair clerk. The output was a simple indication of the line status to guide further corrective action by testers and repairmen. This system was married to a computer-based subscriber line record system in 1975. The resulting loop maintenance operations system streamlines the records as well as the testing, putting both past history and present status of the subscriber line at the fingertips of repair personnel.

For the large and ever-increasing number of private line circuits, access is not readily available for testing purposes as it is in the public network. The late 1960's saw the introduction of systems for gaining test access to nonswitched private line circuits and then making measurements of transmission level, noise, delay distortion, and other parameters essential for maintaining adequate performance. These systems have been improved and expanded until today it is possible to make measurements of many dispersed private line circuits from a centralized location. Most recently, systems for collecting and processing alarm and trouble indications over a wide area of the communications network have been added to round out a rapidly growing family of operations systems for transmission maintenance.

TRANSMISSION IMPROVEMENT

One very important result of the advances in solid-state technology, digital techniques, and operations systems noted above has been a gradual but sure improvement of quality of transmission. In each new transmission system introduced, some of the improvement possible with new technology has been used to achieve better quality as well as lower costs. The overal quality gain is difficult to express on any simple scale. Improvement results from a reduction of many impairments, including not only loss and noise, but also nonlinear distortion, delay distortion, and attentuation distortion. As a rough illustration, the "effective loss" of a connection in a metropolitan area has probably been improved by about 3 dB, and that of a toll connection by

about 9 dB over the past 25 years. The worst connections have been improved by even larger amounts. The bottom line is a much better grade of service for voice transmission today.

Another motivation has been to improve the data transmission capability of the communications network. As indicated in the Introduction, the early 1950's saw the introduction of data transmission over the public network at speeds of about 1500 bits/s. Achieving even these speeds required carefully controlled access to the toll network, and there were problems with whole circuit groups between metropolitan areas being inadequate to support satisfactory data transmission.

Careful detective work uncovered the causes of nonlinear distortion, impulse noise, phase jitter, frequency shift, and other impairments which affected data transmission but not voice. Over a period of time, these problems were cleaned up in existing systems and avoided in new systems by appropriate requirements for data transmission in the design stage. These improvements, together with ever-increasing sophistication of the data transmission terminals, has made it possible to achieve 2400 bits on a routine basis and 4800 and even 9600 bits under special circumstances.

THE FRONTIERS

Today, we have a communications network which has penetrated into all areas of the civilized world. In highly developed countries, telephone service is available virtually everywhere. Furthermore, this network has the capability not only of high quality voice transmission, but high speed data transmission as well. This latter capability has allowed remote computer access systems to become a part of our daily routine.

Direct dialing of telephone calls is now the normal way of life, and more than 70 percent of toll calls in North America are completed in this way. To an increasing extent, people are dialing their own overseas telephone calls. Wire transmission carrier facilities between switching centers have increased from 17 million circuit miles to more than 200 million circuit miles. This represents about 35 percent of long distance circuit mileage, the balance being microwave radio. There is no doubt that the communications network, still heavily based on wire transmission techniques, is the foundation on which much of our technical, economic, and social progress is based.

What, then, are the principal challenges for the future in guided transmission? As the demand for communications capability increases both in growth of existing services and in the introduction of new and more exotic services, the capital required to meet this demand is becoming an ever-larger percentage of the total capital available for investment. Furthermore, the necessary return on investment together with operating expense results in tariffs which limit the availability of service and inhibit growth. Therefore, there is a constant pressure to reduce the life cycle cost of communications facilities.

As indicated previously, much progress during the last 25 years has resulted from the introduction of solid-state technology. During the latter part of the period, large-scale integrated circuits, like those in modern microprocessors, have played an increasingly important role in the development of transmission systems. In particular, large-scale integrated circuits have provided a further impetus to the evolution of digital transmission facilities.

Digital systems have been developed separately to perform the functions of loop transmission, trunk transmission, and switching. Today, these functions are beginning to merge, and the traditional boundaries between transmission and switching are disappearing. The result of this synergy will be an even greater acceleration of digital techniques into the trunk transmission plant and a really significant penetration into the loop transmission plant. This will open up the possibility of lower costs for a wide variety of services other than speech transmission, including such things as high speed graphics.

From a technical standpoint, an evolution to digital services up to, say, 56 kbits could be based on the same transmission media used almost universally today, namely, multipair cables. Digital transmission techniques make it possible to exploit the inherent information transmission capability of multipair cables economically. However, it is likely that other transmission media will eventually be capable of providing such services in a more optimum way.

Already we have seen the development of special low capacitance cables for digital transmission over intermediate distances, and the first experimental installations of optical fibers have been placed in service by several manufacturers. There is little doubt that optical fibers will play an important role in the future of "wired" communications. Not only will a single fiber support digital signals that would otherwise require a much larger number of individual wire pairs with intermediate repeaters, but fibers will make possible the efficient transmission of wideband signals of the kind that will be required for true video communications. The question is simply when system economics and the demand for such wideband services will make the transition to optical fibers practical on a wide scale basis.

To those of us who have been involved in communications business for the past 25 years, progress seems very great indeed. We entered this field when the electron tube was king, and the most advanced wire line system provided 600 voice channels on a coaxial cable. Today, single silicon chips are being used to provide the functions of hundreds of electron tubes in both analog and digital transmission systems. Coaxial cable systems can now carry more than 13 000 voice channels per tube, and millimeter waveguides offer the possibility of many times this capacity. We are doing things today which we never dreamed of in the early 1950's.

However, there is little doubt that the changes in communications will be even greater in the years ahead. Changes in services, in systems concepts, and in transmission media are likely to be of a much more fundamental nature than we have experienced in our technical lifetimes. This all adds up to a tremendous challenge and opportunity for the communications engineer entering the field today.

DIGITAL COMMUNICATIONS— THE SILENT [R]EVOLUTION?*

M. Robert Aaron, Member, IEEE

Digital telecommunications has already made significant strides and the pace will be accelerated with very large-scale integration, optical fibers, and digital switching.

PRELIMINARY COMMENTARY

"Do you speak digitease?" (Or digits with ease?) You probably have "spoken in *digitease*" many times without being aware of the fact. With *digital facilities* your speech signal is *sampled*, digitized in an *analog-to-digital* (A/D) converter and *interleaved* in time with those digits representing your neighbors' speech (or data). This composite digital signal is transmitted over the ordinary telephone wires that may have been under the streets for a number of years. *Sharing* a pair of wires over many conversations increases the system capacity in a cost-effective way. By "mining" additional circuits from the existing copper we conserve a precious resource and we avoid tearing up the concrete. For these, and other reasons to be developed, many countries in the world have introduced digital systems. Thus we can answer our lead-off question in the affirmative. Virtually, all of us have spoken "digitease" (some since 1962) and all of us will speak in bits (1's and 0's) and bytes (groups of bits) in the future. Thus, in a real sense, the evolution to digital has been quiet—no trip-hammers, no bulldozers.

Of course, until we are fitted with a digital ear, it is necessary to convert "digitease" back to the domain of continuous time and amplitude before presentation to the human receiver. For this translation to succeed, we must make sure that almost all of the pulses that enter the underground conduits make it unchanged to the receiving terminal where the conversion to analog takes place. Remember, many of these wire pairs were originally designed to handle a single conversation confined to a band of frequencies of less than 4 kHz. With digital transmission, it is necessary to expand bandwidth to achieve satisfactory performance. As a compensation, transmission is possible at a much lower signal-to-noise ratio than that required in analog systems. Narrow transmitted pulses are attenuated and spread out in time as they travel in the wire pairs. This is due to the fact that the higher frequency components of the signal travel on the outer skin of the copper, and thereby suffer greater attenuation than the lower frequency components which have greater penetration. In addition, the dielectric (pulp, plastic) is lossy yielding high-frequency attenuation. Crosstalk from pair to pair also increases with frequency. Fortunately, these and other effects can be combated by introducing a regenerative repeater about every mile on paired cable to reconstruct the original digital signal from the dispersed, noisy waveform.

This ability to regenerate the signal and strip away noise and crosstalk introduced in transmission is the "payoff for digital" [1]. Performance becomes virtually independent of distance as contrasted with analog transmission where noise and other impairments accumulate with distance. More than 99 percent of the time the errors made in distinguishing pulses (1's) from spaces (0's), and vice versa, is less than one in a million bits in a repeatered line. This error rate has no effect on a phone conversation (and very little impact either on voiceband or direct digital data transmission). Consequently, sufficient information is delivered to the receiver at the end of the cable to permit conversion of the digital signal back to the analog domain. Here it is so like the original acoustic

The author is with Bell Laboratories, Holmdel, NJ 07733.

*Reprinted from *IEEE Communications Society Magazine*, January 1979, Vol. 17, No. 1, pp. 16–26.

Processes in Digitization of Analog Signals

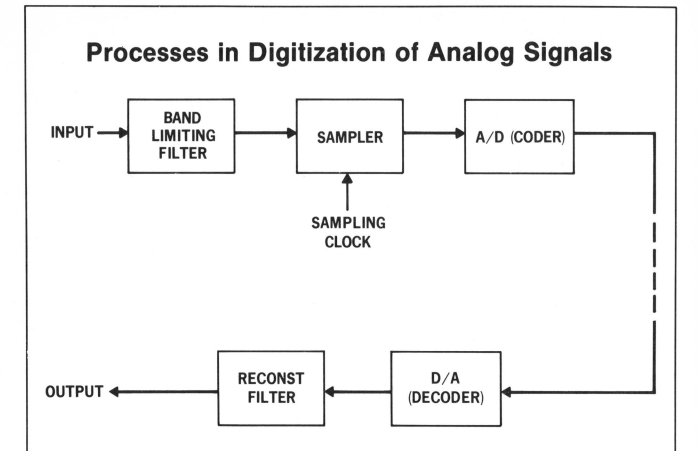

Conversion of an analog signal to digital form (1's and 0's) involves sampling a signal in time followed by representation of the signal samples by groups of 1's and 0's. Prior to sampling, the analog signal must be filtered to limit the amount of out-of-band energy reflected in-band as a result of sampling. It is convenient to consider the process of sampling as multiplying the signal by a train of narrow pulses. This sampling pulse train can be represented by a Fourier series—a series of sine waves. Each of these sine wave carriers—at dc, the sampling frequency and its harmonics—results in translates of the signal spectrum which can overlap unless suitably filtered. For a voiceband signal, the band-limiting filter cuts off at about 3.2 kHz and has an out-of-band attenuation of about 30 dB. Such a filter is easily realized with two operational amplifiers and an assortment of resistors and capacitors. Thin film, thick film, or monolithic forms are available. A sampling frequency of 8 kHz is used worldwide for the voiceband channel.

Once the train of samples has been created, they are "measured" by a quantizer with a stair-step characteristic [18]. The smaller the number of steps, the smaller the error (quantizing error) in the measurement process. Of course the finer the gradations (steps) of this "ruler", the larger the number of bits required to represent the signal. To accommodate the wide dynamic range characteristic of speech, while conserving bits, the step-sizes are arranged approximately logarithmically. This produces small steps for low-level signals and larger steps for higher level signals, making the roundoff (quantizing) error approximately a fixed percentage of the signal independent of its power.

Each of the steps of the quantizer is encoded into a binary number. In digital telephony, 8 bits are used giving 2^8 levels to represent samples of the signal. Thus, with 8 kHz sampling and 8 bits/sample, the bit rate per voice channel is 64 kbits/s. The process of quantizing and coding is performed in an A/D converter.

The 1's and 0's are transmitted over a medium with regenerative repeaters. At the receiver the digital signal is converted back to pulses and smoothed out by a low-pass filter to give (barring many errors in transmission) a close facsimile of the original signal. With 8-bit quantization, the quantizing noise is undetectable for speech and for voiceband data. This permits many tandem A/D/A conversions in a mixed analog/digital network.

waveform at the transmitter (prior to digitization) that one can rarely distinguish the two.

The process of conversion to pulses, their regeneration, and the reconversion back to analog, is known as pulse-code modulation (PCM). These *elementary* terminal operations performed on a single channel in a digital transmission system are summarized in the boxed inset. Fig. 1 depicts the time-division multiplexing (TDM) of such digitized analog signals (speech and video) with data signals that are already in digital form. As we see from this figure, "all signals look alike," independent of their origin (or language!). This dilution of sources into the bubbling bit-stream is another important advantage that digital has over analog (again purchased at the expense of bandwidth!).

The dilution of diverse information sources into the bubbling bit stream is an important advantage that digital has over analog.

As long as the composite pulse stream can be regenerated, there is no interaction between the original signals! With analog transmission by frequency division multiplexing (FDM), the original signals are allotted (ideally) separate frequency bands. Crosstalk from one signal band into another arises, to a limited extent, from the lack of sufficient attenuation in filter skirts. (This accounts for the relatively high cost of channel separation filters.) However, in analog systems, the main source

of undesirable interaction between channels arises from the nonlinear behavior of the nominally linear repeaters spaced throughout the transmission medium to overcome the cable loss. Since the nonlinearity occurs at every repeater, its effects are cumulative; and in long repeatered systems, the linearity requirements imposed on individual repeaters are most stringent—a small fraction of one percent intermodulation is typical and achievable. Furthermore, television, which requires a much higher overall signal-to-noise ratio than speech, must be placed on the line at a relatively high level. This raises the level of crosstalk from the television signal into the voiceband signal above that associated with FDM voiceband signals alone. To date, this factor has limited the combined use of FDM voiceband channels and television on the same broadband analog medium.

The principal objective of this paper is to go beyond the above preliminaries to indicate the present status of digital communications. First, a simple view of a nationwide communications network is given. Next, the applications factors that control digital penetration of this network are enunciated. Statistics for the worldwide use of digital communications by telecommunications administrations are summarized.

Emphasis throughout is on common carrier communications, especially that characterized by the Bell System to which this author's firsthand experience is confined. Furthermore, the emphasis is on digital transmission rather than on switching since an excellent overview of digital switching has recently appeared in this magazine [2]. Where the two areas coalesce (this promises to be almost everywhere!), we inevitably touch upon their interdependence.

A subsidiary objective of the paper is addressed in the final section. There, a brief glimpse of the near future is attempted. The long-term end point may be easy to state, namely, an all-digital network. How to get there is very complex and is left as an exercise for the reader!

THE MESSAGE NETWORK

To better appreciate the factors that have led to the widespread use of digital common carrier communications, one should have a rudimentary understanding of the widely available "message network." We develop such an oversimplified view in Fig. 2. Connections from the subscribers' premises to the local central office (CO) switch are called *lines* or *loops*. Switching machines are connected together by *trunks*—*local* or *exchange* trunks between local offices, *toll connecting trunks* from local to toll switches and *intertoll trunks* between toll switches. The switching hierarchy beyond the toll office, not shown, contains three levels making an overall five-level hierarchy for the North American network [3].

Digital carrier systems can be applied to trunks as well as loops. Due to reduced terminal costs, as compared with analog carrier, penetration of digital carrier has

DIGITAL
TRANSMISSION

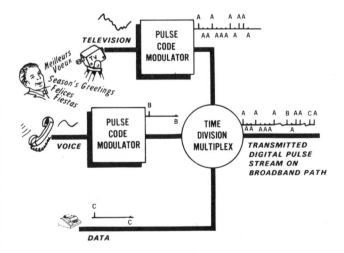

Fig. 1. Time-division multiplexing of digital signals. After the analog signals have been digitized and combined with inherently digital signals "all signals look alike." Of course since television occupies a much wider band than speech, its sampling rate is much higher, resulting in many more bits on the line for video than for speech.

THE "MESSAGE" NETWORK

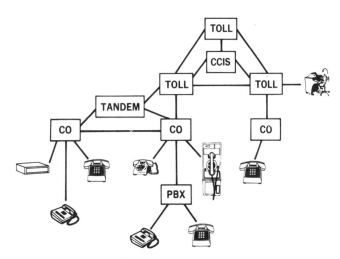

Fig. 2. "The message network." The message network accommodates all forms of services and permits the interconnection of more than 100 million subscribers without human intervention.

been extensive in local- and toll-connecting trunks where terminal costs dominate. This is further examined below. Though there has been only limited application of digital carrier to long subscriber loops, it is expected that the

---

Penetration of digital carrier has been extensive in local and toll connecting trunks where terminal costs dominate.

---

introduction of local digital switching will foster significant growth of digital carrier in the subscriber loops and vice versa. This is proving to be the case in toll-connecting trunks. Here, toll-digital (time-division) switching introduced in 1976 retains the signal in digital form for switching, saving much of the cost of conversion of the digital signal to analog, as required in conventional analog space-division switches [4], [5]. It is to be emphasized that the large digital toll switches [4], even though best matched to digital trunks, prove to be cost effective in an all-analog environment. In long toll trunks where line haul costs dominate, analog systems have been and continue to be cheaper than digital systems. This is a consequence of the bandwidth penalty associated with digital transmission.

APPLICATIONS FACTORS

We have alluded to some of the technical features of digital communications in the above commentary. Here we introduce the economic factors, correlate them with the technical, and formalize some of the earlier material.

It is well known, and it has been repeated *ad nauseam*,

that the transistor was a necessary (though not sufficient) vehicle for the realization of digital terminals in which logic and switching circuits abound. For FDM, where filters and linear modulators predominate, solid-state device technology has certainly had an impact, but not on the same scale as in digital systems. An even more important feature of digital communications is the ability to easily time-division multiplex (TDM) signaling with the customers' pulse stream! Signaling consists of both supervisory information—idle/busy; on hook/off hook—and address information. With FDM in the U.S.A., this information is communicated by modulating a carrier in the signal band (actually at 2600 Hz). Care must be exercised to prevent speech from "talking off" the connection. Compared to the simple TDM approach, the FDM "in-band signaling" is much more expensive. *In toto* we find, as depicted in Fig. 3, that digital terminals are cheaper than their analog counterpart, particularly when signaling is included as it is in all short-haul carrier. Recently, Common Channel Interoffice Signaling (CCIS) has been introduced in the long-haul network [6]. With this approach, signaling information from many channels is TDM and transmitted over a separate channel dedicated to signaling. This is particularly cost-effective and gives improved performance over in-band signaling.

Integrated circuits make digital channel banks cheaper than their analog counterparts.

When CCIS becomes economically competitive for short distances, then signaling will no longer be a significant burden for FDM. However, the economic advantages of digital short-haul carrier will remain due to the decreasing cost of large-scale integration (LSI) on which modern digital communication feeds.

Note from Fig. 3 that, for carrier, the *prove in* distance (the distance at which the per circuit cost of carrier equals that of a single circuit on wire pairs) is dependent upon several factors. For example, if the capacity of the existing wire pairs is running out and it is necessary to put in new conduits and/or new manholes in a metropolitan area, then this capital expense can be avoided by introducing carrier instead of new cable. Thus, the economics for direct relief differ markedly from those where space exists to introduce new paired cable. If the trunk is toll-connecting from an analog local office to a digital toll office, then the cost of digital carrier is reduced, due to the missing multiplex at the digital toll switch. As shown in Fig. 3, the intercept cost of digital carrier is lower than that of analog carrier, due to lower terminal costs. Conversely, line-haul costs are lower for analog carrier on wire pairs or coaxial cable as evidenced by its lower slope. This is due to the bandwidth economy of analog for voice transmission. This results in a *prove-out*

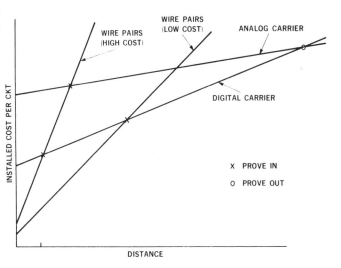

Fig. 3. Cost picture for wire line digital transmission.

> Though digital terminals are cheap,
> Wire line costs rise too steep.
> So that ere long,
> Digits aren't strong
> Till fibers oer distances leap!

distance beyond which analog carrier is cheaper than digital. For example, consider transmission on coaxial cable in which the repeater spacing is approximately one mile for either analog or digital transmission. In the analog case it is possible to get 13,200 two-way voiceband circuits per pair of coaxial tubes [7]. With binary (2-level) transmission at 274 Mbits/s [8], [9], only 4032 circuits are obtained (64 kbits/s per circuit plus additional housekeeping bits). To achieve approximately equal capacity, 8-level transmission is required at the same symbol rate. Realization of such a system is difficult, to say the least. We have greatly oversimplified matters by looking at line capacity rather than total system cost. The comparisons are perhaps more meaningful, though not complete, for *new* long-haul installations.

Specific *prove-in* and *prove-out* distances depend on many factors, including those noted above, and they vary from country to country. For example, in England, analog-signaling costs are quite high and the geography of the country is such as to move the prove-out distance off-shore. (One man's short haul is another man's long haul.) These are but two reasons for the plans in the United Kingdom to install digital carrier in their long-haul

---

Today's long haul analog systems on cable provide three times the circuit capacity of binary digital systems with the same repeater spacing. But tomorrow, long haul fibers . . .

---

network in the 1980's [10]. In the Bell System all new carrier being installed in metropolitan areas are digital.

With digital toll-switching, digital toll-connecting trunks prove-in at zero length! On the other hand, long-haul analog systems on cable provide more than three times the 4 kHz circuit capacity of binary digital systems with the same repeater spacing.

Consider an extreme situation where the 274 Mbits/s *digital data* is aggregated for exchange between two quite remote points. Using T4M or LD4 [8], [9] this can be achieved on a pair of coaxial cables (two "tubes"). To achieve this capacity on an analog system with existing voiceband data modems, feeding each of the 13 200 channels with 4.8 kbits/s data would require more than four pairs of tubes. Doubling the data capacity of an ordinary analog voiceband channel would still necessitate two pairs of L5E tubes to match the data capacity of a single pair of tubes with T4M. A bit rate of more than 20 kbits/s per voice channel on the analog cable system would yield equal data capacity with the 274 Mbit/s digital system. Such a capacity on an analog voice channel in the direct distance dialing (DDD) network might be quoted by either a science fiction writer or the most naive information theorist! On the other hand, even the most ardent digital advocate would have to agree that there are no two points in the world where there is a demand to aggregate or exchange data at such a rate either now or in the near future. (A simple and more realistic example follows from the recognition that a single voice channel in the T1 digital carrier system can carry up to 64 kbits/s of digital data—more than 10 times the 4.8 kbits/s of today's phone channel.) Data traffic, though small today, is growing faster than plain old telephone service (POTS) and it is necessary to factor some requirements for data transmission into the continuing studies of long-haul digital transmission. As seen above, the bandwidth penalty associated with digital transmission of POTS becomes an asset for direct digital data transmission.

Digital transmission on line-of-sight microwave radio links is attractive for intermediate lengths at both 6 GHz and 11 GHz, where a capacity of 1344 circuits may be achieved [11]. (2 GHz, at lower capacity is also of interest.) For long-haul, comparison of digital with analog single sideband (SSB) radio [12], which has a capacity of 6000 circuits, is not favorable for speech signals alone. In addition, the increase in capacity of analog radio can be achieved by modifying existing systems that are in place, at incremental costs much less than that required for complete new systems.[1] The cost-effective increase in capacity of FDM-FM radio systems over the years [14] has been a major engineering achievement and is responsible for the fact that more than two-thirds of long-haul traffic in the U.S.A. is carried on 4 GHz radio. Analog is not dead—yet!

---

[1]In Canada, the availability of 8 GHz for common carrier communication has permitted overbuilding existing radio with 8 GHz digital systems for their long-haul facilities [13].

Both optical fibers and millimeter waveguide are well-suited to digital communications for a variety of reasons —the principal reasons being that the solid-state devices at the frequencies of interest are well-matched to on-off modulation and there is bandwidth to burn. With the waveguide, extremely large capacity is required to make the system economical. This introduces problems of restoring service when such a facility fails, even when the needs for the large capacity materializes. Furthermore, as noted above, overbuilding existing radio with SSB is cheaper. At this writing it appears that fibers will enter the "message network" first in short trunks.[2] In this application, the 4-5 mi repeater spacing for fiber regeneration as opposed to 1 mi for wire-pair regenerators can eliminate outside plant electronics on many routes. Several field experiments have been reported [15]. All new guided media—waveguide, fiber, and coaxial cable systems—are at a disadvantage with respect to microwave radio and satellite for long-haul because of high "right-of-way" costs; i.e., costs of acquiring a corridor of *terra firma* in which to bury the medium. Of course when the airwaves become congested and/or existing right-of-way can be reused, then this disadvantage of guided media will vanish.

Another medium, well-matched to digital transmission, is the satellite employing time-division multiple access (TDMA). Several digital satellite systems have been tested and are scheduled for use in the early 80's. One of the major problems with satellite is that of far-end echo returned to the speaker via the long satellite path [16]. Cost-effective echo cancellers are now becoming available to control this problem for voice service for a single-satellite hop.

SUMMARY OF CHARACTERISTICS AND OPPORTUNITIES

Before we go on to give a quantitative picture of the extent to which digital communications has grown, we pause to recap its "raison d'etre" and its problems (opportunities). Obviously it exists and is growing "because it meets a need at a justifiable cost" [17]. Underlying the obvious we find the following keys to the digital doors of the present and future!

1) **Digital terminals are cheaper than analog.**
   —cheaper filters, time-shared digital circuitry, time multiplexed signaling.

2) **Performance is determined by the terminals.**
   —quantizing noise [18] determines performance, line errors are not controlling even at signal-to-noise ratios much less than those required by analog systems.

3) **Mixing signals is easy—A pulse is a pulse is a pulse . . .**
   —all signals look alike after conversion to digital form; therefore they can be interleaved and regenerated without fear of crosstalk.

4) **Interference limited media.**
   —the ability to regenerate permits the exploitation of crosstalk—dominated media such as wire pairs already in place. In a sense, coaxial cable is too good for long-haul digital transmission.

5) **Digital signal processing is well-matched to device technology.**
   —yesterday the transistor made digital transmission viable, today LSI is making digital switching attractive and further reducing transmission costs. In addition, signal processing techniques such as echo control, service circuits (tone detection, etc.), bit-rate reduction techniques [19] and encryption are amenable to digital device technology evolution.

6) **Nonlinear optical and millimeter devices favor digital.**
   —optical modulators and demodulators operate most efficiently in the nonlinear mode. This is also true of millimeter wave devices for circular waveguide. In passing we ask— will light leapfrog waveguide? If so, where?

7) **Maintenance is simplified.**
   —if the error rate is low, the systems clearly go! In-service performance monitoring is easy, protection switching and off-line fault isolation are affordable, due to item 5) above.

8) **Improved performance.**
   —in the transition from analog to digital communications, performance will improve due to the robustness of tandem A/D conversion,[3] *their ultimate reduction* with digital connectivity, and the virtual elimination of transmission facility noise (item 2 above) [20].

9) **Promise of making new services viable.**
   —this tune has been playing on the digital violins (God forbid!) for a long time. Most of these so-called new services can be achieved with the stored program control (SPC) network—analog or digital coupled with "skinny" (small capacity) digital connectivity on existing analog facilities [21] or on satellite. Ultimately, the promise will be fulfilled —but when? When bonafide high-speed (>4.8 kbits/s) digital data demands proliferate! But when?

---

[2]Fibers are also expected to be used in office "fibering" (wiring) applications where their small size and resistance to electromagnetic interference are of paramount importance.

[3]In a network with both digital and analog links, multiple conversions take place between the two modes. As analog links fade away, noise accumulated in these links will disappear as will noise associated with multiple A/D's.

10) **Long-haul transmission will be dominated by analog techniques for "some time" to come.**

—this follows from the opportunities to derive additional circuits from the existing analog plant at low cost (SSB AM microwave radio, higher capacity coaxial systems on the same cables) and the bandwidth penalty associated with digital transmission. Items 5), 6), and 9) above will determine the exact definition of "some time." With new guided media, bandwidth is not a problem, but right-of-way costs may be—today!

11) **A strong synergy exists between digital switching and transmission.**

—with the device catalyst (item 5) above) it seems clear that local digital switching and loops will have the same mutual regenerative effects on one another that toll-switching and toll-connecting trunks are experiencing. It should be noted that with interconnected digital switching, synchronization is required [20]. Several approaches to synchronization are being considered around the world.

CRYSTAL BALL GAZING—THE TRANSITION

Since I am unsure of the details of my own future (other than its ultimate disposition) I can hardly speak authoritatively for the future of digital communications in the Bell System, let alone the world. Though I plead ignorance of specific milestones and dates, I nonetheless believe in the ultimate dominance of the digital approach. Its takeover in transmission in the exchange plant is already assured

BITS AND BUCKS AROUND THE WORLD

Digital communications has become a very big business since the T1 carrier system was introduced in 1962 [22]. Investment in digital in the Bell System is measured in billions. Western Electric has shipped millions of regenerative repeaters and millions of channel units used in digital channel banks. More than one-third of the local- and toll-connecting trunks in the Bell System are now digital. Significant penetration of digital in Independent Telephone Company networks has also taken place. More than 50 percent of the trunks in the Continental network are digital; T-carrier channels doubled from 1973 to 1976 and local, as well as toll, digital switches have been introduced to accelerate this pace. By 1985, Continental's transmission facilities will be 80 percent digital. About 32 percent of the General Telephone trunks are digital.

In addition to the above brief recital, detailed worldwide statistics have been obtained. Many of the early returns arrived two years ago, some have come in more recently, while others have yet to arrive. For these reasons, most of the data covers information through 1976 and is not all encompassing. Data is confined to what has been called the primary multiplex. In North America and Japan, this means the output of digital channel banks consisting of 24 TDM channels with a transmission rate (including framing and signaling) of 1.544 Mbits/s. European channel banks handle 30 TDM channels (plus two 8-bit time slots per 125 ms frame for framing and signaling) with an output bit rate of 2.048 Mbits/s. (This format is called the CEPT ⟶

TABLE I
Primary Multiplex
Ckt-km Installed (Cumulative-Thousands)

Year	Can.	F.R.G.	Italy	Japan	Nor.	Sp.	Swed.	Swit.	U.K.	U.S.A. Bell	U.S.A. GTE
1962										8	
1963										240	
1964										1240	
1965	110		2.3	5						2700	
1966	150		5.2	125						4530	
1967	180		11.7	480						7210	
1968	220		19.5	1040					11	10 500	
1969	280		35.6	1670	0.4			0.3	155	15 400	
1970	380		57.5	2470	1.3			3.5	400	22 000	720
1971	520		89.3	3350	3.7		6.1	6.7	700	30 000	1380
1972	810		141	4500	10.7	31	11.1	13.9	1120	39 900	1920
1973	1180		269	5800	28.4	78	21.6	31.3	1620	51 600	2580
1974	1680		481	7400	39.2	104	36.9	44.3	2200	60 700	3250
1975	2180		712	9400	75.9	212	39.2	50.1	2810	80 000	3860
1976	3130	137	980	10400		387		72.8		88 000	
1977			1380			710				98 000	

system.) Exceptions to these two standards will be noted. Tables I and II contain the available data on installed circuit-kilometers and circuits, respectively. There may be some discrepancy between this data and circuits in service due to installed spare capacity. On this basis it is perhaps more accurate to refer to these figures as *installed capacity*. Each circuit contains two channel ends. In some cases, the circuit may be jointly owned by two different telephone "families." Thus there may be a small amount of duplication in the U.S.A. figures. Finally, a few circuits from newly introduced higher order multiplexes and subscriber loop carrier may have "crept in" to the numbers. Remarks on each country and/or carrier are given below.

Canada—(Courtesy of Richard P. Skillen, Northern Telecom)

Growth of digital carrier from its introduction in 1965-1976 has been at a rate of about 40 percent; several times that of overall circuit growth. More recently the growth rate has been about 50 percent. It is projected that in the Ontario Region of Bell Canada, 50 percent of the *toll* trunks will be digital in the early 80's and that urban areas in that region will be almost exclusively digital at that time [23]. The average system length, excluding toll, is about 38 km.

Federal Republic of Germany—Courtesy of Theodor Irmer, Deutsche Bundespost)

After several years of extensive field trials, about 8000 digital circuits were introduced into the network of the Deutsche Bundespost in 1976. The total number of circuit-kilometers at that time amounted to 136 800, yielding an average circuit length of 16 km. In the network of the Federal Republic of Germany, distances between local switching exchanges are short, thus accounting for the fact that digital carrier has not achieved widespread use to date.

Italy—(Courtesy of Aldo Mascioli, STET)

Digital communications were born in Italy in 1965, making this the first country in Europe to use PCM. The early installations used the 24 channel format. Today less than 9000 such circuits are in operation. In 1969, a 30-channel system of early design was introduced, but since 1973, the standard CEPT systems are being installed. For 1975, a very small number of PCM systems were installed by ASST (State Agency for long-distance communications). This small quantity is included in the data presented. With the exception of this minor perturbation, all the circuits tabulated are under the jurisdiction of SIP, the operating company for STET (Society Finanziaria Telefonica) which is the Italian equivalent of AT&T. The average circuit length from Tables I and II is 16 km. Presently 25% of the carrier systems are digital. By 1985 it is expected that about 45% of installed carrier channels will be digital.

Japan—(Courtesy of Noriyoshi Kuroyanagi, NTT)

Starting with 10 systems in 1965, Nippon Telephone Telegraph (NTT) has increased their digital capacity 1000-fold in a decade. The average circuit length is about 20 km. We note in passing that in 1977, the Japanese introduced the highest speed PCM system on coaxial cable in the world—the 400 M system transmitting up to 5760 channels at a bit rate of 397.2 Mbits/s [24].

The Netherlands—(Courtesy of P.J.C. Hamelberg, The Netherlands Postal and Telecommunication Service)

Prior to 1976, The Netherlands Administration experimented with digital systems. Starting in 1976 it was anticipated that 10-25 systems would be installed per year over the succeeding few years. Today 30 channel systems are operational. In 1983, 11% of total capacity (650 of the 2 Mbit/s systems) should be digital.

TABLE II
Primary Multiplex
Ckts. Installed (Cumulative-Thousands)

Year	Can.	F.R.G.	Italy	Japan	Nor.	Sp.	Swed.	Swit.	U.K.	U.S.A. Bell	Cont.	GTE
1962										0.25		
1963										7.2		
1964										37		
1965	2.7		0.12	0.24						81		
1966	4.0		0.26	6.7						130		
1967	5.0		0.60	23.8						220		
1968	6.0		1.4	52.1					0.41	310		
1969	7.3		2.7	83.3	0.02			0.03	5.7	460		
1970	9.5		4.3	124	0.05			0.38	14.9	660		22.4
1971	13.5		6.7	168	0.20		0.48	0.77	26	900	5.2	46.8
1972	21		10.5	227	0.59	1.8	0.87	1.8	41.3	1200	5.4	71.3
1973	33		14	291	1.6	4.6	1.7	2.5	60	1600	6.9	106.1
1974	50		32	371	2.2	7.2	2.9	3.8	81.3	1800	9.1	149.3
1975	75		44	471	4.2	14	4.7	5.3	104.2	2300	11.9	201.3
1976	98	8.6	58	520		22		8.3		2700	15.3	
1977			79			36				3200		

Norway—(Courtesy of E. Heggelund, Norwegian Telecommunications Administration)

From 1969 through 1973 a few 24-channel systems were operated. Since 1971, 30-channel systems have been installed. The average system length is 18 km.

Spain—(Courtesy of P. Pastor Lozano, Compania Telefonica Nacional de Espana)

In 1972, 24-channel PCM systems were first introduced and were installed through 1975. No new 24-channel systems are being installed. Thirty-channel systems were first introduced in small numbers in 1973 and in 1975 they constituted 57 percent of the installed PCM capacity. Initial average system length was about 18 km. Circuit kilometers have almost doubled from 1974 to 1977 at which time 20% of interoffice trunks were digital.

Sweden—(Courtesy of Anders Olsson, Swedish Telecommunications Administration)

In 1971, 20 of the 24-channel systems were installed. In succeeding years, 30-channel systems have been deployed. By the end of 1978, 230 of these systems will be in service representing 3% of the trunks. Average system length is 20 km. Digital build up is rapid.

Switzerland—(Courtesy Albert Kuendig, Swiss PTT)

Essentially all systems are of the 30-channel variety. Of these, 53 percent are 30-channel systems with a bit rate of 2.56 Mbits/s. The others are the standard CEPT system which has been employed exclusively for the last several years. The vast majority of the systems have been employed in the Swiss short-haul network at an average system length of 12 km. As of 1976, the Swiss trunk (short-haul) network consisted largely of analog baseband systems (77 percent) with the remaining trunks 14 percent FDM and 9 percent PCM. However, the growth in PCM in 1976 was 43 percent as compared with 5 percent each for baseband and analog carrier, respectively.

United Kingdom—(Courtesy of W. G. Simpson, United Kingdom Post Office)

From 1968 through 1978, the British Post Office (BPO) installed PCM systems of a 24-channel variety that differed markedly from the T1 system. By the end of 1978, there will be some 7000 such systems in service. From 1978 onward, 30-channel CEPT systems are being installed and they will be compatible with a family of digital switching exchanges known as System X. The average length of a 24-channel system is about 25 km with a range from about 10 to 70 km [25].

**United States of America
Bell System**

The T1 system introduced in 1962 marked the beginning of the digital communications era for commercial carriers. Growth has been phenomenal—1000-fold in the first five years and another 10-fold in the next 10 years. Average circuit length is about 30 km. Dense islands of T1-carrier exist in all metropolitan areas. Extension to outstate areas is taking place.

Continental—(Courtesy of Herbert F. Haug, Continental Telephone Service Corporation)

T-carrier was introduced into the Continental network around 1967, but the growth to 1972 was slow for two reasons—1) physical plant in place had not exhausted, and 2) need for coordination with connecting companies on established analog carrier routes. As previously noted, T-carrier "took off" starting in 1973 on Superior Cable Company screened cable (cable made up of wire pairs with a shield to minimize crosstalk between the two directions of transmission). Further acceleration of this pace is expected with the recent and continuing introduction of toll- and local-digital switching centers.

General Telephone & Electronics—(Courtesy of C. George Griffith, GTE Communication Products Group)

GTE has introduced considerable digital carrier over the years. From the data in the tables it can be shown that the length of an average digital circuit has dropped from 30 km to about 20 km. Further digital expansion of both loops and trunks is anticipated with the advent of digital switching.

Other Countries

Though direct inputs from other countries are not available, a few comments can be made based upon the open literature.

It is well-known that France has been the leader in the introduction of digital switching at the local level. Hundreds of thousands of lines have been installed on digital switches since 1970 largely in rural areas. Both digital subscriber loop and digital trunks terminate on these switches. Urban areas also have digital trunk systems of the CEPT variety.

In the U.S.S.R., digital 12-channel systems were introduced in rural areas in the early 70's. Though other system formats have been tried, present emphasis is on the 30-channel CEPT systems. Systems are also manufactured in Czechoslovakia, Hungary, and in the German Democratic Republic. Penetration of digital trunks between local switches does not appear to be extensive to date.

A Graphic View

A portion of the data base of Table I is shown on Figure 4 for those situations where installed ckt-kms exceeds one million in 1976. An initial "explosion" followed by steady growth is typical. Most of those countries of Table I not shown on Fig. 4 are in the rapid initial expansion stage.

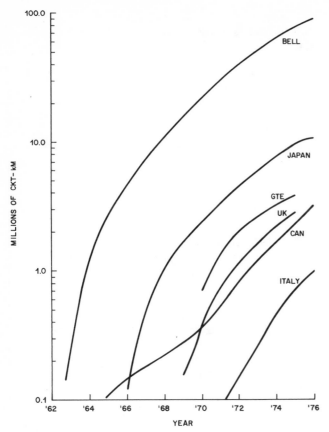

Fig. 4. Growth in ckt-km installed. Rapid initial introduction followed by steady growth is most common.

as seen from Tables I and II. Indeed by the mid-80's more than half of the short-haul trunks in the Bell System should be digital. Toll-switching is becoming digital with 4 ESS machines at a faster pace than earlier projections anticipated. This will drive toll-connecting trunks to be almost exclusively digital well before 1990. In the United Kingdom it is expected that the entire budget for transmission by about 1985 will be devoted to digital facilities [26]. Plans for the evolution to an all-digital network (90 percent digital) in Japan before the turn of the century have been presented [27]. Most telephone administrations have or are developing plans for the process of converting from analog to digital that are peculiar to their own environment.

Indeed, it is the transition period and how to manage the change that is at issue, not whether digital will prevail over analog. Both approaches will continue to be used where they make economic sense. Details of *prove-in* depend upon a variety of factors, including: geography,

---

New services promise to accelerate the pace to digital.

---

the local state of technology, balance of payments, support for local industry, growth of new services,

political, legal and regulatory policies, depreciation strategies, and social factors. A few of these facts of life have been alluded to earlier. One can conjure up scenarios using these factors to support or combat almost any view. However, detailed economic studies relevant to these issues must be made to provide a basis for action. As always, the major problems arise from uncertainties surrounding inputs to these studies. It is much easier to dream up new services than it is to *quantify* their impact.

Much of the controversy revolves about these new services. Some believe that initial demands for new services will be modest—well within the capability of an existing SPC, basically analog, network. Others have the gut feeling that the existence of digital connectivity will stimulate new services. These and other arguments can be heard at any meeting with the word communication or computing in its title. Though the digital (r)evolution literally started out in relative silence, it is now quite vocal. "Digitease" is spoken here, there, and everywhere even without the "digital larynx" and the "digital ear"!

ACKNOWLEDGMENT

I am indebted to all of the gentlemen from around the world who so kindly provided the data on digital carrier. Blame for misinterpretations of this information should be laid at my door.

REFERENCES

[1] B. M. Oliver, J. R. Pierce, and C. E. Shannon, "Philosophy of PCM," *Proc. IRE*, vol. 36, p. 1324, Nov. 1948.

[2] J. C. McDonald, "Techniques for digital switching," *IEEE Commun. Soc. Mag.*, vol. 16, no. 4, p. 11, July 1978.

[3] A. E. Joel, Jr., "What is telecommunications circuit switching," *Proc. IEEE*, vol. 65, p. 1237, Sept. 1977.

[4] A. E. Ritchie and L. S. Tuomenoksa, "No. 4 ESS, system objectives and organization," *Bell Syst. Tech. J.*, vol. 56, p. 1017, Sept. 1977.

[5] J. C. McDonald and J. R. Baichtal, "A new integrated digital switching system," *Nat. Telecommun. Conf. Rec.*, p. 3.2-1, 1976.

[6] A. E. Ritchie and J. Z. Menard, "Common channel interoffice signaling, an overview," *Bell Syst. Tech. J.*, vol. 57, p. 221, Feb. 1978.

[7] J. W. Mulcahy, "The L5E coaxial system," *Int. Commun. Conf. Rec. 1977*, p. 18.1-1.

[8] P. E. Rubin, "The T4 digital transmission system—Overview," *Int. Commun. Conf.*, p. 48-1, 1975.

[9] D. Cowan, "The LD4 system," *Int. Commun. Conf.*, 1974.

[10] W. G. Simpson, "Review of trunk transmission planning in the United Kingdom Post Office," *Proc. IEE*, vol. 123, May 1976.

[11] M. Ramadan, "Practical considerations in the design of minimum bandwidth 90 Mbits/s, 8-PSK digital microwave system," *IEEE Int. Conf. Commun.*, p. 29-1, 1976.

[12] R. E. Markle, "The AR6A SSB long haul radio system," *Int. Commun. Conf. Rec. 1977*, p. 40.1-78.

[13] I. Godier, "DRS 8, a digital radio for long haul transmission," *IEEE Int. Conf. Commun. 1977*, p. 5.4.

[14] U. S. Berger, "The old TD2 became the new TD2C," *Bell Labs. Rec.*, p. 278, Oct. 1973.

[15] S. D. Personick, "Fiber optic communication—A technology coming of age," *IEEE Commun. Soc. Mag.*, vol. 16, p. 12, Mar. 1978.

[16] S. B. Weinstein, "Echo cancellation in the telephone network," *IEEE Commun. Soc. Mag.*, vol. 15, p. 2, Jan. 1977.

[17] M. R. Aaron, "PCM in the exchange plant," *Bell Syst. Tech. J.,* vol. 41, p. 99, Jan. 1962.

[18] A. Gersho, "Quantization," *IEEE Commun. Soc. Mag.,* p. 16, Sept. 1977.

[19] J. M. Tribolet, P. Noll, B. J. McDermott, and R. E. Crochiere, "A study of complexity and quality of speech waveform coders," *IEEE Int. Conf. Acoust., Speech, Signal Processing 1978,* p. 586.

[20] J. E. Abate, L. H. Brandenburg, J. C. Lawson, and W. L. Ross, "The switched digital network plan," *Bell Syst. Tech. J.,* vol. 56, p. 1297, Sept. 1977.

[21] K. L. Seastrand and L. L. Sheets, "Digital transmission over analog microwave radio systems," *IEEE Int. Conf. Commun. 1972,* p. 29-1.

[22] K. E. Fultz and D. B. Penick, "The T1 carrier system," *Bell Syst. Tech. J.,* vol. 44, p. 1405, Sept. 1965.

[23] D. A. Chisholm, "The ABCs of digital telephone technology," *Tel. Eng. Manag.,* p. 33, Sept. 15, 1976.

[24] T. Soejima, "Digital transmission systems in NTT," *Jap. Telecommun. Rev.,* p. 97, Apr. 1977.

[25] J. F. Boag, "The end of the first pulse code modulation era in the U.K.," *The Post Office Elec. Eng. J.,* vol. 71, part 1, p. 2, Apr. 1978.

[26] J. S. Whyte, "The United Kingdom telecommunication strategy," *Intelcom 1977,* p. 145.

[27] T. Mirrami, T. Murakami, and T. Ichikawa, "An overview of the digital transmission network in Japan," *Int. Commun. Conf.,* 1978, p. 11.1.

M. R. Aaron joined Bell Laboratories in 1951 after receiving the BS (1949) and MS (1951) in Electrical Engineering from the University of Pennsylvania. Initially, he was concerned with the design of networks for a variety of transmission systems. He made fundamental contributions to computer-aided design and applied these techniques to the development of the first repeatered transatlantic cable system. Since 1956 he has been involved in the development of digital systems. He was a key contributor to the design of the T1 carrier system, the first practical digital system in the world designed to meet the needs of the exchange telephone plant. At present, he has responsibility for exploratory development of digital signal processing terminals and techniques.

Mr. Aaron has had more than 50 papers and patents published in the fields of circuit design, control, and communications. Several of these have been chosen for republication in collections of benchmark papers. He has given many technical talks and several of his poems(?) have cropped up in technical journals.

DIGITAL TRANSMISSION BUILDING BLOCKS *

S. D. Personick

A variety of hardware systems are now being developed and manufactured to provide for the rapidly expanding networks of digital telecommunications.

Digital transmission in **telecommunications** applications began with the introduction of "T1" by the Bell System in 1962. In the "early days," T1 line repeaters cost the equivalent of roughly $500 in today's money. A present day T1 line repeater costs less than $100; it consumes significantly less power; it is significantly smaller and more reliable. Today the appetite of the telecommunications market for digital transmission and switching equipment often exceeds the ability of even the largest equipment manufacturers to respond.

An article by M.R. Aaron [1] in the January 1979 issue of this MAGAZINE traces the history and the growth of digital telecommunications. This paper will provide an overview of some of the specific hardware subsystems and systems used to provide digital transmission between telephone buildings and to interconnect customers to telephone central offices.

THE MARKET AND THE REQUIREMENTS

The simplest way to provide telephone connections between customers is via "pairs" of copper wires contained in telephone cables as shown in Fig. 1. The basic function of the connection is to provide a talking path with a quality acceptable to the customer. Subjective testing by telephone system designers provides an input to the tradeoff which must be made between the cost and the quality of service. Generally, circuits are designed so that the transmission performance is rated good or excellent by a majority of users. Quantitative factors which affect the subjective quality of service are the available bandwidth, added noise and "crosstalk," end-to-end acoustical attenuation, delay, and echo.

In addition to providing a talking path, certain other functions have been designed into the telephone system. Customers lines are constantly and automatically scanned to determine whether a customer has gone "off hook." Customers expect to receive a dial tone in a very short interval (typically a few hundred milliseconds) after lifting their handset. The telephone switch located in a *central office* (or remote housing) must be prepared to receive dialing information communicated by rotary and Touchtone® dialing techniques. For incoming calls, a low-frequency high-voltage signal is applied to the customers line to activate a ringer. A battery must also be applied. On "trunks" between telephone buildings, similar on-hook, off-hook, and dialing information (signaling) must be exchanged.

When metallic connections are replaced by electronic carrier systems as shown in Fig. 2, all of the functions described above must be implemented in a transparent fashion. That is, existing customer equipment and telephone office equipment must interconnect and operate with the carrier system as if it consisted of parallel wire pair paths.

In the real world of telecommunications, there are survivability requirements which must be met in addition to the functional requirements outlined above. Lightning strikes on cables and towers, short circuits caused by cable damage or maintenance actions, and large extraneous voltages and currents are everyday occurrences which must be accomodated with minimal damage to equipment—in many cases without disrupting service.

The author is with TRW/Vidar, Mountain View, CA 94040.

*Reprinted from *IEEE Communications Society Magazine,* January 1980, Vol. 18, No. 1, pp. 27–36.

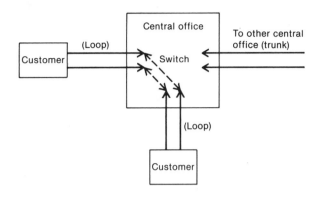

Fig. 1. Wire pair interconnections.

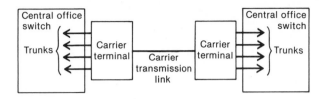

Fig. 2. Carrier trunk system.

These survivability requirements can significantly increase the cost of telecommunications equipment. and can significantly complicate the designs.

T1: WHERE IT ALL BEGAN

The telephone circuits which go between telephone buildings are called *trunks*. These circuits are shared among customers on a request basis using switching machines to assign available trunks as required. When the distance between telephone buildings is sufficiently large (and for some digital switching machine applications at any distance) it is uneconomical to allocate separate physical pairs of copper wires for each voice circuit. Instead, voice circuits are first multiplexed together (combined) to form a composite signal which is then transmitted. Unless the switching machine itself processes signals in multiplexed form, there is a cost at each office for a terminal to do this multiplexing operation. However, if the transmission cost per circuit is

lower for the multiplexed signal, then there is a "prove-in distance" where multiplexed transmission savings are greater than the terminal costs. Before 1962, multiplexing of analog circuits was accomplished by stacking voice channels in 4 kHz frequency bands. A terminal which combines individual voice circuits in such a frequency division multiplexed (FDM) format is called an analog "channel bank." In addition to providing multiplexed voice circuits which can be demultiplexed without crosstalk, the channel bank must accept the signaling information (on-hook off-hook, and dialing) of each channel, and transmit it to the distant end in a transparent fashion. This is typically done in analog channel banks with audio tones. The composite broadband analog signals can be transmitted on wire pair transmission systems and radio transmission systems.

The T-carrier system introduced in 1962 uses digital channel banks to convert up to 24 incoming analog voice circuits to a 1.544 Mbit/s time division multiplexed (TDM), pulse code modulation (PCM) format.

Fig. 3 shows a block diagram of a conventional *D* channel bank. Incoming analog voice circuits can be in either "two-wire" or "four-wire" format, that is, either both directions of transmission travel over the same pair of wires, or the transmission directions are separated.

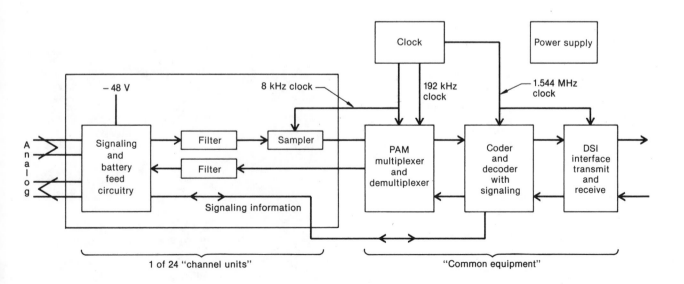

Fig. 3. *D* channel bank.

Two-wire incoming circuits must be converted to four-wire format using a "hybrid" as shown in Fig. 4. In order for the hybrid to function well, the balancing impedance must match that of the incoming two-wire circuit. If it does not various echo paths can be formed as shown.

The "channel unit" is the interface between the incoming analog signal and the "common equipment." In a traditional D channel bank, in addition to two-wire to four-wire conversion, the channel unit detects and generates the appropriate signaling information (e.g., on-hook off-hook transitions, E and M signaling); provides –48 V office battery to the analog wire pairs; filters and samples incoming analog signals; and filters outgoing samples from the common equipment. The interfaces between the channel units and the common equipment are: transmit and receive pulse amplitude modulation (PAM) samples; signaling information; and an 8 kHz clock signal for the sampler. The common equipment accepts the PAM and signaling interfaces of 24 channel units. Analog PAM samples are time division multiplexed (interleaved), and the resulting composite PAM signal is converted to PCM using a shared coder–decoder. "Framing bits" are added to identify individual channels. Signaling information is also added by occasionally "robbing" bits from the voice information. The resulting composite digital interface is a 1.544 million bit/s waveform suitable for transmission over two pairs of wires. This precisely specified interface is called a DS1 interface.

Recently, the architecture of some D channel banks has been modified because of advances in digital LSI technology. The PCM coding function has been placed on the channel unit card rather than shared in the common equipment. Thus 24 PCM signals (64 kbits/s each) derived from individual voice circuits are multiplexed in the common equipment instead of multiplexing

analog samples and using a shared codec. The tradeoff is predominantly economic, depending on the relative costs of shared and per-channel codecs. However, there are some technical and system differences as well. With a per-channel codec, the channel unit interface to the common equipment is digital. Thus, to provide high speed (up to 64 kbit/s) data interfaces, a voice channel unit can be replaced by a data channel unit. Also, under some circumstances, crosstalk between channels can be lower if per-channel codecs and digital multiplexing are used.

Fig. 5 shows a photograph of a typical modern D channel bank with 24 channel unit cards (vertical) and 4 common equipment cards (horizontal). Different types of channel unit cards are used for the multiplicity of analog trunk interface types (2 wire, 4 wire, dial pulse signaling, E and M signaling, foreign exchange, etc.). A fully equipped channel bank currently costs around $4000.

DS1 signals are transmitted over two pairs of wires for distances of up to 6000 feet depending upon the wire guage and other factors. The attenuation of the wires and crosstalk between pairs periodically necessitates a digital repeater in order to keep error rates at acceptable levels (less than 10^{-6} from end to end).

Fig. 6 shows a block diagram of a typical T1 digital repeater. The repeater is powered directly over the wire pairs carrying the DS1 signals using a constant current "simplex" powering scheme.

The function of the "automatic line buildout" (ALBO) is to compensate for the various frequency dependent losses of different cable lengths and cable types. A typical ALBO can adapt to cables having between 7.5 and 35 dB of loss at the "half baud" (772 kHz). The ALBO makes the equalizer input appear to be a result of passing a DS1 output signal down 6000 feet of standard 22 guage cable. The "equalizer" compensates for the frequency depen-

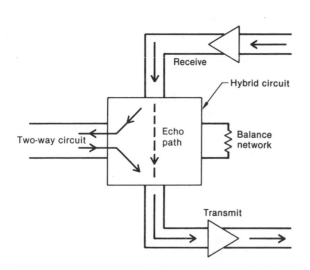

Fig. 4.   Two wire–four wire hybrid.

Fig. 5. Typical *D* channel bank.

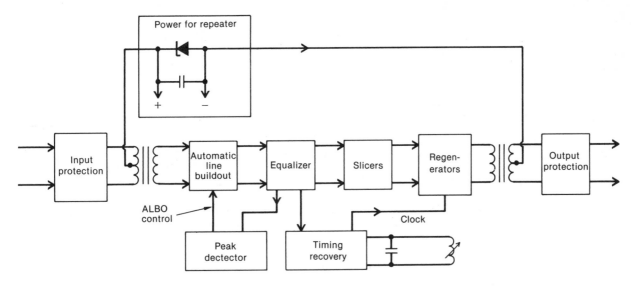

Fig. 6. T1 repeater (one way shown).

dent loss of the cable—removing the associated inter-symbol interference and restoring the signal to a standard level. The timing recovery circuit derives a stable 1.544 MHz clock from the incoming signal by nonlinear means. This clock is used to sample and regenerate the equalized waveform to a new DS1 output. Input and output protection circuits consist of surge resistors and zener diode arrays to protect the repeater from lightning surges and power crosses.

Fig. 7 is a photograph of a typical two-way T1 minirepeater. It requires 7.5 V at 60 mA nominal and it costs less than $100.

Fig. 8 shows a typical T carrier system including channel banks and repeaters. The possibility of automatic transfer to recover from the failure of a T1 transmission line is indicated. Automatic transfer will be treated later in the section on maintenance and protection.

Outside plant repeaters must be housed in protective containers. A typical repeater housing is shown in Fig. 9. It is suitable for pole and manhole applications. Just as the electronic equipment must meet stringent functional and survivability requirements, so must the repeater housing. Crosstalk between repeaters due to wiring and connectors must be extremely low. Grounding must be adequate to protect against lightning strikes to the incoming cables. Some applications require the housing to hold air pressure to keep out water. For outside above ground installations, a standard shotgun survivability test must be passed to meet REA (Rural Electrification Administration) requirements.

The popularity of the T carrier is phenomonal due to its low first cost and its excellent maintainence record. Today, in metropolitian areas it is used almost exclusively to provide interoffice trunks. The prove-in distance between analog switching machines is around 4 miles. Between digital switches, T carrier proves in at zero miles, since channel banks are not needed.

HIGHER SPEED SYSTEMS FOR WIRE PAIR AND COAXIAL CABLE

The remarkable success of T1 carrier has spurred interest in higher capacity systems which can reduce transmission costs. By combining several DS1 signals together in a terminal called a multiplexer, one can share the transmission medium among more circuits. Once again, the cost of the multiplexer must be offset by the savings in transmission. In the United States, several standard transmission rates have been established. The DS2 rate of 6.3 Mbits/s allows four DS1 signals to be multiplexed. A transmission system called T2 has been introduced by the Bell System to carry DS2 signals over wire pairs. Unfortunately, the increased cable loss and crosstalk at the half baud (3.15 MHz) of T2 does not

Fig. 7. Typical two-way T1 minirepeater.

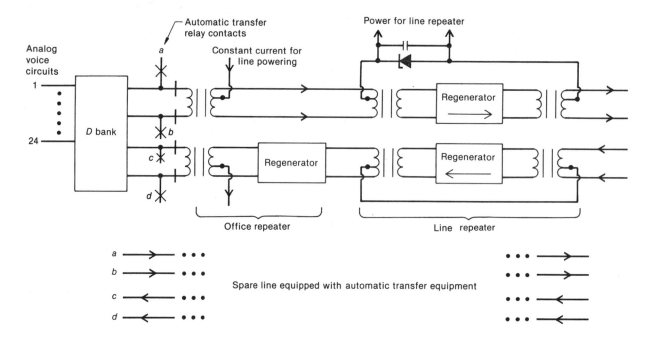

Fig. 8.   T carrier system.

allow its use on standard exchange grade cables with standard 6000 foot repeater spacing. In T2 applications, a special cable is used which is generally compatible with 12 000 foot spacing. Since T2 cannot be used with existing exchange cables, its popularity is much less than that of T1.

The DS4 rate of 274 Mbits/s allows 168 DS1 signals to be transmitted simultaneously. The large bandwidth requires the use of coaxial cables. The T4M and LD4 systems have been used in the United States and Canada to provide large capacity superhighways between telephone offices, with 1 mile repeater spacing. Although the T4M repeater is much more complex and expensive than a T1 minirepeater, its cost is shared among 168 times as many circuits. Thus, when circuit requirements justify, it is economically competitive with

T1. Unfortunately, the large circuit capacity of T4M limits its application to extremely high density routes in selected metropolitan areas. Fig. 10 is a photograph of a T4M repeater.

In recent years there has been considerable effort made to develop systems which allow twice the DS1 rate to be transmitted over existing T1 cables with the same repeater spacing and other engineering rules. Increased cable loss, increased crosstalk, and the use of twice the data rate reduce the T1 signal-to-noise margin by about 25 dB when one tries to scale a T1 system up to a double rate "T1C" system. To accommodate this, special engineering rules are required which limit the applications severely. Improved repeater housings and reduced-bandwidth repeater designs have been used with moderate success to increase the number of instances where T1 repeaters can be "retrofitted" with a double rate type. No perfect retrofit system has yet been developed although the "duobinary" approach introduced by Lenkurt has come very close.

In anticipation of DS1C transmission systems, and to minimize terminal costs for multiplexers, special *D* channel banks have been introduced which generate DS1C signals directly from 48 voice circuits with a minimum of interface circuitry.

OTHER TRANSMISSION MEDIA

The widespread acceptance of digital transmission for metropolitan trunking over wire pairs has led to a proliferation of DS1 signals in the telephone network. Indeed, the transmission of telephone conversations is

Fig. 9.   Nonmetallic repeater case for miniature T1 repeaters.

Fig. 10. Bell T4M repeater (shown open to illustrate inside parts).

rapidly being reduced to the transport of these 1.544 Mbit/s DS1 signals. In applications where cable is inappropriate, or suboptimal, other media can be used to carry individual or multiplexed DS1's. Digital radio has been recently introduced into the telephone plant. Digital satellite and fiber optic systems will soon follow. A digital radio accepts signals directly in digital format (DS1 signals or otherwise) and uses these signals to modulate a radio carrier.

Traditional microwave radio systems have used analog multiplexed (FDM) signals to frequency modulate a microwave carrier. In this process a certain efficiency of spectrum utilization has been achieved. Since available radio channels are at a premium, FCC regulations require digital radios to carry roughly the same number of voice conversations per unit bandwidth as the analog systems they compete with. The requirements vary with the frequency band. For 2 GHz radios, the available spectrum is broken into 3.5 MHz slots. The required spectral efficiency is 1.8 bits (of standard PCM encoded voice information) per second per hertz of bandwidth. In the 30 MHz wide 6 GHz slots and the 40 MHz wide 11 GHz slots, the requirements are 2.6 bits/s · Hz and 1.9 bits/s · Hz, respectively. In the 2 GHz band, several

approaches have been used to transmit four or more DS1 signals in each 3.5 MHz wide slot.

One approach is to divide the 3.5 MHz slot into two pieces. A separate RF carrier is used in each half. Two DS1 signals independently modulate the two quadrature phases of each carrier. This method eliminates the need to synchronously multiplex the DS1 signals. Order wire and alarm signals must be added separately, perhaps using superimposed frequency modulation of the carrier signals.

Another approach is to synchronously multiplex 4, 6, or 8 DS1 signals and to use the composite high data rate signal to modulate a single carrier. Order wire and alarm signals in digital format can be multiplexed in with the main payload.

In the higher frequency broadband 6 and 11 GHz slots synchronous multiplexers are typically used to combine up to 56 DS1 signals together with order wire and alarm information. The high-speed signal (90 Mbaud) modulates a single carrier using one of several modulation techniques (PSK, QAM, etc.).

In addition to meeting the FCC requirements on spectral efficiency and out of band radiation, all radio systems must contend with multipath fading and rain

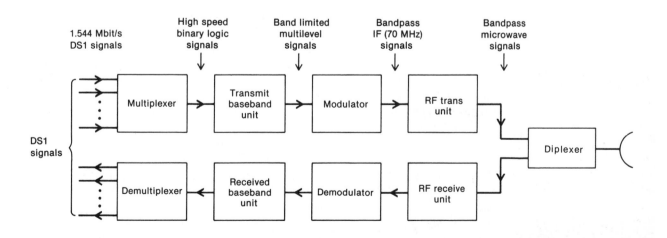

Fig. 11. Digital radio (protection switching circuitry not shown).

attenuation. As the name implies, multipath fading occurs when several microwave paths of differing delays exist between the microwave transmitter and receiver antennas. If the delay difference at a particular frequency is an odd multiple of 180° and if the signals have comparable amplitudes, then significant cancellation can occur resulting in tens of decibels of signal drop. During a fade, the attenuation may be a function of frequency across the bandwidth occupied by the digitally modulated radio carrier. This "frequency selective fading" distorts the modulation, typically causing intersymbol interference. Thus for 6 and 11 GHz signals occupying a wide bandwidth, multipath selective fading is more severe than flat attenuation. It is difficult to simulate multipath selective fading in the laboratory. Thus early systems in the wide 6 and 11 GHz slots may produce some surprises. In the narrower 2 GHz slots, the problems of selective fading are essentially negligible.

Even without selective fading, flat fading will cause system outages if adequate margin is not engineered into the design of a radio link. During a fade, not only will the desired signal be reduced by tens of decibels relative to noise, but also interference from other radio signals becomes a serious problem.

One solution is to use shorter radio hops. This significantly reduces the fading, which is a rapidly increasing function of the path length. With short hops and minimal fading, lower power transmitters and smaller antennas can be used thereby significantly reducing radio system costs.

Another approach to reducing the interference problems during fades is to transmit low-power signals except during fading. The receiving end of a link will call for more power, as needed, from a distant transmitter over the reverse direction transmission.

Fig. 11 shows the block diagram of a typical 2 GHz digital radio exclusive of protection switching and service channel circuitry. Up to six incoming DS1 signals are synchronized and multiplexed together to form a 9 Mbaud data stream.

The transmit baseband unit separates this into two 4.5 Mbit/s two-level streams and converts each of these to a 2.25 Mbit/s bandlimited four-level signal. The 2.25 Mbit/s signals amplitude modulate the quadrature phases of a 70 MHz IF carrier. The IF carrier is mixed with a stable local oscillator to produce a 2 GHz signal which is amplified to the desired level in a linear fashion (to avoid the generation of excessive levels of out-of-band components). Transmit and receive signals, which are at different frequencies, interface to the same antenna via a diplexer. The allowed loss between the transmitter output and the distant receiver input is in excess of 100 dB. Since six 1.544 Mbit/s signals share a 3.5 MHz frequency slot, the bandwidth efficiency is 2.7 bits/Hz · s. Fig. 12 is a photograph of a 1:1 protected two-way radio rack including the power supplies, multiplexer,

Fig. 12. Two-way digital radio rack.

RF transmitter-receiver, order wire, and alarm equipment.

The newest media to be candidates for digital transmission are satellites and fiber optics. Satellites offer transmission with a cost independent of distance. Traditionally satellite channels have been shared by analog voice circuits on an FDM basis. Recently much interest has been focused on time division multiple access satellites, where many ground stations share the same satellite transponders on a carefully coordinated time interleaved basis. To further increase the efficiency of satellite spectrum usage, the transmit and receive beams of the satellite can be scanned periodically over a large area to provide both spatial and temporal sharing.

Fiber optic systems are most attractive in digital modulation applications, because they offer an excellent tradeoff between bandwidth and available signal-to-noise ratio. Fiber costs (present and projected) are significantly more than those of wire pairs, so fibers must be shared by multiplexed DS1 signals, in the same way as coaxial cable is shared. Because of this, fibers are best suited for high density routes of sufficient length to justify the multiplexing cost and the high capacity. The advantages of fiber systems are long repeater spacings (several miles) small cross section of fiber cables, immunity to interference, and with proper cable design, immunity to lightning and ground faults.

Fig. 13 shows a block diagram of a typical fiber optic system. The optical transmitter containing an LED or a laser diode is on–off modulated by the binary digital signal emanating from the multiplexer. The optical receiver contains a photodiode detector to convert light back to a current for amplification, filtering, and regeneration.

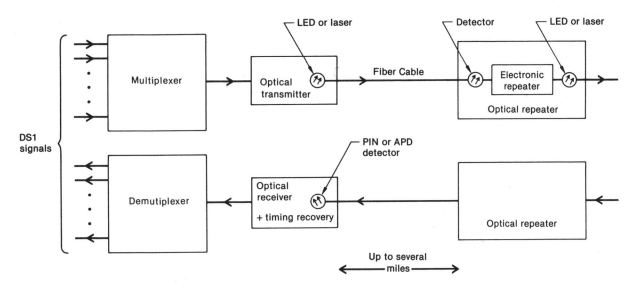

Fig. 13. Fiber optic system.

Fig. 14 shows (left to right) a fiber optics transmitter card, a timing recovery regenerator card, and a fiber optics receiver card. The square module on the transmitter card contains an LED. An armored fiber "pigtail" cable terminating in an optical connector is clearly visible. This module interfaces with a TTL or balanced ECL logic signal to produce optical pulses at up to a 20 Mbit/s rate. The module on the receiver card contains an avalanche photodiode. The receiver sensitivity allows for up to 40 dB of attenuation between it and the transmitter. Automatic gain controls on the receiver card provide for a 1 V peak-to-peak output (unregenerated) for transmitter-to-receiver losses of between 0 and 40 dB.

SPECIAL TERMINALS

In addition to the standard D channel banks that interface 24 analog trunk circuits to a DS1 transmission system, there are several types of terminals for other applications which also interface to DS1 signals.

Fig. 15 shows a subscriber carrier system. Telephone customers who are located far from a telephone central office can have their lines terminating on a nearby "remote" terminal. This terminal converts their two wire analog signals to digital form and multiplexes 24 subscribers together to form a 1.544 Mbit/s DS1 signal. The quality of the customer's connection is governed by the

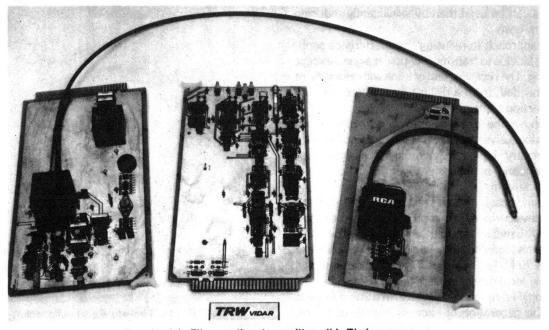

Fig. 14. (a) Fiber optics transmitter. (b) Timing recovery regenerator. (c) Fiber optics receiver.

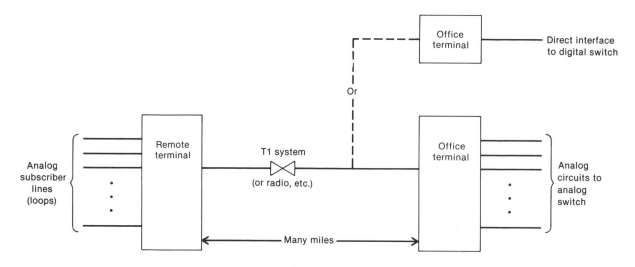

Fig. 15.   Subscriber carrier system.

distance of the customer from the remote terminal and by the quality of this subscriber terminal. Service quality to distant customers can thus be dramatically improved. In addition, the transmission cost to the office is shared among many customers.

At the central office end, the DS1 signal can be converted back to analog form or can interface directly with a digital switch. The subscriber carrier terminal must provide the same functions as a central office termination. It must detect when customers go off hook, it must ring telephones, operate with coin phones and multiparty lines, etc. Fig. 16 is a photograph of a remote subscriber terminal in an outside plant housing. The terminal contains a −48 V battery which provides for continuous service during a failure of local ac power. The terminal can also test subscriber lines and the quality of the overall connection, under the remote control of the central office. Commands to perform these tests are sent over the DS1 link using signaling bits built into the DS1 format.

Recently there has been a great deal of interest in taking advantage of the 'redundancy of speech to increase the number of circuits which can share a DS1 line. One approach is to code the voice signals more efficiently, using less than 64 kbits/s per voice conversation. Another approach is to take advantage of the silent periods of normal speech (typically only one person in a conversation is talking at a given time) to interleave more conversations. This later approach is used on analog ocean cable links and is called TASI (time assignment speech interpolation). Advances in various adaptive modulation techniques, improved "companding" techniques, and digital implementations of TASI will lead to a variety of bit-rate-efficient terminals in the near future. Indeed some terminal vendors already offer special channel units that share a 64 kbit/s interface between two voice circuits.

Fig. 16.   TRW Vidar D3 SCT subscriber carrier system.

The terminals described above interface individual analog circuits to DS1 lines. There are applications where it is desirable to interface FDM analog signals to digital lines without the expense of back-to-back digital and analog channel banks. For instance one might try to interface two analog "super groups" of 60 voice circuits each to five DS1 signals.

One approach is to essentially build back-to-back terminals into a single terminal which minimizes interface circuitry, jacks, power supplies, etc. Another approach is to sample the FDM analog signal at a high rate, encode it with a very high accuracy A/D converter, and then to use digital processing to convert the encoded signal to multiple DS1's. Terminals which perform this FDM to TDM interface are called transmultiplexers.

Other terminals worth mentioning are those which accept a combination of voice and data signals. Also

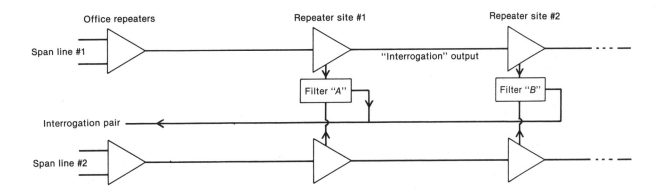

Fig. 17.   T span fault location system.

noteworthy are the new terminals which include simple encryption devices on the digital side to provide reasonably economical privacy.

MAINTENANCE AND PROTECTION

Telephone equipment—particularly outside plant equipment—must be simple to maintain, and to fault locate, and in some cases must automatically protect itself from single point failures.

Easy maintenance of T1 lines is provided by the audio tone fault locating technique. Referring to Fig. 17, separate outputs are provided on T1 repeaters which essentially duplicate the main output signal. These outputs can be connected in parallel at a given repeater site to a narrow-band audio filter. The filter outputs from various sites bridge on to an auxiliary fault locating pair. The audio filter at each repeater location along a T span is tuned to a different frequency (12 frequencies are used). To test a T line for problems, a special signal is used to drive the office repeater. This signal has an imbalance of positive and negative pulses which varies in

T carrier spans one extra span line can act as a spare for several active lines. Special circuits monitor each active line for faulty operation (indicated by code restriction violations), and automatically transfer both ends to standby if appropriate. Obviously such transfer equipments must be more reliable than the lines they protect. Sophisticated handshaking routings help assure that proper "transfers" and "resets" actually occur. In many cases, transfers will activate remote alarms to alert craftspeople to the occurrence of a failure.

The revolution in integrated technology has made remote electronic terminals, like the subscriber carrier terminal described above, economically feasible. Such terminals provide improved service and reduced cost to distant customers. However, they create maintenance challenges. One can imagine a subscriber carrier terminal several tens of miles distant from an attended office, interconnected by T carrier, radio, etc. Trouble reports and routine maintenance of the subscriber's lines must not necessitate costly travel to the remote terminal, except when necessary to perform repairs. Even then, it is desirable to diagnose the problem from the attended

---

The appetite of the telecommunications market for digital transmission and switching systems in many cases exceeds the ability of the giant manufacturers to develop and produce the equipment.

---

polarity at an audio rate. The level of audio return from a given location (via the local filter and fault locate pair) is used to measure the health of the repeater at that location. To "interrogate" a specific repeater on a chosen line, one varies the pulse imbalance at the audio frequency corresponding to the filter at the desired site.

In many applications, the desire to maintain uninterrupted service necessitates automatic protection switching equipment.

Such equipment must detect faulty operation of a digital transmission link and switch to a standby link. On

office, so that the appropriate personnel and equipment can be dispatched. Therefore, sophisticated remote testing techniques have been devised which allow remote instruments to make measurements at unattended equipment locations under central office control, and to return the results of those measurements back to the central office. The automatic testing of subscriber loops in a subscriber carrier terminal (described above) is an example. The challenge is to make these maintenance features economical enough so as not to undermine the advantages of the remote terminal.

CONCLUSIONS

Since the introduction of T1, digital transmission and switching of telephone signals has grown at an astounding rate. To a large extent its growth is limited in many cases only by the ability of the giant telecommunications manufacturers to develop and produce the equipment. In the future, two-way digital voice and data links directly into the customer's premises, digital transmission of two-way and entertainment video, and a proliferation of new services made economical by digital processing will provide many opportunities for reduced business and personal travel, increased and enhanced leisure activities, enhanced security and property protection, and increased educational opportunities.

ECHO CANCELLATION IN THE TELEPHONE NETWORK*

Stephen B. Weinstein

Abstract　Most long-distance telephone connections generate echoes, which must be heavily attenuated in order to obtain satisfactory transmission quality. Voice-actuated switches (echo *suppressors*) are widely used to eliminate echoes but have an unfortunate tendency also to cut out part of the desired signal from the other end of the line. Because the distortion caused by echo suppressors is particularly noticeable on satellite-routed connections, the advent of telephone communication via satellite, including the recent introduction of satellite circuits into the U.S. domestic network, has motivated the search for a better way to eliminate echoes. The answer may be the echo *canceler*, an adaptive filter which selectively eliminates echoes. Advanced echo canceler designs have been undergoing field trials in recent years. This article explains why echo cancelers are advantageous and how they work.

INTRODUCTION

When we speak, we like to hear what we are saying, but only if we hear it right away. Studies of subjective reaction to echo in the telephone network have shown that it is difficult to carry on a conversation when one's own voice returns with a delay of more than a few tens of milliseconds. Delayed echoes as much as 40 dB below the outgoing voice level will cause some speakers to characterize the transmission quality as unsatisfactory, particularly since the bulk of our telephone experience is with connections on which distant echoes have been suppressed. It must be emphasized that only long-delayed echoes are irritating; shorter echoes, called "sidetone," are actually desirable and are intentionally inserted to keep the telephone from sounding "dead."

Data communication equipment is just as sensitive to echo interference as are human beings. There is, however, an important difference. Modems are disturbed by echo energy of *any* delay, which encourages the consideration of echo elimination at station locations rather than at points within the telephone networks. More will be said about this later.

Long-delayed echoes are observed only on long-distance connections. Fig. 1 is an illustration of a typical long-distance telephone circuit; although physically simplified, it is a good model as far as echoes are concerned. It represents a long-distance connection as a tandem sequence of a two-wire portion (including the local loop which connects a customer to the telephone office), a four-wire carrier facility, and another two-wire portion at the other end of the line. Signals travel in both directions on the two-wire segments, but are transmitted through two distinct channels, one for each direction, in the four-wire segment.[1]

Echoes are the result of impedance mismatches in the communication circuit. A substantial amount of echo energy comes from mismatching of impedance at the *hybrid couplers* [1] which, as Fig. 1 indicates, are located at the two-wire/four-wire interfaces in the telephone circuit. These devices,

The author is with Bell Laboratories, Holmdel, NJ 07733.

Reprinted from IEEE Communications Society Magazine, January 1977, Vol. 15, No. 1, pp. 9–15.

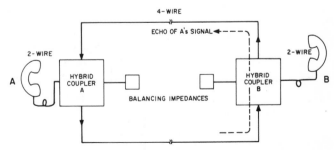

Fig. 1.　Model for long-distance telephone circuit. The four-wire carrier section may possibly include a satellite link.

[1] These channels are likely to be among many telephone channels multiplexed into broad-band radio or cable signals.

made up of inductive elements in a balanced bridge configuration, are passive four-port directional couplers. The ideal hybrid coupler (Fig. 2) passes an incoming signal from its "in" port to its "two-wire" port, attenuating it by 3 dB, and does not pass anything to the "out" port. Conversely, it passes (again with 3-dB attenuation) signals from the "two-wire" port to the "out" port, without reflecting any energy back into the two-wire line. However, this ideal performance is obtained only if the balancing impedance, shown in Fig. 2, is equal to the impedance presented by the two-wire line. Since many different local loops have access to a given hybrid coupler, it is unlikely that a fixed balancing impedance will always be satisfactory, and it should come as no surprise that considerable "leakage" and reflection signal may be generated by a hybrid coupler.

The attenuation from "in" to "out" port of the hybrid coupler, ideally infinite, is actually observed to have an average value of about 15 dB with a standard deviation of 3 dB (if we neglect the 3-dB losses associated with any transfer from one port of the hybrid coupler to another). The echoes that result are generated at the hybrid itself and at all other impedance mismatches in the telephone circuit, including those at the telephone set or other terminating device. No echoes will return from within the four-wire part of the circuit since the channels here are unidirectional. It can be seen from Fig. 1 that a long-delayed echo will result from speaker A's signal "leaking" through hybrid coupler B and returning to A's location.

Most of the remainder of this article will concern the elimination of voice echoes and will first describe the currently implemented technique of echo *suppression* before introducing the new proposals for echo *cancellation*. A discussion of the somewhat different proposals for cancellation of echoes of data signals is provided toward the end of the article.

ECHO SUPPRESSORS

Telephone companies realized early that some way had to be found to deal with echoes. It was noted that most con-

versations consist largely of *single talking,* when one person speaks at a time, in contrast to *double talking* when both parties are speaking simultaneously. This means that, at least for connections not characterized by long delay, returning echoes almost always return by themselves, and not mixed with a desired distant signal. Under these conditions, it is reasonable to install a complementary pair of switches, one near each four-wire/two-wire interface, such that the single active speaker sets the switches to pass his signal and block all returning signals. Special provision for closing both switches or inserting a moderate attenuation in both transmission directions can be made for those brief periods when double talking does occur. This voice-actuated switch, illustrated in Fig. 3, is called an *echo suppressor,* and has been in general use in the United States since the late 1920's [2].

The echo suppressor, when operating properly, has been a quite satisfactory solution to the echo problem on circuits with moderate (perhaps less than 100 ms) round-trip delay. It does, of course, have a tendency to "chop" during back-and-forth conversation [3], and it may clip out a noticeable segment at the beginning of an utterance, but telephone users have had a favorable reaction to the echo suppressor. Considerable effort continues to be put into improving these units [1], [4].

On circuits with long (more than 200 ms) round-trip delay, e.g., satellite-routed circuits, the distorting effects are much more pronounced. The reasons are not completely clear, but Brady and Helder [3] have suggested how the suppressors can disrupt a very interactive conversation. Because of the long delay in the circuit, a quick response by speaker B to something said by speaker A may cause suppression of something said by speaker A at a later time. The deletion is noticed by B, encouraging this speaker to stop and wait for A to get through. The resulting confusion, which may stop conversation entirely while each party waits for the other to say something, is only partially alleviated by restraining suppression during double talking.

SELECTIVE ECHO ELIMINATION
BY ECHO CANCELLATION

The impedance mismatches which cause echoes also tend to distort them. The problems associated with the echo suppressor could be largely avoided if a way could be found to simulate the transformation undergone by speaker A's signal as it travels from point a in Fig. 3 to point b where it is an echo on its way back to speaker A. If this transformation were known, it could be simulated in a filter with input from point a, thus generating a copy of the echo signal which could be subtracted from the total signal at point b to eliminate the echo without disturbing any desired signal which might also be present at point b. Fig. 4 shows how the filter would be connected. The combination of filter and subtractor is an *echo canceler*.

Unfortunately, the echo channel, although largely linear and unchanging with time (with an important exception to be described later), varies considerably with the connection of

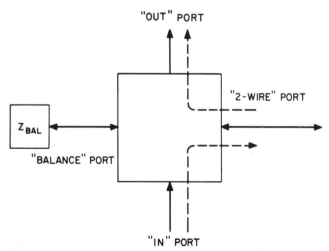

Fig. 2.   Hybrid coupler as a four-port directional coupler.

different two-wire circuits to hybrid coupler *B*, so that a fixed filter cannot be used in the canceler of Fig. 4. Attention, therefore, has been concentrated on adaptive filter structures which can automatically *learn* the echo channel characteristic, take on this characteristic, and adapt to changes in the characteristic. An echo canceler built around such a filter will align itself automatically, without human intervention, to produce the minimum residual uncanceled echo possible within its structural limitations.

The adaptive filter structure which underlies almost all echo canceler designs is the nonrecursive transversal filter (Fig. 5)

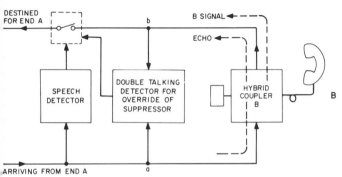

Fig. 3.   Echo suppressor passing speaker *A*'s signal and blocking echo of speaker *A*'s signal. A signal originating from speaker *B* will also be blocked.

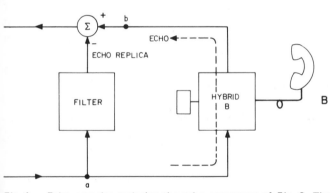

Fig. 4.   Echo canceler replacing the echo suppressor of Fig. 2. The filter simulates the echo channel from point *a* to point *b*.

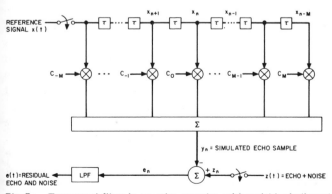

Fig. 5.   Transversal filter in an echo canceler, with variables indicated for time *t* = *nT*. Samples are taken at rate >1/2*f*max. The reference signal is found at point *a* in Fig. 4, and the echo + noise signal at point *b*.

adapted according to a gradient algorithm which minimizes mean-square output error. Although the structure is generally credited to Norbert Wiener, the minimum mean-square error (MMSE) adaptation strategy is largely due to Bernard Widrow, who is one of the authors of an excellent recent tutorial paper [5] on the theory and broad range of applications of this filter. The use of the MMSE strategy specifically for voice echo cancellation was first proposed in two papers published in the *Bell System Technical Journal* in 1966, one by Becker and Rudin [6] and the other by Sondhi and Presti [7]. Much work was done subsequently at Bell Laboratories [8]-[12], at COMSAT Laboratories [13]-[15], and by workers in Australia [16], Germany [17], France [18], and Japan [19].

MINIMIZING MEAN-SQUARE ERROR VIA A GRADIENT ALGORITHM

Although the reader should consult [5] or [12] for a more extensive derivation and discussion of the MMSE cancellation algorithm, a heuristic development at this point may be helpful in understanding the remainder of this article. Referring to Fig. 5, assume we have a delay line tapped at time intervals τ (or in digital implementation, a shift register shifted at intervals τ), where τ is less than or equal to the reciprocal of twice the highest significant frequency component f_{max} of the echo waveform. At any instant of time, delayed samples of the reference signal appear at the various taps and are multiplied by *tap weights* $c_{-M}, \cdots, c_0, \cdots, c_M$. These products are summed to form the filter output.

A transversal filter of infinite length can realize any transfer function on the frequency interval $(-f_{max}, f_{max})$. This can quickly be seen by writing down the filter impulse response (which is nothing more than a sum of delayed impulses) and taking its Fourier transform:

$$g(t) = \sum_{-\infty}^{\infty} c_m \delta(t-m\tau) \qquad (1a)^2$$

$$G(f) = \int_{-\infty}^{\infty} g(t)\exp[-j2\pi ft]\,dt = \sum_{-\infty}^{\infty} c_m \exp[-j2\pi mf\tau]$$

$$= \sum_{-\infty}^{\infty} c_m \exp[-jm\pi f/F], \qquad (1b)$$

where $F = 1/2\tau$. The last summation is a Fourier series for any well-behaved function on the interval $(-F,F)$, and can obviously be designed, by proper choice of the tap weights, to realize the desired echo transformation.

It is, however, our desire to get by with a transversal filter of finite length, and to have its tap weights adjust themselves to generate the best possible approximation to the echo chan-

---

²For simplicity of presentation, a nonrealizable impulse response is assumed. The actual impulse response of a finite-length filter will exhibit a delay of approximately half the length of the delay line.

nel. The MMSE approach to this problem is to establish and attempt to minimize a cost function.

$$\langle e_n^2\rangle = E\ [(\text{residual echo plus noise at time } n\tau)^2]\qquad(2)$$

where the residual echo is that part of the echo signal sample entering the subtractor of Fig. 5 which is *not* canceled by the transversal filter output y_n. It can be shown [20] that $\langle e_n^2\rangle$ is a convex (cup-shaped) function of the set of tap weights, so that the possibility exists of sliding down the side of the cup to reach the tap weight settings which produce the minimum $\langle e_n^2\rangle$. A *gradient algorithm* is a procedure designed to do just this. The movements in tap-weight space are illustrated in Fig. 6 for a mythical single tap weight; in the actual case of multiple tap weights, the convex function is a surface in a higher dimensional space.

The gradient of $\langle e_n^2\rangle$ (with respect to the tap weight set) will point in the direction of maximum rate of increase of $\langle e_n^2\rangle$, so the gradient algorithm seeking a *minimum* of $\langle e_n^2\rangle$ must move in the negative gradient direction. Using vector notation, the tap adjustment formula is

$$c_{n+1} = c_n - \beta\,\nabla\langle e_n^2\rangle.\qquad(3)$$

where

c_n tap weight vector $(c_{-M},\cdots c_0,\cdots,c_M)$ at time $n\tau$

$\nabla\langle e_n^2\rangle$ gradient of $\langle e_n^2\rangle$ at time $n\tau$ with respect to c_n

β step size.

In practice, rather than compute $\nabla\langle e_n^2\rangle$, it is usual to assume β very small and take the gradient of the *unaveraged* square error. The components of this gradient vector are, of course, partial derivatives of the unaveraged square error:

$$\nabla e_n^2 = \left\{\frac{\partial e_n^2}{\partial c_{-M}},\cdots,\frac{\partial e_n^2}{\partial c_0},\cdots,\frac{\partial e_n^2}{\partial c_M}\right\}.\qquad(4)$$

Although ∇e_n^2, after multiplication by β, will exhibit random fluctuations in addition to a small but consistent tendency in the right direction, the net result over many iterations will be

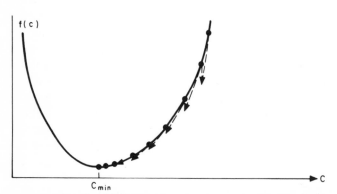

Fig. 6.   Convex function of a single variable, showing how the minimum value of the function can be reached by a series of movements on the negative gradient direction.

equivalent to use of the gradient $\nabla\langle e_n^2\rangle$ of the averaged error.[3]

The gradient vector in (4) can be evaluated by noting from Fig. 5 that

$$e_n = z_n - y_n = z_n - \sum_{m=-M}^{M} c_m x_{n-m},\qquad(5)$$

where

$$x_n = (x_{n+M},\cdots,x_n,\cdots,x_{n-M})$$

is the vector of reference signal samples stored in the transversal filter, z_n is the incoming sample (echo plus noise), and y_n is the output of the transversal filter. Then

$$\frac{\partial e_n^2}{\partial c_j} = 2\,e_n\,\frac{\partial e_n}{\partial c_j} = -2 e_n x_{n-j},\qquad(6)$$

or in vector form,

$$\nabla e_n^2 = -2 e_n x_n.\qquad(7)$$

Using the gradient ∇e_n^2 of the unaveraged error in place of $\nabla\langle e_n^2\rangle$ in (3), and absorbing the factor of 2 into β, we have

$$c_{n+1} = c_n + \beta e_n x_n.\qquad(8)$$

This is the basic gradient adjustment algorithm, and it is so robust that almost any direction-preserving variation on the correction term $\beta e_n x_n$ will work almost as well. For example, it suffices to use sgn (e_n) (the algebraic sign of e_n) in place of e_n. This particular substitution can greatly reduce the implementation complexity [7], is guaranteed to work [10], and does appear in commercially oriented designs [13].

The performance demanded of the echo canceler depends on how big the echo is in comparison with nominal signal levels at the canceler location. Telephone jargon refers to the relative size of the echo (in dB) as echo return loss. The additional suppression provided by the echo canceler is called echo return loss enhancement, and is rarely required to be greater than 30 dB. Some further improvement can be obtained from center clipping, as will be discussed later on.

An important parameter of the echo canceler, which affects its ability to thoroughly cancel an echo, is its length or time span. If the echo channel disperses signals over a time interval longer than the total delay of the transversal filter, there will be some echo energy which the canceler will be unable to reach.[4] The dispersion in most echo paths (defined as the signal path from point a to point b in Figs. 3 and 4) is

---

[3]There is a three-way tradeoff relating the degree of cancellation achieved, the speed of convergence and/or tracking, and the level of the noise (or any other signal) added to the echo. If noise is present and relatively large tap corrections are made (i.e., β is relatively large), then the canceler will be fast, but the large random fluctuations in the tap weights will reduce the degree of cancellation which can be achieved [5]. If, on the other hand, the step size β is relatively small, then it will take longer to converge but the result will be a higher degree of cancellation. Increasing or decreasing the noise level correspondingly worsens or improves the degree of cancellation.

[4]Gitlin and Thompson [30] have proposed placing an additional recursive-type filter in the echo path in order to shorten the echo dispersion.

probably less than 8 ms, but there may be considerable delay in the four-wire path between the echo canceler and the hybrid in those cases (such as in satellite ground stations) when the canceler cannot be located near the four-wire/two-wire interface. When there is large delay, many more transversal filter taps than the 64 (at 0.125-ms intervals) needed to span 8-ms dispersion will be called for, but many will be superfluous on any particular call. It would be desirable to limit the number of active taps (defined as those having weights which are allowed to adapt), both to hold down complexity and because the effects of noise become worse as the number of active taps grows, but the practical difficulty of determining which taps should be allowed to be active has discouraged work in this area, except for that related to cancellation of data echoes at station locations.

It is not the purpose of this article to develop the theory of echo cancellation any further than (8), but it is appropriate to note that a considerable body of analysis exists of the conditions required for convergence of the MMSE gradient algorithm and on the transient and asymptotic statistics of the tap weights and the residual error [5], [20]-[24]. New work in this area can be expected to appear in the literature in connection with proposals for new canceler structures and applications.

THE PERILS OF DOUBLE TALKING

It was noted in footnote 2 that an additive noise contained in the incoming waveform $z(t)$ (Fig. 5) will perturb the tap weights from their correct settings [5]. If the noise is uncorrelated with the reference waveform, as one hopes is the case, then the settings are correct on the average, but fluctuate about their true values. The power of the residual (uncanceled) echo resulting from this fluctuation is more or less in proportion to the noise power and is also a sensitive function of the step size β in (8).

These facts would be academic for a canceler used on a telephone channel, where the random noise is likely to be very small, were it not for the activity of the speaker at the other end of the line (and the possible phase roll discussed in the next section). From the point of view of echo canceler 2 (Fig. 7), which is trying to cancel echoes of speaker A's signal, speaker B's signal is a gigantic additive noise. When double talking occurs, it seems unavoidable that the tap weights will diverge from their proper settings and the echo will not be adequately canceled.

In order to avoid this possibility, the tap weights of currently proposed voice echo cancelers are "frozen" as soon as double talking is detected.[5] A double-talking detector (Fig. 8) must therefore be provided in the echo canceler. The design of a good double-talking detector is difficult and much more of an art than the design of the adaptive filter itself. Even with the assumption of fast-acting double-talking detector, there is

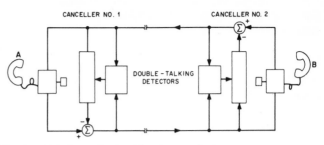

Fig. 7.   Telephone circuit with echo cancelers.

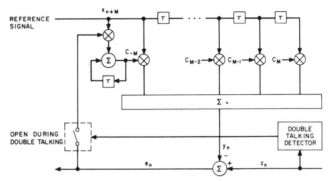

Fig. 8.   Echo canceler with tap-adjustment circuit shown for one tap only (C_{-M}). The double-talking detector stops adjustment when the "other" talker is active.

still the possibility of changes occurring in the echo channel during the time that the canceler is frozen, which will lead to increased uncanceled echo. Fortunately, the duration of a period of double talking is usually very short. A more serious difficulty is presented by the inactivity of the canceler during periods of single-talking in the reverse direction. A properly designed canceler must be capable of extremely rapid adaptation immediately following the end of an inactive period.

PHASE ROLL AND NONLINEAR DISTORTIONS

To add to the problems of effective echo canceler design, it sometimes happens that part of the echo energy returns with distortions which cannot be replicated in a fixed or slowly adapting transversal filter. A form of distortion which is becoming less common in the United States but is cause for concern in some other countries is *frequency offset,* which refers to a shift (rarely more than 1 or 2 Hz) in all frequency components of a signal. Frequency offset, which is associated with the modulation and demodulation operations at the end of a carrier link, is perceived by an adaptive canceler as a constantly increasing phase, or "phase roll," in the echo signal, and the canceler can effectively cancel only if its adaptation can keep up with the phase roll. Unfortunately, as we have already emphasized, faster adaptation in the presence of additive noise implies less cancellation.

Rather than attempt to compromise between speed and effectiveness, one can provide a separate phase-tracking unit. One implementation [16] makes successive computations of tap correction during each sampling interval, but more advanced structures based on a frequency tracking loop (second-

---

[5]An alternative approach, which presents some implementation difficulties in digital signal processing equipment and is not needed for normal voice conversation, is to "gearshift" to a small step size β and continue adaptive operation.

order phase-locked loop) are also possible [31]. It is conjectured that the frequency variable can be "locked" during brief periods of double talking (when, as discussed earlier, tap-weight adaptation is turned off), thus avoiding loss of synch. However, it will be impossible to hold synch over long periods of single-talking in the reverse direction.

Nonlinear distortions are much harder to compensate. An effort in this direction was made by Thomas [9], but the difficulty of identifying and characterizing low-level non-linearities has constrained the development of practical designs.

CENTER CLIPPING

Even an echo canceler which is working well will leave some residual uncanceled error. Limitations on the achievable cancellation ratio are imposed by the presence of additive noise, by nonlinear distortions, by echo dispersion beyond the length of the transversal filter, and by digital resolution constraints. Under single-talking conditions, when only uncanceled echo and random noise are returning to the speaker, these small undesired signals can be blocked by a nonlinear device which only passes signal magnitudes above a certain (small) threshold value. Unlike an echo suppressor, which can be viewed as a center clipper with a very high threshold, the center clipper associated with an echo canceler will distort, but not block the double-talking signal. Fig. 9 indicates the input-output relationship of a center clipper and gives an example of a center-clipped waveform.

Berkley and Mitchell [25] have determined that a center clipper which is designed to operate separately on several subsets of the voice spectrum will not only effectively eliminate a substantial echo, but will also not noticeably degrade the desired signal which is present during activity of the speaker at the other end of the line, and thus does not have to be removed during double talking. The center clipper appears to be a way around the problem of eliminating virtually uncancellable small residual echoes, and may be part of any commercially successful echo canceler.

ECHO CANCELLATION FOR TWO-WAY DATA TRANSMISSION

Most of the technical effort in echo cancellation, and correspondingly most of this article, is devoted to voice echo cancellation within the telephone network. These cancelers, if present at or near all internal network two-wire/four-wire interfaces, would still not eliminate the near echoes (sidetone) which, as noted earlier, are actually desirable in voice conversation. Such echoes are the consequence of impedance mismatches at the end station itself and in the network facilities up to the first two-wire/four-wire interface.

For full-duplex (simultaneous two-way) data transmission, however, near echoes are just as damaging to successful receiver operation as are distant echoes. For this reason, and also because echo cancelers are *not* deployed throughout the telephone network, it has been proposed [26]-[29], as illustrated in Fig. 10, that hybrid couplers and echo cancelers be provided at *station locations* to create an echo-free interface between the four-wire modem and the available two-wire telephone line.[6] If this capability could be provided at moderate cost, dialed lines would become more attractive as backups in computer communication networks and as links to networks in which reverse-channel signals for synchronization and other purposes are provided.

Several special problems are associated with echo cancellation for full-duplex data transmission. In contrast to voice conversation, long sessions of double talking (full-duplex operation) will take place. It will not be practical to freeze the echo canceler during double talking because some small changes in the echo channel are likely to take place during a lengthy communication session. Adaptation during double talking is, happily, possible through use of an averaged-gradient algorithm [21], although it is very slow.

Another serious difficulty unique to echo cancellation at station locations is the very large delay which can be exhibited by the distant echo (that returning from the far end of the circuit rather than deriving from the hybrid coupler at the local station or from the impedance mismatches in the circuit

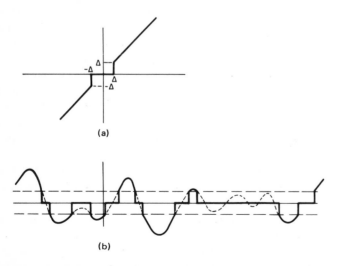

Fig. 9.   Center clipping. (a) Input-output characteristic. (b) Example of center-clipped waveform.

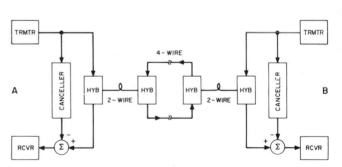

Fig. 10.   Echo cancelers at station locations for full-duplex data transmission over a two-wire line.

[6]This proposal is relevant only for dialed connections. Four-wire private lines are leased at very little more than two-wire private lines, so there is no point in leasing a two-wire private line and going to the trouble of canceling echoes. The cost of two dialed calls, in contrast, is usually twice as much as the cost of one.

up to the nearest two-wire/four-wire network interface). If the circuit includes a satellite link in its four-wire part, the distant echo will be delayed by 500 ms or more. It is clearly impractical, because of complexity constraints and the large tap fluctuation noise, to employ the 4000 or so taps[7] which would be required in a transversal filter spanning dispersions of this magnitude. A reduction by a factor of 4 in the number of taps can be achieved by the use of a symbol-interval transversal filter [27], [28] rather than a voice-type Nyquist-interval transversal filter, but the number is still unreasonably high for satellite-routed calls.

A possible solution is to provide a bulk delay between sections of active transversal filter [26], as illustrated in Fig. 11. This approach will reduce the number of active Nyquist interval taps, regardless of the transmission delay of the circuit, to something of the order of 100. It is, however, necessary to sound the echo channel at the beginning of a call to determine the delay of the distant echo group. In data communications, the allocation of a little time at the beginning of a call to a setup protocol is not an overwhelming difficulty.

ALTERNATIVE STRUCTURES FOR DATA ECHO CANCELERS

Although the transversal filter structure of Fig. 5, which is the essential component of a voice-echo canceler, can be used directly in the system of Figs. 10 and 11, data signals have a peculiarity which suggests modified structures of reduced complexity. This peculiarity is that although a modulated data signal may require sampling at Nyquist intervals $\tau = 1/2f_{max}$ in order to satisfy the sampling theorem, the signal (and hence an echo of the signal) is completely determined by data symbols at substantially larger intervals[8] T. This means that the echo canceler can accept as input data symbols at intervals T, rather than samples of the transmitted signal at intervals τ. Two possible arrangements are shown in Fig. 12.

Cancellation at Nyquist intervals τ [Fig. 12(a)] is possible as before, and the transversal filter in this case will be sparsely filled: symbols will appear at spacings T on the delay line, although the delay line will be tapped at intervals τ [28].

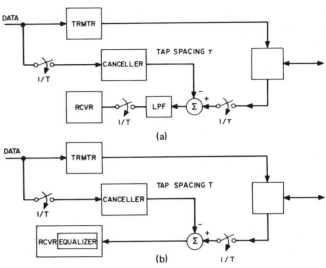

Fig. 12. Data-driven echo canceler structures. (a) Producing samples at Nyquist intervals. (b) Producing samples at symbol intervals $T > \tau$. (The subtractor can be located after the channel equalizer rather than before, implying somewhat different performance.)

Alternatively, canceler outputs can be generated only at symbol intervals T [Fig. 12(b)], since this is the sampling interval of most modems; the transversal filter in this case has taps (and updates tap weights) only at intervals T [27], [28]. The number of operations per symbol interval in the latter case is considerably smaller than that of the former case, but in the latter case operation of the two transmitters at opposite ends of the communication circuit must be fully synchronized.

THE BRIGHT FUTURE OF ECHO CANCELLATION

To this writer's knowledge, no echo canceler has yet been used commercially anywhere in the world. However, members of the family of echo cancelers developed over the last five years at COMSAT Laboratories have been tested on international satellite telephone circuits [15] and, more recently, on U.S. domestic satellite circuits. Comparisons have been made between customer reaction to satellite circuits with echo suppressors and satellite circuits with echo cancelers, and the preference has generally been for the circuits with cancelers.

It is, to be honest, remarkable that a technique which has been studied, favorably appraised, and implemented in experimental hardware for more than 10 years has still not been commercially applied. The obvious reason is that echo cancelers which perform better than echo suppressors have tended to cost much more than echo suppressors; it is intrinsically more difficult to discover and separate an echo from a mixture of signals than to block everything. Despite these difficulties, the combination of increasingly skillful design and new technology has brought the expected cost of effective echo cancellation down to a reasonable level. During the next few years it is conceivable that echo cancelers will be introduced not only into the telephone network, but also into communication station equipment and other systems where an interference derived from a known reference signal must be modified or eliminated.

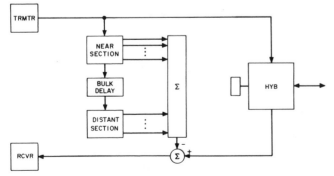

Fig. 11. Echo canceler in which transversal filter incorporates a bulk delay between "near" and "distant" sections.

[7]At Nyquist intervals.
[8]Typically $T = 4\tau$.

ACKNOWLEDGMENT

I am genuinely grateful to R. D. Gitlin, J. F. Hayes, G. K. Helder, J. S. Thompson, and S. J. Campanella for advice and assistance in compiling this article.

REFERENCES

[1]   R. G. Gould and G. H. Helder, "Transmission delay and echo suppression," *IEEE Spectrum*, pp. 47-59, Apr. 1970.

[2]   A. B. Clark and R. C. Mathes, "Echo suppressors for long telephone circuits," *Proc. AIEE*, vol. 44, pp. 481-490, Apr. 1925.

[3]   P. T. Brady and G. K. Helder, "Echo suppressor design in telephone communications," *Bell Syst. Tech. J.,* vol. 42, pp. 2893-2917, 1963.

[4]   "Session on digital echo suppressors," in *Proc. 1976 Int. Conf. Commun.,* vol. III, pp. 36-1 - 36-18.

[5]   B. Widrow, *et al.,* "Adaptive noise cancelling: Principles and applications," *Proc. IEEE,* vol. 63, pp. 1692-1716, Dec. 1975.

[6]   F. K. Becker and H. R. Rudin, "Application of automatic transversal filters to the problem of echo suppression," *Bell Syst. Tech. J.,* vol. 45, pp. 1847-1850, 1966.

[7]   M. M. Sondhi and A. J. Presti, "A self-adaptive echo canceller," *Bell. Syst. Tech. J.,* vol. 45, 1966.

[8]   J. R. Rosenberger and E. J. Thomas, "Performance of an adaptive echo canceller operating in a noisy, linear, time-invariant environment," *Bell Syst. Tech. J.,* vol. 50, no. 3, pp. 785-813, 1971.

[9]   E. J. Thomas, "An adaptive echo canceller in a nonideal environment (nonlinear or time variant)," *Bell Syst. Tech. J.,* vol. 50, no. 10, pp. 2779-2795, 1971.

[10]   M. M. Sondhi, "An adaptive echo canceller," *Bell Syst. Tech. J.,* vol. 46, pp. 497-511, Mar. 1967.

[11]   D. Mitra and M. M. Sondhi, "Adaptive filtering with nonideal multipliers: Applications to echo cancellation," *Proc. 1975 Int. Conf. Commun.,* vol. II, pp. 30-11 - 30-15.

[12]   M. M. Sondhi and D. Mitra, "New results on the performance of a well-known class of adaptive filters," *Proc. IEEE,* vol. 64, Nov. 1976.

[13]   S. J. Campanella, H. G. Suyderhoud, and M. Onufry, "Analysis of an adaptive impulse response echo canceller," *COMSAT Tech. Rev.,* vol. 2, Spring 1972.

[14]   H. G. Suyderhoud and M. Onufry, "Performance of a digital adaptive echo canceller in a simulated satellite circuit environment," *Progress Astronaut. Aeronaut.,* vol. 33, pp. 455-477, 1974.

[15]   H. G. Suyderhoud, S. J. Campanella, and M. Onufry, "Results and analysis of a worldwide echo canceller field trial," *COMSAT Tech. Rev.,* vol. 5, pp. 253-274, Fall 1975.

[16]   N. Demytko and L. K. Machechnie, "A high speed digital adaptive echo canceller," *Austral. Telecommun. Rev.,* vol. 7, no. 1, pp. 20-28, 1973.

[17]   H. Hoge, "Analysis of an adaptive echo canceller with optimized gradient gain," *Siemens Forsch. Entwicklungsberichte,* vol. 4, no. 3, pp. 127-131, 1975.

[18]   A. Castanet, "Auto-adaptive echo canceller for long distance telephone circuits," *L'Onde Electrique,* vol. 55, pp. 14-20, Jan. 1975.

[19]   Y. Kato, S. Chiba, T. Ishiguro, Y. Sato, M. Tajima, T. Ogihara, S. J. Campanella, H. G. Suyderhoud, and M. Onufry, "A digital adaptive echo canceller," *NEC Res. and Dev.,* no. 31, pp. 32-41, Oct. 1973.

[20]   R. W. Lucky, "Techniques for adaptive equalization of digital communication systems," *Bell Syst. Tech. J.,* vol. 45, Feb. 1966.

[21]   ——, "Automatic equalization for digital communication systems," *Bell Syst. Tech. J.,* vol. 44, pp. 547-588, Apr. 1965.

[22]   R. W. Lucky, J. Salz, and H. Weldon, *Principles of Data Communication.*   New York: McGraw-Hill, 1968.

[23]   R. D. Gitlin, J. E. Mazo, and M. G. Taylor, "On the design of gradient algorithms for digitally implemented adaptive filters," *IEEE Trans. Circuit Theory,* vol. CT-20, pp. 125-136, Mar. 1973.

[24]   A. N. Gersho, "Adaptive equalization of highly dispersive channels for data transmission, *Bell Syst. Tech. J.,* vol. 48 pp. 55-61, Jan. 1969.

[25]   O. M. M. Mitchell and D. A. Berkley, "Full duplex echo suppressor using center clipping," *Bell Syst. Tech. J.,* vol. 50, pp. 1619-1630, 1971.

[26]   V. G. Koll and S. B. Weinstein, "Simultaneous two-way data transmission over a two-wire line," *IEEE Trans. Commun.,* vol. COM-21, pp. 143-147, Feb. 1973.

[27]   K. M. Mueller, "A new digital echo canceller for two-wire full-duplex data transmission," *IEEE Trans. Commun.,* vol. COM-24, pp. 956-967, Sept. 1976.

[28]   D. D. Falconer, K. M. Mueller, and S. B. Weinstein, "Echo cancellation techniques for full-duplex data transmission on two-wire lines," in *Proc. Nat. Telecommun. Conf.,* Dallas, TX, Dec. 1976.

[29]   H. C. Van den Elzen, P. J. van Gerven, and W. A. M. Snijders, "Echo cancellation in a two-way full-duplex data transmission system with bipolar encoding," in *Proc. Nat. Telecommun. Conf.,* Dallas, TX, Dec. 1976.

[30]   R. D. Gitlin and J. S. Thompson, "New structures for digital echo cancellation," in *Proc. Nat. Telecommun. Conf.,* Dallas, TX, Dec. 1976.

[31]   ——, patent applied for.

Stephen B. Weinstein (S'59-M'66-SM'76) was born in New York City on November 25, 1938. He received the B.S. degree from the Massachusetts Institute of Technology, Cambridge, in 1960, the M.S. degree from the University of Michigan, Ann Arbor, in 1962, and the Ph.D. degree from the University of California, Berkeley, in 1966, all in electrical engineering.

He worked for Philips Research Laboratories, Eindhoven, The Netherlands in 1967-1968, and since May 1968 he has been with Bell Laboratories, Holmdel, NJ. During the year 1973-1974 he was Bell Labs Visiting Professor at Howard University, Washington, DC. His technical interests include data communication, statistical communication theory, and information retrieval.

Dr. Weinstein was the Editor of *Communications Society Magazine* from 1974 to 1976.

OPTICAL FIBERS, A NEW TRANSMISSION MEDIUM*

S. D. Personick

I. INTRODUCTION

Recent announcements of extremely low loss optical fibers ($\sim$ 1 dB/km), and the introduction of fibers and associated electronic components into the market, have served to generate much interest in this new transmission medium as a potential competitor in applications where metallic and radio media are currently used. This interest among communication engineers and users is accompanied by a number of excellent tutorial papers describing the fiber art [1], [2].

The present tutorial paper is directed specifically to the communication theorist, and emphasizes signals, noise, and distortions associated with digital fiber systems. It is written for a reader who is familiar with such concepts as intersymbol interference, equalization, timing, random noise, etc., which occur in other digital systems. As will be shown, fiber systems are similar in many ways to conventional metallic cable systems, and much of the theory of these systems carries over easily to fibers.

II. A BLOCK DIAGRAM OF THE FIBER SYSTEM

Fig. 1 outlines a typical fiber system. The information arrives at the upper left in binary digital form. The transmitter emits pulses of optical *power* at a specified rate, *B*, bits/s, each pulse being "on" or "off" according to the corresponding data. The reader is reminded that power is a baseband quantity that varies at the data rate and not at optical frequency rates (power is the square of the complex envelope and *does not* contain an $\exp(i\Omega t)$ dependence, where Ω is the optical frequency). The transmitter contains a driver for the light source (typically a high-current low-voltage device) and an injection laser diode or light-emitting diode (LED). (Other light sources such as miniature Nd-YAG lasers pumped by LED's are possible, but the preceding appear most practical for presently envisioned systems.) These light sources can be turned on and off directly by turning the drive current on and off.

The author is with Bell Laboratories, Holmdel, N. J. 07733.

[1] It is possible to use very small core diameter fibers which propagate only a single mode (ray), and therefore avoid this pulse spreading [1]. Such fibers can only be used with lasers and may present splicing difficulties.

*Reprinted from *IEEE Communications Society Magazine*, January 1975, Vol. 13, No. 1, pp. 20–24.

Fig. 1(a).

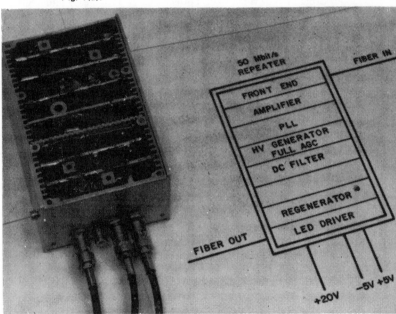

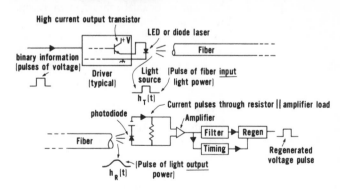

Fig. 1. The fiber system (digital).

The power emitted by the transmitter is captured in part by the fiber and propagates according to the principles of total internal reflection [1], as shown in Fig. 2. We see from Fig. 2 that different "rays" travel different paths of different lengths and experience different delays per unit length. Thus, pulses of

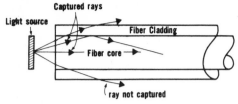

Core index of refraction > cladding index of refraction

Fig. 2.   Total internal reflection.

light which enter at one end of the fiber become wider in duration as they propagate along the fiber.[1] In addition, some light sources emit light simultaneously over a broad band of wavelengths which travel at different speeds in the fiber (dispersion). This also results in pulse broadening in propagation. Furthermore, pulses of light power traveling in the fiber are attenuated (lose *area*, which corresponds to energy) because of absorption (conversion of optical energy into heat) and scattering (loss of light from the fiber caused by fiber imperfections). Thus the pulses of power exiting the fiber have less area (energy) and are spread out in time relative to those that enter.

The receiver serves three functions: to convert pulses of optical power impinging upon it into pulses of electrical current (detection), to amplify, filter, and equalize (shape) the current (linear processing), and to decide whether each pulse is "on" or "off" (regeneration). We see from Fig. 1 that the detector (typically a photodiode) is essentially a converter of light power to electrical current. Since it responds to light power, we are tempted to think of it as a square law device. However, since we modulate the power at the transmitter, and since (as we shall see better below) the fiber operates linearly on the propagation power (spreads and attenuates the pulses of voltage), we can think of the detector more as a linear device which converts *power* to *current*. Once the pulses of current are produced by the detector, the rest of the processing is identical in nature to that of a wire medium repeater. Fig. 1(a) shows a picture of a one-way 50 Mbit/s optical repeater with receiving and transmitting fibers attached.

In summary, the block diagram of a digital fiber system resembles that of a wire pair or coaxial cable system except that power, rather than current (both baseband quantities), propagates in the fiber system. At each end of the fiber medium there is a current-to-light converter—an LED or laser at the transmitting end and a photodiode at the receiving end.

III. MATHEMATICAL MODEL

In the following section, the mathematical model of the fiber transmission system will be outlined. The purpose is to give the reader a rough idea of the factors involved in limiting the system performance (repeater spacing). After this outline, we shall merely state the results of detailed analysis.

Referring to Fig. 1, we can write (approximately) the transmitter power $p_{in}(t)$ as follows:

$$p_{in}(t) = \sum_k a_k h_T(t-kT) \text{ (watts)} \tag{1}$$

where $h_T(t)$ is the transmitted pulse shape, normalized to have unit area, $a_k = E$ or 0 where E is the energy (Joules) in pulse k if it is "on," and T is the spacing between pulses = 1/bit rate.

Generally, it is assumed that the transmitted pulses do not overlap. Directly modulated optical sources have nonlinear memory characteristics (e.g., shape and height of present pulse depends upon previous pulses), particularly at high bit rates. If care is taken in the design of the driver, and if the device is not driven to the limits of its allowable response times, then we can assume that the pulses are identical in shape and either "on" or "off," as described in (1) above. Typical average powers which can be coupled from LED or injection laser sources into low loss fibers are from a few microwatts to a few milliwatts, depending upon the source, fiber, and quality of the connection between the two.

The output of the fiber can be written as

$$p_{out}(t) = \sum_k a_k e^{-\alpha \ell} h_R(t-kT) \tag{2}$$

where $h_R(t)$ is the shape of the received pulse normalized to have unit area, α is the fiber loss per unit length (nepers/kilometer), and ℓ is the fiber length (kilometers).

It is generally assumed that the fiber behaves "linearly in power" such that the output pulses add even if they overlap. A discussion of the validity of "power linearity" is too involved for this paper [7], but it appears from analysis and experiments that this assumption is valid (for performance calculations) for systems of interest. In the rest of this paper we shall assume "power linearity."

From the "power linearity" assumption and (1) and (2), we have the convolution relation

$$h_R(t) = \int h_T(t-\tau) h_F(\tau) \, d\tau \tag{3}$$

where $h_F(\tau)$ is the fiber impulse response, a positive function having unit area.

Pulse Spreading from Group Velocity Spread and Dispersion

The shape of the fiber impulse response (whose finite width corresponds to pulse spreading in propagation) depends upon the spread of group velocities of the propagating fiber modes (rays), fiber mode coupling (hopping of energy from one fiber mode to another), and the dispersion in propagation associated with the bandwidth of the optical source (typically much larger than the modulation bandwidth). Pulse spreading due to group velocity spread and dispersion act approximately as two impulse responses in tandem, as follows:

$$h_F(\tau) = h_{F \text{ mode}}(\tau) * h_{F \text{ dispersion}}(\tau) \tag{4}$$

where the asterisk denotes convolution, $h_{F \text{ mode}}$ is the fiber impulse response with group velocity spread and without dispersion (narrow band optical source), and $h_{F \text{ dispersion}}$ is the fiber impulse response if all modes (rays) traveled at the same group velocity but dispersion were present.

Typically, we can neglect dispersion for laser sources. For GaAs LED sources $h_{F \text{ dispersion}}$ is roughly Gaussian in shape:

$$h_{F \text{ dispersion}}(t) = \left[\frac{1}{\sqrt{2\pi} \, \sigma_d} \right] e^{-t^2/2\sigma_d^2} \tag{5}$$

where for fused silica fibers $\sigma_d \cong (9 \times 10^{-12} \text{ s}) \times$ rms optical source bandwidth (in angstroms) $\times$ fiber length (in kilometers).

The portion of the fiber impulse response due to the spread

of mode group velocities depends upon the type of fiber being used [1]. For the sake of simplicity, we can assume that it is Gaussian in shape:

$$h_{F\ mode}(t) = \left[\frac{1}{\sqrt{2\pi}\ \sigma_m}\right]e^{-t^2/2\sigma_m 2} \qquad (6)$$

For fibers without mode coupling, σ_m is proportional to the fiber length with a proportionality constant which *can be* as small as a fraction of a nanosecond/kilometer in practical low-loss fibers. It can be much larger as well. For fibers *with* mode coupling, σ_m is proportional to the square root of the fiber length for sufficiently long fibers.

Detector Operation

When the fiber output power given in (2) falls upon a photodiode (see Fig. 1), current flows at a rate which *on the average* is proportional to the incident optical power. This current consists (heuristically) of electrons generated at an *average rate* (electrons/second) proportional to the incident optical power. However, the *exact* number of electrons generated in a given time interval, and their generation times, depends only statistically upon the optical power. For a detector with internal gain, such as an avalanche detector, each "primary" photoelectron produced by the incident light can produce many "secondary" electrons by the mechanism of collision ionization. We can write the detector output current as follows:

$$i_{det}(t) = e \sum n_k h_{det}(t-t_k) \qquad (7)$$

where $h_{detector}(t)$ is the displacement current impulse response to a detector generated electron, t_k is the generation time of the kth electron, e is the electron charge, and n_k is the number of "secondary" electrons produced by primary photoelectron k if the photodiode is of the avalanche type. If the photodiode is not an avalanche type, $n_k = 1$ for all k.

The statistics of primary electron generation correspond to a Poisson process. The average rate of "arrivals" of primary electrons $\lambda(t)$ is given by

$$\lambda(t) = \lambda_0 + \frac{\eta}{\hbar\Omega}p_{incident}(t) \qquad (8)$$

where λ_0 is the electrons per second produced even in the absence of incident light (dark current), η is the detector quantum efficiency (of order unity), $\hbar\Omega$ is the energy in a photon $\sim 2.2 \times 10^{-19}$ Joules at 0.9 μm (GaAs wavelength), and $p_{incident}(t)$ is the incident optical power (watts).

The average current flowing is given by averaging (7) over the Poisson arrival statistics [3]:

$$\langle i_{det}(t) \rangle = Ge \int \lambda(\bar\tau)h_{det}(t-\tau)d\tau \qquad (8a)$$

where G is the average value of n_k (secondary electrons/primary electrons) = 1 for nonavalanche detectors.

If $h_{det}(t)$ is "broad band" compared to the pulse shape $h_R(t)$ in the received power waveform $p(t)$, then the average detector current essentially follows the optical power variations (neglecting the dc term due to dark current). We can write the output current as the sum of its average value and the deviations from the average [combining (2), (8), and (8a)]:

$$i_{det}(t) = \langle i_{det}(t)\rangle + n_s(t)$$
$$= \left[\lambda_0 eG + \frac{\eta eG}{\hbar\Omega}\sum a_k e^{-\alpha\ell}\ h_{det\ out}(t-kT)\right] + n_s(t) \qquad (9)$$

where $n_s(t)$ is commonly referred to as "shot noise" or "quantum noise" and $h_{det\ out}(t)$ is the detector output pulse shape = $\int h_R(\tau)h_{det}(t-\tau)d\tau$.

It can be shown [3] that $n_s(t)$ has the following correlation function:

$$\langle n_s(t)n_s(t+\tau)\rangle = FG^2 e^2 \int \lambda(\tau')h(t-\tau')h(t+\tau-\tau')d\tau' \qquad (9a)$$

where $FG^2 = \langle n^2 \rangle$ is the mean-squared number of secondary electrons per primary electron (=1 for nonavalanche detectors).

Post-Detector Noise

In order to process $i_{det}(t)$, electronic amplification is required. This amplification process adds additional "thermal" noise. The amplified current, available for further processing, is given by

$$i_{amp\ out}(t) = A[i_{det}(t) + i_n(t)] \qquad (10)$$

where A is the (mathematically irrelevant) amplifier gain and $i_n(t)$ is the amplifier noise.

The amplifier noise is not necessarily "white." Typically [4], [5], it has a spectral density $S_n(f)$ of the form

$$S_n(f) = \sum_{\ell=0}^{\infty} (f)^{2\ell} S_{n\ell}$$

where at least the first two terms ($\ell=0$, $\ell=1$) are non negligible.

Once $i_{amp\ out}(t)$ is available, it can be processed in the same manner as any digital baseband signal with added noise and (possibly) intersymbol interference. For example, it can be filtered and equalized to reach a reasonable compromise between noise and intersymbol interference. The resulting noisy pulse stream can then be regenerated by a clocked comparator. Timing information can be recovered from the pulse stream itself using a phase-locked loop or filter.

The performance (error rate) is governed by both the shot noise described by (9) and the amplifier noise.

IV. SYSTEM PERFORMANCE-THEORY AND PRACTICE

A detailed analysis of the system described in Section III is too tedious to duplicate here [4], [5]. Typical calculated receiver sensitivity curves (for a 10^{-9} error rate) are shown in Fig. 3 for a well-designed front-end amplifier following the detector, and for detectors with and without avalanche gain. The finite width of the "curves" accounts somewhat for component uncertainties and practical design tradeoffs. Also shown on those curves are some experimentally achieved receiver sensitivities [5], [6]. [2]These curves assume that the received optical pulses $h_R(t)$ are half duty cycle rectangular

---

[2]The 6.3 Mbit/s points have been adjusted downward 1.5 dB (with avalanche gain) and 3 dB (without avalanche gain) to reflect "non-return-to-zero" equalization which is represented by these curves. The actual repeater used return-to-zero equalization to facilitate dc restoration.

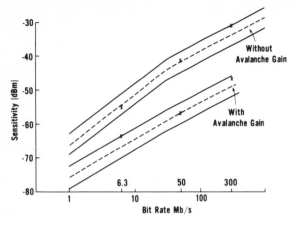

Fig. 3.

pulses. We see, for example, that for 1 mW of average transmitted power (typical for a laser source), and a 50 Mbit/s transmission rate, as much as 57 dB of attenuation can be placed between repeaters (neglecting possible pulse spreading in propagation).

If the received pulses $h_R(t)$ [see (2)] are very narrow, then the average optical power required to achieve a desired error rate is minimized. For other received pulse shapes, the receiver sensitivity is lower. If, for example, we assume that the receiver "equalizes" the received pulses to produce a fixed output pulse shape (input to the regenerator), then we can calculate [4] the loss in sensitivity (relative to a narrow received pulse) as a function of the received pulse width. Define the rms width of the received pulse $\sigma_{received}$ as[3]

$$\sigma_{received}^2 = \int t^2 h_R(t)dt - \left[\int t\, h_R(t)dt\right]^2.$$

Fig. 4 shows a typical plot of the loss in receiver sensitivity versus $\sigma_{received}$ for a receiver using "zero forcing equalization."

The experimental points [8] show that the assumption of Gaussian-shaped received pulses is probably conservative. We see that for optimum sensitivity, it is desirable that the rms

width of the received pulses be less than 0.25 times the spacing between pulses.

As in other digital systems, "zero forcing" equalization is not necessarily optimum or practical. The amplifier output current [see (8) and (9)] could be processed in other ways to obtain a better or more robust tradeoff between noise and intersymbol interference. This is still a topic for further research.

V. CONCLUSIONS

We have seen that in many ways, optical fiber systems are similar to baseband wire pair systems, with optical power rather than voltage as the information carrier. Fiber systems are limited by attenuation and pulse spreading, as are their wire pair analogs, with the exception that the equivalent fiber baseband attenuation may have a significant low frequency value, whereas in wire pair systems, the attenuation is typically much larger near the half bit rate than at low frequencies. It appears that repeater spacings of 50 dB (of attenuation) or more, corresponding to a few miles of fiber, are reasonably projected from available laboratory results for systems transmitting tens of megabits per second of data.

REFERENCES

[1] S. E. Miller, E. A. J. Marcatili, and T. Li, "Research toward optical fiber transmission systems," *Proc. IEEE*, vol. 61, pp. 1703-1751, Dec. 1973.

[2] R. D. Maurer, "Fibers for optical communication," *Proc. IEEE*, vol. 61, pp. 452-462, Apr. 1973.

[3] E. Parzen, *Stochastic Processes*, San Francisco: Holden-Day, 1962, p. 156.

[4] S. D. Personick, "Receiver design for digital fiber optic communication systems," *Bell Syst. Tech. J.*, vol. 52, pp. 843-886, Aug. 1973.

[5] J. E. Goell, "An optical repeater with high impedance input amplifier," *Bell Syst. Tech. J.*, vol. 53, pp. 629-644, Apr. 1974.

[6] P. K. Runge, "A 50 Mb/s repeater for a fiber optic PCM experiment," in *Proc. Int. Conf. Commun.*, 1974, pp. 17B1-17B3.

[7] S. D. Personick, "Baseband linearity and equalization in fiber optic digital communication systems," *Bell Syst. Tech. J.*, vol. 52, pp. 1175-1194, Sept. 1973.

[8] D. M. Henderson, "Dispersion and equalization in fiber optic communication systems," *Bell Syst. Tech. J.*, vol. 52, pp. 1867-1876, Dec. 1973.

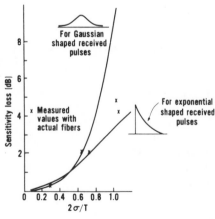

Fig. 4.

S. D. Personick received the B. E. E. degree from the City University of New York, New York, N.Y., in 1967 and the Sc. D. degree from the Massachusetts Institute of Technology, Cambridge, in 1969.

Since then he has been engaged in studies of optical communication systems and in communication theory. He is currently a member of the Technical Staff at Bell Laboratories, Holmdel, N.J.

[3]Remember that $h_R(t)$ has unit area [see (2)].

FIBER OPTIC COMMUNICATION, A TECHNOLOGY COMING OF AGE *

S. D. Personick, Senior Member, IEEE

Abstract   In a little over ten years, lightwave communication using optical fibers has progressed from a laboratory proposal to a near commercial reality. Losses in optical fiber waveguides have been reduced from hundreds of dB/Km in the early seventies to less than 1 dB/km at some wavelengths today. Bandwidths of multimode fibers can now exceed 1 GHz in km lengths. Strengths in kilometer length fibers have been increased to hundreds of KPSI (more than steel). Cables containing hundreds of fibers, multiple fiber splices, and single fiber connectors have been developed. Lasers which had lifetimes measured in minutes or hours in the early seventies now have extrapolated lifetimes of over a million hours. New material systems which can use the lower loss longer wavelength regions of the optical spectrum are evolving. Meanwhile, prototype systems carrying voice, data, and video services have been placed in service for commercial telephone and military applications. We can anticipate the widespread use of optical fibers on a routine basis beginning in the early eighties. Along with this will come reduced costs for existing services and the introduction of new services made more economical by this new transmission medium.

I. INTRODUCTION

With the announcements this year of a sizeable number of commercial demonstrations of optical fiber transmission systems carrying live traffic, it is clear that this new technology is rapidly being transformed from a laboratory proposal into a commercial reality. Experimental systems carrying voice data and video signals for telephone and military applications are being tested by AT&T, GTE, and ITT in the United States (to name a few), and by numerous European and Japanese concerns.

In a little over ten years dramatic progress has been made in reducing fiber losses, increasing multimode fiber bandwidths, developing practical cable and connector structures, developing reliable optical sources and detectors compatible with the fibers, and developing systems to combine these components [1]–[7].

In the pages that follow, we shall review some of this progress and summarize characteristics of some of the prototype systems which are being tested.

II. A TYPICAL OPTICAL FIBER TRANSMISSION SYSTEM

Fig. 1 shows a typical digital fiber optic transmission system. It is assumed for illustration that the information originates as analog telephone conversations and is converted to digital form by more or less conventional terminal equipment (channel banks and multiplexers). Since fibers cost more than wires and have relatively large bandwidths, for most applications economics generally dictate a fairly high data rate (consistent with point-to-point communications requirements). The digital signals emitted by the terminal equipment are used to modulate the current of an injection laser or light emitting diode. The laser is a threshold device which begins to turn on at currents of about 100 mA and is fully on with an additional 20 mA or so. It can be switched on and off at up to gigabit rates with careful design of the driver circuitry. It can couple a few milliwatts of power into an optical fiber. The LED is a roughly linear light power versus current device which can couple up to a few hundred microwatts of power into a multimode fiber with drive currents of a few hundred milliamps. Modulation speeds for the LED are limited to around a few hundred MHz or less.

The fiber carries the light with losses typically less than 10 dB/km in what are presently considered good cables. In

The author is with Bell Laboratories, Holmdel, NJ 07733.

*Reprinted from *IEEE Communications Society Magazine*, March 1978, Vol. 10, No. 2, pp. 12–20.

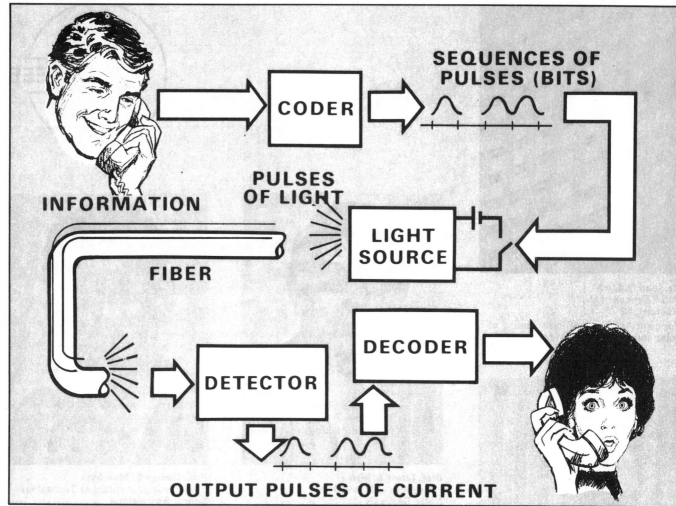

Fig. 1.   Digital fiber optic transmission system.

addition to attenuating the pulses, some pulse dispersion (spreading in time) occurs because of multipath phenomena. Eventually the fiber must terminate on a detector which converts the light into a current for amplification and regeneration. Repeaters can typically be spaced more than four miles apart for present optical component and light cable technology.

III. OPTICAL FIBER CHARACTERISTICS

The most important characteristic of an optical fiber is its attenuation. In the time which has passed since Kao first proposed that low loss optical fibers might be possible to fabricate [1], dramatic progress has been made in improving this critical parameter. Fig. 2 shows a curve of best reported loss at 0.82 μm wavelength (the wavelength of typical gallium arsenide laser diodes) as a function of time. The major breakpoint for practical fiber long distance applications was the announcement by Corning Glass Works of 20 dB/km losses in 1970. We see that, at this wavelength, losses approaching 1 dB/km have now been achieved.

In early fibers, the main source of attenuation was absorption (the conversion of light into heat). To reduce this effect concentrations of metallic impurities like iron, cobalt, and copper had to be reduced to below one part per billion.

Water in the form of OH radical had to be reduced to below one part per million concentrations. When absorption is made sufficiently small, Rayleigh scattering from frozen in molecular density and compositional inhomogeneities becomes the dominant source of loss. This scattering follows a $(\lambda)^{-4}$ dependence (where λ is the wavelength). Thus in the best fibers made today, losses are lower as one moves to wavelengths beyond 1 μm. Two loss versus wavelength curves are shown in Fig. 3 for two exceptionally low loss fibers. Because of the desire to operate with the minimum of fiber losses, intensive effort is being expended to develop optical sources and detectors at these longer wavelengths.

After loss, dispersion is the next transmission property of concern. In multimode fibers not all rays (modes) travel at the same group velocity. Thus narrow pulses of light injected into a fiber (which in general excite many modes) will spread out in time as they travel down the fiber. By careful control of the refractive index profile (index of refraction as a function of distance from the fiber axis), in order to equalize the group velocities of the modes, group delay effects of this kind have been reduced from about 50 ns/km of spreading down to below 1 ns/km in the best fibers. When wide optical bandwidth sources are used, a

REDUCTION OF FIBER LOSS

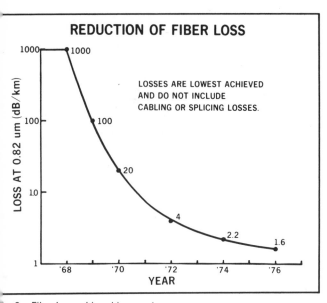

LOSSES ARE LOWEST ACHIEVED AND DO NOT INCLUDE CABLING OR SPLICING LOSSES.

Fig. 2.   Fiber loss achieved in recent years.

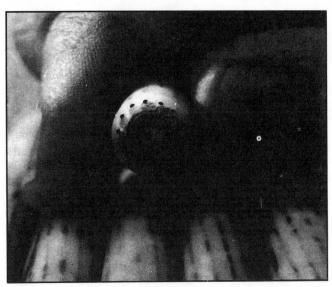

Fig. 4.   End of a 144 fiber cable.

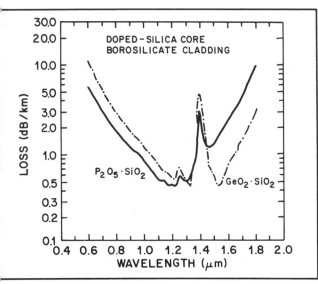

DOPED—SILICA CORE BOROSILICATE CLADDING

Fig. 3.   Fiber loss versus wavelength.

second source of pulse spreading comes into play. Different wavelengths within the optical spectrum of the source travel at slightly different group velocities. The spreading at 0.82 μm wavelength is about 0.1 ns/nm/km. Thus an LED with 50 nm of optical bandwidth would produce about 5 ns/km of dispersion. If this is too much, one must use the injection laser (which typically has less than 2 nm of bandwidth) or one must move to longer wavelengths of around 1.3 μm where the dispersion due to wavelength variations is a minimum (and where, fortuitously, losses are also a minimum). Since long wavelength sources do not yet exist in commercial form and long wavelength detectors are also generally unavailable, one must use the first option with today's technology.

In order to fabricate practical cables one requires strong fibers. Glass fibers are inherently very strong. So called pristine fibers have strengths exceeding that of steel (more than a million pounds per square inch). The problem is that microscopic flaws in the fiber surface dramatically reduce the strength. Also these flaws tend to grow with time in the presence of water and tension, producing a long-term failure phenomenon called static fatigue. To eliminate these surface flaws, the fibers must be protected by special coatings during the fiber drawing process before the fiber can come in contact with any abrasive surfaces. The intensive research efforts which have been expended in the area of fiber strength, drawing techniques, and coatings have now produced long fibers with tensile strengths of over 750 KPSI. In production, fibers with proof test strengths of 25–50 KPSI are routinely made with high yield.

To facilitate splicing and to keep losses from radiation down, one must carefully control the fiber diameter. By using in line fiber diameter measuring equipment, operated in a feedback fashion with the drawing machine, diameter variations of less than 1 percent (3 "sigma") have been reported. In practice, production variations are presently somewhat larger.

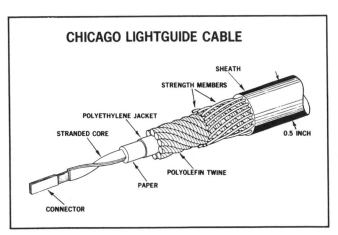

CHICAGO LIGHTGUIDE CABLE

Fig. 5.   Schematic of a 24 fiber cable.

Cost is the bottom line for applications purposes. To reduce fiber manufacturing costs both materials useage efficiency and production speed must be maximized. Present fiber prices are around $1–3 per fiber meter. This is an artificial price reflecting the small present demand. As demand picks up, prices will be limited by the cost of the raw materials for making fibers (silicon tetrachloride, dopants, carrier gases, fused silica tubes) and by the rate at which fibers can be drawn and performs fabricated. Projections for future fiber costs are around 10¢ per fiber meter dropping toward a few cents per fiber meter in the longer run as fabrication technology improves.

IV. CABLING AND SPLICING

Fibers are too fragile to handle as isolated entities in a field environment. In order to put them to practical use one requires cables and connectors which minimize the necessity of dealing with individual unprotected fibers. In recent years much progress has been made in these directions. Fig. 4 shows the end view of a 144 fiber cable manufactured by Bell Laboratories and Western Electric Company. A 24 fiber version of the cable is shown schematically in Fig. 5. It consists of ribbons of twelve contiguous fibers stacked to form a 12 x 2 array (12 x 12 for a 144 fiber cable). The 12 x 12 array is approximately ⅛ inch square. The ribbons are fabricated by sandwiching fibers between plastic tapes held together by an adhesive. Fig. 6 shows some recent loss measurements in cables of this type. When interpreting this figure one should keep in mind that it represents hundreds of kilometers of nominally similar fibers and not just a selected kilometer or two. Also, since the cabling process introduces loss (through short period low amplitude bending of the fibers) these results are even more impressive. The cabling process accounts for about 1 dB/km of the losses shown.

The ribbon cable can be terminated in an array splice as shown in Fig. 7. This splice consists of grooved chips in which fibers can be laid in precise registration. Chips and fiber ribbons (stripped back to expose the fibers) can be alternated to produce a structure where as many as 144

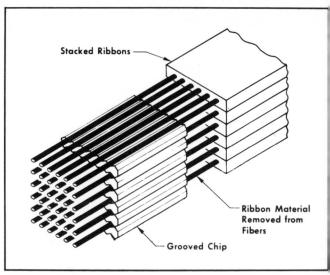

Fig. 7.   Array splice termination for a ribbon cable.

fibers have been held in precisely controlled positions. The array splice can then be potted and polished; and two half splices mated in the field to interconnect up to 144 fibers simultaneously. Losses in such an array splice can average below 0.25 dB, implying tolerances below 0.0001 inch in the axial offsets between fibers in the two halves.

The array splice is designed to be put together once with careful cleaning before assembly. For practical systems one also needs a demountable connector which can be used on cables containing a single fiber. Fig. 8 shows such a connector which can be molded from plastic materials. This connector can produce sufficiently small axial and longitudinal offsets in the fiber cores to obtain losses well below 1 dB. At present splice losses for single fiber connectors are limited by fiber diameter variations, ellipticities, and the lack of a suitable index matching material with physical properties appropriate for a demountable connector.

V. OPTICAL SOURCES

For fiber optics applications one requires a source which is simple, cheap, compatible with the fiber dimensions and with the wavelengths where fibers have low losses and dispersion. GaAs LED's and injection lasers are presently the most interesting candidates. LED's can couple up to a few hundred μW into a typical multimode fiber, although in practice 50 μW is a more reasonable design number. They can be directly modulated by varying the drive current at up to a few hundred MHz, with 50 MHz being a more conservative figure. They require peak drive currents of about 100–300 mA and have roughly linear light output versus drive current characteristics up to the point of saturation of the light output due to device heating. GaAlAs LED's can be fabricated at wavelengths between 0.8 and 0.9 μm. They typically have optical bandwidths of 40-50 nm at room temperature. This results in material dispersion (pulse spreading in transmission) of 3–5 ns/km as mentioned earlier. Fig. 9 shows the structure of a typical Burrus type LED. The various layers of semiconductor material (with varying amounts of aluminum) serve to

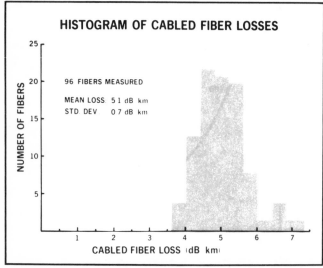

Fig. 6.   Recent loss measurements for a fiber cable.

confine the holes and electrons to a thin region near the p-n junction. The dot contact insures that only a small diameter column of current is injected. The etched well allows the fiber to be physically close to the junction where holes and electrons combine to emit light. This structure insures that 1) holes and electrons tend to combine radiatively rather than nonradiatively, 2) that reabsorption of emitted light before it reaches the fiber is minimized, and 3) that light is generated only in a small area under the fiber core. Together these precautions maximize the conversion efficiency between electrical drive power and power captured by the fiber. With a drive current of about 100 mA at 2 V about 50 μW peak can be captured by a typical multimode fiber. Thus the conversion efficiency (overall) is about 0.025 percent.

An injection laser can be fabricated as shown in Fig. 10. The cleaved front and back faces reflect light with about 30 percent reflection coefficient. Because of the large gain per unit length in the pumped semiconductor, this is sufficient to give an overall round trip gain. Once again the layered semiconductor structure serves to confine electrons to a region close to the junction, and also forms an optical waveguide that confines the light as well. The laser is a threshold device which turns on at about 100 mA of drive current and reaches maximum output levels (limited by mirror damage) with perhaps 20–30 mA more current.

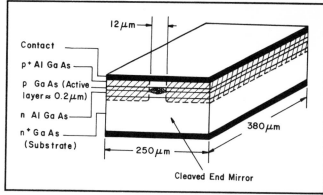

Fig. 10.   Injection laser.

(Some Japanese companies have reported lasers with thresholds of about 10 mA using a different structure). It can couple a few mW of light into a fiber and thus is about 10–50 times more efficient than an LED. It can be modulated by varying the drive current at up to GHz rates. Linearity, even around an operating point above threshold in the light output versus current characteristic, is not good. Thus lasers with simple driver electronics are considered most appropriate for digital applications. Since the optical bandwidth is much less than that of an LED, material dispersion is not a problem.

The more serious drawbacks of the laser with respect to the LED are lifetime and driver complexity. Fig. 11 shows the progress which has been made in laser lifetime over the past few years. Early devices had lifetimes measured in minutes or hours. Now extrapolated room temperature lifetimes of more than 1 million hours are being reported. However, as the temperature increases, laser lifetimes drop. Also lifetime variations amongst devices are still fairly large. Thus lifetime continues to be a concern for many applications where large numbers of devices may be used.

The laser threshold varies with temperature and age and this leads to a requirement for a feedback controlled driver to monitor the light output locally and to keep the device

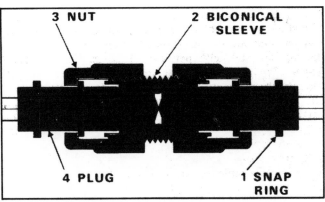

Fig. 8.   Molded plastic connector for a single fiber.

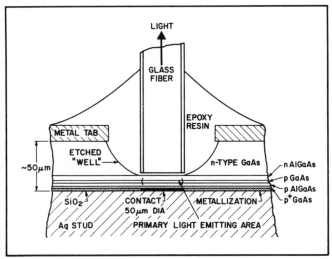

Fig. 9.   Structure of a typical LED.

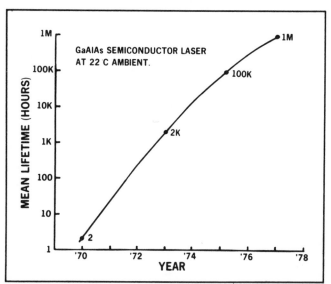

Fig. 11.   Improvement in laser lifetime.

biased near threshold. This driver is more complex than that required for the LED. (On the other hand one must switch only tens of mA with the laser versus hundreds of mA with the LED making high speed modulation easier with the laser.)

Since fiber loss and material dispersion are lower at longer wavelengths effort is being focused on producing sources and detectors at these wavelengths. The so called quaternary material system, indium, gallium, arsenide, phosphide, appears very promising. Lasers with 10 000 hour lifetimes have already been reported using these materials. High radiance LED's of these compositions have also been produced.

VI. DETECTORS

For 0.8–0.9 μm wavelengths silicon pin and avalanche photodiodes are used. These detectors convert light to current with excellent efficiency and low noise. For avalanche detectors multiplication statistics are important in determining both the useful gain, and the ultimate sensitivity of the receiver in which they are employed. Today detectors with excellent uniformity, low multiplication noise, low dark current, and useful gains beyond one hundred are readily available.

For longer wavelengths, germanium and indium gallium arsenide detectors are potential candidates. Germanium detectors have been around for a while, but were deemphasized in favor of silicon when work centered on the 0.8–0.9 μm region of the spectrum. Work on germanium detectors is now being revived. Other materials are being studied, particularly with respect to electron and hole collision ionization properties which affect multiplication statistics in avalanche devices.

VII. SUBSYSTEMS

In order to make optical sources and detectors useable in the field they must be packaged for easy interfacing both electrically and optically. One approach which has been almost universally adopted is to attach a connectorized fiber pigtail to the source or detector. The detector diode is placed on a hybrid integrated circuit containing a low noise preamplifier designed specifically to work with a photodiode. The laser or LED is combined with a discrete or integrated driver circuit with feedback control (for lasers) provided by a local photodiode detector. Fig. 12 (left) shows a photodiode package on a complete receiver card used in the Bell System Chicago project [5]. The black plastic package (in the upper left portion of the card) contains an avalanche photodiode and preamplifier. The black cable attached to the package is an armored single fiber which is attached to the detector in the package and which terminates in a single fiber connector at the other end (lower center of card). When this board is inserted in a card frame, electrical and optical connections are made simultaneously. The electrical output of the preamplifier is between 1 mV and 1 V depending upon the optical signal level and the avalanche gain setting. The package can accommodate 40 dB of optical range and has a 45 Mbaud digital capability. Fig. 12 (right) shows a transmitter card containing a packaged feedback controlled laser and an auxiliary control circuit. The fiber pigtail and connector are again visible. The laser package accepts standard ECL levels at a 45 Mbaud rate. Laser output is held at 1 mW peak by the feedback control circuitry over a 0–60°C temperature range and over the laser lifetime threshold variations.

In addition to optical sources and detectors suitably packaged for easy interfacing, one needs an interface

Fig. 12.   Optical repeaters.

between the single fiber cables, used to interconnect these devices, and the multifiber cable. Fig. 13 shows a prototype "fanout" used in the Bell System Chicago experiment. The fiber cable enters the fanout box and attaches to the fanout via an array connector described previously. Individual fibers exit the box as armored single fiber cables terminated in demountable single fiber connectors. Fig. 14 shows a fiber distribution panel. The fanout box is mounted behind this panel with individual fiber cables attached to connector sockets on the back of the panel. (A socket mounted in the panel accepts a single fiber connector on each side, front and back, to make a demountable splice.) Single fiber cables from shelves containing transmitters and receivers also

terminate on the back of the panel (bay "wiring"). Interconnections are made on the front of the panel by armored single fiber jumper cables with single fiber connectors on both ends.

VIII. PUTTING IT ALL TOGETHER

In the last couple of years a number of companies in the United States, Europe, and Japan have installed optical fiber systems in the "field." These systems range in sophistication from laboratory equipment transplanted to the field to bring attention to the new technology, to near prototype systems which for the most part are manufacturable and capable of being installed by operating craftspeople. Some systems are digital, operating at data rates varying from a few to hundreds of Mbits/s. Some systems are analog, generally carrying television signals, using a variety of modulation schemes and offering a range of performances. Below we shall review the characteristics of a system, carrying "live" customer traffic, installed by Bell Laboratories, Illinois Bell, Western Electric Company, and AT&T in Chicago in early 1977 [5]. I choose this system to describe because I am familiar with its details. Other companies have installed their own systems as mentioned above, with varying degrees of sophistication.

Fig. 15 shows a route map of the "Chicago Project." Cables containing twenty-four fibers each (two ribbons of twelve fibers) were installed between the Franklin Central Office, the Wabash Central Office, and the Brunswick office building (a commercial building). Manholes and splices are shown. These cables carry 44.7 Mbaud digital information for voice loops (4 kHz telephone connections between the central office and a customer's premises), voice trunks (4 kHz telephone connections between switching machines in

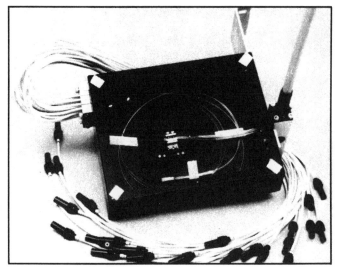

Fig. 13.   Prototype "fanout."

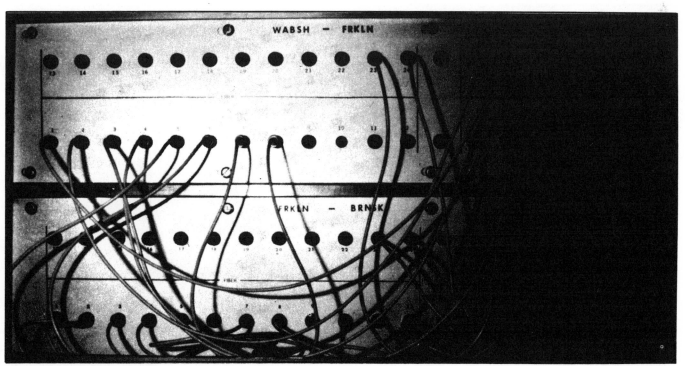

Fig. 14.   Fiber distribution panel.

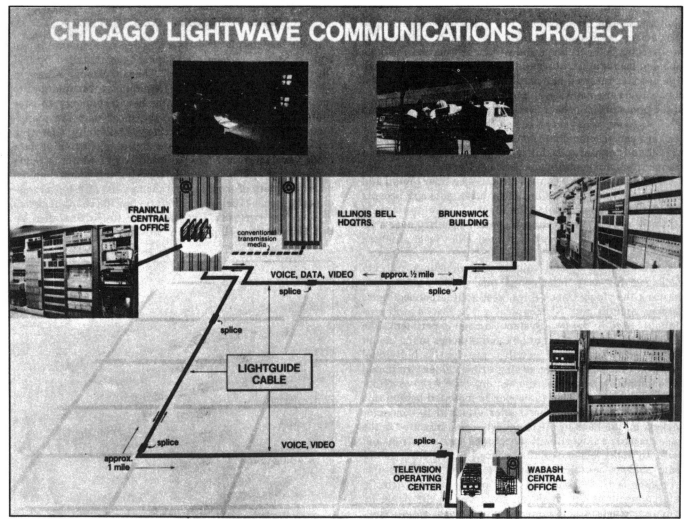

Fig. 15.   "Route map" of the Chicago project.

central offices), Dataphone Digital Service® loops, and Picturephone Meeting Service® loops (a 4 MHz bandwidth black-and-white conference service).

The cable design was as shown in Figs. 4 and 5 except that only two ribbons were used. The ducts chosen for installation of the fiber cable were old and plagued by bends and offsets. In manholes, level shifts between the incoming and outgoing ducts could be more than ten feet. These problems complicated the cable installation and were among the considerations which led to the choice of short sections (a few hundred meters) of cable interconnected by array splices in splicing manholes. To protect the cable an inner duct of plastic hose was installed in the duct system, and the fiber cable pulled through this inner duct. The inner duct was made air tight and pressurized to keep water away from the fiber cable. The cable and inner duct were installed in the middle of winter at night and on weekends by Illinois Bell craftspeople and Bell Labs engineers under Bell Laboratories supervision. Fig. 16 shows a photo of a cable being pulled off of a reel and into a manhole.

®Registered service mark of the American Telephone and Telegraph Company.

Fig. 17 is a photo of a typical repeater shelf in a bay. Optical repeaters were built on the three cards shown in Fig. 12. The receiver card (left) contains the photodetector module and amplifier circuits. The decider (center) card contains a phase locked loop for timing recovery and a regenerator. The transmitter card (right) contains a laser module (shown) or LED. The laser module optical output from the fiber pigtail is –3 dBm average, the LED pigtail output is –18 dBm average, and the minimum power level required at the receiver pigtail for a 10^{-9} error rate is –55 dBm average. The receiver optical dynamic range is >35 dB.

As mentioned above, when these cards are inserted in a card frame, electrical and optical connections on the back plane are automatically made.

Fig. 18 shows a typical end-to-end connection between the Wabash and Franklin central offices. One should note the large number of single fiber connectors required to implement a system with plug in capability and a fiber cross connect. Typical splice losses for single fiber demountable connectors are less than 1 dB, and for array splices less than 0.5 dB. Lower splice losses are possible with improved fiber dimensional tolerances. Fig. 19 shows a time domain reflectometer used to inject pulses of light into the fibers and

Fig. 16.   Cable being pulled off of a reel and into a manhole.

Fig. 17.   Typical repeater shelf in a bay.

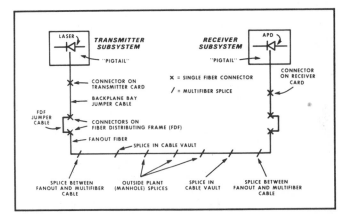

Fig. 18.   Typical end-to-end connection between Wabash and Franklin central offices.

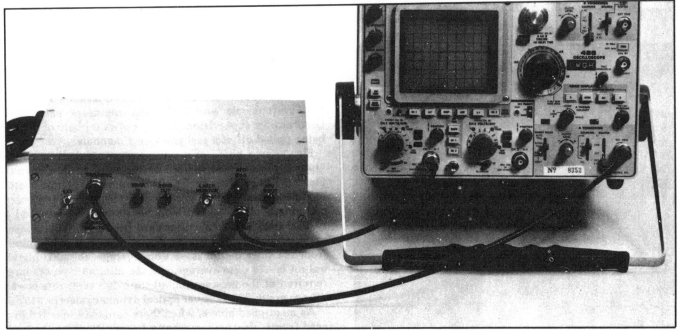

Fig. 19.   Time-domain reflectometer.

to look for echoes from breaks or splices. (No breaks occurred in the Chicago Project cables.) Fig. 20 shows the "echo scan" from a cable between the Franklin and Wabash Companies. One can see reflections from the launch point in the Franklin bay, the Franklin cable vault, three splicing manholes, the Wabash cable vault and the Wabash bays. The time scale is 2000 ns/cm, corresponding to 200 meters of one-way distance per cm on the scale.

The "Chicago" project provided valuable information about the installation of a fiber system which is being used to refine these techniques for anticipated introduction of fiber systems on a standard basis in the early eighties.

IX. CONCLUSIONS

In the next few years we will see the widespread introduction of fiber optic technology into the communications industry. Early systems will provide increased repeater spacings, and broader bandwidths than what is obtainable with wire pairs. Fiber cables will be cheaper, lighter in weight, and smaller in size than coaxial cables. For telephone applications fiber systems will of course be used where these properties lead to reduced cost. For specialized applications the extraordinary properties of fibers such as immunity from interference will help solve serious engineering problems which exist with metallic media. As familiarity with the technology increases and demand picks up, the economies of scale will bring down prices and further accelerate the growth. Beyond this, progress at longer wavelengths, where losses are lower, will lead to increased repeater spacings, lower costs, and even more applications. Thus, all indications are that we are at the turning point for the long anticipated fiber optics revolution in communication.

REFERENCES

[1] K. C. Kao and G. A. Hockham, "Dielectric fiber surface waveguides for optical frequencies," *Proc. Inst. Elec. Eng.,* vol. 113, pp. 1151–1158, July 1966.
[2] S. E. Miller *et al.,* "Research toward optical fiber transmission systems," *Proc. IEEE,* vol. 61, pp. 1703–1751, Dec 1973.
[3] D. Gloge, Ed., *Optical Fiber Technology.* New York: IEEE Press, 1976.
[4] M. L. Barnoski, Ed., *Fundamentals of Optical Fiber Communication.* New York: Academic, 1977.
[5] A. R. Meier, "Real world aspects of Bell fiber optics systems begin test" *Telephony,* pp. 35–39, April 11, 1977.
[6] T. L. Maione and D. D. Sell, "Experimental fiber optic transmission system for interoffice trunks," *IEEE Trans. Commun.,* vol. COM-25, pp. 517–523, May 1977.
[7] I. Jacobs, "Lightwave communications passes its first test," *Bell Lab. Rec.,* pp. 291–296, Dec. 1976.

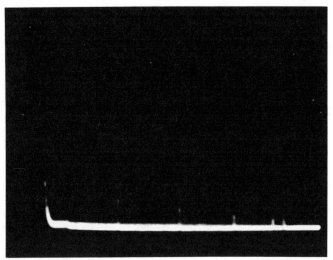

Fig. 20. "Echo-scan" from a cable.

S. D. Personick (S'69–M'69–SM'77) was born in New York, NY, on February 22, 1947. He received the B.E.E. degree from the City College of New York, New York, NY, in 1967 and the Sc.D. degree from the Massachusetts Institute of Technology, Cambridge, in 1969.

Since 1967 he has been an employee of Bell Laboratories, working in the area of transmission systems. From 1969 to 1975 he did research in the fiber optics area. In 1975 he became a Supervisor, doing development work for fiber optic trunk transmission systems. His current responsibility is repeater development. He is chairman of the IEEE Communications Society Transmission System Subcommittee on Fiber Optics.

Section 5

Signal Processing

EDITORS' COMMENTS

Signal processing has a rich history, and its importance is evident in such diverse fields as communications, acoustics, sonar, radar, seismology, nuclear science, biomedical engineering, and many others. In some applications we may wish to remove interference, such as noise, from the signal or to modify the signal to present it in a form which is more desirable or more easily interpreted. In other applications, a signal transmitted over a communications channel is generally perturbed in a variety of ways, including channel distortion, fading, and the insertion of background noise. One of the objectives at the receiver is to process the signal to compensate for these disturbances. The six papers in this section cover the recent advances in both digital and analog signal processing techniques, as well as specific devices used in communications.

The first paper, entitled "Tutorials on Signal Processing for Communications: Part I—Digital Techniques for Communication Signal Processing," by K. V. Mina, V. B. Lawrence, and J. J. Werner, is theoretical in nature and explains the principles of digital signal processing. It also explains why, where, and how such processing is used in communications. Almost all the theoretical considerations involved in the design of digital signal processors would be of little value if a good understanding of the issues involved in practical implementations were not included. Recent advances in integrated circuit technology promise economical implementations of very complex digital signal processing systems. A companion paper, entitled "Tutorials on Signal Processing for Communications: Part II—Digital Signal Processing Architecture," by S. K. Tewksbury, R. B. Kieburtz, J. S. Thompson, and S. P. Verma, reviews the design of digital signal processors, concentrating on the overall architecture. The major design considerations, i.e., cost, real-time performance, and reliability are clearly presented. The tradeoffs between software and hardware, and between custom and programmable chips, are enunciated in detail.

The evolution of the practical application of digital signal processing was accelerated in 1965 by the invention of an efficient algorithm for computing Fourier transforms, the fast Fourier transform (FFT). The third paper in this section, "The Discrete Fourier Transform Applied to Time Domain Signal Processing," by F. J. Harris, presents the develop-

ment and the understanding of the fundamental ideas of the FFT algorithm. It provides examples to show how the versatility of this algorithm is reflected in the numerous ways in which it is applied to signal processing tasks.

The fourth paper, "TDM/FDM Translation as an Application of Digital Signal Processing," by S. L. Freeny, traces the history and growth of transmultiplexers. Freeny outlines the theory, explains the implementation, and summarizes the architecture of various prototype TDM/FDM translators. Transmultiplexers are essential systems for connecting analog and digital networks, and will be very important during the long transition from analog to an all digital communications network.

The fifth paper, "Charge-Coupled Devices: The Analog Shift Register Comes of Age," by A. Gersho, is an introduction to charge coupled devices (CCDs). Gersho outlines the theory and the physics of CCDs, and explains the fundamental limitation on their performance. He discusses the implementation of fixed and adaptive filters, as well as recursive and nonrecursive filters. He includes the noise sources, dynamic range, and nonlinearity of CCDs, and gives examples of systems that use such devices.

In the sixth paper, entitled "Surface Acoustic Wave Devices," L. B. Milstein and P. K. Das outline the theory and technology of surface acoustic wave (SAW) devices. Research and development on SAW devices, (in the context of potential applications in electronics), has received a flurry of recent activity. The growth of interest in the subject has been enormous. The subject has developed to such a point that many sophisticated devices are now available on the commercial market. In this paper, Milstein and Das explain why SAW devices are ideal for very high frequency signals, typically in the range 10 MHz to 1 GHz. They show how SAW devices can be used as delay lines, fixed and programmable bandpass filters, matched filters, correlators, and oscillators. They also show how a SAW device can easily perform Fourier transforms. Finally, they give examples showing how these devices are making significant impact on various electronic systems, particularly radar, spread-spectrum communications, Electronic Counter Measures (ECM), and domestic television.

TUTORIALS ON SIGNAL PROCESSING FOR COMMUNICATIONS: PART I— DIGITAL TECHNIQUES FOR COMMUNICATION SIGNAL PROCESSING *

K. V. Mina, V. B. Lawrence, and J. J. Werner

Abstract   Digital processing of communication signals is now a practical alternative to analog processing. This application of a well-understood theory as a viable alternative is largely due to the availability of low cost arithmetic and storage LSI circuits; the availability of low cost A/D and D/A circuits; the commonplace occurrence of digital signals due to the ever increasing use of digital transmission and routing; and the significant advantages offered by the flexibility of the digital approach. This paper is an attempt to present in an introductory manner the what, why, where, and how of digital signal processing.

WHAT IS DIGITAL SIGNAL PROCESSING (DSP)

To the unfamiliar engineer, the two basic concepts crucial to understanding the digital processing of signals are sampling and quantization. Uniform sampling is the periodic measurement of a continuous signal. The resulting time series of numbers is an accurate representation of the original signal provided the sampling rate F_S is at least twice the bandwidth of the signal. Failure to sample at a sufficiently high rate results in a distortion called frequency folding or aliasing. Digital processing of the sampled data cannot occur until the data are quantized. In this process continuous amplitude ranges of the sample are each assigned a finite binary number. In uniform quantization all internal ranges are of equal size and the largest number represents all larger signals (clipping). The time series of numbers that result from quantized sampled signals can now be processed using any or all of the available digital

computer type capabilities. The digital processing of communication signals may be defined as any set of calculations on a real-time digital sampled signal that improves its communicated value. Typical examples are more informative. Linear digital filtering can be accomplished by multiplying, adding, subtracting, and storing samples. All the familiar continuous (analog) filtering capabilities carry over to the digital world with the implied constraint of a frequency domain limited to $Fs/2$.* In addition several new aspects are now practical as in Hilbert transformers [1], multipath networks, sample rate changers [2], and FFT's [3]. Many nonlinear processing capabilities exist, and the reproducibility of these digitally controlled nonlinear calculations is especially important. The μ-law or A-law logarithmic compression is simple to accomplish since this is essentially a floating-point rather than a fixed-point quantization operation.

The real-time processing of digital signals can of course be accomplished by a general purpose digital computer with

The authors are with Bell Laboratories, Holmdel, NJ 07733.

*Reprinted from *IEEE Communications Society Magazine*, January 1978, Vol. 16, No. 1, pp. 18–22.

*The equally good concept of repeated frequency images may be more common to the reader.

the proper input/output interface. Significant economies are to be gained by using specialized digital signal processors where a number of specific arithmetic operations are carried on in parallel. Typically the arithmetic operations are time multiplexed over a number of communications signals so as to operate the arithmetic unit (AU) at the highest possible efficiency. An appropriate amount of programmability is often included in the signal processor. The various aspects of system architecture are discussed in the companion paper by Tewksbury *et al.*

WHY USE DSP

The basic motivation for the use of digital signal processing is cost savings. Three technological factors dominate the reasons for this possibility. First, cheap, low power, high speed, LSI circuits now exist for performing arithmetic, storage, timing, and miscellaneous bit picking operations. Second, signals are available in digital format in ever increasing amounts. A third factor is important but more subtle. Digital signal processors offer a flexibility in their ability to be programmable in a time multiplexed environment.

The reduced cost and power and increased speed and complexity of LSI circuits are quite spectacular. In the storage of past values of the input and output signals, as is often required in signal processing, bulk RAM's and shift registers are very useful [4]. The register approach is applicable whenever the signal processor structure is relatively fixed. The use of RAM's is preferred when the structure of the processor is highly programmable. Multipliers are available in LSI chips that can be connected in modular fashion to obtain a parallel array multiplier or a serial multiplier [5]. It is to be expected that the full arithmetic function required, e.g., for a second-order recursive section, will be available on a chip in the near future.

The major reason for increased availability of signals in digital format is due mainly to a preference for the transmission and routing of digital signals. The growth of digital transmission is spectacular. In the Bell System, e.g., the 1.544 Mbit T-1 transmission system has grown as shown in Fig. 1. Even more important are proposed digital systems that operate at a 274 Mbit/s channel rate (equivalent to 4032 voiceband circuits). These include a radio system, a coaxial system, a millimeter wave system, and in all likelihood fiber optic systems [6]. The result will be an ever increasing population of digital signals. The ease with which digital signals can be routed is also a source for the increased use of digital signals. Long distance telephone calls are switched by a time-division switch with a digital signal format [7].

Another important reason for cost savings in digital signal processing is the inherent flexibility of using standard arithmetic operations and storage to perform a multiple of operations. On the lowest level, e.g., a single multiplier can be time shared to perform all the multiplications, as shown in the typical second-order filter (Fig. 2). On a somewhat higher level, a complete second-order section can be constructed consisting of five multipliers and four adders. This unit can then be programmed to operate as a cascade of K different second-order sections, typically operating on N time-multiplexed channels. On an even higher level the same arithmetic capability can, e.g., be programmed to function at times as a digital filter, at times as a tone generator, and at times as a tone detector. Thus the same hardware arithmetic circuits could perform all the required communication processing and simultaneously handle the required routine self-maintenance. In this way circuits that normally have low usage in a communications system can be avoided. This implies that with shrewd allocation of resources, the arithmetic circuits in digital signal processors can be made to function in virtually the most efficient power-speed condition.

HOW TO APPLY DSP

This section will discuss some of the basic tools employed in DSP, namely: digital filters, sample rate changers, tone generators, and detectors, spectral analyzers, and phase-locked loops.

Linear Digital Filtering: Theoretical but practical design approaches for digital filtering now exist. Typically a designer has filter requirements in the frequency, and or in the time domain. It is necessary to set the sampling rate of

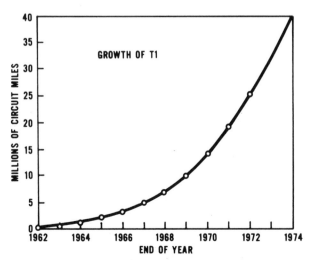

Figure 1.   (Source: W. E. Danielson, *Bell Labs Record*, 1975).

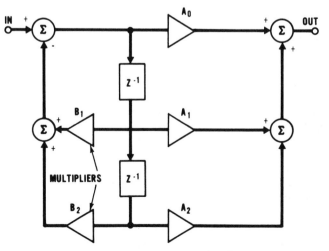

Figure 2.   Second Order Section (IIR).

the filter, the basic type, the order, and the detailed structure of the filter including: data and coefficient word lengths, number format, and quantization techniques. The sampling rate of the filter is often identical to that of the input data, although this is not always the case. The filter is of two basic types, infinite impulse response (IIR) and finite response (FIR). For FIR filters the zeros are distributed through the Z plane but the poles are all at z = 0. The choice between the FIR or IIR filters involves many factors and definite situations exist where each type is preferred [8]. The major differences are: FIR filters can easily be designed with perfectly linear phase [1] while IIR filters cannot. IIR magnitude designs often have a much lower order and therefore a more efficient hardware realization. Finite arithmetic effects are also substantially different.

The design of standard bandpass IIR filters is usually accomplished by using analog filters with well known S domain poles and zeros and using transformations to obtain the Z domain singularities. This approach is documented [9] for the Elliptic (Cauer), Chebyshev I and II, Butterworth, Bessel, and Transitional filters using three different transformations. In addition, all-pass filters which can phase compensate the basic IIR design are readily designed when composite magnitude and phase characteristics are required [10]. When nonstandard magnitude design is required, optimization techniques are usually applied directly on the singularities in the Z plane to meet the necessary requirements. The design of FIR filters has typically been carried out directly in the digital domain. Usually the finite impulse response is derived directly since many implementations make direct use of these coefficients. Although FIR filters can be designed with arbitrary phase the majority of applications involve exactly linear phase since this is easily accomplished. Appropriate design methods and software codes have been documented and these programs provide extremely versatile methods for generating low-pass, high-pass, bandpass, and multiple bandpass and bandstop filters [1], [11].

The implementation of IIR filters often occurs as cascades of the second-order section shown in Fig. 2. The implementation of linear phase FIR filters is often done in the direct form shown in Fig. 3 although cascades of second- and fourth-order sections are also useful and efficient. Most real hardware filters have been constructed with linear rather than floating-point number formats for

arithmetic operations. Floating point (μ-law) formats are possible but have not been used because of a reduced modularity of the hardware. It may become more useful in the future. The final design of a filter requires setting the exact topology, the coefficient values, the data word length [12], and the control of the time-multiplexed environment. These factors interact so that often several iterations are required to obtain an acceptable solution. Although the cascade organization can be regarded as a reliable approach, other totally new structures have been found that, e.g., can be implemented with reduced coefficient accuracy [13].

One special consideration for IIR implementation is stability. Since nonlinear operations, product quantization, and adder overflow exist in a feedback loop, oscillations will occur [14], [15] but can be controlled. Two examples of filtering techniques that are practical in the digital world are associated with phase shifters and with the Hilbert transforms. An IIR all-pass multipath filter that provides approximately $\pi/2$ phase difference between a pair of outputs over a frequency band of interest can be easily designed and implemented using continuous s-plane design [16] and the bilinear z-transform. An FIR linear phase filter whose amplitude approximates the sign function over a desired frequency band can be easily designed and implemented [1].

Sample Rate Changers: The sampling rate of a digital signal can often be advantageously increased or reduced by filling in additional (interpolating) samples or discarding (decimating) unnecessary samples [17]. The motivation for this is to keep sampling rates at the most practical rate in terms of reducing system hardware. The main effect is to cause multiple signal images during sample rate increase or aliasing during decrease. This rate changing procedure has been found useful in a number of processing situations. When, for example, narrowband low-pass filtering is done in stages, the sampling rate can be reduced in each stage. In meshing systems operating at different rates both increase and decrease of rates is often employed. These are especially beneficial when direct form FIR filters are employed since unnecessary calculations need not be done. One particularly practical application involves cascades of FIR filters with simple coefficients that are easily designed [18]. In addition to simple rate change situations, asynchronous systems can often be managed by using appropriate interpolation.

Spectral Analysis: The use of the DFT [19] (Discrete Fourier Transform) for spectral analysis of digital signals is quite commonplace. This can be used for tone detection systems, or can be used to accomplish real-time linear filtering by employing both the DFT and its inverse.

The major reason for employing the DFT is the particularly efficient methods for calculating it by using the FFT algorithms [3]. FFT processors are now marketed that offer significant advantages over a general purpose computing approach. It should be mentioned that although the DFT is analogous to the continuous Fourier transform, there are several unique properties that must be understood before maximum advantages can be obtained. In addition, attention must be paid to finite word length effects due to overflow of sums or quantization of products and coefficients [3].

Tone Detection and Generation: The generation of

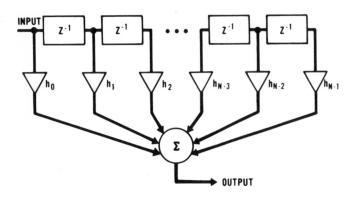

Figure 3.   FIR Filter.

sinusoids is necessary in many communication systems. One method applicable is the use of ROM to store the necessary sampled values. The number of bits stored determines harmonic content. This method is particularly useful when several frequencies that are harmonically related must be generated. If increased accuracy is required it is possible to use interpolation techniques on the stored values. An alternate approach is to use second-order feedback sections with poles on the unit circle. The frequency is controlled by the value of the feedback coefficient and is therefore easily programmable.

Multifrequency (MF) signaling is used extensively in the telephone network. The advent of telephone offices (No. 4 ESS) that switch directly time-multiplexed digitally encoded voice channels provides the motivation and environment for digital service circuits [24]. In this type of MF signaling the receiver has to be able to decide which two out of the six possible different tones (700, 800, 1100, 1300, 1500, and 1700 Hz) have been transmitted. Tone amplitude detection can be achieved by quadrature detection which consists in generating in-phase and quadrature components of narrowband signals centered around each of the possible frequencies [23], [24]. If a tone is present at a given frequency, its amplitude can be computed by taking the square root of the sum of the squares of the in-phase and quadrature components. A simple, approximate algorithm for the computation of the square root is given in [22]. A threshold detector is then used to decide which two tones are present. An alternate way to detect the two tones consists in computing the discrete Fourier transform (DFT) at each signaling frequency [21]. The received power at a given frequency is simply the squared magnitude of the DFT evaluated at this frequency. In addition to tone detection, the MF receivers must also have the ability to detect the presence of excessive noise, to bridge signal interruptions, and to recognize signal spacings. All these functions can be implemented digitally. Hardware efficiency through greater multiplexing can be achieved by using subsampling techniques [23], [24].

Another example of MF signaling is TOUCH-TONE® or pushbutton signaling. Each dialing digit is represented by a pair of frequency components selected from a low frequency set and a high frequency set. The CCITT recommendations for the nominal frequencies are (697, 770, 852, 941 Hz) for the low-frequency group and (1209, 1336, 1477, 1633 Hz) for the high-frequency group. An experimental all-digital push-button receiver is described in [26]. An MOS/LSI implementation is reported in [25]. The first operation performed at the receiver consists in separating the low-frequency group from the high-frequency group. Since only one tone is present in each group there is no need for quadrature detection and the frequency can be measured, for example, by counting pulses between adjacent zero-crossings [26]. The algorithm proposed in [25] is based on a digital range filter concept.

Digital Phase Lock Loops (DPLL): DPLL's can be viewed as time-varying nonlinear digital filters. They are used for synchronization and demodulation. [33] describes a DPLL which emulates an analog PLL used for demodulation of FM

®Registered service mark of American Telephone and Telegraph Company.

signals. However, most DPLL's used in practice do not emulate analog PLL's but rather take advantage of the peculiar nature of DSP. One such popular DPLL locks the loop to the incoming sinewave by advancing or retarding the sampling epoch of the A/D under the guidance of a control algorithm [34], [38].

WHERE IN COMMUNICATION

SSB Modulation: An example of a complex digital system is the generation of the digital group band signal, that is twelve 4-kHz voice channels from 60 to 108 kHz, from the set of digital baseband signals. This function as well as its inverse are useful because of the interface between existing analog FDM transmission systems and new digital systems. Several approaches have been tried illustrating the flexible nature of digital systems. In [27]-[29], the Weaver two-path method is used with IIR/FIR filters, rate changing, and modulation techniques. In [30] and [31], a novel approach uses a "polyphase" network and time interleaving to accomplish rate changing. An approach that uses all pass IIR filters with two outputs that provide approximately $\pi/2$ phase difference is given in [32].

Voiceband Data Communications has provided DSP with its most successful applications. Recently, several manufacturers have advertised all-digital modems using LSI devices [36], [46]. The analog modems used in the early 1960's were operating at speeds equal to or lower than 2400 bits/s. Reliable transmission at 4800 and 9600 bits/s became feasible in the late 1960's and early 1970's because of an increasing use of digital technology, particularly in the implementation of digital adaptive equalizers [35], [46]. Modems incorporate a great number of the functions that are usually encountered in communications. References [38]-[42] present several different techniques for implementing digital transmitters. An all-digital VSB transmitter implemented on an I²L chip is described in [39]. Microcoded transmitters in which the modulation scheme and/or speed of operation can be changed by a simple software modification have been proposed in [41]. It has also been shown that complete modems can be implemented on a digital signal processor [37], [38]. The recent availability of fast bipolar microprocessors could well make such an approach cost-effective and lead to the development of more flexible and perhaps faster modems. An important consideration in the design of high-speed digital modems is the fact that most of the functions can be performed at the baud rate which is usually lower than 2400 Hz. Therefore baud processing can be performed with slower, i.e., less expensive, devices than processing which has to be done at a rate which is dictated by the sampling theorem (>6 kHz). Algorithms for demodulators, equalizers, carrier recovery, and timing recovery as well as the description of the digital implementation of some of these functions, are described in [43]-[54]. The cost of the A/D can generally not be justified for the implementation of medium and low-speed modems (≤2400 bits/s). However, the modulation schemes that are used at these speeds, such as FSK or DPSK, carry all the useful information in their zero-crossings. Therefore, the incoming signal can be passed through a simple hard-limiter without loss of information and all the rest of the processing can be done on binary waveforms [44], [45].

DSP is particularly well suited for the implementation of functions that require precise delays and accurate arithmetic operations. Two such applications are *adaptive equalization* and *echo cancellation*. Adaptive equalizers and echo cancelers usually consist of a transversal filter whose taps (coefficients) are adaptively updated by using a gradient type of algorithm. Taps inhibit, taps preset and other ancillary tasks are easily implemented digitally. Digital adaptive equalizers are used extensively in high-speed voiceband modems. References [46], [48]-[53] present digital implementations and algorithms for adaptive equalizers. Digital echo cancelers have potential applications in the telephone network especially in view of the increasing availability of digital transmission lines and switching systems. They may replace the present analog echo suppressors which introduce an unavoidable mutilation of the speech signals and whose performance might not be considered good enough for domestic satellite service. It should be noted however, that digital processing can also be used for the implementation of echo suppressors. Digital echo suppressors overcome many of the problems associated with conventional analog echo suppressors [60]. References [55]-[59] describe some implementations of digital adaptive echo cancelers suited for use in satellite communications. Echo cancelers might also find some applications in data communications for 2-wire full-duplex transmission [61].

Radar Systems: In most radar systems, performance and accuracy considerations, rather than cost, dictate the type of technology. The salient features of DSP devices such as weight, size, and reliability make them suitable particularly for aircraft radar. Presently, most modern receivers [62]-[64] incorporate digital signal processors either as banks of time-varying digital filters or FFT's. MTI radar receivers using custom-designed DSP are currently available on the market [5]. Programmable DSP radar receivers using LSI microprocessors have been built [64]. These processors incorporate an economical redundancy, that together with processor programmability, also provide fault tolerance.

Complex Terminal Equipments: Continual advances in DSP may lead to better and cheaper terminals. Such terminals may well include coders and decoders for digital transmission of audio and video signals. Tradeoffs exist in the overall cost of a communication system between transmission and terminal costs. Complex terminal equipment, such as vocoders, can transmit intelligible speech with a bandwidth saving of 10:1 or more. Despite this high potential in bandwidth saving, they have only been applied in special communication situations. It is hoped that recent advances in the implementation of vocoders [65] using DSP will offer further improvements in speech quality as well as reduce the cost of terminal equipment. Predictive and transform coding [67] can cut down the channel capacity required for video signals. The implementation of some transform coding schemes using DSP, such as Hadamard transforms, require a lot of memory but no multiplication. Efficient transmission of color TV signals may require the separation of its three primary components before encoding. Such separation of the primary components may best be done using DSP.

REFERENCES

[1] L. R. Rabiner and B. Gold, *Theory and Application of Digital Signal Processing,* Prentice-Hall Inc., 1975.

[2] E. P. F. Kan and J. K. Aggarwal, "Multirate Digital Filtering," *IEEE Trans. Audio Electroacoust.,* Vol. AU-20, No. 3, 8/72.

[3] L. R. Rabiner and C. M. Rader, *Digital Signal Processing,* IEEE Press, 1972, pp. 223-260.

[4] *MSI-LSI Memory D.A.T.A. Book,* Spring 1976, Edition 10.

[5] J. R. Mick, *Digital Signal Processing Handbook 1976,* Advanced Micro Devices, Inc.

[6] N. E. Snow and N. Knopp, Jr., "Digital Data Network System Overview," *BSTJ,* Vol. 54, No. 5, 5/75.

[7] G. E. Watson, "Getting No. 4 ESS on Line on Time," *Bell Labs Record,* April, 1976, pg. 82.

[8] L. R. Rabiner et al., "Some Comparisons Between FIR and IIR Digital Filters," *BSTJ,* Vol. 53, No. 2, 2/74.

[9] J. F. Kaiser, "Computer Aided Design of Classical Continuous System Transfer Functions," *Proc. Hawaii Int. Conf. Sys. Sciences,* Univ. Hawaii Press, 1968.

[10] J. P. Thiran, "Equal-ripple Delay Recursive Digital Filters," *IEEE Trans. Circuit Theory,* Vol. CT-18, No. 6, Nov. 1971, pp. 664-669.

[11] J. F. Kaiser, "Nonrecursive Digital Filter Design Using the Io-Sinh Window Function," *Proc. 1974 IEEE Symp. Circuits and Sys.*

[12] B. Liu, "Effect of Finite Word Length on the Accuracy of Digital Filters—A Review," *IEEE Trans. Circuit Theory,* Vol. CT-18, No. 6, 7/71.

[13] A. Fettweis et al., "Design of Wave Digital Filters for Communication Applications," *IEEE Proc. ISCAS '75.*

[14] P. M. Ebert et al., "Overflow Oscillations in Digital Filters," *BSTJ,* Vol. 48, 11/69.

[15] L. B. Jackson, "Analysis of Limit Cycles Due to Multiplication Rounding in Recursive Digital Filters," *Proc. Seventh Annual Allerton Conf. Circuit and System Theory,* 1969.

[16] S. D. Bedrosian, "Normalized Design of 90° Phase-Difference Networks," *IRE Trans. on Circuit Theory,* Vol. CT-7, No. 2, 6/60.

[17] R. W. Schaefer et al., "A Digital Signal Processing Approach to Interpolation," *Proc. IEEE,* Vol. 61, No. 6, 6/73.

[18] D. J. Goodman, "Digital Filters for Code Format Conversion," *Electronic Letters,* Vol. 11, 6/75.

[19] A. V. Oppenheim and R. W. Schafer, *Digital Signal Processing,* Prentice-Hall, 1975, Ch. 3.

[20] A. D. Proudfoot, "Simple Multifrequency Tone Detector," *Elec. Letters,* pp. 525-526, 10/19/72.

[21] I. Koval and G. Gara, "Digital MF Receiver Using Discrete Fourier Transforms," *IEEE Trans. on Commun.,* Vol. COM-21, No. 12, pp. 1331-1335, 12/73.

[22] F. G. Braun and H. Blaser, "Digital Hardware for Approximating Amplitude of Quadrature Pairs," *Electronic Letters,* Vol. 10, No. 13, pp. 255-256, 6/27/74.

[23] F. G. Braun, "Nonrecursive Digital Filters for Detecting multifrequency Code Signals," *IEEE Trans. Acoust. Speech, Signal Processing,* ASSP-23, No. 3, pp. 250-256, 6/75.

[24] C. R. Baugh, "Design and Performance of a Digital Multifrequency Receiver," *IEEE Trans. Commun.,* June 1977.

[25] D. R. Gibson, "LSI in Communications: A Tool for Effecting Product Leadership," *Conf. Rec. ICC '76.*

[26] K. Niwa and M. Sato, "Multifrequency Receiver for Pushbutton Signaling Using Digital Processing Techniques," *Conference Record, ICC '74,* Minneapolis.

[27] S. L. Freeny et al., "Design of Digital Filters for an all Digital Frequency Division Multiplex-Time Division Multiplex Translator," *IEEE Trans. Circuit Theory,* Vol. CT-18, No. 6, 11/71.

[28] S. L. Freeny et al., "System Analysis of a TDM-FDM Translator/Digital A-Type Channel Bank," *IEEE Trans. Commun.,* Vol. COM-19, No. 6, 12/71.

[29] S. Singh et al., "Digital Single-Sideband Modulation," *IEEE Trans. Commun.,* Vol. COM-21, No. 3, 3/73.

[30] M. G. Bellanger and J. Daguet, "TDM-FDM Transmultiplexer: Digital Polyphase and FFT," *IEEE Trans. Commun.,* Vol. COM-22, No. 9, 9/74.

[31] M. G. Bellanger et al., "Digital Filtering by Polyphase

Network: Application to Sample-Rate Alteration and Filter Banks," *IEEE Trans. Acoust. Speech, Signal Processing,* Vol. ASSP-24, No. 2, 4/76.

[32] R. Boite and H. Leich, "Use of Numerical Phase-Difference Filter for the Digital Channel Bank," *Proc. IEEE,* Vol. 122, No. 4, April 1975.

[33] J. Garodnick et al., "Response of an All Digital Phase-Locked Loop," *IEEE Trans. Commun.,* Vol. COM-22, No. 6, June 1974.

[34] A. Weinberg and Bede Liu, "Discrete Time Analyses of Nonuniform Sampling First- and Second-Order Digital Phase Lock Loops," *IEEE Trans. on Commun.,* Vol. COM-22, 2/74.

[35] E. R. Kretzmer, "The New Look in Data Communication," *Bell Labs Record,* pp. 258-265, 10/73.

[36] H. Harris et al., "An All Digital 9.6K BPS LSI Modem," *NTC Record,* San Diego, December, 1974

[37] D. N. Sherman and S. P. Verma, "System Description of a Programmable Multiple Data Set," *NTC Record,* New Orleans, December, 1975.

[38] A. C. Salazar et al., "Implementation of Voiceband Modems on a Digital Signal Processor," *NTC Record,* New Orleans, December, 1975.

[39] P. J. Van Gerwen et al., "A New Type of Digital Filter for Data Transmission," *IEEE Trans. Commun.,* February, 1975, pp. 222-234.

[40] F. A. M. Snijders et al., "Digital Generation of Linearly Modulated Data Waveforms," *IEEE Trans. Commun.,* 11/75, pp. 1259-1270.

[41] M. F. Choquet and H. J. Nussbaumer, "Microcoded Modem Transmitters," *IBM J. Res. Develop.,* 7/74.

[42] K. H. Mueller, "A New Approach to Optimum Pulse Shaping in Sampled Systems Using Time-Domain Filtering," *BSTJ,* May-June, 1973, pp. 723-729.

[43] D. A. Spaulding, "A New Digital Coherent Demodulator," *IEEE Trans. Commun.,* 3/73.

[44] V. D. Heutinen et al., "A Digital Demodulator for PSK Signals," *IEEE Trans. Commun.,* 12/73, pp. 1352-1360.

[45] D. D. Falconer and J. J. Werner, "Comparison Detection of Hard-Limited Digital PM Signals," *ICC Conf. Record,* Minneapolis, 1974.

[46] H. L. Logan and D. G. Forney, "A MOS/LSI Multiple Configuration 9600 bps Data Modem," *ICC Conf. Record,* Philadelphia, 1976.

[47] K. H. Mueller and M. Muller, "Timing Recovery in Digital Synchronous Data Receivers," *IEEE Trans. Commun.,* May, 1976, pp. 516-531.

[48] M. G. Taylor, "A Technique for Using a Time-Multiplexed Second-Order Digital Filter Section for Performing Adaptive Filtering," *IEEE Trans. Commun.,* March, 1974, pp. 326-330.

[49] S. Horvath, "Adaptive IIR Digital Filters for On-Line Time-Domain Equalization and Linear Prediction," *IEEE Workshop on Dig. Sig. Proc.,* Arden House, 2/76.

[50] K. H. Mueller and D. A. Spaulding, "Cyclic Equalization—A New Rapidly Converging Equalization Technique for Data Comm.," *BSTJ,* 2/75.

[51] D. D. Falconer, "Analysis of a Gradient Algorithm for Simultaneous Passband Equalization and Carrier Phase Recovery," *BSTJ,* 4/76.

[52] D. D. Falconer, "Jointly Adaptive Equalization and Carrier Recovery in Two-Dimensional Digital Comm. Systems," *BSTJ,* 3/76.

[53] A. Lender, "Decision-Directed Digital Adaptive Equalization Technique for High-Speed Data Transmission," *IEEE Trans. Commun. Technol.,* 10/70.

[54] D. R. Morgan, "Analysis of Digital AGC Amplifiers Using the Discrete-Time Volterra Series," *ICC Conf. Record,* Minneapolis, 1974.

[55] S. J. Campanella, H. G. Suyderhoud, and M. Onufry, "Analysis of an Adaptive Impulse Response Echo Canceller," *Comsat Technical Review,* Vol. 2, No. 1, Spring, 1972, pp. 1-36.

[56] H. G. Suyderhoud and M. Onufry, "Performance of a Digital Echo Canceller in a Simulated Satellite Circuit Environment," AIAA 4th Communications Satellite Systems Conference, Washington, D.C., April 24-26, 1972, AIAA Paper No. 72-539.

[57] H. G. Suyderhoud, S. J. Campanella, and M. Onufry, "Results and Analysis of Worldwide Echo Canceller Field Trial," *Comsat Technical Review,* Vol. 5, No. 2, Fall, 1975, pp. 253-274.

[58] M. Demytko and L. K. Mackechnie, "A High-Speed Digital Adaptive Echo Canceller," *A.T.R.,* Vol. 7, No. 1, 1973, pp. 20-28.

[59] Y. Kato et al., "A Digital Adaptive Echo Canceller," *A.T.R.,* Vol. 7, No. 1, 1973, pp. 20-28.

[60] Digital Echo Suppressors, *ICC Conf. Record,* Philadelphia, 1976 (5 papers).

[61] K. H. Mueller, "A New Digital Echo Canceller for Two-Wire Full-Duplex Data Transmission," *IEEE Trans. Commun.,* 9/76.

[62] C. E. Muehe et al., "New Techniques Applied to Air Traffic Control Radar," *Proc. IEEE,* Vol. 62, No. 6, June, 1974.

[63] *Proc. of IEEE 1974 National Aerospace and Electronic Conference.*

[64] *Proc. of IEEE 1975 Electronics and Aerospace Systems Convention.*

[65] Special Issue on Man-Machine Communication by Voice, *Proc. IEEE,* April, 1976.

[66] Special Issue on Digital Picture Processing, *Proc. IEEE,* Vol. 60, 7/72.

TUTORIALS ON SIGNAL PROCESSING FOR COMMUNICATIONS: PART II—DIGITAL SIGNAL PROCESSING ARCHITECTURE *

S. K. Tewksbury, R. B. Kieburtz, J. S. Thompson, and S. P. Verma

Abstract   **A number of the basic considerations involved in the selection of an overall digital signal processor architecture are reviewed.**

INTRODUCTION

We review below several aspects of digital signal processor design, concentrating on the overall architectural approach. First, we consider the major design constraints—real-time processing, cost, performance, and reliability. Second, we discuss some of the alternatives regarding the overall processor architecture, and finally we briefly describe some aspects of processor partitioning. References have been included to provide the detail which space limitations prevented being included in this report.

GENERAL DESIGN CONSTRAINTS

Real-Time Data Processing

In the preceding paper [1] the digital signal processing algorithm was represented by a signal flow graph consisting of functional operators (e.g., multipliers, adders, limiters, scalers, etc.), sample period delay operators (Z^{-1}), and signal links showing the signal flow among the various operators. Input signal samples are entered into the graph at the uniform rate f_s, i.e., the signal sample rate. Since the data

The authors are with Bell Laboratories, Holmdel, NJ 07733.

*Reprinted from *IEEE Communications Society Magazine*, January 1978, Vol. 16, No. 1, pp. 23-27 and 32.

are generally applied for an indefinite period of time, the input data file is, in effect, infinite. As a result, the digital signal processor executing the algorithm must complete its program in the sample period $\tau_s = f_s^{-1}$. In particular, the digital signal processor must be a real-time processor.

The minimum rate at which the processor must execute operations of a given type is readily determined from the signal flow graph. For example, the required multiplication rate f_m is merely $f_m = N_m \cdot f_s$ where N_m is the number of multiplication operators in the signal flow graph and f_s is the signal sample rate. In the case of a processor executing digital processing algorithms for a number of input signal channels (i.e., a shared processor or Type I multiplexing in the terminology of Jackson et al. [2], the required processor multiplication rate is the sum of the required rates for each input signal channel.

There are several ways that the operation rate of a processor can be matched to the requirements of the given digital signal processing algorithm. Given a device technology, a functional module can be designed to complete its operation in one machine cycle (requiring considerable hardware) or in a number of machine cycles (requiring less hardware). Multiplier designs are a good example of various designs giving various multiplication rates using the same speed technology [3]-[5]. Alternatively, given the logical design of the functional module, technologies ranging from low speed (and low power) IC technologies to high speed (and high power) IC technologies can be used to realize the module [6]. Finally, given the module's logical design and the device realization, the processor can be designed to use only one such module or a number of such modules [7] (with a net increase in processor operation rate). These three approaches allow great flexibility in the choice of processor architectures, hardware modules, and IC devices capable of realizing the given signal processing algorithm with reasonable efficiency.

The real-time data processing requirement can be characterized as a resource threshold, defined as the minimum processing capability for which the given processor is capable of executing the specified algorithm. This processing capability includes the required operation rate, the required data storage, and the ability to execute all the operations specified in the algorithm.

Cost Threshold

The high cost of digital signal processors has been the primary reason why they have not seen more widespread use [8]. It is only recently that the economies of large-scale integration (LSI) device technologies have made digital signal processors competitive in a number of practical applications. However, in some applications such as radar, the superior performance of digital signal processors is more important than cost [9]. It is in such applications that digital signal processing has had the greatest practical impact [10]. It is therefore convenient to use the concept of a cost threshold, defined as the minimum cost of a digital signal processor for which it remains an acceptable approach.

The cost depends critically on all aspects of the digital signal processor design and includes both the fixed costs (development, materials, manufacture, etc.) and the recurring costs (maintenance, power, building space, cooling, etc.) [11]. When even the most efficient designs are close to the cost threshold, economies must be sought in each phase of the processor design. The result is little flexibility in the design of the overall processor which often emerges as a highly specialized processor. However, when the cost of highly specialized processors lies well below the cost threshold, a wide variety of processor architectures become available. More lofty objectives such as flexibility in the use of the processor, superior performance, and superior reliability may then be considered. Perhaps most important, the long-term impact of the current design (particularly in communications) can be considered in addition to the immediate costs incurred.

As the cost of digital processor components continues to decrease, more and more applications will be less constrained by cost considerations. The result will undoubtedly be the development of new processor architectures particularly suited to digital signal processing problems and less dependent on trends in general-purpose data processors.

Performance Threshold

As discussed in the preceding paper, finite data accuracy effects lead to imprecise computations that generate noise and distortion in the processed signal [12]. Much of the current literature on the synthesis of digital signal processing algorithms concerns such finite accuracy effects and a number of "preferred" (i.e., low noise and distortion), algorithms have been identified or suggested. Of course, the noise and distortion due to finite data accuracy may be decreased indefinitely merely by increasing the processor's internal data accuracy, although this obviously increases the cost of the processor [13].

We introduce the concept of performance threshold defined as the minimum performance (in terms of noise and distortion) for which the digital signal processor remains acceptable. When near the cost threshold, the minimum data accuracy will generally be chosen, with the processor therefore operating near the performance threshold. On the other hand, when well below the cost threshold, the processor's internal data accuracy may be chosen much better than the minimum accuracy required. Not only does this provide greater flexibility in the use of the processor for a variety of applications with different performance requirements, but it also allows the use of nonpreferred algorithms despite their higher noise and distortion. As a result, the processor architecture becomes less dependent on any single algorithm.

Reliability Threshold

In general, the application will also impose reliability requirements on the mean time between failures (MTBF) or other down-time criteria [14]. The reliability of hardware is our major consideration but software verification may also be needed in some programmable processor applications [15].

Without special provisions the reliability of a processor is worse than the reliability of its components. However, a number of techniques have been developed to improve system reliability even to the point that the system is more reliable than its components. The occurrence of a failure can be determined by fault detection [16] using modular redundancy voting [17], system functional tests, coding techniques [18], etc. The fault can be located to within a replaceable portion of the system using fault diagnosis techniques [19]. Although fault detection and diagnosis can considerably reduce system down time, fault correction must be used to increase the MTBF. Fault-tolerant computers and digital signal processors use fault-correction techniques to switch a defective module out of the system and to switch in a working ("hot") spare module or use fault-correcting codes [20], [21]. Digital signal processors generally require that only those parts of the processor performing data-dependent decisions (e.g., tone detection) be reset when fault correction is performed. Furthermore, immediate detection of a fault is generally not necessary, the only requirement being that the fault be detected in a reasonable period of time. For such reasons, the characteristics of a fault-tolerant digital signal processor may be quite different from those of a fault-tolerant data processor.

We define the reliability threshold as the minimum reliability (in terms of appropriate criteria such as MTBF) for which the digital signal processor remains acceptable. When near the cost threshold, the processor will typically be designed near the reliability threshold associated with the specific application. The result is limited flexibility in the overall processor design. On the other hand, when well below the cost threshold fault-tolerant processors allow considerable flexibility in the choice of the processor architecture, hardware design of the processor components and the IC devices used.

MAJOR ARCHITECTURAL CONSIDERATIONS

Parallel Versus Sequential Processors

For our purposes, we define parallel processing [22] as the concurrent execution of several functional operations using a number of distinct functional (i.e., processing) units under the coordination of a common control structure. Sequential processors, on the other hand, sequentially execute all the functional operations of the algorithm using a single functional module [9], [15], [23]-[25]. Between these extremes lie a great variety of architectures that use a mixture of these two approaches. (The same statement applies to the other two topics discussed in this section. Space permits only a general summary of the alternatives as defined.) The reader should, however, recognize that the "hybrid" approaches often achieve advantages of both alternatives while retaining some of the disadvantages of each. The tradeoffs leading to the selection of any specific approach are usually application dependent and are best understood by referring to the references.

First, consider some of the advantages of the parallel processor [26]. 1) The functional operation rate of a parallel processor may be increased, without requiring new functional module designs, merely by adding more functional modules. The operation rate of a sequential processor, however, is limited by the operation rate of its single functional module. In this sense, the parallel processor offers the advantage of expandability although the expansion is ultimately limited by the increasing communication rate with the common control hardware [27]. 2) The incremental cost of fault detection and fault tolerance may be less with parallel processors [28] than with sequential processors. First, consider fault detection. If the parallel processor uses several identical modules, special machine cycles may be used to perform majority voting among the already existing modules. Even if redundant modules are added for majority voting, one set can sequentially serve all modules of that type in the processor. Sequential processors, on the other hand, require full replication of the processor for each unit increase in the processor's redundancy. In the case of "hot" spares for fault tolerance, a single "hot" spare module can service all the modules of that type in the parallel processor whereas the sequential processor usually requires a full spare processor. 3) The parallel processor also allows the use of efficiently designed functional units (such as multipliers) with limited functional capabilities to execute commonly encountered functions. Less common functions such as overflow correction can be executed by a more flexible but less efficient functional module. The sequential processor, on the other hand, requires a versatile functional unit capable of executing all the operations in the algorithm with the usual consequence that each specific function is executed inefficiently. A familiar example is a sequential processor which performs multiplications using a shift-and-add sequential process. An important topic in parallel pro-

cessing concerns the extent to which an algorithm can be executed using parallel, concurrent operations, i.e., the inherent degree of parallelism of the algorithm [29]. However, digital signal processing algorithms generally lack data dependent control and the signal flow graphs exhibit a high degree of parallelism. In this sense, digital signal processing algorithms are well suited to the parallel processing approach.

Sequential processors dominated early data processor architectures [30], [10], [31], [32], [12], [24]. The great range of IC device speeds available among the various technologies often permits the capabilities of a sequential processor to be closely matched to a digital signal processing problem. Some advantages of sequential processing are as follows: 1) The sequential processor offers the advantage of simplicity in a number of ways. Conceptually, the architecture is less complex than for a parallel processor and efficient hardware designs are usually more straightforward. Sequential processors are often easier to program since the software need not specify which functional module performs which operation as in the case of parallel processors. 2) Since several functional modules communicate with the common control and other common support facilities of a parallel processor, different communications rates are generally found in different parts of a parallel processor. Time division multiplexing and demultiplexing is therefore generally required to interface parts of the processor operating at different rates. The sequential processor, on the other hand, tends to have common communication rates throughout the processor making the multiplexing/demultiplexing hardware unnecessary. 3) Different functional units of a parallel processor may have different latencies (i.e., throughput delays). Execution of the algorithm often requires careful consideration of the time ordering of operations to ensure correct operation at points where parallel processed signals are merged. The fully sequential execution of an algorithm generally avoids such problems. 4) Finally, it should be noted that when an overall system function test indicates a functional unit failure, the fault diagnosis is less complex in a sequential processor with one potentially faulty functional module than in a parallel processor with several.

Centralized Versus Distributed Support
Hardware in Parallel Processors

In addition to the functional and sample period delay operations shown in the algorithm's signal flow graph, the processor will have to perform a variety of additional operations such as data path switching, scratch pad memory, and generation and distribution of both the control information defining the state of the processor and of timing information defining movement of data and execution of functions and of program constants, etc. These additional operations are included here under the heading of support functions [27].

By a centralized support facility [33] we mean that all or part of some support function is concentrated in a single hardware module whose resources are time-shared among the several modules of the parallel processor. By a distributed support facility [34] we mean that all or part of some support function required for some given processor module is realized with hardware that is permanently dedicated to that processor module. Obviously, most processors will contain both centralized (also called common) and distributed support hardware.

Centralized support facilities have a number of advantages. 1) Since the support hardware can be assigned to a variety of modules, a more flexible processor usually results than with distributed support systems. For example, the data memory space assigned to a given module may not be adequate. In a distributed memory system, the module could not be used with full efficiency whereas a common memory system could assign a larger memory block to that module. 2) A centralized support facility is also conceptually simpler than a distributed support facility. The problem of partitioning the support function during the hardware design is avoided, although the problem does reappear in the software design. The highly specialized "hard wired" digital signal processor [35], however, must resolve the partitioning problem for the hardware design for other reasons and the efficiencies and low cost of distributed support are often taken advantage of in such designs. 3) However, to a considerable extent centralized (and consolidated) support functions are more compatible with existing standard LSI digital devices than distributed support systems. In particular, large and efficiently fabricated memory devices are currently available. A random access memory can perform all the support functions listed earlier (even in the same memory unit by time-sharing its operation over several support functions) as well as provide the main data storage requirements. The impact of LSI memories on digital signal processors (as well as on data processors) has been considerable and tends, at this time, to favor centralized support facilities.

Despite the trend toward centralized support facilities in parallel processors, distributed support will continue to play an important role for several reasons. 1) The major difficulty with centralized support facilities is that its operation rate increases as more processor modules communicate with it. The result is that the expandability of the parallel processor is severely limited by the capabilities of the common support facility which, in turn, are technology limited. With distributed support, the addition of another module is accompanied by the addition of its dedicated support and less incremental demands are placed on the common support hardware. Complete distribution of the support function without greatly increasing the complexity of the processor's hardware and software is, however, unlikely. A typical result is to judiciously choose the distributed support to minimize the incremental demands placed on any common support facilities by the addition of more processor modules. 2) The decreased interaction rate of the modules with common support facilities which results from distributing support often permits the functional modules to be operated at higher efficiency since unnecessary delays in the interfaces to a centralized support facility are avoided.

Fixed-Program Versus Reprogrammable Processors

A fixed-program processor is defined here as a processor designed to execute a single specific algorithm with the possible exception of changes in the program constants by changing memories storing program constants [35]-[37]. A reprogrammable processor is defined here as a processor whose sequence of operations as well as program constants can be changed either by software or by replacing hardware components [38], [27], [13]. In particular, we avoid rewiring the processor in both cases since the result would, in fact, be a new processor.

The limited flexibility of fixed-program processors is often compensated by significantly lower cost processor realizations. Otherwise, it is doubtful whether there would be much interest in this approach. Fixed-program processors eliminate all unessential and redundant hardware, often using custom-designed IC devices and always using custom-designed processor modules. To a large extent, the program is hardwired—the reason for the lack of versatility. By necessity, a fixed-program processor incurs a higher hardware development cost. Therefore, fixed-program processors are usually found in applications where the development costs can be distributed over a large production run. When program constants are stored in read-only memories (ROMs), the limited flexibility of fixed-program digital signal processors is not as great as one might expect. For example, although the processor may be capable of performing only second-order sections, changing filter coefficients by replacing ROMs allows a wide range of filtering functions to be performed. The problem, of course, is that one may wish to do more than just filtering.

Despite their generally higher cost, there may be compelling reasons why programmable processors might be preferable. Perhaps the greatest advantage is the potential to avoid an overwhelming proliferation of highly customized processors designed for each application, with the resulting maintenance problems [39]. These maintenance problems include unfamiliarity with the specific hardware, the difficulty of maintaining a supply of spare parts, and the limitations on replacing old components that have failed with current state-of-the-art IC devices. However, this potential advantage of programmable processors will not emerge necessarily. In particular, it is possible that a unique programmable processor may be designed for each application with again a resulting proliferation of highly customized processors.

Programmable processors offer other important advantages. Some applications may require a flexible processor (e.g., test equipment, etc.). A single programmable processor can then replace a set of fixed-program processors. In some applications, experience with the digital signal processor may suggest a superior algorithm (in terms of performance) for performing the signal processing function. A programmable processor can, if it had adequate processing capacity, be reprogrammed to perform the desired algorithm. Finally, especially in large systems such as communications, the evolution of the system may modify the required signal processing function. Again, the programmable processor has obvious advantages.

PARTITIONING

The previous section reviewed some gross characterizations of the digital signal processor design. At the other end of the design problem is the selection of specific IC devices and their organization, according to some physical design plan, into a realization of the processor. This aspect of the processor design is called partitioning and the partitioning problem encompasses virtually all aspects of the physical design.

In early processors using SSI digital devices, logic gates were the basic hardware building blocks used to realize the processor. A single step partitioning of the entire processor is extremely complex in such cases, and most designs used a hierarchy of partitioning levels. The IC logic gates were organized into more complex hardware units (e.g., adders, multipliers, etc.) which were used as building blocks for still more complex hardware units [40]. It is convenient to define these hardware units as modules and design the overall processor using modules at various partitioning levels [41]. Today, however, the basic hardware building blocks are MSI and LSI digital devices with >100 gate complexity, corresponding to relatively complex processor modules.

Designers seeking complete freedom in the processor architecture and efficient realizations of that architecture can partition the processor, using logic gates as building blocks, to define digital modules which can be realized as custom MSI or LSI devices [42]. Considerable computer-aided design software has been developed to simplify the design and fabrication of custom MSI and LSI devices once the logical function has been defined. Typically, the custom device approach yields efficient realizations with the penalty of high development cost. As a result, the custom approach is used most often where a large production run of the processor reduces the impact of development costs.

On the other hand, designers wishing to avoid the development time and cost associated with custom processors have available to them a wide variety of standard MSI and LSI devices. Standardization may, in fact, extend beyond the IC devices used and include standard processor modules containing several IC devices. Although standardization probably offers the greatest long-term benefits, the continued use of custom designs suggests that standardization has important limitations. Perhaps the greatest limitation is the tradeoff between flexibility and efficiency of the processor realization. We illustrate this limitation using as an example the impact of LSI on processor designs.

The primary advantages offered by LSI include 1) less physical space occupied by the processor, 2) less power dissipation due to fewer and shorter device interconnections, and 3) higher reliability due to a reduced number of interconnections (often the primary cause of failures). However, these advantages are not obtained automatically. For example, combining two 16-pin MSI functions to obtain a single 32-pin LSI device gives the same board area for mounting and the same number of external connections as the two MSI devices. Indeed, the physical design may suffer since the two functions on the single LSI device cannot be located at different physical positions when desirable.

The advantages of LSI are obtained largely to the extent that interdevice connections are replaced by intradevice connections [43]. However, the flexibility with which the logic on a single LSI device can be used decreases as the number of external leads available for interconnections decreases. In the case of custom LSI devices, the processor architecture (chosen *a priori*) is partitioned to reduce the number of interdevice connections required. The conse-

quence is generally LSI devices which are suited only for that specific processor organization. On the other hand, design of standard LSI devices with few external connections must presume, to some extent, the use of the device in the processor realization. The result is that standard LSI tends to dictate the architecture. A good example of this effect is the large random-access memories available which favor large centralized memories over smaller distributed memories.

There are at least three approaches to maintaining flexibility in a processor realized with efficient LSI devices. Flexible software can be used in conjunction with inflexible hardware (both hardware devices and data paths). Inflexible software and an inflexible data path network can be used along with flexible (i.e., programmable) devices. Finally, inflexible software and nonprogrammable devices can be used in conjunction with a flexible data path network.

The greatest flexibility in efficiently realized processor architectures designed using standard devices appears to result when the flexibility is provided in the interconnection paths (rather than dedicated paths using flexible devices). Time-shared communications paths are an important characteristic of contemporary processor designs. A second major characteristic of contemporary processors is the use of large random access memories to implement many of the support functions, as noted earlier.

However, it must be recognized that contemporary processors using standard LSI devices are subject to a major constraint. The standard LSI devices available are designed for use in general-purpose data processors and are not necessarily well suited to digital signal processors. More research will be necessary before it will become clear whether a set of standard LSI devices, specifically designed for digital signal processors, will lead to overall architectures distinct from the general-purpose data processors.

REFERENCES

[1] Mina, Lawrence, and Werner, "Digital Techniques for Communication Signal Processing," this issue, pp. 18-22, 32.
[2] Jackson et al., "An Approach to the Implementation of Digital Filters," *IEEE Trans. Audio and Electroacoustics*, Vol. AU-16, pp. 413-421, 1968.
[3] McIver, et al., "A Monolithic 16x16 Digital Multiplier," *Digest of Technical Papers, 1974 IEEE Solid State Circuits Conf.*
[4] Kane, "A Low Power Bipolar Two's Complement Serial Pipeline Multiplier Chip," submitted to *IEEE Journal on Solid State Circuits*, Oct. 1976.
[5] Peled, "On the Hardware Implementation of Digital Signal Processors," *IEEE Trans. ASSP*, Vol. ASSP-24, pp. 76-86, 1976.
[6] Hoeneisen and Mead, "Fundamental Limitations in Microelectronics—I. MOS Technology," *Solid State Electronics*, Vol. 15, pp. 819-829, 1972. "— II. Bipolar Technology," *Solid State Electronics*, Vol. 15, pp. 891-897, 1972.
[7] Flynn, "Very High Speed Computing Systems," *Proc. IEEE*, Vol. 15, pp. 1901-1909, 1966.
[8] Newell and Verma, "When to Use Real-Time Digital Signal Processors," *NTC 1975 Record*.
[9] Gold et al., "The FDP—A Fast Programmable Digital Signal Processor," *IEEE Trans. on Computers*, Vol. C-20, pp. 33-38, 1971.
[10] Shay, "Radar Arithmetic Processing Elements as an MTI Filter," *NTC 1974 Record*.
[11] Berg and Wald, "The Impact of Implementation Techniques on System Parameters," *NTC 1974 Record*.

[12] Digital Signal Processing Committee, *Selected Papers in Digital Signal Processing, II,* pp. 335-524, IEEE Press, NY, 1976.

[13] Allan, "Computer Architecture for Signal Processing," *Proc. IEEE,* Vol. 63, pp. 624-633, 1975.

[14] Naresky, "Reliability Definitions," *IEEE Trans. Reliability,* Vol. R-19, pp. 198-200, 1970.

[15] Cook and Flynn, "System Design of a Dynamic Microprocessor," *IEEE Trans. Computers,* Vol. C-19, pp. 213-222, 1970.

[16] Friedman and Menon, *Fault Detection in Digital Circuits,* Prentice-Hall, 1971.

[17] Mathus and Avizienis, "Reliability Analysis and Architecture of a Hybrid Redundant Digital System," *AFIPS Proc., 1970 Spring Joint Computer Conf.,* Vol. 36, pp. 375-383, 1970.

[18] Wakerly, "Partially Self-Checking Circuits and Their Use in Performing Logical Operations," *IEEE Trans. on Computers,* Vol. C-23, pp. 658-666, 1974.

[19] Chang, Manning and Metze, *Fault Diagnosis of Digital Systems,* John Wiley and Sons, 1970.

[20] Peterson, *Error Correcting Codes,* MIT Press, 1961.

[21] Larsen and Reed, "Redundancy by Coding vs Redundancy by Replication for Fault-tolerant Sequential Circuits," *IEEE Trans. on Computers,* Vol. C-21, pp. 130-137, 1972.

[22] Hobbs et al. (Eds.), *Parallel Processor Systems, Technologies and Applications,* Spartan, 1970.

[23] Gold, "Parallel vs. Sequential Tradeoffs in Signal Processing Computers," *1974 NTC Record.*

[24] Hornbuckle and Arcona, "LX-1 Microprocessor and its Application to Real-Time Signal Processing," *IEEE Trans. on Computers,* Vol. C-19, pp. 216-220, 1970.

[25] Kratz et al., "Microprogrammed Approach to Signal Processing," *IEEE Trans. on Computers,* Vol. C-23, pp. 808-817, 1974.

[26] Ihrat et al., "The Use of Two Levels of Parallelism to Implement an Efficient Programmable Signal Processing Computer," *Proc. 1973 Sagamore Computer Conf. on Parallel Processing.*

[27] Sherman and Verma, "System Description of a Programmable Multiple Data Set," *NTC 1975 Record.*

[28] Avizienis and Parhami, "A Fault-Tolerant Parallel Computer System for Signal Processing," *Proc. 1974 Int. Symp. on Fault-Tolerant Computing.*

[29] Salazar et al., "Implementation of Voiceband Modems on a Digital Signal Processor," *NTC 1975 Record.*

[30] White and Nagle, "Digital Filter Realization Using a Special Purpose Stored Program Computer," *IEEE Trans. Audio and Electroacoustics,* Vol. AU-20, pp. 289-294, 1972.

[31] Blackenship et al., "LSP/2 Programmable Signal Processor," *NEC 1974 Record.*

[32] Costio, "A High Speed Microprocessor for Real Time Processing of Small Dedicated Jobs," *NTC 1974 Record.*

[33] Agrawala and Rauscher, *Foundations of Microprogramming,* Academic Press, 1976.

[34] Comfort, "A Modified Holland Machine," *AFIPS Proc. 1963 Fall Joint Computer Conf.,* pp. 481-488.

[35] Freeny et al., "Design of Digital Filters for an All Digital Frequency Division Multiplex-Time Division Multiplex Translator," *IEEE Trans. Circuit Theory,* Vol. CT-18, pp. 702-711, 1971.

[36] Baugh, "Design and Performance of a Digital Multifrequency Receiver," *IEEE Trans. on Communications,* June 1977.

[37] Tierney et al., "A Digital Frequency Synthesizer," *IEEE Trans. Audio and Electroacoustics,* Vol. AU-19, pp. 48-57, 1971.

[38] Thompson, "Digital Signal Processor for Voiceband Processors," *NTC 1974 record.*

[39] Thompson, "Some Architectural Implications of Stored Program Control in DSP," *EASCON 1975,* pp. 216 A-F.

[40] McDonald, "Impact of Large-Scale-Integrated Circuits on Communications Equipment," *Proc. NET, 1968.*

[41] Nordmann, "Design of Digital Modules for Signal Processing," Naval Ordnance Lab. Technical Report 74-170-1974.

[42] Neugebauer and Doyle, "Approaches to Custom LSI," *NTC 1973 Record.*

[43] Russo, "On the Tradeoff Between Logic Performance and Circuit-to-Pin Ratio for LSI," *IEEE Trans. Computer,* Vol. C-21, pp. 147-153, 1972.

GENERAL REFERENCES

[1] Freeny, "Special Purpose Hardware for Digital Filters," *Proc. IEEE,* Vol. 63, No. 4, pp. 633-648, April 1975.

[2] Hill and Peterson, *Digital Systems: Hardware Organization and Design,* John Wiley and Sons, 1973.

[3] Peatman, *The Design of Digital Systems,* McGraw-Hill, 1972.

[4] Chu, *Digital Computer Design Fundamentals,* McGraw-Hill, 1962.

THE DISCRETE FOURIER TRANSFORM APPLIED TO TIME DOMAIN SIGNAL PROCESSING *

Fredric J. Harris

Abstract The discrete Fourier transform (DFT), implemented as a computationally efficient algorithm called the fast Fourier transform (FFT), has found application to all aspects of signal processing. These applications include time domain processing as well as frequency domain processing. The proper noun "Fourier" may elicit images of frequency domain data, and by these images may restrict the vista of applications of this important tool. We will review herein the DFT with emphasis on perspectives which facilitate time domain processing. In particular we will review a number of applications to communications.

A review of the discrete Fourier transform, emphasizing the use of DFT in direct and indirect methods of time domain signal processing.

THE discrete Fourier transform (DFT), implemented as a computationally efficient algorithm called the fast Fourier transform (FFT), has found application to all aspects of signal processing. These applications include time domain processing as well as frequency domain processing.

The proper noun "Fourier" may elicit images of frequency domain data and, by these images, restrict the vista of applications of this important tool. We will review the DFT with emphasis on perspectives which facilitate time domain processing. In particular, we will review a number of applications to communications.

INTRODUCTION

The fast Fourier transform [1,2] is the generic name of computationally efficient algorithms which implement the discrete Fourier transform. Computational efficiency is one obvious reason for the great interest in the DFT based

algorithms. Another reason is the great versatility of the DFT, which is reflected in the numerous ways in which it is applied to time domain signal processing tasks [3]. This versatility is related to the development of a number of equivalent interpretations of the DFT algorithm. Through these interpretations we have been able to augment the FFT with pre- and post-processing algorithms to achieve efficient realizations of many signal processing tasks. The basic block diagram of a real-time FFT based time domain signal processor is shown in Fig. 1.

Examples of pre- and post-processing operations include quadrature heterodynes, low pass filtering, re-sampling, windowing, overlapping, zero extending, synchronous sampling, synchronous averaging and coherent averaging. Many of these operations are performed to convert the block processing structure of the FFT to a continuous processing time domain structure.

We emphasize that it is the development and understanding of these different perspectives that brings great versatility to the FFT. It is these perspectives that provide the necessary steps to bridge the gap between the FFT as an algorithm and a signal processing tool. It is these perspectives we review in this paper.

BACKGROUND

To better describe the DFT in the context of modern sampled data signal processing, we need the following

*Reprinted from *IEEE Communications Society Magazine,* May 1982, Vol. 20, No. ?, pp. ??.

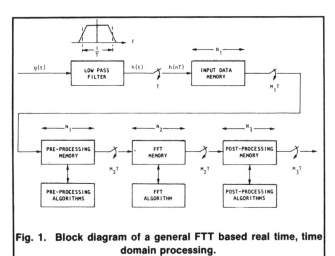

Fig. 1. Block diagram of a general FTT based real time, time domain processing.

background. A bandlimited signal, band limited to B Hz, is uniformly sampled at a rate which satisfies the Nyquist criterion (2BT samples per second or 1/T Hz) [4]. This rate, equal to the two-sided bandwidth of the signal, is required to assure that the samples truly represent the signal. If the samples do indeed represent the signal, then the original continuous signal can be reconstructed from the samples. Because of real world constraints of cost and complexity in the pre-sample analog filter, the actual sample rate is always greater than the desired two-sided bandwidth. To distinguish between the two bandwidths, we will call them processing bandwidth (sample rate related) and analysis bandwidth (data related). The ratio of processing bandwidth to analysis bandwidth is typically greater than 1.2. For instance, a single baseband voice channel has a two-sided bandwidth of 6.6 KHz and is usually sampled at 8.0 KHz [5].

A preselected number of these samples, N, are collected for

processing. Since the sample rate has already been established, the selection of N samples is equivalent to defining the time duration over which the samples are taken (NT seconds). This total time duration defines the longest period that can be identified in the data. This in turn defines the smallest nonzero frequency or the smallest frequency difference which can be resolved in the data ($\frac{1}{NT}$ Hz). We recognize that the two-sided processing bandwidth is $\frac{1}{T}$ Hz and that the minimum resolution is $\frac{1}{NT}$ Hz; thus concluding that we can resolve exactly N distinct frequencies in the spectrum of the data sequence. In a sense we have come full circle; not surprisingly, if N points are required to describe the data in the time domain, then N points are also required to describe the data in the frequency domain. Note that the sample rate is determined by the highest frequency content of the data, while the sample duration is determined by the lowest frequency content.

CLASSICAL INTERPRETATION OF INVERSE DFT (IDFT)

The IDFT is the mathematical expression of the following description. Imagine a collection of N signal generators. The generators are paired, and each pair outputs a cosinusoidal and a sinusoidal sequence of k cycles per observation interval. An observation interval is defined by N equally spaced sample intervals. The different pairs of generators are identified by the index k. They output, in concert with a time index n, the ordered pair of sequences $\cos[\frac{2\pi}{N} kn]$ and $\sin[\frac{2\pi}{N} kn]$ respectively. The outputs of each generator

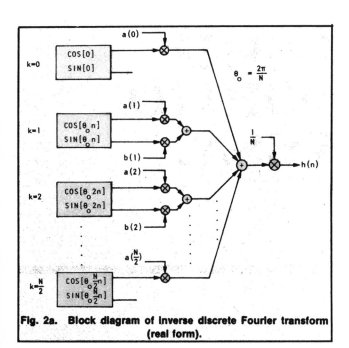

Fig. 2a. Block diagram of inverse discrete Fourier transform (real form).

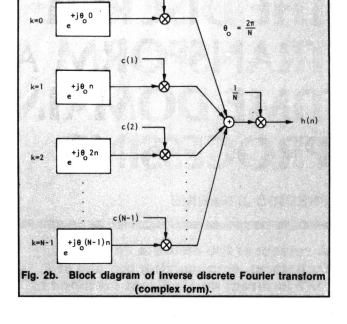

Fig. 2b. Block diagram of inverse discrete Fourier transform (complex form).

are then scaled by the ordered coefficients a(k) and b(k). At each time index n, an output is formed by the summation of the weighted generator outputs. A block diagram reflecting this description is shown in Fig. 2(a) and a concise rendering of this description is given in (1a).

$$h(n) = \frac{1}{N} \sum_{k=0}^{\frac{N}{2}-1} \left[a(k)\cos[\frac{2\pi}{N}kn] + b(k)\sin[\frac{2\pi}{N}kn] \right] \quad (1a)$$

If we treat the generator outputs as ordered pairs identified by

$$c(k) = [\frac{a(k)}{2}, \frac{b(k)}{2}] = \frac{a(k)}{2} + j\frac{b(k)}{2}$$

and

$$e^{j\frac{2\pi}{N}kn} = \left[\cos[\frac{2\pi}{N}kn], \sin[\frac{2\pi}{N}kn] \right]$$

$$= \cos[\frac{2\pi}{N}kn] + j\sin[\frac{2\pi}{N}kn]$$

then we can present an even more concise version of the above description as (1b). Equation (1b) is the definition of IDFT. The block diagram interpretation of (1b) is shown in Fig. 2b.

$$h(n) = \frac{1}{N} \sum_{k=0}^{N-1} c(k) e^{+j\frac{2\pi}{N}kn} \quad (1b)$$

It is easy to visualize that (1b), the IDFT, describes for every choice of weighting coefficients a periodic sequence of length N. The real significance of (1b) is that every finite amplitude sequence of length N can, by the appropriate selection of these weighting coefficients, be represented in this manner. These weighting terms are of course called the Fourier coefficients of the sequence. We note that for these finite summations there is no question of convergence.

We comment as an aside that (1) represents a special case of a more general expression involving the weighted summation of arbitrary orthogonal sequences [6]. The attraction of the sinusoidal sequences is of course related to the fact that the steady state response of linear systems to sinusoidal excitation is again sinusoidal. We say that linear systems are invariant operators for the complex exponentials. It is this property that makes the sinusoidal expansions such a useful tool.

All that remains now is to examine the method by which we can extract (or compute) the Fourier coefficients associated with a given sequence. That process is classically called Fourier analysis.

CLASSICAL INTERPRETATION OF DFT

In preparing for the Fourier analysis, we examine an important property of the separate sequences that contribute to the summed output sequence. The periods of the sin and cos sequences have been selected in a very special way. The periods are such that each sinusoid has precisely an integer number of cycles in the total interval. For instance, the first frequency has exactly one cycle in the total interval, the second frequency has exactly two cycles, etc. As a result, the summation of the sample values of any of the sinusoidal sequences over the total N point interval will lead to a zero sum. The zero frequency sequence (the "dc" sequence) is the only exception, and the summation over the sample values of that sequence will sum to N. See Fig. 3 for examples.

Thus within a known scale factor (N), the weighting term of the zero frequency sequence can be obtained from an arbitrary sequence by simply summing over all the sample values of the sequence. The sample values of each of the contributing sinusoids in the arbitrary sequence will separately sum or average to zero. We say that these other contributors are orthogonal to the zero frequency sequence.

The knowledge that only the sample values of the zero frequency sequence survive the summation is the clue to determining the coefficients of the arbitrary k-th frequency sequence. We simply invoke the modulation theorem to heterodyne (or shift) the k-th frequency sequence to zero frequency. We then perform the summation or average to determine the new zero frequency coefficients. There are generally two zero frequency terms as a result of the complex heterodyne: one from the cos heterodyne and average, and one from the sin heterodyne and average. These are proportional to the amplitude of the k-th cos and sin sequences respectively. An equivalent block diagram of this operation, in both real and complex forms, is presented in Fig. 4(a) and (b). The expression for the heterodyne and average operation is presented in (2).

$$c(k) = \sum_{n=0}^{N-1} h(n) e^{-j\frac{2\pi}{N}kn} \quad (2)$$

The total DFT process can be visualized as a collection of N such heterodynes and averages. This is shown in Fig. 5. It is instructive to compare (1) and (2), the IDFT and the DFT, which are so similar in form, while also comparing their

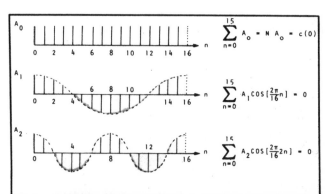

Fig. 3. Summation over sample points of sinusoidal sequences.

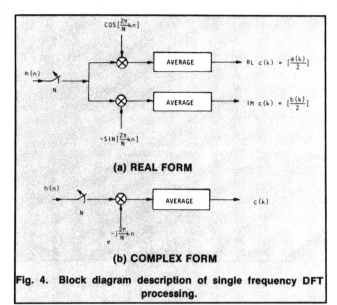

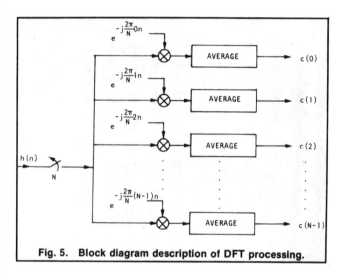

Fig. 4. Block diagram description of single frequency DFT processing.

Fig. 5. Block diagram description of DFT processing.

interpretations in Figs. 2 and 5, which are so completely different.

When we consider classical transform-based analysis, it is these interpretations that first come to mind. They supply the answer to the questions, "What are the amplitudes of the sinusoids which have been combined to form this specific signal?" and "What is the signal that corresponds to this specific collection of sinusoidal amplitudes?"

We note that the addition of harmonically related periodic signals, as visualized in Fig. 2, suggests that the resultant signal must also be periodic. In fact, the summation is periodic and is called the periodic extension of the observed signal. Hence, spectrum analysis based on the DFT is often called periodogram analysis. For many applications a number of pre- and post-processing algorithms are applied along with the DFT to reduce any undesired artifacts created by the truncation and periodic extension of the original signal. These include windowing [7], overlapping [8], synchronous sam-

pling [9], synchronous averaging [9], and incoherent averaging [8]. Since our concern here is time domain processing, we will now leave the classical spectrum analysis-directed interpretation of the DFT.

FAST CONVOLUTION AND CORRELATION

One important time domain process of interest in communication signal processing is that of replica correlation [10]. Correlation, often performed as a matched filter, is used to maximize signal to noise ratio, achieve pulse compression, clock synchronization or frame synchronization. When implemented as a filter, the processing task is simply that of determining the response of a known filter to a given input sequence. The filter is described in the time domain by its finite duration impulse response, i.e., its response to a single sample. Such a filter is depicted in Fig. 6.

The M point running weighted summation indicated in Fig. 6 has to be performed for each output point of the filter. The computational burden of a direct convolution is approximately M multiplies and adds per output point. In many applications, during signal acquisition for instance, an output point is computed for each input point. In such a case the computational workload is M multiplies and adds per input point.

We observe here for later use, that the response to a single input sample is of length M, the response to two consecutive input samples is of length $M + 1$, to three successive input samples of length $M + 2$, etc. By extension, it follows that the response to an input sequence of length P is of length $M + P - 1$.

There is an indirect method for determining the response of a given filter to a particular input. The indirect method requires a description of the filter's sinusoidal steady state frequency response. We know that linear stable filters when driven with a sinusoidal signal will exhibit a steady state

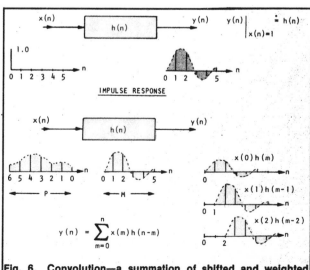

Fig. 6. Convolution—a summation of shifted and weighted responses.

response which is a sinusoid of the same frequency as the input signal. The response can differ from the input in only magnitude and phase shift. At each frequency we define the ratio of output to input amplitudes as the magnitude gain, and the difference between the output and the input phase angles as the phase gain. The magnitude and phase gains are the polar representation of the complex gain at that frequency. A

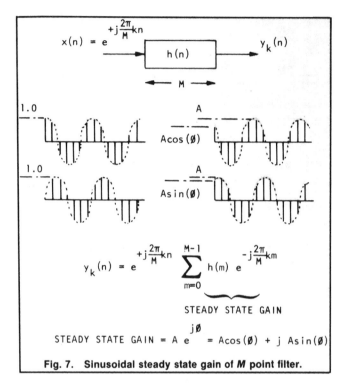

Fig. 7.   Sinusoidal steady state gain of _M_ point filter.

list of these gains at each frequency is one description of the filter's frequency response. Conceptually, we could determine a filter's gain at a single frequency by the method depicted in Fig. 7.

We note from examining Fig. 7, that the sinusoidal steady state gain of the filter is simply the DFT of the filter impulse response. Thus we can determine the collection of sinusoidal steady state gains by simply performing a DFT on the filter impulse response.

The DFT based equivalence to convolution will now be described. A block diagram equivalent to this description is shown in Fig. 8. The input sequence x(n) is processed by a DFT to determine the amplitudes of the sinusoids comprising that sequence. The impulse response of the filter h(n) is similarly processed by a DFT to determine the sinusoidal steady state gains at each of the frequencies associated with the input sequence. These gains are then applied multiplicatively to the amplitudes of the corresponding input sinusoidal amplitudes and thereby construct a new set of scaled amplitudes. These amplitudes are the sizes of the sinusoids comprising the output sequence y(n). The output sequence is formed by an IDFT applied to the scaled amplitudes. This technique is called indirect convolution because the time function was obtained indirectly through the frequency domain; when indirect convolution is performed with the use of an FFT, it is called fast convolution.

A word of caution: the length of the filtered output sequence is $P + M - 1$, the sum of the filter sequence length (M) and the input sequence length less one ($P - 1$). We know that if we require $P + M - 1$ data samples to describe the output sequence in the time domain, we must also require this many sample val-

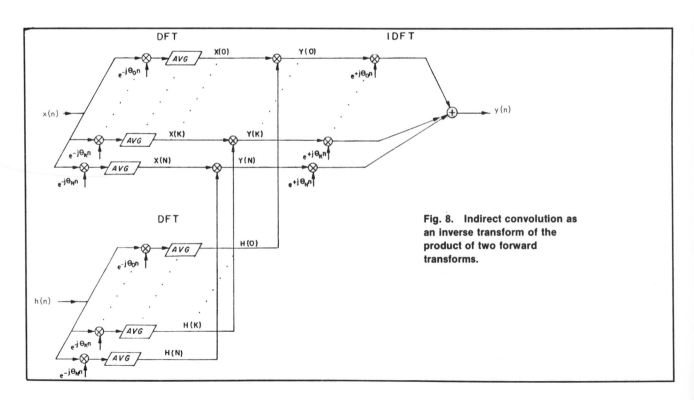

Fig. 8.   Indirect convolution as an inverse transform of the product of two forward transforms.

ues in the frequency domain. To satisfy this requirement, the frequency response (or the DFT) of the filter and of the input data must be computed at least in that many locations. We accomplish this by zero extending the two sequences to some convenient length N, where $N \geqslant P + M - 1$. The length N, to which we zero extend the data, is simply the size of the transform we will be using. Failure to zero extend the data will result in circular wrap around or aliasing of the time domain response. This wrap around is called circular convolution, and it is the zero extending which allows us to convert circular convolution to linear convolution. Fig. 9 is a block diagram of the zero-extended fast convolution algorithm.

In many applications, the length P of the input data sequence is essentially unlimited. In such a case we can not process the data stream by the zero-extended fast convolution technique. In addition, even for merely long sequences, transforms of the required size may not be available or the

delay may be unacceptably large until the end of the sequence prior to processing. In these cases, we must slightly modify our fast convolution technique. The modification entails the partitioning of the input data into time intervals or segments called blocks and then sequentially processing each block as it is formed. We must exercise care that the artificial boundaries induced by the block partitioning do not generate artifacts in the processed data.

We avoid the edge effects by overlapping adjacent blocks by the length $M - 1$ where M is the length of the filter impulse response. Assume we have a transform of length N. Then during each successive block processing we select and process $N + 1 - M$ new input sample points and obtain the same number of new output points. The intervals of $M - 1$ points which were overlapped contain the processing artifacts induced by the partitions. There are two classic techniques for handling these edge effects. They are called overlap and add, and overlap and discard [11]; the latter is sometimes called overlap and save.

In the overlap and add technique, $N + 1 - M$, new sample points are identified for processing and are zero-extended to the right for a total of N points (Fig. 10). The data is then processed by a conventional fast convolution routine and the results are transferred to the segment recombination routine. Remember that each output segment overlaps the previous segment by $M - 1$ terms. We recombine the processed segments in two steps. Adjacent segments overlap in precisely $M - 1$ sample points. These overlapped points are merged by simple addition at each sample location. The

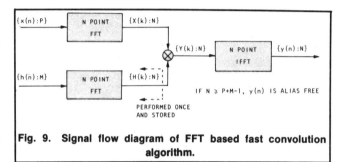

Fig. 9. Signal flow diagram of FFT based fast convolution algorithm.

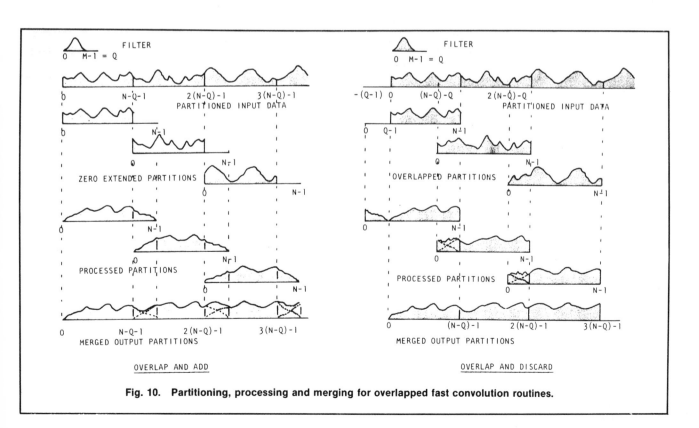

Fig. 10. Partitioning, processing and merging for overlapped fast convolution routines.

remaining sample points from the nonoverlapped intervals are merged by simple juxtaposition (Fig. 10).

In the overlap and discard technique, a total of N sample points are identified for processing. Of these N samples, the rightmost $N + 1 - M$ points are new data points and the leftmost $M - 1$ points are the overlapped points from the previous data block. This data is processed by a conventional fast convolution routine. The leftmost $M - 1$ terms in the resultant data are circularly aliased data points and are simply discarded (hence the name!). The remaining $N + 1 - M$ points represent adjacent nonoverlapped data samples. The recombination routine merges these adjacent segments by simple juxtaposition.

THE DFT AS A BANK OF MATCHED FILTERS

In the previous section, we described how the DFT could perform time domain processing by indirect means. The DFT was used to move our descriptions of the signals between the two domains, time and frequency. The attraction of using these transformations is that operations which are computationally burdensome (convolution) in one domain are significantly less so (weighting) in the other domain. The important aspect of these comments is that the DFT has performed the time domain processing indirectly through the frequency domain.

The DFT can also be used to perform certain time domain processes directly. By directly, we mean that both the input and the output data are in the time domain. We will not require the second transformation to return to the time domain; we will not have left it. We will now demonstrate that the outputs of a DFT are equivalent to the outputs of a bank of filters matched to tone burst sequences.

The block diagram of an N point finite impulse response filter is shown in Fig. 11. The impulse response of this filter is the sequence of weighting coefficients $h(n)$. The output of this filter is known to exhibit the maximum signal to noise ratio (SNR) for the sequence $h(N - 1 - n)$ in additive white Gaussian noise (AWGN)* [12]. If we select the weighting coefficients $h(N - 1 - n)$ to be the ordered pair of sequences $\cos[\frac{2\pi}{N}kn]$ and $\sin[\frac{2\pi}{N}kn]$ then the filter is the optimal filter for detecting sinusoidal tone bursts in AWGN.

We note by examining Fig. 11 that when the data completely fills the filter memory, which happens when the output index $n = N - 1$, the computed output term $y(N-1)$ is identical to the DFT computation of (2). Thus, the matched filter can be used to compute the value of a given DFT output point.

On the other hand is the crux of this discussion: the DFT can be used to compute the output of the matched filter. In

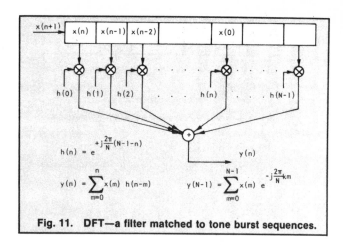

Fig. 11.   DFT—a filter matched to tone burst sequences.

fact, the DFT will simultaneously compute the output of N such filters processing the same data. The DFT computation can be performed at other index times besides the index value $n = N - 1$. This is accomplished by a technique known as overlapped or sliding transforms. We can think of the transform memory sequentially shifting or sliding through a list of input data samples. The DFT can be performed at any location during the sequential shifting. For instance, we can compute a DFT each time the memory shifts one new data point. Then the sequence of successive transform output points from a given index k is a time series corresponding to the output data from the selected matched filter in the filter bank. Note that by sequentially applying the DFT to overlapped blocks of data, we have generated a collection of time series, one sample point per filter per transform performed.

We have just observed that by proper overlap (or shifting or sliding) of the data segments into the DFT memory, the DFT can synthesize the outputs of a bank of matched filters. We now ask the question, "What is the frequency response of each of these filters?" The frequency response of any filter is simply the Fourier transform of its impulse response. The impulse response of these filters is simply the sequence of coefficients defining the weighting terms. For instance, the impulse response and the corresponding frequency response of the 4-th filter is shown in Fig. 12. The frequency response is the well known Dirichlet kernel with its classic $\frac{\sin(x)}{(x)}$ behavior. The successive zeros of this function are separated by the reciprocal of the data interval, i.e., $\frac{1}{NT}$. The function has its maximum gain, N equal to the number of data points in the interval, at the frequency of the tone burst. The filter response centered at this frequency and bounded by the adjacent zeros is called main lobe response. The oscillatory behavior between successive zeros is called sidelobe response.

THE DFT AS A BANK OF NARROWBAND FILTERS

Note in Fig. 12 that the frequency response of the matched filter is zero at those frequencies corresponding to the other

---

*We remark that for the sampled noise sequence, white implies independence of samples. This is assured by the bandlimiting of a wideband noise source by the anti-aliasing filter and subsequent sampling at the Nyquist rate.

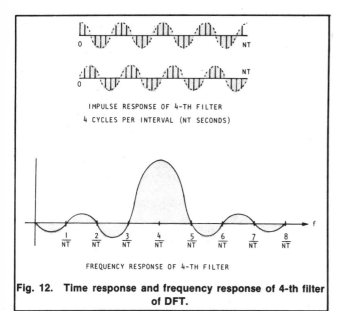

Fig. 12. Time response and frequency response of 4-th filter of DFT.

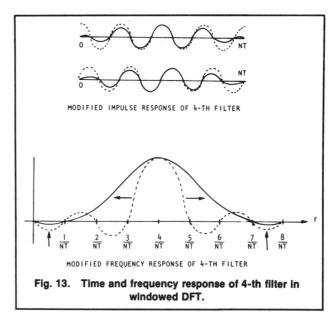

Fig. 13. Time and frequency response of 4-th filter in windowed DFT.

filters in the filter bank. This again reflects the mutual orthogonality of the signal sets used in the DFT. The filters do respond however with nonzero gain, in fact with very large gain, at frequencies remote from the mainlobe location. For instance, with respect to Fig. 12, consider a tone corresponding to the frequency of 7.5 cycles per interval applied to the filter with peak response at 4.0 cycles per interval. This tone is attenuated by only 0.0909 or approximately 21 dB relative to the peak gain. If the frequency of the input tone were changed to 7.0 or to 8.0 cycles per interval, the attenuation would be arbitrarily large. This significant change in response behavior caused by small changes in input frequency is undesirable for many applications. The effect is called spectral leakage and is obviously related to the sidelobe behavior of the filter response. The sidelobe structure of the filter can be controlled by modifying the impulse response of the equivalent matched filter. Multiplicative modification is known as windowing, and is one well known method of filter design; non-multiplicative techniques include the Remez multiple exchange algorithm and frequency sampling algorithms [13]. Any of these techniques can be used to alter the filter frequency response shape. Regardless of the technique used to design the new filter coefficients, a common thread runs through the results. Lower sidelobe response is always achieved in exchange for mainlobe width in the frequency domain, and envelope smoothing in the time domain. A typical modification of the filter impulse response and the corresponding frequency response is shown in Fig. 13. We note that the filters with the suppressed sidelobe response are no longer matched to the tone bursts of the DFT; we now consider them simply as banks of narrowband contiguous filters.

In the previous discussion, we visualized sidelobe control as multiplicative modification of the filter impulse response. In the DFT based filter bank, we do not have access to the impulse responses, only the input data and an efficient algorithm. However, we achieve the same result by applying the multiplicative modifications to the separate data blocks as they are moved into the transform memory. This is commonly called windowing of the data, and its function in this context is to shape the frequency responses of the filter bank. This process is demonstrated in Fig. 14. Note that the sequential output data points from any given DFT filter is a narrowband sequence located at the filter center frequency. In this application, we have not extracted the complex envelopes; the envelopes still reside on their carriers.

The attractiveness of the DFT based filter bank is the economy of computation when realized as an FFT. When realized in the direct form, each of the narrowband filters synthesized by the transform requires approximately $\frac{N}{2}$ multiplies and N adds per output point. To construct a bank of M equally spaced equal bandwidth filters by direct computation, the total computational load would be $\frac{MN}{2}$ multiplies and MN adds for the M outputs. This is equivalent to $\frac{N}{2}$ multiplies and N adds per output point per filter.

If the same filter bank is synthesized by a windowed and pruned* FFT, the computational load for the M filters is approximately $[N + 2M \log_2 M]$ multiplies and $[N + 3M \log_2 M]$ adds. This is equivalent to $[\frac{N}{M} + 2 \log_2 M]$ multiplies

---

*Often all N output filters of an N-point FFT are not needed for a given processing task. Processing speed advantages can be realized by selectively eliminating those computations leading to these non-used outputs. Such modifications to the FFT algorithm are called pruning (to reflect the removal of undesired branches in a tree). Pruning can be considered the fine tuning of the algorithm to the particular processing task.

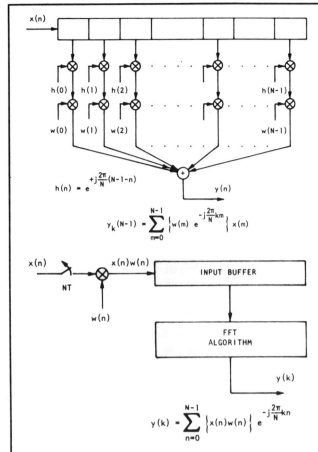

$$h(n) = e^{+j\frac{2\pi}{N}(N-1-n)}$$

$$y_k(N-1) = \sum_{m=0}^{N-1} \left\{ w(m) \ e^{-j\frac{2\pi}{N}km} \right\} x(m)$$

$$y(k) = \sum_{n=0}^{N-1} \left\{ x(n)w(n) \right\} e^{-j\frac{2\pi}{N}kn}$$

Fig. 14. Comparison of modified impulse response of direct convolution filter with equivalent windowed DFT response.

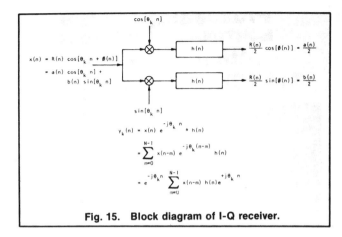

Fig. 15.   Block diagram of I-Q receiver.

and $[\frac{N}{M} + 3 \log_2 M]$ adds per output point per filter. As an example, suppose we wish to construct a bank of 32 adjacent narrowband filters each of length 128 points. By direct techniques, we would require 64 multiplies and 128 adds per output point per filter. By the FFT based technique, we would require 14 multiplies and 19 adds per output point per filter. In a sense, the attractiveness of the FFT based process for narrowband filtering is the price break we get by constructing the entire set of filters in a single algorithm. Think of it as a price break for large quantity purchases!

THE DFT AS AN I-Q DEMODULATOR

Another interpretation of the DFT will allow us to synthesize a bank of complex demodulators. This process is also commonly known as a Weaver (single sideband) heterodyne, an I-Q (In phase and Quadrature phase) receiver and a complex envelope receiver. The attractiveness of the DFT to this application is the computational advantage available when synthesizing a bank of such demodulators. The computational savings will be comparable to those described in the previous section.

We note that the block diagram for the I-Q filter (Fig. 15) is

strikingly similar to that of the DFT (Fig. 4a). The difference between the two diagrams is that one is for a block process which is exercised once for N input points, and the other is for a continuous process which is exercised upon the arrival of each new input sample point. Of course a block process can always synthesize a continuous process by sequential shifting of overlapped data blocks into the process. In fact, that is how we used the DFT to synthesize the bank of narrowband filters in the previous section.

The difference between the DFT and the I-Q heterodyne can be found in the sin and cos sequence generators used for the heterodynes. In the DFT, these generators reset their time index upon the start of each new block of data. In the I-Q heterodyne, on the other hand, these generators operate on a sequential time index as does the input data. To make the DFT appear to be an I-Q heterodyne, a phase correction has to be applied to account for this difference in phase of the two sets of generators.

The phase correction can be seen in the expression for the I-Q output shown in Fig. 15. Notice that the summation part of this expression is identical to the windowed summation which performed the narrowband filtering as demonstrated in the previous section. Here the impulse response of the filter, h(n), replaces the window sequence. Thus the summation part of the I-Q heterodyne can be implemented as a windowed and overlapped DFT. This is how we synthesized the narrowband filters in the previous section. Thus, the phase correction terms are simply a heterodyne of the narrowband filter output from the center frequency to baseband. This is indicated in Fig. 16.

We recognize that in basebanding and filtering of the input signal, we have significantly reduced the output bandwidth relative to the input bandwidth. We can, consequently, reduce the output sample rate relative to the input sample rate. Remember that we can always reduce the sample rate after processing as long as we satisfy the Nyquist criterion relative to the analysis bandwidth.

We can visualize performing this sample rate reduction on the output of the separate I-Q filter banks (Fig. 16a). Here we conceptually transfer every R-th data point out of the process

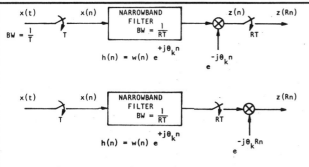

Fig. 16. I-Q receiver as the basebanded output of a narrowband filter basebanding before and after desampling.

and in doing so reduce the output sample rate by the factor R. But note that even though only every R-th point is transferred, every point is phase corrected as it is outputted from the filter. By a simple modification of the block diagram, we slide the desampler to the input side of the phase correcting heterodyne. We thus only phase correct the points we transfer from the process. As a result, the angle increment on the phase correction heterodyne is increased by the factor R. This new incrementing angle can be interpreted as the aliased version of the original center frequency, where the aliasing is induced by the desampling.

In reality, the desampling is performed as part of the overlapped DFT algorithm. The desampled rate is identical to the transform rate. Thus, if we perform a new transform for every R new input data samples, the output rate is $\frac{1}{R}$ of the input rate. A total FFT based I-Q filter bank with spectral shaping, desampling and phase correction is shown in Fig. 17.

CONCLUSIONS

We have reviewed, in what we hope is an intuitive manner, the structure of the discrete Fourier transform. The intent of this review was the underscoring of perspectives which emphasize use of the DFT in time domain signal processing; these perspectives have included both direct and indirect methods.

The indirect methods, or fast convolution, perform their processing in the frequency domain. Here the FFT is merely the conveyance for moving the description of the signals between the two domains. We note that when the fast convolution technique is augmented by overlap and segmenting capabilities it becomes the core of very efficient general purpose filtering algorithms.

The direct methods of DFT based signal processing achieve time domain outputs by the sequential processing of overlapped blocks of input data. This technique can be used to construct the output of narrowband filters. The advantage of this method is the ability to economically realize the output of a large bank of narrowband filters. In fact, this technique has found wide use in the communications sector in the synthesis of banks of I-Q receivers.

REFERENCES

[1] J. W. Cooley and J. W. Tukey, "An algorithm for the machine calculation of complex Fourier series," *Math. Computation*, vol. 19, pp. 297-301, 1965.

[2] J. W. Cooley, P. A. W. Lewis and P. D. Welch, "Historical notes on the fast Fourier transform," *IEEE Trans. Audio Electroacoust.*, vol. AU-15, pp. 76-79, June 1967.

[3] A. V. Oppenheim, Ed., *Applications of Digital Signal Processing*, Englewood Cliffs, NJ: Prentice-Hall, 1978.

[4] Mischa Schwartz, *Information Transmission, Modulation, and Noise*, 2nd ed., New York: McGraw-Hill, ch. 3.2-3.4, pp. 117-131.

[5] K. E. Fultz and D. B. Penick, "T1 Carrier System," *Bell Syst. Tel. J.*, vol. 44, pp. 1405-1451, Sept. 1965.

[6] J. W. Wozencraft and I. M. Jacobs, *Principles of Communication Engineering*, New York: Wiley and Sons, ch. 4.3, pp. 223-228, 1965.

[7] F. J. Harris, "On the use of windows for harmonic analysis with the discrete Fourier transform," *Proc. of the IEEE*, vol. 66, no. 1, pp. 51-83, Jan. 1978.

[8] F. J. Harris, "On overlapped fast Fourier transforms," *Int. Telecommun. Conf.*, Nov. 1978.

[9] Application Note DSP-035, "Measuring frequency response functions on operating systems," *Scientific Atlanta Spectral Dynamics*

[10] F. J. Harris, "Convolution, correlation, and narrowband filtering with the fast Fourier transform," IEEE—ESIME Conf., *Semana de la Ingenieria en Comunicaciones Electricas*, Mexico City, July 1980.

[11] A. V. Oppenheim and R. W. Schafer, "*Digital Signal Processing*," Englewood Cliffs, NJ: Prentice-Hall, pp. 110-115, 1975.

[12] J. W. Wozencraft and I. M. Jacobs, *Principles of Communication Engineering*, New York: Wiley and Sons, ch. 4.4, pp. 239-242, 1965.

[13] Rabiner and Gold, *Theory of Applications of Digital Signal Processing*, Englewood Cliffs, NJ: Prentice Hall, pp. 187-204, 1975.

Fredric J. Harris was born in Brooklyn, New York on April 6, 1940. He received a B.E.E. from the Polytechnic Institute of Brooklyn and an M.S.E.E. from San Diego State University. He has also completed all course requirements for a Ph.D. at the University of California at San Diego.

He is currently a Professor in the Electrical and Computer Engineering Department at San Diego State University. In addition, he holds a one day per week appointment with the Signal Processing and Display group at the Naval Ocean Systems Center in San Diego. His major fields of interest are the development and application of digital signal processing and modern communications.

Harris is a Senior Member of the IEEE and is a member of the Audio Speech and Signal Processing Society and the Communication Society.

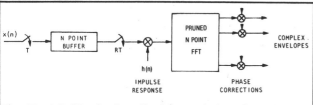

Fig. 17. I-Q filter bank realized by a windowed, overlapped, phase corrected FFT.

TDM/FDM TRANSLATION AS AN APPLICATION OF DIGITAL SIGNAL PROCESSING *

Stanley L. Freeny

Transmultiplexers utilize sophisticated digital processing to convert between time and frequency division multiplexed signals.

The general-purpose digital computer had not been in existence very long before it occurred to someone that it could be used not only for bulk calculations, but also "on-line," so to speak, for real-time control and signal processing. These early applications were 1) slow, and 2) sufficiently complex to require the inherent precision and sophistication afforded by digital control. However, as we all know, a veritable revolution has taken place over the last two decades in the amount of computing power that can be obtained in a given size for a given cost. With the availability of computing entities that are orders of magnitude faster, smaller, and cheaper, it is now feasible to consider using real-time *digital signal processing* (DSP) techniques in a large variety of applications, including many that were heretofore done by conventional analog methods.

One such application will be described in this article. It arises in the field of telecommunications—a field in which signal processing is pervasive and in which opportunities consequently abound for using DSP techniques. The particular application to be discussed concerns the *transmultiplexer*, an interface between the two standard signal multiplexing formats used in telephony: *frequency division multiplex* (FDM) and *time division multiplex* (TDM). In FDM, voiceband signals are stacked into

contiguous 4 kHz channels in the frequency domain by means of single sideband (SSB) amplitude modulation. FDM has been in existence for over 50 years and is used universally throughout the world for transmitting long distance telephone calls. In TDM, a technique of much more recent general usage, voice signals are each digitized (pulse code modulated) at an 8 kHz sample rate and the resulting pulse streams are interleaved in time. Even though TDM requires as much as fifteen times the bandwidth of FDM to transmit a given number of channels, the signals are very robust and the hardware takes full advantage of the steadily decreasing cost of digital integrated circuits. Thus its application to both switching and exchange area trunk transmission is rapidly growing. It is primarily the first application that gives rise to the subject of this article.

In 1976 the first class 4 digital toll switching office was placed into service in Chicago and new class 4 offices are presently being installed in the United States at the rate of about 20 per year. Among the chief characteristics of these offices is the fact that they switch signals exclusively in the TDM format. This means that all analog signals entering and leaving the office—including FDM signals which the office, by its nature, must handle—have to be converted to and from digital form. Thus arises the TDM/FDM interface problem.

Of course, DSP techniques are not necessary to implement this interface. Indeed, at present it is satisfactorily accomplished by demultiplexing the signals in one format down to baseband and remultiplexing them again

The author is with Bell Laboratories, Holmdel, NJ 07733.

*Reprinted from *IEEE Communications Society Magazine*, January 1980, Vol. 18, No. 1, pp. 5–15.

248

into the other format, all with conventional analog and digital multiplexing techniques. Nevertheless, as an application of digital processing, TDM/FDM translation is particularly appealing. The frequency range of the signals involved makes efficient use of present-day technology. The interface specifications are sufficiently stringent to benefit from the inherent precision of the digital approach. Furthermore, the problem is well defined yet intricate enough to be intellectually challenging. Beginning with the early work of Darlington [1] it has attracted much attention and many ingenious solutions have been proposed [2]-[22]. Because of their diversity, these solutions encompass most important aspects of the digital processing field. Hence, in addition to being worthwhile in its own right, a survey of this problem provides a good introduction to the practical application of DSP and the techniques peculiar to it.

DIGITAL SIGNAL PROCESSING

During the past decade, the literature on DSP has grown enormously [23],[24]: the subject has become a standard offering in the curricula of many universities and several textbooks have been written [25]-[29]. Nevertheless, on the assumption that there are readers not all that familiar with the field, the following very concise review is offered.

In digital signal processing we deal, naturally enough, with digital signals, which are nothing more than sequences of numbers that occur at a uniform rate. One usually thinks of such a sequence as representing the amplitudes of a uniformly sampled analog wave, but in many cases this association is not necessary. A digital processor is any computing entity which performs real-time "number crunching" on one or more input sequences. As with computers of any type, the discussion of digital processors divides naturally into the topics of hardware and software.

DSP Software

Although the more complex, fully programmable digital processors are complicated enough to involve their designers in most aspects of modern computer science, the essential element of DSP software is simply the *processing algorithm*. In its basic form this is just an equation or set of equations that relates the processor output to the input. (An important and relevant example is given by (1) in the inset on digital filtering.) In most DSP applications of interest the algorithm is not unique, i.e., there exists more than one that will do the job. Hence, it is important to have some means of comparing the efficiency of different algorithms for performing a given task.

DSP Hardware

The ultimate figure-of-merit for any processing algorithm is the cost of the hardware required to realize it. Now, the hardware implementation of a DSP algorithm can be accomplished in either of two basic ways: 1) the

custom approach in which large-scale integrated circuits are designed and built for one application only; and 2) the "building block" approach in which existing modular components are used. Approach 1) is followed when production volume is sufficient to justify the extra time and cost involved. Method 2) is generally less efficient for any given application but has the appeal of shorter design time and lower startup cost. The transmultiplexer application is actually a candidate for either approach; however, we will consider only the building block method since it is difficult to say anything meaningful about the custom approach in this simplified exposition.

Insofar as the modular approach is concerned, much useful hardware has appeared in the last few years, including microprocessors, multiplier/accumulators, and bit-slice devices. But the future most probably lies with the present trend toward producing single chip devices containing: 1) enough logic to perform a basic arithmetic operation plus such auxiliary tasks as scaling, overflow control, etc.; and 2) memory for storing coefficients, the results of interim calculations, some number of data samples, and the program. In other words, each will be self-sufficient in the sense that a given application will require one or more identical digital processor chips, plus (in some cases) standard blocks of memory to augment that on the chip. In operation, each chip will perform a well-defined arithmetic function in a specified time called the *machine cycle*. Given a specific arithmetic function, it is possible to compute how many such functions must be done per second to realize a particular DSP algorithm. Comparing this to the *computation rate* (reciprocal of the cycle time) of our building block chip gives an idea of how many chips are required—and thus an estimate of the relative hardware cost—for implementing that particular algorithm (to this must be added any extra memory needed). Hence the nature of the basic arithmetic function is fundamental to how broadly applicable a building block chip is to a variety of DSP applications. The following discussion of arithmetic functions for digital filtering should illustrate the point.

Perhaps the simplest useful arithmetic operation is a multiply-and-add, i.e., during each machine cycle there are two inputs and one output related by

$$\text{output} = (\text{input}_1) \cdot (\text{coefficient}) + (\text{input}_2). \quad (2)$$

During each cycle, the appropriate coefficient is read from an internal memory location under control of the program. Suppose we wish to implement the second-order digital filter of Fig. 3 using a processor chip with this basic arithmetic operation. Moving left to right through the filter, it is obvious that each succeeding variable can be computed from the previous one by (2); hence five machine cycles of the simple multiply/add type are required to compute each output sample of a second-order section.

Filters of higher order can be obtained by cascading

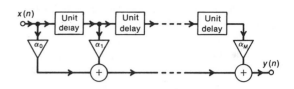

Fig. 2.   FIR digital filter.

DIGITAL FILTERING

An important subset of DSP is *digital filtering*. In a digital filter the output sequence $y(n)$ is related to the input sequence $x(n)$ by

$$y(n) = \sum_{i=0}^{N} \alpha_i \, x(n-i) - \sum_{i=1}^{M} \beta_i \, y(n-i). \qquad (1)$$

This equation is the general form of a *linear difference equation* and plays the same role in digital filtering as the linear differential equation does in describing analog filters. Thus the familiar analog concepts of amplitude and phase response and poles and zeros of transmission are equally applicable to digital filters. In (1), the x's and y's are the *data*, the α's and β's are the *coefficients*, and M and N together define the *order* of the filter.

Without further qualification, (1) actually describes a *discrete-time* or *sampled-data* filter, as no restriction is implied upon the accuracy with which the data and coefficient words are expressed. By their nature, digital filters—and in fact digital processors in general—must deal with these and other numerical constraints. In addition to the necessity for quantizing the data and coefficient parameters, data words must be requantized (rounded) after various arithmetic operations and the overflow characteristics of these same operations taken into account. These finite accuracy considerations are important and introduce impairments in the form of quantizing noise, small and large scale limit cycles and perturbations from the ideal transfer function. Much of the DSP literature is devoted to a study of these impairments and ways to control them. Nevertheless, important as these quantization matters are, in the TDM/FDM application (and other applications as well) they have a lesser effect on the hardware than more basic parameters such as sample rate and filter order. Therefore, to simplify our discussion of an intricate subject, we will deal primarily with the sampled-data form of the processing algorithm.

To lay further groundwork for the accompanying discussion, we review some additional digital filter terminology. If one were to draw from (1) a block diagram composed of adders, multipliers, and unit delay elements, the result would look like Fig. 1. This is known as the *direct form* of a digital filter. Were it not for the quantization effects discussed above, the order in which the computations of (1) are performed would be irrelevant and this form would suffice for all digital filters. Such is not the case,

however, and in fact a surprising variety of forms different from Fig. 1 have been proposed and studied. Again, in the interests of simplicity, we will confine our attention to the basic digital filter forms that appear most frequently in the work on TDM/FDM translation (and in most other DSP applications as well). These are the *finite impulse response (FIR)* form, depicted in Fig. 2 and

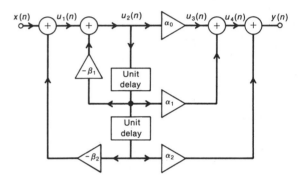

Fig. 3.   Recursive second-order digital filter section.

obtained by setting all the β's in (1) to zero; and the *recursive second-order section,* shown in Fig. 3 and described by (1) when $M = N = 2$. Filters of any order are obtained with the latter by cascading two or more sections. (Filters of general order can also be obtained by connecting second-order sections in parallel; however, this approach is generally inferior to the cascade in regard to quantizing noise.) The cascade form falls into the general class of *infinite impulse response (IIR)* digital filters and is sometimes referred to that way.

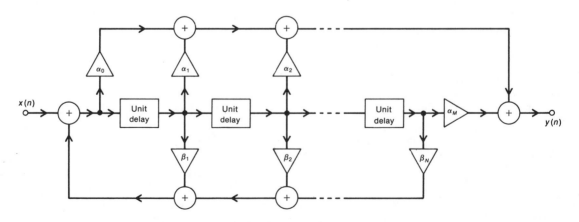

Fig. 1.   Direct form digital filter.

second-order sections. Using the above reasoning, a general Nth-order (N even) filter would require $5N/2$ multiply/add cycles. However, it is possible to factor a term from each section (thus normalizing one of the coefficients, say α_0, to one) and gather all these terms into a single scaling multiply which would appear at the beginning or end of the overall filter. In this case, the number of multiply/adds becomes $(2N+1)$ for an Nth-order filter, and the formula is true for N odd as well as even.

Suppose we wish to realize the FIR filter of Fig. 2 using the same machine cycle of (2). From the figure it should be clear that there are $(M+1)$ multiplies and M adds; hence we conclude immediately that $(M+1)$ cycles are required for each output sample of an Mth-order FIR filter.

Now suppose we wish to realize a filter belonging to the important subclass of FIR filters that have symmetrical coefficients. The transfer functions of such filters have exactly linear phase and, in fact, most FIR designs considered in the literature possess this property. For these filters, coefficients spaced symmetrically about the center tap have equal values and, by pairing terms appropriately, each output sample thus requires only $(M/2+1)$ multiplies and M adds (for M even). With the basic function defined by (2), the pairing is impossible and M cycles are necessary; whereas if

$$\text{output} = \left[(\text{input}_1) + (\text{input}_2)\right] \cdot (\text{coeff}) + (\text{input}_3) \quad (3)$$

$(M/2+1)$ cycles suffice. Hence, for symmetrical FIR filters a processor with basic cycle (3) has essentially double the computing power of one using (2).

However, the incremental cost of implementing this function over that of (2) arises not just from the hardware for the extra add but, perhaps more importantly, from the increased program overhead required to deal with three inputs rather than two. These tradeoffs are still being studied and debated by designers and little can be said at this time about where the point of diminishing returns lies regarding arithmetic unit complexity. Nevertheless, the issue is important and will arise again in the section on TDM/FDM processing algorithms.

DIGITAL TDM/FDM TRANSLATION

Before proceeding further it is necessary to be more specific about certain details of the TDM/FDM interface, including aspects which are peculiar to a digital implementation. By international agreement, the standard FDM hierarchy assembles voiceband channels first into *groups* (twelve 4 kHz channels), then into supergroups (five groups or 60 channels), and then into larger bundles which need not concern us here. In the frequency domain, the groupband occupies the interval from 60 to 108 kHz and the supergroup occupies that from 312 to 552 kHz. As to the TDM hierarchy, there is unfortunately less widespread agreement regarding its construction. In North America and Japan, the basic unit is 24 time slots

(channel positions) while in Western Europe it is 32. In both cases, however, the per-channel sample rate is 8 kHz. In the first system, all time slots are available to users while in the second there are 30 active time slots, one signaling time slot, and one framing time slot.

The fundamental elements of a digital TDM/FDM transmultiplexer are shown in Fig. 4. FDM-to-TDM conversion is accomplished by: 1) removing any unwanted out of band components from the FDM signal with an analog filter; 2) passing this cleaned-up signal through an analog/digital converter to produce a digital word stream; and 3) "unscrambling" the individual channels in this stream via the real-time processing algorithm. In the TDM-to-FDM direction, operations inverse to these are performed (except for the analog filter which performs essentially the same function in both directions).

In designing an interface between the TDM and FDM hierarchies, the first issue to be settled is the level at which the translation should take place. Of the various possibilities, only two have received serious considerations: 1) translation between two 12-channel groupbands and a 24-channel North American TDM unit; and 2) translation between the 60-channel supergroup and two 30-channel European TDM units. Important issues which bear on the choice are the relative expense of A/D and D/A converting a group signal versus a supergroup signal, the relative expense of the digital processing in each case, and the overall way in which the telephone network is administered (in North America there are not as many opportunities for interfacing at the supergroup level as there are in the European network). Conversions at both the group and supergroup levels have received roughly equal attention in the literature; however, the same principles apply to both and for the most part, we will confine subsequent discussion to the groupband case.

Another question to be answered is what sample rate to use for the digital signal $v(m)$ in Fig. 4. The theoretical minimum rate is equal to twice the bandwidth of the FDM group signal which is equal to the word rate of the TDM signal $r(n)$, viz., 96 kHz for the groupband case. But there is no reason why the sample rates of $v(m)$ and $r(n)$ should be the same. In fact, making them identical

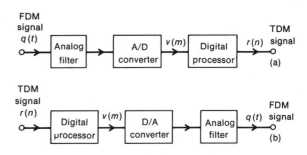

Fig. 4. Digital TDM–FDM translation.

places very severe demands on the cutoff skirts of the analog reconstruction filter of Fig. 4(b). On the other hand, the costs of the A/D and D/A converters and of the digital processors increases with the sample rate of $v(m)$ and there is nothing to be gained by making it larger than necessary. Now, the group format is "bandpass" in that a gap exists between dc and the lower band edge of the spectrum (Fig. 5). Moreover, the band edge frequencies are such that bandpass sampling techniques can be used to sample $q(t)$ directly without resorting to analog modulating steps either before A/D conversion or after D/A conversion. A little reflection should convince one that any sampling rate between the upper band edge (108 kHz) and twice the lower band edge (120 kHz) will work without causing any overlap of channels due to aliasing. Add to this the desirability of using an integer multiple of the basic 8 kHz PCM rate (this considerably simplifies the processing algorithm) and the choice is quickly reduced to $8 \times 14 = 112$ kHz. This rate provides 8 kHz guard bands between signal images (thus minimizing the complexity of the analog reconstruction filter) and is sufficiently close to the 96 kHz theoretical minimum to be a very acceptable compromise. Similar reasoning can be applied to the supergroupband case.

The remaining issues to be considered in this section have to do primarily with allowed signal impairments—noise, crosstalk, phase and frequency distortion, and nonlinear distortion. In principle it should be possible to present specifications on these impairments in a straightforward way. In practice it is not that easy because: 1) different telephone administrations throughout the world allocate the impairments in different ways; 2) the amount of each impairment allocated to the transmultiplexer depends on exactly where it appears in the network; and 3) the specifications are in a constant state of evolution, being subject to change as modifications and improvements are made to other parts of the telephone plant. Nevertheless, to form the basis of further discussion, a somewhat simplified set of "typical" numbers is offered in Table I. These are baseband parameters which apply to each channel and would in practice be measured by connecting a transmitter [Fig. 4(b)] and a receiver [Fig. 4(a)] back-to-back.

In the type of system under consideration, the above constraints are dealt with as follows. Sources which contribute idle channel noise are the A/D and D/A converters and numerous roundoff points within the digital processors. The requirement is met by using con-

verters of sufficient accuracy (typically 13 bits) and a sufficiently large data word (typically 20 bits) in the digital processing. The only important sources of nonlinear distortion are the A/D and D/A converters and, as it happens, converters which meet the idle channel noise requirement easily meet the nonlinearity requirement. Performance in regard to the remaining categories—crosstalk and frequency response—is more intimately associated with details of the digital processing algorithm and will be discussed in the next section.

Finally, mention should be made of a performance characteristic which is often ignored but which is of increasing importance. This is the amount of flat delay imposed by the transmultiplexer. Echoes are a fact of life in telephone systems and, as is well known, the subjective annoyance value of an echo increases as the round trip delay increases. Anything which contributes unnecessarily to this delay is therefore to be avoided. There is as yet no agreed upon upper limit to allowable round-trip delay and thus no way of allocating amounts to various network components such as a transmultiplexer. Nevertheless, some proposed digital transmultiplexer algorithms—such as a block FFT approach in which the block size is large—introduce large amounts of flat delay and this property should be taken into account in their evaluation.

DIGITAL TDM/FDM ALGORITHMS

We now turn our attention to specific algorithms for performing TDM/FDM conversion. As will become apparent, this is an intricate subject and thus no attempt will be made to cover exhaustively all the work that has been done. Instead, two approaches of historical importance will be used to illustrate basic principles.

The key element in digital TDM/FDM conversion is digital single sideband modulation and demodulation. As many have noted, these operations can be viewed in turn as variants of digital sample rate interpolation and decimation. We therefore begin this section with a brief review of the latter subjects.

Fig. 5.   Group band spectrum.

TABLE I

Maximum idle channel noise referenced to peak signal point	−80 dB
Minimum crosstalk attenuation between any two channels	60 dB
Frequency response:	
within $\begin{cases} +0.5, -0.5 \text{ dB} \\ +0.5, -3 \text{ dB} \end{cases}$	500 to 3000 Hz 200 to 500 and 3000 to 3400 Hz
Maximum rms nonlinear distortion referenced to peak signal point	−40 dB

Interpolation and Decimation

Digital interpolation is the process of taking a given sequence $x(n)$ and creating a new sequence $y(m)$ whose samples occur R times as fast. Decimation refers to the inverse process of sample rate reduction. Conceptually at least, interpolation can be accomplished by passing $x(n)$ through an ideal D/A converter to produce an analog signal $x(t)$ and then resampling this at the higher rate. In practice, of course, it is not necessary to create an analog signal; the process can be carried out by purely digital means as follows.

Let $x(n)$ be a sequence with sample period T seconds and consider the spectrum of this signal, depicted in Fig. 6(a). A basic feature of this spectrum is that it is periodic with period $1/T$ Hz. Let $H(f)$ be a lowpass digital filter with cutoff frequency of $1/2T$ Hz and operating at the high sample rate of R/T Hz [Fig. 6(b)]. If we apply the low-rate signal to this filter (intervening input samples are considered zero) it will, by its nature, produce an R/T rate output whose samples are basically the same as those that would be obtained in the conceptual interpolation process described in the preceding paragraph. As shown in Fig. 6(c), the filter simply removes unwanted signal images and leaves a spectrum that would have resulted from sampling an analog signal at the high rate to begin with (at least to the degree that $H(f)$ approximates an ideal lowpass filter).

In the dual process of decimation, one starts with a high-rate sequence, applies this to the filter $H(f)$ to remove any signal energy in the interval $1/2T$ Hz to $R/2T$ Hz and then retains only every Rth output sample, discarding the rest. Mathematically, the last operation performs a step-and-repeat process (with step interval $1/T$ Hz) on the spectrum of Fig. 6(c), to produce that of Fig. 6(a), as desired.

Digital SSB Generation

In the previous section the interpolation filter $H(f)$ was used to extract a portion of the signal spectrum centered about the origin, i.e., it acted as a standard lowpass filter. In a straightforward way, the concept of interpolation can be generalized to allow the extraction of any portion of the Fig. 6(a) spectrum, including single rather than double sideband portions.

This leads immediately to the *bandpass* method of digital SSB generation, which is a direct counterpart of analog SSB generation used in long-distance telephone networks throughout the world (except that in analog systems, a modulation step, i.e., multiplication by a sinusoidal wave, is necessary, whereas in the digital case the carrier is "built in" because of the periodic nature of the digital spectrum). It is instructive to consider the computational complexity of this method. It can be shown that a recursive digital bandpass filter of order 16 (i.e., eight cascaded second-order sections) will approximate the desired transfer function to the degree neces-

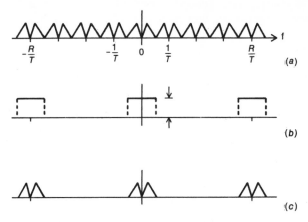

Fig. 6. Interpolation spectra.

sary to satisfy the requirements of Table I. From previous discussions we know that $1/T = 8$ kHz and $R/T = 112$ kHz (therefore $R = 14$). From the formula derived in the section on DSP hardware, and assuming the processor described by (2), the number of machine cycles/second required to compute the digital SSB stream is

$$\Big[(2)\,(16) + 1\Big]\,(112\ 000) = 3\ 696\ 000.$$

To generate a groupband signal, the bandpass interpolation process described above is performed separately on each of 12 baseband channels and the results added to produce one composite 112 kilosample/second FDM signal ($v(m)$ of Fig. 4). Although every channel has a different bandpass filter, the order of all filters is the same and the above number is the computation rate for each channel. As we shall soon see, this is a large per-channel computation rate and the bandpass method has therefore never received serious consideration. Nevertheless, it is easy to understand and makes a good starting point in a discussion of TDM/FDM algorithms.

For the bandpass demodulation process the same steps are performed in reverse. The composite FDM stream $v(m)$ is applied simultaneously to 12 bandpass filters whose outputs are then decimated by a factor of 14. The per-channel computation rate for demodulation is thus identical to that for modulation. In fact, this duality is so universal that continuing to discuss the inverse operation is pointless. We henceforth confine our attention to the TDM-to-FDM direction only.

Before going further, one small loose end should be tied up. If one were to implement the bandpass method exactly as described, the resulting FDM signal would contain a mixture of upper and lower sidebands because of the nature of the spectrum of Fig. 6(a). Sidebands of like parity are desired and, fortunately, can be produced quite easily. It is merely necessary to alternate the sign of the input sequences of every other channel, say the odd numbered ones. This is equivalent to multiplying each of these sample sequences by the sequence

$$\cos(n\pi) = \cdots -1,1,-1,1,\cdots.$$

Recalling an elementary property of Fourier transforms, this has the effect of shifting the spectra of each of these inputs by exactly half the sampling rate (i.e., 4 kHz). In any given frequency slot an upper sideband is exchanged for a lower or vice versa, thus achieving the desired end. In the bandpass demodulator, a similar sign alteration must be performed on half the output channels following decimation.

The Weaver Method

In 1956, D. K. Weaver proposed a so-called "third" method [30] of analog SSB generation, which was designed to overcome certain disadvantages of other analog approaches. As it happens, Weaver's method has not been used much in analog telephony but, in the early work on digital TDM/FDM translation, the approach proved quite useful because it lent itself to two important techniques for reducing complexity [1].

Fig. 7 depicts a digital version of Weaver's original proposal. Starting at the left, an 8 kHz baseband input signal is split into two paths and each is modulated by midchannel (2 kHz) sine waves 90° apart in phase. (The accompanying spectral plots are pictorial representations of Fourier transforms in which overlapping sidebands are depicted as being distinct. This artifice helps one visualize how various sidebands cancel at the appropriate points.) Two identical lowpass 112 kHz sample rate filters then remove energy above 2 kHz. In a second quadrature modulation step, the sideband combinations produced thus far are translated to the desired channel position in the groupband. The phases of the various modulating carriers are such that when the two branches are summed, the unwanted sideband cancels out.

The computation rate necessary to implement the above process for the groupband case can be reckoned as follows: The low-speed modulation involves multiplication by the factors zero and ±1. In most DSP designs such degenerate multiplications can be handled without using separate machine cycles and so these will

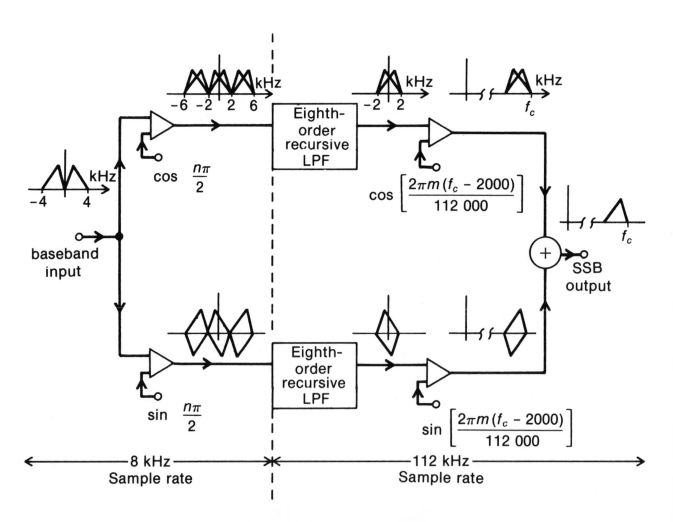

Fig. 7.   Digital Weaver SSB generator.

not be counted. The interpolation filter is lowpass rather than bandpass, which means the filter order is 8 and not 16. However, two identical filters are needed to accommodate both branches, so the equivalent order per channel is still 16. Adding the multiplies required for the high-speed modulation to those required for the filters gives an overall per-channel computation rate of

$$(2) \left[(2)(8) + 1 \right] (112\ 000) + (2)(112\ 000) = 4\ 032\ 000.$$

As it stands, this number is actually greater than that for the bandpass case (because of the extra modulation steps involved). Fortunately, means exist to reduce this number drastically.

In most interpolation (and decimation) applications of interest there is much to be gained by doing the operation in more than one step [31]. For the case at hand, several combinations have been studied. An efficient approach, which still meets all applicable specifications, is to replace the eighth-order high rate recursive filter with a similar one operating at 8 kHz, followed by a high rate FIR filter to remove the unwanted signal images centered about multiples of 8 kHz. An FIR filter is used since advantage can be taken of the fact that most of its input values are zero due to the sample rate increase. A block diagram of this new arrangement is shown in Fig. 8. It turns out that a 55th-order FIR filter is required to adequately remove the unwanted signal energy. Ordinarily this filter would need 56 multiply/adds to produce each output; however, since 13 of every 14 inputs are zero, each output point requires only $56 \div 14 = 4$ machine cycles. The overall computation rate thus becomes

$$(2) \left[(2)(8) + 1 \right] (8000) + (2)(4)(112\ 000)$$
$$+ (2)(112\ 000) = 1\ 392\ 000$$

which represents a reduction of about a factor of three.

The key to further computation rate reduction takes advantage of a special relationship between the multiplicity of SSB signals that must be added together at the output. As it happens, the expression relating the output $v(m)$ to the 24 inputs to the FIR filters bears much resemblance to a discrete Fourier transform (this arises primarily from the fact that the 24 modulating frequencies are all integer multiples of a single frequency—in this case 2 kHz). This suggests the use of fast Fourier transform (FFT) computing techniques. Such techniques can, in fact, be used to considerable advantage. The details [1] are beyond the scope of this discussion; however, the net

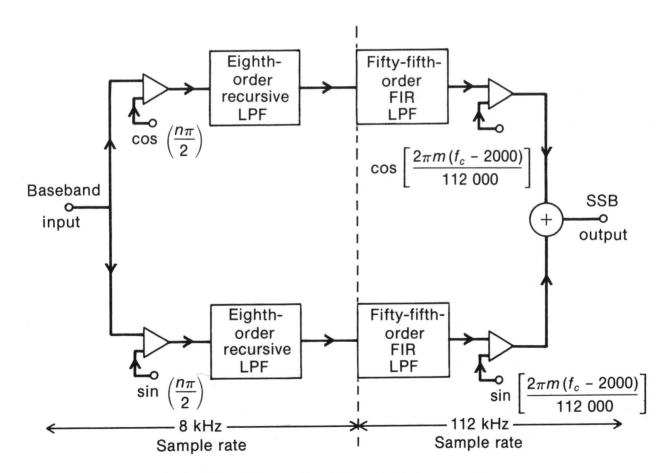

Fig. 8. Weaver SSB generation with two-step filtering.

result is to allow the second modulation step and the summing of channels to be done before the FIR filter. We now have a single FIR filter (identical to the ones used before) whose 8 kHz input, for the particular case at hand, requires 196 multiplies per sample value to be computed from the 24 outputs of the eighth-order recursive filters. Under these circumstances, the per-channel computation rate becomes

$$(2) \quad \left[(2)(8) + 1 \right] \quad (8000) + \frac{(196)(8000)}{12}$$

$$+ \frac{(4)(112\,000)}{12} = 440\,000$$

which represents a further reduction by about a factor of three. However, this second reduction comes at a price. Because of the special nature of the computations that occur between the 24 recursive filters and the FIR filter, the simple arithmetic unit of (2) will not suffice. The reduction can be achieved only with a basic function which, among other things, performs a number of pre-adds along with each multiply. Also, considerably more scratchpad memory is required for temporary storage of interim calculations.

The Polyphase Method of Bellanger *et al.* [7],[10],[12],[16]

This method represented the first significant departure from the original ideas of Darlington and paved the way for a more general understanding of digital FDM generation. The following explanation differs somewhat from that in the references but, in this author's opinion, makes the method easier to understand and to compare with the other ideas put forth in this discussion.

We return once again to the concept of sample rate interpolation. Consider the process depicted in Fig. 9, in which an 8 kHz sequence $x(n)$ is applied simultaneously to a bank of R digital filters, where R is the ratio of high to low sample rates as before. Assume that the ℓth filter imposes a flat delay of $\ell/8R$ ms on its input (this transfer function cannot be realized exactly—except for the degenerate case where ℓ is an integer multiple of R—but can be approximated to any desired accuracy by using filters of sufficiently high order).

Since all R filters operate in synchronism from the same input, their outputs occur at the same time instants. If we pass these outputs through a second bank of purely digital delays as shown in Fig. 9, they can be distributed evenly over the 125 μs sample interval and then combined to produce a single $8R$-kHz sequence $y(m)$. (In Fig. 9, the analog envelope $x(t)$ is shown to help clarify what is being done.)

Now, since this bank of filters and delays produces an interpolated sequence $y(m)$ from the original sequence $x(n)$, it is clearly equivalent to the basic interpolation filter discussed earlier and must therefore have an overall

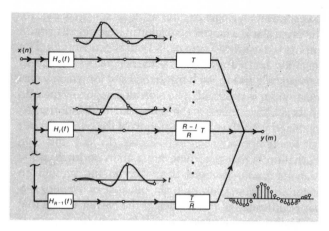

Fig. 9.   Polyphase interpolator.

input-to-output transfer function identical to that of Fig. 6(b). Moreover, the computational complexity of this interpolation procedure is the same as the previous one. It can be shown [12] that if, say, an eighth-order recursive approximation to the transfer function of Fig. 6(b) suffices, then each of the $H(f)$'s in Fig. 9 will be eighth-order also, and one $8R$-kHz eighth-order filter requires the same computation rate as R 8-kHz eighth-order filters. The network of Fig. 9 performs interpolation with a multiplicity of flat delays (phase shifts) and it is therefore often referred to as a *polyphase interpolator*.

Such networks can be used for very efficient digital FDM generation. To see this, first consider using the arrangement of Fig. 9 to realize each of the lowpass filter/interpolators in the Weaver method of Fig. 7. In this case the bank will consist of $R = 112 \div 4 = 28$ filters, each operating at 4 kHz. (The baseband input has rate 8 kHz, but the interpolator inputs are effectively only 4 kHz since the low-speed modulation step multiplies alternate sample values by zero.) By the same reasoning used in the previous paragraph, the computation rate for this scheme is identical to that for the Weaver method before the two rate reduction techniques were applied. As we have seen, this rate is anything but efficient; however, we are again fortunate in that an elegant and equally effective reduction technique applies here as well.

Consider that portion of each signal branch that contains the 28-filter interpolator followed by the high-speed modulation step. Now, the multiplications that effect this modulation can be performed directly on the 112 kHz sample stream, or they can be done on each of the 28 branches individually, before the samples are interleaved. If the latter course is followed, an interesting thing happens. As noted before, the carrier for each frequency slot is an integer multiple of 2 kHz. Since the sample rate for each filter branch is 4 kHz, the phase of the carrier associated with a particular branch changes by an exact multiple of 180° each time around; hence, the numerical value of the carrier stays fixed and there is at most a sign alternation. Ignoring the sign change for the moment, what we have done is to convert the high-speed modulation to a fixed scalar multiplication on each filter branch. Moreover, this scalar multiply can be performed anywhere on the branch, including in front of the flat delay filter.

As matters now stand, our new version of the Weaver FDM generator consists of twenty-four 28-filter networks connected in parallel, each operating at 4 kHz. We can immediately reduce this to 12 parallel 28-filter networks by combining the split channel paths together, but each network now operates at 8 kHz, thus saving nothing as far as computation rate is concerned. The next big step is to realize that these 12 networks can be coalesced into a single 28-filter array as shown in Fig. 10. The prescalar is a matrix operation in which each of its 28 outputs is a weighted sum of all 12 baseband inputs. The weighting factors are just the carrier values unique to each branch and to each frequency slot for which a particular input is destined. The sign alternation of the odd inputs is what remains of the low-speed modulation when combined with the sign fluctuation of the carrier values that we temporarily ignored above. (In fact, this sign alternation is doing nothing more than performing the sideband parity reversal described in the section on digital SSB generation.) Finally, because of the very regular relationship of the phases of the carrier values that weight the input signals on each branch, the prescalar operation turns out to be mathematically identical to the discrete Fourier transform, thus opening the door again to the use of FFT techniques.

When all of the above is taken into account, the per-

channel computation rate for the polyphase FDM generator is

$$\frac{(28)\,[\,(2)\,(8)+1\,]\,(8000)+(28)\,(7)\,(8000)}{12}=448\,000.$$

The first term is the rate for 28, 8-kHz, eighth-order recursive filters, the second reflects the fact that the prescalar/DFT operation requires an average of 7 multiplies/output branch to compute, and the whole is divided by 12 to give the per-channel rate. In a recent paper on the polyphase method [16], Bellanger has noted how certain symmetries in the coefficients of the flat delay filters can be used to further reduce the rate to roughly half the above value. However, this reduction requires even more specialized hardware than has been discussed so far.

The Generalized Scheme of Classen and Mecklenbräuker

Although the Weaver and polyphase methods were arrived at originally by quite different reasoning, there appear to be definite points of similarity, especially after various complexity reduction techniques have been applied. The groupband Weaver method passes 12 input signals through a bank of 24 8-kHz, eighth-order filters, then performs an FFT-like operation on the outputs, followed by a single 112 kHz filtering operation distributed over the 12 channels. In the polyphase method, an FFT operation is performed first, followed by a bank of 28 eighth-order, 8-kHz filters (there is no high-rate filtering required). The two methods thus perform similar operations in inverse order. This suggests that both methods might be special cases of something more general.

Such a generalized scheme has been proposed by T. Classen and W. Mecklenbräuker [21]. Although the details are beyond the scope of this discussion, a block diagram would resemble Fig. 10. In this general scheme, the prescalar is no longer confined to be a discrete Fourier transform matrix and, except for special cases, the filters H_l operate at rate $8R$ kHz, rather than 8 kHz. A particular combination of prescalar and filter bank must satisfy certain conditions in order to be a valid FDM generator. Further conditions are imposed under which "fast" realizations are possible, e.g., prescalars which allow the use of FFT-like algorithms and filter banks which can operate at 8 kHz and not $8R$ kHz. Among the new forms uncovered by the approach are the use of a Hadamard prescalar (i.e., one whose coefficients are ± 1 and whose implementation, assuming the appropriate hardware, requires no multiplications at all), and the use of filter banks with a tree-like structure that serves to reduce computation rate. The Classen/Mecklenbräuker paper does not provide much information about the relative computational complexity of the multitude of new forms it discusses. Nevertheless, it offers a great deal of

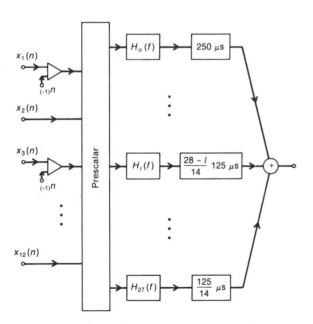

Fig. 10. Polyphase FDM generation.

insight and food for thought to any who would delve further into the subject.

Other Work

A glance at the references will indicate a large body of work not explicitly mentioned here. However, the great majority of it is built on the principles and techniques discussed above. The aim has been simply to prepare an interested reader to better understand and evaluate various schemes that have been proposed. Omission of a particular piece of work thus implies nothing about its relevance or importance.

CONCLUSIONS

This article has included a definition and general discussion of the TDM/FDM translation problem, a concise review of digital signal processing techniques and some indication of how the latter can be applied to the former. Particular emphasis was placed on the importance of measuring and reducing computational complexity in processing algorithms.

Digital signal processing has a mature theoretical base but is still in its hardware infancy.

The field of digital signal processing is in an interesting state. It has a mature theoretical base but is still in its hardware infancy. The voices proclaiming the coming digital revolution have been with us for some time but the relentless increase of device packing density will sooner or later prove them right. If recent experience with microprocessors is any indication, the forthcoming general purpose DSP chips will quickly give way to new generations of more sophisticated chips as architectures evolve with the increase in our understanding and experience. Hopefully, this discussion has conveyed some flavor of one of the more interesting applications of a field whose time has finally come.

REFERENCES

[1] S. Darlington, "On digital single-sideband modulators," *IEEE Trans. Circuit Theory*, vol. CT-17, pp. 409-414, Aug. 1970.

[2] C. F. Kurth, "SSB/FDM utilizing TDM digital filters," *IEEE Trans. Commun.*, vol. COM-19, pp. 63-71, Feb. 1971.

[3] S. L. Freeny, R. B. Kieburtz, K. V. Mina, and S. K. Tewksbury, "Design of digital filters for an all digital frequency division multiplex-time division multiplex translator," *IEEE Trans. Circuit Theory*, vol. CT-18, pp. 702-711, Nov. 1971.

[4] ——, "Systems analysis of a TDM-FDM translator/digital A-type channel bank," *IEEE Trans. Commun.*, vol. COM-19, pp. 1050-1059, Dec. 1971.

[5] C. Y. Kao, "An exploratory TDM-FDM SSB generator," in *Proc. Nat. Electron. Conf.*, Oct. 1972, pp. 47-50.

[6] S. Singh, K. Renner, and S. C. Gupta, "Digital single-sideband modulation," *IEEE Trans. Commun.*, vol. COM-21, pp. 255-262, Mar. 1973.

[7] M. G. Bellanger and J. L. Daguet, "TDM-FDM transmultiplexer: Digital polyphase and FFT," *IEEE Trans. Commun.*, vol. COM-22, pp. 1199-1205, Sept. 1974.

[8] P. M. Terrell and P. J. W. Rayner, "A digital block-processor for SSB modulation and demodulation," *IEEE Trans. Commun.*, vol. COM-23, pp. 282-286, Feb. 1975.

[9] R. Boite and H. Leich, "Use of the numerical phase-difference filter for digital channel banks," *Proc. IEE*, vol. 122, pp. 340-344, Apr. 1975.

[10] J. L. Daguet, M. G. Bellanger, and G. Bonnerot, "Méthode simplifiée de multiplexage en fréquence de signaux numériques réels," *Câbles et Transmission*, vol. 29, pp. 259-265, July 1975.

[11] R. Lagadec, "Digital channel translators based on quadrature processing," *AGEN-Mitteilungen*, pp. 3-35, Nov. 1975.

[12] M. G. Bellanger, G. Bonnerot, and M. Coudreuse, "Digital filtering by polyphase network: Application to sample-rate alteration and filter banks," *IEEE Trans. Acoust., Speech, Signal Processing*, vol. ASSP-24, pp. 109-114, Apr. 1976.

[13] A. G. Constantinides and I. Colyer, "Digital phase-splitting network design for digital FDM applications," *Proc. IEE*, vol. 123, pp. 1313-1315, Dec. 1976.

[14] M. Tomlinson and K. M. Wong, "Techniques for the digital interfacing of TDM-FDM systems," *Proc. IEE*, vol. 123, pp. 1285-1292, Dec. 1976.

[15] C. R. Williams and B. J. Leon, "The design of digital single-sideband modulators using a multirate digital filter," Purdue University, West Lafayette, IN, Rep. TR-EE, 77-29, May 1977.

[16] G. Bonnerot, M. Coudreuse, and M. G. Bellanger, "Digital processing techniques in the 60 channel transmultiplexer," *IEEE Trans. Commun.*, vol. COM-26, pp. 698-706, May 1978.

[17] A. Peled and S. Winograd, "TDM-FDM conversion requiring reduced computation complexity," *IEEE Trans. Commun.*, vol. COM-26, pp. 707-719, May 1978.

[18] R. Maruta and A. Tomozawa, "An improved method for digital SSB-FDM modulation and demodulation," *IEEE Trans. Commun.*, vol. COM-26, pp. 720-725, May 1978.

[19] F. Takahata, Y. Hirata, A. Ogawa and K. Inagaki, "Development of a TDM/FDM transmultiplexer," *IEEE Trans. Commun.*, vol. COM-26, pp. 726-733, May 1978.

[20] T. Tsuda, S. Morita, and Y. Fujii, "Digital TDM-FDM translator with multistage structure," *IEEE Trans. Commun.*, vol. COM-26, pp. 734-741, May 1978.

[21] T. Classen and W. Mecklenbräuker, "A generalized scheme for an all-digital time-division multiplex to frequency division multiplex translator," *IEEE Trans. Circuits Syst.*, vol. CAS-25, pp. 252-259, May 1978.

[22] D. Pelloni, "Dimensionierung von Digitalen TDM-FDM Transmultiplexern nach der Polyphasenmethode," *AGEN-Mitteilungen*, Mar. 1979.

[23] H. D. Helms, J. F. Kaiser, and L. R. Rabiner, *Literature in Digital Signal Processing*. New York: IEEE Press, 1975.

[24] J. F. Kaiser and H. D. Helms, *Supplement to Literature in Digital Signal Processing*. New York: IEEE Press, 1979.

[25] B. Gold and C. M. Rader, *Digital Processing of Signals*. New York: McGraw-Hill, 1969.

[26] A. V. Oppenheim and R. W. Schafer, *Digital Signal Processing*. Englewood Cliffs, NJ: Prentice-Hall, 1975.

[27] L. R. Rabiner and B. Gold, *Theory and Application of Digital Signal Processing*. Englewood Cliffs, NJ: Prentice-Hall, 1975.

[28] A. Peled and B. Liu, *Digital Signal Processing*. New York: Wiley, 1976.

[29] R. W. Hamming, *Digital Filters*. Englewood Cliffs, NJ: Prentice-Hall, 1977.

[30] D. K. Weaver, "A third method of generation and detection of single-sideband signals," *Proc. IRE*, vol. 44, pp. 1703-1705, Dec. 1956.

[31] R. E. Crochiere and L. R. Rabiner, "Optimum FIR digital filter implementations for decimation, interpolation and narrow-band filtering," *IEEE Trans. Acoust., Speech, Signal Processing*, vol. ASSP-23, pp. 444-456, Oct. 1975.

Stanley L. Freeny received the B.E.E. degree from the Georgia Institute of Technology, Atlanta, in 1958, and the M.E.E. degree from New York University, New York, NY, in 1960.

Since 1958, he has been associated with Bell Laboratories, Holmdel, NJ, where he has worked on timing and regeneration problems in long chains of digital repeaters, on hybrid multilevel PCM transmission systems, and most recently, on digital filters. In 1971, he was appointed Head of the Signal Processing Research Department, whose primary mission is to investigate various applications of digital signal processing to the telephone plant.

CHARGE-COUPLED DEVICES:
THE ANALOG SHIFT REGISTER
COMES OF AGE *

A. Gersho

In recent years delay lines, particularly tapped delay lines or transversal filters, have become ubiquitous components of communication systems. In theoretical studies, the transversal filter continually arises in optimal receiver structures. Yet when it comes to hardware implementation, the delay line or transversal filter is not a trivial component to be pulled out of a parts bin. While old-fashioned and bulky lumped filter networks have been used in the past to approximate a distortionless delay line, i.e., one with flat amplitude and linear phase characteristics, the dominant technique today uses costly A-D conversion followed by multibit parallel shift registers. A new device that achieves analog delay without the need for A-D conversion is now emerging, the charge-coupled device (CCD).

The CCD is essentially a discrete-time or sampled-data device that acts like an analog shift register, shifting analog samples of a signal along an array of cells under the control of a clock. Based on standard metal-oxide semiconductor (MOS) technology, the CCD consumes almost negligible power, can be clocked at rates up to 100 MHz, and can achieve time delays as long as a tenth of a second. A CCD chip can be fabricated as a complete transversal filter. In fact, CCD chips have been built containing two 500 tap transversal filters with prescribed weights, and occupying an area much less than a square centimeter.

CCD's have already been developed and marketed for two other applications—image sensing for TV cameras and digital memories. This paper will restrict itself to CCD's for analog signal processing. Their potential in this field is still speculative, but it appears not unlikely that CCD's will revolutionize analog signal processing.

THE CHARGE TRANSFER CONCEPTS

The CCD works by transferring packets of charge from cell to cell along a linear array of cells. An input signal voltage is converted into a charge packet, usually a bunch of electrons, whose size (total charge) represents the sample value. This packet is stored initially in the first storage element of the array. After one clock period the packet is transferred, with its size intact, to the second element, and a new packet representing the next input sample is deposited into the first element. After another period has elapsed, the first packet is transferred intact into element three, the second packet sample moves into element two and a new sample of the input is used to create a third packet which is stored in element one. Continuing in this way, charge packets travel along the array without changing in size, and therefore storing the analog amplitudes of the sampled input signal. The successive charge packets appearing in the last element of the array are used to generate analog output pulses which, when smoothed, form a delayed replica of the original continous-time input signal. Of course, the input is assumed to be suitably band-limited to avoid aliasing effects. Weighted taps are achieved by nondestructively sensing the size of a packet in a particular cell and generating an output component proportional to the stored charge value, with a specified constant of proportionality.

It is, of course, not possible to deposit a new charge packet into a cell at the same time as an old packet is being removed from that cell without losing the separate identities of the two packets. Therefore, each storage element consists of two or more distinct cells only one of which is storing a charge packet any instant of time. In this way a packet can always be transferred into an already emptied cell. For example, in a *two-phase* transfer scheme, every other cell is empty during a storage period and two transfers occur per clock period, one moving packets from, say, even-numbered to odd-numbered cells and the next moving packets from odd-numbered to even-numbered cells. In a 3-phase transfer scheme, two out of every three cells are empty during any storage instant and three transfers occur every period.

PHYSICAL BASIS OF CHARGE-COUPLING

There are other ways of achieving charge transfer (notably via bucket-brigade devices) but the *charge-coupling* technique dominates because of its simplicity and quality of operation. A layer of doped semiconductor, usually p-type silicon, is coated

The author is with Bell Laboratories, Murray Hill, N.J. 07974.

*Reprinted from *IEEE Communications Society Magazine*, November 1975, Vol. 13, No. 6, pp. 27-32.

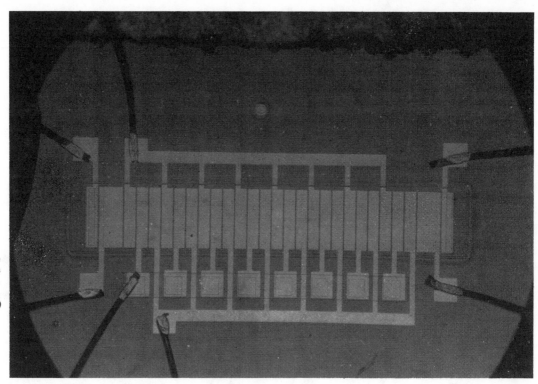

First charge-coupled device comprising 8 delay elements with 3-phase transfer scheme. (Tompsett *et al., Appl. Phys. Lett.,* vol. 17, pp. 111-115, 1970.)

with an oxide insulator, and then an array of closely spaced metal electrodes is placed on the oxide surface. When a positive voltage appears on an electrode, the majority charge carriers (holes) in the silicon are pushed away from the vicinity of the electrode, creating a *depletion layer* near the oxide with a net negative charge. If a bunch of free electrons are injected into this layer close to the oxide, they will be attracted to the interface between oxide and silicon. Assuming the two adjacent electrodes are substantially less positive, this will be a stable location for the packet of electrons, which implies a minimum of potential energy. This *potential well* keeps the charge packet safely ensconced. (Thermal activity in the depletion layer will occasionally free a bound electron which then jumps into the well and joins the packet. This *dark current* eventually masks the stored packet, and limits the useful storage time of the device to a figure in the 10-100-ms range.)

By raising the voltage at the next $k+1$th electrode to match the voltage on the kth electrode, a new potential well is created. But because the two electrodes are so closely spaced, the old and new potential wells combine almost as if the two electrodes have become one double width electrode. This *coupling* of potential wells causes some of the charge to move along the interface to the region below the $k+1$th electrode. The voltage on the kth electrode is then gradually reduced and the remainder of the packet of electrons diffuses over to the new well as the old well disappears. During this transfer process the $k-1$th electrode is kept at a low voltage to prevent the backflow of electrons. In a three-phase CCD, three clock-controlled drive waveforms as shown in Fig. 1 must be generated off-chip to supply the electrode voltages.

Finally, we note that charge is injected into the first cell of the array by applying suitable voltages to a p-n junction built into the input end of the device.

CHARGE TRANSFER INEFFICIENCY

With charge-coupling, or any other method of charge transfer, there is an inevitable imperfection that places a fundamental limitation on performance. This imperfection is called charge transfer inefficiency (CTI), and consists of the phenomenon of some electrons not getting transferred with their packet, but being left behind to subsequently join the next packet to come along. In each storage site at the oxide-silicon interface there are *traps*, quantum mechanical states, that grab hold of some of the electrons as soon as the packet arrives. Then, after the rest of the packet has moved on, these electrons begin to be released. Although captured very quickly, they are released more slowly with different release times depending on the particular energy level of the traps. As a result, many of the trapped electrons are released too late to rejoin their parent packet, and are attracted to the next packet to come along.

A simplistic but graphic analogy is to picture each potential well as a ship moving along with its crew of sailors (charge packet). During the transfer process the ship is moving from one port to the next adjacent port. When the ship is in port, some of the sailors go gamboling around, quickly filling up the available traps (pubs or what have you). When the ship starts to move on, some sailors are left behind in the traps and do not get out in time to catch the ship. As a result, the laggards end up in the next ship to come into port.

This and other physical effects produce transfer inefficiency. The net result is that a small fraction ϵ of the signal amplitude stored in a particular packet is left behind and added to the following packet.

Let $x_k(j)$ denote the signal amplitude stored in the kth delay stage of the CCD at the discrete-time instant $t = jT$. Then transfer inefficiency is described by the relation

$$x_{k+1}(j+1) = \rho x_k(j) + \epsilon x_{k+1}(j) \qquad (1)$$

where $\rho \approx 1-\epsilon$. That is, the sample value passed on to the next (k+1th) stage at the next (j+1th) time instant is the sample slightly attenuated from the previous stage at the previous time plus the left-behind trace of the sample previously stored in the jth stage. This is a very accurate (although a linearized) approximation.

The actual quantity of charge in a packet cannot be negative and is always kept well above zero by adding a fixed background charge to the input packet (10 to 25 percent of the maximum packet size) even when the input sample is zero. This so-called *fat zero* technique helps to reduce the transfer inefficiency. The coefficient ϵ of CTI is in the range 10^{-3} to 10^{-5} depending on the particular CCD design.

Fig. 2(a) shows an equivalent linear model for one delay stage of a CCD, which can be called a "unit transfer element." If ρ and ϵ are treated as independent parameters, it is clear that ρ simply attenuates the signal while ϵ introduces a frequency dependence. A CCD delay line thus can be modeled as in Fig. 2(b).

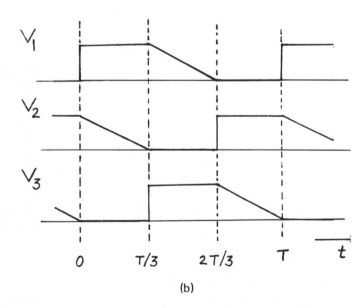

(a)

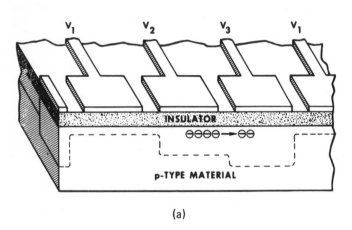

(b)

Fig. 1   (a) Section of 3 phase CCD during charge transfer. (b) Drive waveforms.

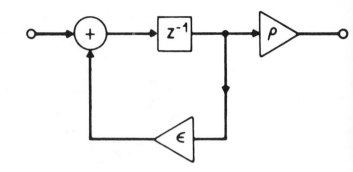

Fig. 2   (a) Model of one CCD delay element incorporating effect of transfer inefficiency.

EFFECT OF TRANSFER INEFFICIENCY

For any discrete-time filter made of CCD delay elements, i.e., any interconnection of summer weights (to scale signal amplitude) and unit transfer elements, the transfer function attained will differ from that obtained with ideal delay elements. Using z transforms, we see that the unit transfer element has the transfer function

$$\lambda^{-1}(z) = \frac{\rho z^{-1}}{1 - \epsilon z^{-1}} . \qquad (2)$$

Thus, if $H(z)$ is the transfer function of a filter based on ideal-delay elements with transfer function z^{-1}, then the actual transfer function attained using unit transfer elements, replacing z^{-1} by λ^{-1}, is

$$G(z) = H(\lambda^{-1}(z)). \qquad (3)$$

In particular, a delay line with N stages of delay (total delay NT where T is the clock period) has the transfer function

$$H(z) = z^{-N} \qquad (4)$$

for ideal delay elements. With CCD delay elements the actual transfer function becomes

$$G(z) = \left(\frac{(1-\epsilon)z^{-1}}{(1-\epsilon z^{-1})}\right)^N \approx e^{-N\epsilon(1-z^{-1})}z^{-N} \qquad (5)$$

where we have used the approximation $1 - \delta \approx e^{-\delta}$. Hence the frequency-dependent magnitude and phase of a CCD delay line are respectively $e^{-N\epsilon} \cos \omega T$ and $N\epsilon \sin \omega T$. Because ϵ is very small, it is clear that very many stages N of delay can be used before the $N\epsilon$ product is large enough to cause substantial high-frequency attenuation and phase distortion. Typically $N\epsilon = 0.1$ is considered an upper limit for reasonably good quality delay lines. Note that we are talking about discrete-time filtering so that the frequencies of interest here lie between zero and $1/2T$ hertz, assuming the analog signal is sampled at $1/T$ hertz.

More generally, for filters implemented with CCD's, so-called *charge-transfer filters*, we can readily see the effect of CTI on filter performance. Thus, if p is a pole or zero of a desired transfer function $H(z)$, from (2) and (3) it follows that the actual pole or zero attained using unit transfer elements is

$$p^1 = \rho p + \epsilon . \qquad (6)$$

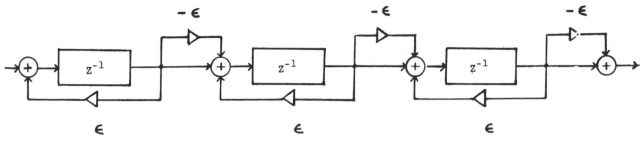

Fig. 2 (b) Model of three element charge-coupled delay line.

That is, the pole is radially attenuated by $\rho(= 1-\epsilon)$ and shifted to the right by ϵ (see Fig. 2). Examining the geometry in the unit circle shows that the effect of CTI is more significant for high-frequency poles and for higher order poles. Readers familiar with digital filtering can use their pole-zero plot intuition and the mapping (6) to determine the effect of CTI on particular filter designs.

CCD TRANSVERSAL FILTERS

A very effective way to make a complete transversal filter with a CCD is the *split-electrode* weighting technique. This method is based on the fact that when a charge packet moves under an electrode, a transient current flows through the clock line coming from that electrode, and the integrated current flow is proportional to the size of the electrode times the size of the arriving charge packet. Now suppose the electrode is split into two sections, as shown in Fig. 3, with a separate clock line connected to each segment. Then, by comparing the integrated current flow from the two clock lines, a quantity proportional to the size of the stored charge packet times the difference between capacitances of the two segments is measured. All the electrodes for a particular clock phase (every third electrode in a three-phase CCD) are split so that the two segments of the electrode for the kth delay element have lengths (and capacitances) in the proportion $1 + h_k$ to $1 - h_k$, where h_k is the desired weight value (normalized so that $(h_k) \leq 1$) for each k. The clock lines from all the lower electrode segments are connected together on a common bus line and similarly for the upper electrode segments. The two lines are applied to an on-chip differential current integrator to produce the transversal output. In this way, the CCD is made into a transversal filter with specified weight values essentially by modifying the design of one photomask to incorporate the appropriate electrode splits.

The most striking feature of this design is that the resulting transversal filter cost is hardly more than for a CCD delay line alone. Since the marginal cost of additional tap weights is negligible, it is practical to achieve transversal filters with hundreds of tap weights. Furthermore, as with any transversal filter, a linear phase response can be obtained by choosing the weight values so that $h_i = h_{N-i}$ for $i = 0,1,2,\cdots$ with N odd.

The key limitation of the split-electrode method is weight inaccuracy due to the limited resolution available in making photomasks. With the best current technology, a weight inaccuracy in the range 0.1 to 1.0 percent is inevitable. The more taps used in a transversal filter, the greater is the degradation due to the weight error. For example, for low-pass filter designs there is a best attainable stopband rejection, typically 35-45 dB, for an optimal number N, of taps, where N depends on passband ripple and is inversely proportional to transition width.

Nevertheless, for many applications where filter specifications are not too severe, the CCD transversal filter may be an effective and elegant way to do the job.

RECURSIVE FILTERING

Discrete-time recursive filters can be implemented using CCD delay elements, resistors for weighting, and operational amplifiers for summing. Recursive filters have the advantage of being able to achieve sharp transitions or high Q with only a moderate number of delay elements and weights. Because of stability considerations, coefficient weight values must be accurately realized by such means as laser-trimmed thin-film resistors. The input summing amplifiers which collect feedback signal components must similarly have a precisely determined gain. Consequently, it is not yet practical to use less accurate on-chip amplifiers.

Also, the sensitivity of pole positions to inaccuracies in the coefficients of high-order polynomials requires, as in digital filtering, a structural realization based on parallel or cascade combinations of second-order recursive filters or based on certain ladder configurations. This sensitivity problem prevents the use of a transversal filter imbedded in a feedback loop. Thus, recursive filters cannot so far be as fully integrated as transversal CCD filters.

One technique that may make recursive filters more practical is *multiplexing*, i.e., time-sharing common hardware (the delay line in particular) for several filtering functions. An approach that has received some attention recently uses a single CCD delay line with $2N$ elements, as shown in Fig. 4. By time-multiplexing N input signals, operating the CCD at N times the sampling rate of the individual signals, and periodically switching the weight values, the structure can effectively

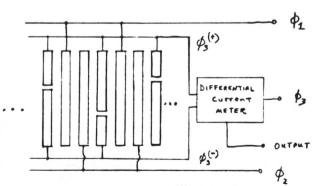

Fig. 3 Split electrode weighting method.

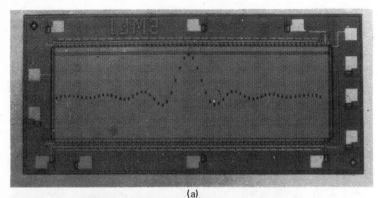

(a)

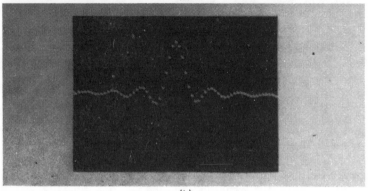

(b)

Fig. 4   (a) Photomicrograph of low-pass split electrode transversal filter, (b) Observed impulse response. (Baertsch, Engeler, Goldberg, Puckette, and Tiemann, in *Proc. CCD-74 Int. Conf.*, 1974, pp. 229-236.)

achieve N distinct second-order recursive filtering operations. Structures of this type can be interconnected to obtain a bank of N distinct higher order recursive filters. An obvious potential application for such structures is channel bank filtering for frequency division multiplexing or time-division multiplexing of voice signals.

ADAPTIVE FILTERING

While the split-electrode technique implies fixed preprogrammed weight values, it is possible to use CCD's for implementing adaptive filters where the weight values are to be varied dynamically. Two approaches have been proposed for this purpose. In one approach the CCD is used as a tapped delay line by modifying the CCD design to nondestructively sense the size of the charge packet in each delay element, and variable conductance MNOS field-effect transistors are used to obtain variable tap weights controlled by an applied voltage. Some tap output nonlinearity arises with this approach. The other approach is based on the binary representation of tap weight values. A parallel bank of tapped CCD delay lines uses only tap weight values 1 or 0 to form M transversal filters. The filter outputs are weighted by successive powers of two and added. In this way, a digitally adapted transversal filter is realized with M—bit coefficients and good linearity. Of course, adaptive filters tend to be relatively insensitive to both tap nonlinearities and transfer inefficiency because of their inherent self-optimizing character.

Finally, it should not be forgotten that since CCD's are clock controlled, the time-delay can be varied dynamically from milliseconds to microseconds, providing a potentially

valuable dimension to adaptive signal processing. For example, a tuned filter with electronically variable resonant frequency can readily be attained by varying the clock frequency. In fact, the first application for charge transfer circuits was to obtain variable delay to compensate for speed variations in audio tape recorders.

NOISE AND DYNAMIC RANGE

The various noise sources in a CCD are too numerous and complex to discuss individually here. In one study of a 256-element device the total rms noise equivalent signal was found to be 90 dB below the maximum signal amplitude in a charge packet. Of greater concern are the effects of clock pickup, harmonic distortion, and dark current. These effects can be substantially suppressed by operating two CCD's in parallel with complementary inputs, and combining the outputs with a differential amplifier.

NONLINEARITY

Another limitation to be considered is nonlinearity at both the input and output of the CCD. Nonlinearity associated with one charge injection technique at the CCD input has been found to introduce second and third harmonics of 45 dB below the amplitude of the fundamental sinusoidal input, and it is believed that 60 dB can be attained with an improved design. Further improvements in linearity apparently would substantially increase input noise and degrade high-frequency performance.

APPLICATIONS

There are, of course, numerous applications for cheap delay lines. For example, charge-coupled delay lines have been effectively demonstrated for delaying video signals. One application is for line interpolation in Europe's PAL type TV receivers. Another application is in frame storage for video bandwidth reduction. Analog delay lines can be used to elec-

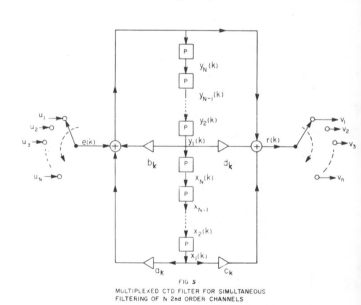

FIG 5

MULTIPLEXED CTD FILTER FOR SIMULTANEOUS FILTERING OF N 2nd ORDER CHANNELS

Fig. 5   Multiplexed second-order recursive charge-coupled filter. Weight values are periodically switched to the appropriate channel samples arriving at the tap positions.

tronically focus and steer antenna arrays. One such application is for ultrasonic imaging in medical diagnostics. Fancier applications involve time-axis compression and expansion, and recirculating loop memories.

In addition to general filtering applications, the charge-coupled transversal filter is very well suited for matched filtering. The optimality property of a matched filter implies that its output SNR is relatively insensitive to first-order perturbations in the tap weight values. Hence, very large numbers of taps can be used with little degradation in performance. Pulse compression radar and spread spectrum communications are likely candidates.

Another application of transversal filters reported recently is for spectral analysis by means of a sliding chirp z transform. A real-time 500 point discrete Fourier transform is performed using a suitable arrangement of four 500 tap transversal filters to perform most of the needed computation. Spectral analysis with CCD's could have numerous applications, a notable example being the generation of speech spectrograms.

Transversal CCD filters are also being explored for SSB signal generation by implementing Hilbert transform filters. If this approach can in fact achieve sufficient sideband rejection, it may turn out to be a very cheap way to do SSB modulation.

CONCLUSIONS

Many other applications for CCD's not mentioned here have also been proposed. While most of the applications are still in the speculative stage, others are already in development. The scope and potential of CCD's is obviously vast, but charge-coupled delay lines and transversal filters are still custom-designed items with a high-production cost, and high volume applications are needed to obtain low unit cost. Yet in its first five years the technology has progressed so rapidly that we may not have to wait very long before charge-coupled delay lines or even complete charge-coupled filters become standard stockroom items.

Most of the material discussed here may be found in much greater depth in the new and comprehensive book [1]. For brevity I have not cited the many individual contributors to the topics briefly mentioned here. A few representative papers of interest are [2] - [6]. Some recent work was reported in a special CCD session at ICC-75. A forthcoming Special Journal Issue on CCD's [7] may also be of interest.

This paper is an admittedly superficial introduction to CCD's. There are many subtleties and complications not dis-

cussed here. There is much more to say about the different kinds of devices being explored, and the work on imaging and digital memories has not been discussed at all. Nevertheless, I hope this introduction stimulates potential users to find out more about this new technology, and perhaps, to discover that CCD's will offer a better way to process their analog signals.

REFERENCES

[1] C. H. Sequin and M. F. Tompsett, *Charge Transfer Devices*, (Supplement to Advances in Electronics and Electron Physics). New York: Academic, 1975.

[2] D. D. Buss, D. R. Collins, W. H. Bailey, and C. R. Reeves, "Transversal filtering using charge transfer devices," *IEEE J. Solid-State Circuits*, vol. SC-8, pp. 138-146, Apr. 1973.

[3] D. A. Smith, W. J. Butler, and C. M. Puckette, "Programmable bandpass filter and tone generator using bucket-brigade delay lines," *IEEE Trans Commun.*, vol. COM-22, pp. 921-925, July 1974.

[4] R. W. Brodersen, H.-S. Fu, R. C. Frye, and D. D. Buss, "A 500-point Fourier transform using charge-coupled devices," in *ISSCC Dig. Tech. Papers*, Philadelphia, Pa. pp. 144-145, 1975.

[5] J. J. Tiemann, W. E. Engeler, R. D. Baertsch, and D. M. Brown, "Intracell charge-transfer structures for signal processing," *IEEE Trans. Electron Devices*, vol. ED-21, pp. 300-308, May 1974.

[6] A. Gersho and B. Gopinath, "Filtering with charge transfer devices," in *Proc. Int. Symp. Circuits and Systems*, Newton, Mass., Apr. 1975, pp. 183-186.

[7] D. D. Buss and M. F. Tompsett, Ed., "Joint Special Issue on Charge Transfer Devices," *IEEE J. Solid-State Circuits* and *IEEE Trans. Electron Devices*, Feb. 1976.

Allen Gersho received the B.S. degree in electrical engineering from the Massachusetts Institute of Technology, Cambridge, Mass., in 1960, and the M.S. and Ph.D. degrees from Cornell University, Ithaca, N.Y., in 1961 and 1963, respectively.

Since 1963, he has been a member of the technical staff at Bell Laboratories, where he is in the Mathematics and Statistics Research Center at Murray Hill, N.J. During the 1966-1967 academic year he was also Assistant Professor of Electrical Engineering at the City College of the City University of New York, New York, N.Y. He has been engaged in research studies on time-varying and nonlinear signal processing, synchronization of digital networks, adaptive equalization for data transmission, and the statistical approach to digital filter design. Recent work includes adaptive digital encoding of analog signals, social systems modeling, and an examination of the effect of transfer inefficiency on the performance of multiplexed CCD filters. He has just completed a theoretical study of photoacoustic spectroscopy, a new technique for the investigations of solid materials. □

SURFACE ACOUSTIC WAVE DEVICES *

Laurence B. Milstein
and Pankaj K. Das

SAW devices are particularly suited for high frequency and wide bandwidth signal processing applications.

The increasing use of higher frequency and wider bandwidth communication systems, most notably spread spectrum systems, puts greater demand on signal processing devices. For example, processing a 50 MHz bandwidth signal with current microprocessor technology is essentially out of the question due to the tremendously high clock rate that would be needed. *Surface acoustic wave* (SAW) devices have certain general characteristics which make them ideal choices for use in systems which must process such high data rate signals. In what follows, the technology of the device will be briefly surveyed and various applications of SAW devices will be discussed. Also, a comparison will be given between SAW technology and its two prime competitors, charge-coupled devices and digital electronics (i.e., microprocessors).

SAW TECHNOLOGY

To understand why SAW devices are so potentially valuable in signal processing applications, it is first necessary to review some basic principles of the underlying physics. The properties of SAW devices have been described in detail in many references, with those of [1] and [2] being typical. By definition, a SAW is an elastic wave that travels along the surface of a solid and is confined to the vicinity of that surface. This is in contrast to bulk waves which occupy the entire cross section of the medium. While SAW technology makes use of all waves which can propagate through a solid and which are confined to be near the surface of that solid, the most common of these are known as Rayleigh waves because Lord Rayleigh first studied this form of wave propagation in connection with earthquakes. The ordinary sound waves that the human ear can detect (at frequencies less than 30 kHz) are in general bulk waves traveling in air. For frequencies higher than 30 kHz, these waves are called ultrasound waves. A SAW is simply an ultrasound wave that is confined to the surface of a solid.

Fig. 1(a) illustrates the propagation of the bulk sound wave in a solid represented by grid points. The surface of the solid is the plane $z = 0$, and an increasing value of z measures increasing depth of the solid from the surface. Each of the grid points on the solid is uniformly spaced before the wave propagates. While the wave is propagating (in the y direction), one observes compression and elongation along the y axis and uniformity along the x axis. In contrast, Fig. 1(b) shows the SAW propagation. Note that for large values of z, the wave barely exists.

To make this SAW useful for signal processing, we must be able to generate and detect such a wave using electrical signals. The transducer commonly used to convert electrical energy to mechanical energy and vice versa is a piezoelectric material which when compressed or elongated produces an electric field inside the material. Alternatively, the application of an electric field to such a material produces a mechanical stress. Thus by connecting metal contacts to a piezoelectric plate (trans-

L. B. Milstein is with the Department of Electrical Engineering and Computer Sciences, University of California at San Diego, La Jolla, CA 92093.

P. K. Das is with the Electrical and Systems Engineering Department, Rensselaer Polytechnic Institute, Troy, NY 12181.

Reprinted from IEEE Communications Society Magazine, September 1979, Vol. 17, No. 5, pp. 25–33.

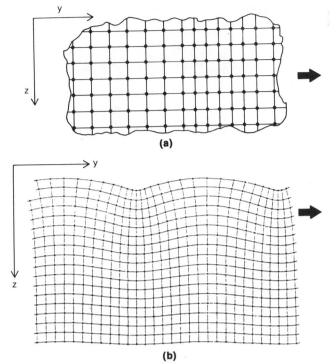

Fig. 1.   Cross section of the displacements of a rectangular grid of material points for a surface acoustic wave on an isotropic material.

ducer) and gluing it to a solid, bulk ultrasound can be generated in the solid (provided the electrical signal is applied at a proper RF frequency). A second transducer can be glued to the other end of the solid to detect the bulk ultrasound. In this way, a so-called bulk ultrasound delay line can be formed. It is possible, however, to achieve a major simplification by using a delay line in which a SAW propagates entirely on a piezoelectric material (commonly referred to as a substrate). In this latter case, one can integrate the entire process of generation, propagation, and detection of a SAW using so-called *interdigital transducers* (IDT) on the piezoelectric substrates.

A typical IDT is shown in Fig. 2. It is simply a set of metal strips placed on the piezoelectric substrate on which the SAW is to be generated. Alternate strips (commonly called fingers) are connected to each other by two other metal strips which form electrical inputs. The usual width of each metal finger is one quarter of a wavelength ($\lambda/4$), with the fingers also separated by $\lambda/4$. When an RF voltage is applied to the two input terminals of this IDT, an electric field is set up between all the adjacent fingers simultaneously. Thus the piezoelectric material alternately gets compressed and elongated between the fingers and generates a SAW which starts traveling with a velocity v, the surface wave velocity. Since the SAW is generated simultaneously by each pair of adjacent fingers, we have a situation in which the sources are distributed. To obtain efficient surface wave generation, all

these excitations must generate waves which add coherently.

Let the period of the RF voltage be T seconds. To propagate a distance of $\lambda/2$ therefore takes $T/2$ seconds. Since this is also the time it takes for the electric field between two adjacent fingers to change polarity, all the generated waves will add constructively. For detection, the reverse is true; i.e., as the elastic deformation passes under the fingers it induces inphase voltages between each finger-pair. This type of operation makes the IDT a truly remarkable and simple SAW generator.

To make a SAW delay line, one uses two IDT's on a piezoelectric substrate. If lithium niobate (LiNbO₃) is used as the piezoelectric material, the surface velocity is $v \approx 3.6 \times 10^3$ m/second so that if $f = 100$ MHz, then $\lambda = 36$ μm. Thus the width of each metal finger is 9 μm and 1 cm of separation between the IDT's will produce a delay of about 2.8 μs.

Why use many finger-pairs instead of only one pair? Actually, an IDT can be (and sometimes is) made with only a single finger-pair. However, with N finger-pairs, greater efficiency is achieved for the same applied voltage, since, in effect, N sources are adding up. However, there is a bandwidth reduction as one uses more and more finger-pairs. This will be explained later.

If we pause here for a moment and compare a SAW delay line with a bulk ultrasound delay line, we immediately see the advantages of the SAW delay line. First, the SAW delay line can be made using a one-step photolithographic process whereas bulk delay lines need the transducer glued to the material through which the wave propagates, a rather cumbersome and expensive process. Second, for a tapped delay line, it is both simpler and less expensive to implement the SAW delay line. For example, to produce a tapped delay line with $N = 10$ taps

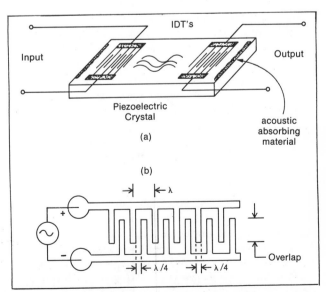

Fig. 2.   The interdigital transducer (IDT) is simply a set of metal strips placed on the piezoelectric substrate.

and a delay time between taps of $T_d = 1$ μs, one needs an input transducer and nine output transducers situated 0.36 cm apart. The fabrication cost for a SAW delay line does not change for this case. However, for a bulk delay line nine more transducers have to be built and glued, making this bulk tapped delay line considerably more expensive. In fact, this latter process is not even feasible if the distance between the transducers becomes too small.

Transversal Filters

To build a transversal filter with transfer function given by

$$H(\omega) = \sum_{n=0}^{N-1} C_n\, e^{-j\omega n T_d} \qquad (1)$$

the delay line output at each of the N taps is multiplied electronically by the corresponding coefficient C_n, and the product is applied to a summing circuit. In general, the multipliers can be switched to different coefficient values so that one can implement a programmable filter. If, however, one is merely interested in a fixed filter response, then the coefficients can be incorporated in the transducer itself. This can be accomplished in many different ways, but we shall discuss the one called *apodization*.

Consider Fig. 3. Suppose that the same SAW is incident on finger-pairs A and B. Note that the finger-pair A overlaps the entire width of the SAW whereas the finger-pair B overlaps only a small fraction (10 percent) of the SAW width. Thus it is expected that the voltage detected at A and B will correspond to coefficients $C_A = 1$ and $C_B = 0.1$. By using a single IDT with such variable overlaps, the overall voltage produced will be the sum of the contributions due to each finger-pair.

This is a very convenient way of making inexpensive filters. The calculated coefficient values are incorporated directly into the mask used to fabricate the IDT pattern. To synthesize some specific transfer function, say $H_1(\omega)$ (assuming $H_1(\omega)$ is suitably bandlimited), one can simply expand $H_1(\omega)$ in a Fourier series and end up with an expression similar to (1). In particular, if the impulse response of $H_1(\omega)$ is $h_1(t)$, then the coefficient C_n in (1) can easily be shown [3] to be given by

$$C_n = h_1(nT_d)$$

where T_d is now chosen to correspond to a correct sampling interval for reconstruction of the specific bandlimited function $h_1(t)$. Thus when we look at a SAW tapped delay line through a microscope, we actually see a sampled version of the impulse response of the device by observing the pattern of overlapping fingers. One can use the well-known techniques of digital filter design for the mask design of a specified filter. As a result, the realization of transversal or finite impulse response filters

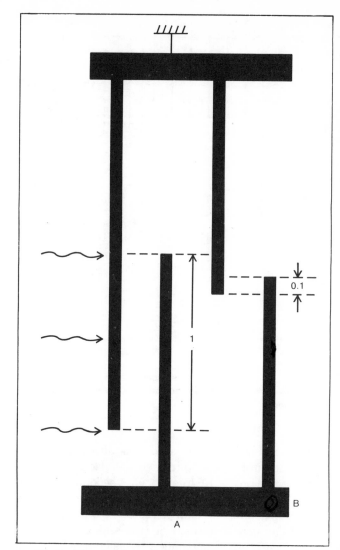

Fig. 3.   Finger-pair overlap determines transversal filter coefficients.

with specified characteristics (within certain limitations) is readily achievable [2].

We now apply the above-mentioned tapped delay line idea to examine the frequency response of the basic IDT with N fingers as discussed earlier. In the previous notation, with all the coefficients $C_n = 1$, we have

$$H(\omega) = \sum_{n=0}^{N-1} e^{-j\omega n T_d} \text{ where } T_d = \frac{1}{f_0},$$

and where f_0 is the center frequency of the design. In the vicinity of f_0, the magnitude of the transfer function can be shown to be approximately

$$N \left| \frac{\sin N\pi\,\Delta f/f_0}{N\pi\,\Delta f/f_0} \right|$$

where $\Delta f = f - f_0$ and $f = w/2\pi$. This "sinc-function" response centered around f_0, has bandwidth inversely proportional to N, and corresponds to the rectangular impulse response where all coefficients are unity over a finite interval. To design a sharp band-pass performance, the impulse response of the tapped delay line should look like the sinc-function as shown in Fig. 4. From the above discussion, it can be seen that there is indeed a bandwidth price that one pays as more and more fingers are used.

Current Technology

We have presented a rather simplified picture of SAW devices. Many other fine details must be considered to obtain a good filter. For example, the input transducer has its own bandwidth limitations. One must certainly take this frequency response into account to attain a good design. Thus it turns out that the initial design of the device may indeed be complex. However, once the mask is designed, the usual process of photolithography (well developed by the integrated circuits industry) can be used to manufacture these filters with low incremental cost. Current technology permits accurate implementations of SAW filters. For example, 100 dB rejection for band-pass filters has been reported and, with proper care, 60 dB is routine [1]. The rejection ratio is limited by the generation of spurious or undesired waves such as bulk waves. (It turns out that whenever you generate a SAW you always generate a small amount of bulk waves.) The device itself is quite small so that a typical filter might be the size of a 25-cent coin, while the number of taps in the delay line can be as large as several hundred.

The design of SAW filters has reached the stage where insertion losses as low as a few decibels, delays of tens of microseconds, and a center frequency which ranges from 10 MHz to a few gigahertz have been achieved [1]. However, at times these specifications conflict with one another, so that, for example, a device with a high center frequency will invariably have a high insertion loss. For routine implementation, the center frequency is now limited to 500 MHz due to fabrication complexity and the insertion loss tends to be in the 20 dB region. The insertion loss depends upon the substrate material used, the electrical matching networks employed, the center frequency of the devices, and the fractional bandwidth. The dynamic range of these devices tends to be in the vicinity of 70 dB [27].

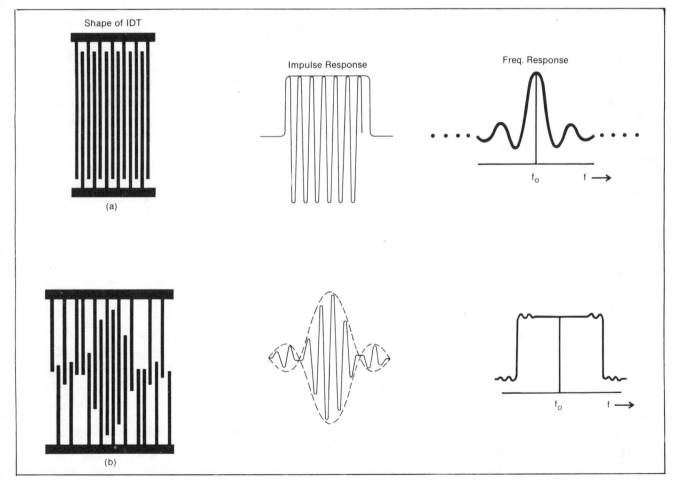

Fig. 4. Two IDT configurations and the corresponding filter characteristics.

SAW filters with center frequencies from tens to hundreds of megahertz are routinely implemented.

In addition to band-pass filters, various other components of a communication system can be implemented with SAW devices. For example, an oscillator [5] can be made using a SAW delay line and an external amplifier in a feedback loop. The Q of this delay line oscillator is generally of the order of a few thousand. For higher Q oscillators [6], a SAW resonator is used. A resonator is made of a piezoelectric substrate at the two ends of which many periodically spaced metal fingers or grooves are constructed. These periodic fingers or grooves reflect the surface waves efficiently at the resonant frequency and a pair of these reflectors forms a planar cavity. The cavity can be connected to the external circuit by interdigital transducers of proper frequencies placed at the middle of the cavity. Oscillator Q's of the order of 30 000 have been reported. With these resonators, oscillators can be made which have stability nearly as good as oscillators using quartz overtone crystals, but at higher frequencies. Thus SAW resonators are finding increasing use as the frequency determining element in UHF and VHF oscillators.

The basic advantage of the SAW devices is that they use planar technology and thus can be cheaply mass produced. For the tapped delay line applications, one major handicap is the rather large temperature coefficient of $LiNbO_3$ or other high coupling material which has low insertion loss. If one is willing to sacrifice efficiency, a substance known as ST-cut quartz is available which has a zero first-order temperature coefficient. By way of comparison, minimum achievable insertion loss for a 30 MHz bandwidth filter with a center frequency of 100 MHz is approximately 2 dB for an $LiNbO_3$ substrate whereas for ST-cut quartz it is 25 dB [4].

Space Charge Coupled SAW Devices

Another class of important SAW device exploits the acoustoelectric interaction of a SAW with the free carriers on a semiconductor surface. A sandwich structure of semiconductor on piezoelectric substrate (e.g., silicon on $LiNbO_3$) is used to form a so-called space-charge coupled SAW device [7],[8]. Being an elastic deformation wave on a piezoelectric substrate, the SAW induces charge separation. Thus it carries an electric field with it which exists both inside and outside the piezoelectric substrate. The outside electric field disturbs the carriers in the neighboring semiconductor and thereby produces "space" charge. This interaction can be effectively utilized to implement real-time signal processing devices known as *acoustic surface wave convolvers*. A silicon-on-lithium niobate structure is the implementation most frequently used.

To understand the operation of such a device,

consider the situation in Fig. 5 where an RF signal[1] $f(t)e^{j\omega t}$ is applied to one input transducer to generate a traveling wave. At the other end of the device, the input applied is $g(t)^{j\omega t}$. These two waves, while traveling under the semiconductor, induce a propagating electric field and a space charge which can be represented at any point x and t inside the medium (to within a multiplicative factor) by

$$f\left(t - \frac{x}{v}\right)e^{j(\omega t - kx)} \quad \text{and} \quad g\left(t + \frac{x}{v}\right)e^{j(\omega t + kx)}$$

where k is the propagation constant of the wave and v is its velocity. If overlapping, these waves interact in a nonlinear manner and the current density inside the semiconductor consists of the fundamental and higher order harmonics of the arguments of the exponentials. The output is proportional to the integral of the current density with respect to x. Thus, by detecting only the second-harmonic term, the output is proportional to

$$\int_{-L/2}^{L/2} f\left(t - \frac{x}{v}\right) g\left(t + \frac{x}{v}\right) dx$$

$$= v \int_{t - L/2v}^{t + L/2v} f(\tau) g(2t - \tau) d\tau.$$

For time-limited signals, with $T = L/2v$, the above expression represents convolution, except for an output time compression factor of two; thus, Fig. 5 depicts a convolver. Using an Si-on-$LiNbO_3$ structure, convolvers have been built with a bandwidth up to 100 MHz, interaction time (ratio of the physical length of the device to the SAW velocity) up to 20 μs, and time-bandwidth product up to 1000. Fig. 6 shows a photograph of an actual device.

An important feature of these real time analog signal processors is that they act as programmable filters or correlators in the sense that one convolver can be used for many types of signals, provided they satisfy the requirements of bandwidth and time limitation. This, of course, follows since the convolver is a two-input device,

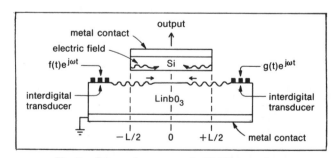

Fig. 5.   Space charge coupled SAW convolver.

[1] The use of complex notation is done for convenience. The actual signal is the real part of the complex signal.

Fig. 7. Chirp filter to obtain Fourier transform.

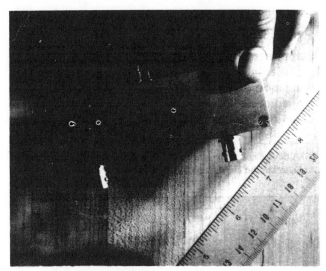

Fig. 6. Packaged experimental SAW convolver.

and hence to change the signal with which a received waveform is to be convolved (i.e., a reference waveform), one merely changes the reference input to the device. The ease of programmability is a great advantage over the transversal filters discussed earlier, as it is much more difficult to make these latter devices adaptive. On the other hand, the transversal filters do have an advantage over convolvers by being self-synchronizing; that is, the shape of the reference signal with which one wants to convolve the received waveform is built into the tap coefficients so that there is no need to establish precise timing information between the received signal and the reference.

FOURIER TRANSFORMATION USING SAW DEVICES

One application of SAW devices that is currently receiving a good deal of attention is the generation of real-time Fourier transforms using the so-called chirp-transform technique [9]. Using complex notation, if a signal $f(t)e^{j(\omega t - \Delta t^2)}$ (that is, a waveform $f(t)$ modulating a linear FM or chirp waveform) is convolved with the signal $e^{j(\omega t + \Delta t^2)}$, the result of that convolution will be the Fourier transform of $f(t)$ modulating a chirp carrier. That is, referring to Fig. 7

$$y(t) = e^{j(\omega_0 t - \Delta t^2)} \int f(\tau) e^{-j\omega\tau} d\tau$$

where $\omega = 2\Delta t$ (i.e., the radian frequency variable ω evolves linearly with time). Therefore, if these two waveforms are used as the two inputs to the SAW convolver described in the previous section, assuming $f(t)$ is time-limited to some value $T \leqslant A$, where A is the interaction time of the device, then the convolver output will be $F(\omega)$, the Fourier transform of $f(t)$. This transform will be accurate over the range $\omega > 2\Delta T$ and $\omega < 2\Delta A$ [10].

Alternately, rather than use the convolver, if one implements a tapped delay line as described in the previous section with an impulse response given by $e^{j(\omega_0 t - \Delta t^2)}$, one can obtain $F(\omega)$ as the output of the delay line when $f(t)e^{j(\omega t + \Delta t^2)}$ is the input.[2] In either case, the radian frequency variable ω will be a linear function of time, so that $F(\omega)$ will be generated at the device output in real time.

The above technique is sufficient only if the waveform is time-limited to a small enough interval. When the waveform is a sequence of contiguous pulses as is typical of most digital communication systems, some means of altering the procedure is necessary, and one such scheme is described in [12].

Basically, the technique consists of dividing the input alternately into two data streams, processing each data stream separately, and then combining the results at the end. Conceptually, it is a straightforward technique, but from an implementation point of view there are a variety of subtleties that have to be addressed involving such things as the accurate generation of constant envelope chirps in the two parallel branches and the minimization of crosstalk effects.

Finally, it must be realized that since one obtains a transform valid only over a finite range in frequency, one can only inverse transform over that range in frequency. In general, one obtains at the output of an inverse transformer the desired inverse convolved with a (sin x)/x type weighting function.

APPLICATIONS

The largest area of application of SAW devices to communications is in linear filtering. SAW band-pass filters, for example, are currently being used as television IF filters [13], and the use of SAW devices as matched filters (or correlators) for detection and/or synchronization in spread spectrum systems is becoming increasingly popular [10],[14],[15], as is their use as code generators (e.g., Barker codes, PN sequences, etc.).

Spread spectrum communications is currently the area providing the greatest impetus for further advances in SAW technology. Specifically, the use of a tapped delay line as a programmable matched filter for a direct sequence spread spectrum system is becoming increas-

---

[2]A device known as a reflective array compressor is at times used to generate the chirp filter [11].

ingly common. Fig. 8 shows a 64-tap programmable delay line (with associated electronics) built by the Hazeltine Corporation. It is designed to operate with either a phase-shifted keyed (PSK) signal or a minimum-shift keyed (MSK) signal [16]. Some of its specifications are tabulated in Table I. It is of interest to note that a SAW filter can be used to generate an MSK waveform from a PSK signal [17].

As mentioned earlier, an alternate way to implement a programmable matched filter is to use a SAW convolver. A detailed description of the application of such a device to spread spectrum communications can be found in [14] and [18].

The use of such devices has found its way to as seemingly distant an area as packet switching in a multiple access packet radio system [19]. As described fully in [19], a SAW convolver has been used to implement a fixed matched filter as part of what was termed an Experimental Packet Radio Unit, and then a programmable matched filter was used in what was termed an Upgraded Packet Radio Unit. This latter more sophisticated unit was designed to allow the packet radio the benefit of using a jam-resistant waveform, and it was the program-

mability of the SAW convolver that made the system feasible.

As a final example, the use of SAW devices as Fourier transformers has led to research on the desirability of so-called "transform domain filtering" [10],[12],[20]. Briefly, this type of system, shown in Fig. 9(a), consists of a Fourier transformer, a multiplier, an inverse Fourier transformer, and a matched filter. In essence, the filtering by the transfer function $H(\omega)$ is done by multiplication followed by inverse transformation rather than by convolution. This multiplication, while ostensibly being performed in the "frequency domain," is of course accomplished by the SAW device in real time. Alternately, the receiver may be implemented as shown in Fig. 9(b), wherein the matched filtering is performed by inversely transforming the product of the transforms of the filtered input waveform and the impulse response of the matched filter.

To illustrate this last technique, Fig. 10 shows results of narrow-band interference removal when $s(t)$ is a 13-bit binary sequence. The interference is a sine wave, and it is filtered out by multiplication in the frequency domain by a rectangular pulse (i.e., an ideal bandpass filter). Note that once the SAW implemented Fourier transform blocks are available, the implementation of ideal "brickwall" filters (within the bandwidth of the system) in the transform domain is completely trivial. This is in marked contrast to the classical situation where these filters are physically unrealizable.

It can be seen from the figure that the interference has been effectively eliminated. The distortion, seen in the final trace, is due in large part to the bandwidth of the final video filter. As an incidental result, if traces 1 and 3 are compared with each other (also traces 4 and 6), one can see the fidelity with which the Fourier transform can be taken.

Fig. 8. Programmable SAW delay line. *Photograph courtesy of Hazeltine Corporation.*

TABLE I
Summary of Hazeltine Programmable Matched Filter Specifications

Number of taps	64
Tap spacing/code rate	0.2 μs/5 Mbit/second
Center frequency	75 MHz
MSK spectral bandwidth	3.1 MHz
Sidelobe level	Within 2 dB of theoretical
Processing gain	Within 1 dB of theoretical
Insertion loss	45 dB
Noise output	−105 dBm
Programming time	8 μs

SAW devices are currently being used as television IF filters.

One very important advantage in the use of SAW devices for signal processing is the possibility of fabri-

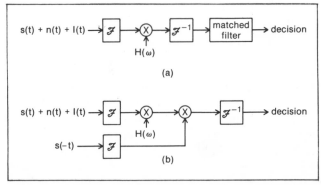

Fig. 9.   Transform domain filtering.

cating entire receivers on a single substrate. Fig. 11 shows a hypothetical SAW implementation of a spread spectrum system described in [21] wherein almost all the components of the system, including the filters, PN code generator, and phase-locked loop are implemented with SAW devices. Notice that both types of SAW devices are utilized in this system. The matched filter is a passive SAW tapped-delay line and this is used to initially acquire the correct phase position of the incoming PN sequence. The correlation on the other hand, implemented with a SAW convolver, is used to coherently remove the PN code from the received waveform once the correct phase position has been identified. Furthermore, because most SAW devices tend to have large dynamic

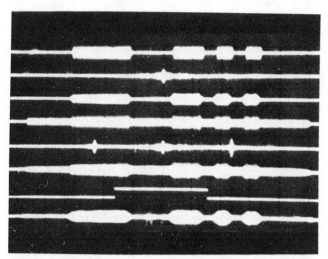

Fig. 10.   Filtering of binary Code Signal (hor. scale—5 μs/div.). Trace 1—binary Code input; 2—Fourier transform of 1; 3—Inverse Fourier transform of 2; 4—binary code plus interference; 5—Fourier transform of 4; 6—Inverse transform of 5; 7—$H(\omega)$-gating signal; 8—Filtered signal.

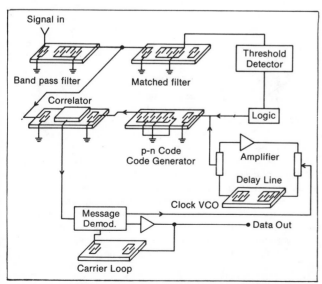

Fig. 11.   SAW Implementation of a spread spectrum system.

ranges, it appears at this point in time that the dynamic range of many of the systems alluded to in this paper will ultimately be limited by other system components such as mixers.

Finally, it should be pointed out that, in principle, any real-time analog convolver or any analog tapped-delay line could be used to perform the same types of signal processing functions. However, in most cases, SAW devices have distinct advantages in terms of compactness and ease of fabrication. For example, due to the very high velocity of electromagnetic waves, a huge length of RF cable would be required to achieve a delay comparable to what a SAW device can achieve.

COMPARISON OF SAW, CCD, AND MICROPROCESSORS

For use in wide bandwidth communication systems, SAW devices have two primary competitors, namely charge-coupled devices (CCD's) [22],[23] and microprocessors. CCD's have an inherent advantage over SAW devices. Being one branch of so-called silicon technology, the CCD can be made to interface much more readily with the final baseband signal processing stages necessary in any digital communication system. However, SAW's and CCD's are not always competitive. As pointed out in [23]-[25], a natural interface exists between the two which allows them to complement one another in various applications (e.g., Doppler radar). For bandwidths in excess of 20 MHz, SAW technology is usually simpler and/or cheaper than either CCD or digital electronics, and for bandwidths in excess of 50 MHz, SAW technology is probably the only realistic choice.

More specifically, digital electronics is typically the best choice at bandwidths less than about 100 kHz, while CCD's are competitive in the range 1 kHz to 20 MHz. As an example, consider the 50 MHz bandwidth 20 μs interaction time SAW convolver shown in Fig. 6. To implement a similar correlation receiver using a microprocessor which employs an FFT algorithm, it would be necessary to perform somewhere in the vicinity of 4×10^4 operations in 20 μs thus requiring approximately 0.5 ns per operation. This is probably three to four orders of magnitude faster than the rate at which microprocessors currently operate.

CONCLUSION

This paper has presented a brief summary of the operation and applications of SAW devices. While the devices emphasized here are among the most common implementations, other devices like the reflective array compressor [11] and the multistrip coupler [26] have become increasingly popular. It would appear that the demand for these devices as signal processors can only increase. Currently, devices with time-bandwidth products of the order of 1000 (corresponding to, say, a bandwidth of 50 MHz and an interaction time of 20 μs) have been achieved using the SAW convolver, and time-bandwidth products in the vicinity of 10 000 are achievable using the tapped delay-line implementation. In the future, it appears the prime use for the device will be in the area of spread spectrum communications. Indeed, as indicated in the previous section, even at this point in time SAW matched filters are competing with the more established technology of digital matched filters using microprocessor technology.

ACKNOWLEDGMENT

The authors would like to thank the Hazeltine Corporation for supplying the photograph and description of the programmable convolver of Fig. 8, and would like to thank the Anderson Laboratory for loaning some of the devices used to make the measurements. Finally, the authors would like to acknowledge the many helpful discussions with D. R. Arsenault and R. T. Webster.

REFERENCES

[1] The following six symposium proceedings and three special issues of journals have enormous papers concerning SAW technology and its applications: (a) *1972 Ultrasonics Symp. Proc.*; (b) *1973 Ultrasonics Symp. Proc.*; (c) *1974 Ultrasonics Symp. Proc.*; (d) *1975 Ultrasonics Symp. Proc.*; (e) *1976 Ultrasonics Symp. Proc.*; (f) *IEEE Trans. Microwave Theory Tech.* (Special Issue on Microwave Acoustics), vol. MTT-17, Nov. 1969; (g) *IEEE Trans. Sonics Ultrason.* (Special Issue on Microwave Acoustic Signal Processing) vol. SU-20, Apr. 1973; (h) *Proc. IEEE* (Special Issue on Surface Acoustic Wave Devices and Applications), vol. 64, May 1976; (i) *Proc. Symp. Optical and Acoustical Microelectronics*, vol. 23, Brooklyn, NY: Polytechnic, 1974.
[2] H. Matthews, *Surface Wave Filters.* New York: Wiley, 1977, ch. 3.
[3] R. H. Tancrell, "Analytic design of surface wave bandpass filters," *IEEE Trans. Sonics Ultrason.*, vol. SU-21, pp. 12-22, Jan. 1974.
[4] R. M. Hays and C. S. Hartman, "Surface-acoustic wave devices for communications," *Proc. IEEE*, vol. 64, pp. 652-671, May 1976.
[5] M. F. Lewis, "Some aspects of SAW oscillators," in *Proc., 1975 Ultrasonics Symp.*, pp. 334-347.
[6] E. J. Staples and R. C. Smythe, "Surface acoustic wave resonators," in *Proc., 1973 Ultrasonics Symp.*, pp. 307-310.
[7] G. S. Kino, "Acoustoelectric interactions in acoustic-surface-wave devices,"*Proc. IEEE*, vol. 60, pp. 724-748, May 1978.
[8] W. C. Wang and P. Das, "Surface wave convolver via space-charge nonlinearity," in *Proc. IEEE Ultrasonics Symp.*, Oct. 1972, p. 310.
[9] J. D. Maines and E. G. S. Paige, "Surface acoustic wave devices for communications," *Proc. IEEE*, vol. 64, pp. 639-652, May 1976.
[10] L. B. Milstein and P. Das, "Spread spectrum receiver using surface acoustic wave technology," *IEEE Trans. Commun.*, vol. COM-25, pp. 841-847, Aug. 1977.
[11] R. C. Williamson, "Properties and applications of reflective-array devices," *Proc. IEEE*, vol. 64, pp. 702-710, May 1976.
[12] L. B. Milstein, D. R. Arsenault, and P. Das, "Transform domain processing for digital communication systems using surface acoustic wave devices," in *Proc. AGARD Symp. on Digital Communications in Avionics*, June 1978.
[13] A. J. DeVries and R. Adler, "Case history of a surface-wave TV IF filter for color television receivers," *Proc. IEEE*, vol. 64, pp. 769-771, May 1976.

[14] J. H. Cafarella, W. M. Brown, Jr., E. Stern, and J. A. Alusow, "Acoustoelectric convolvers for programmable matched filtering in spread spectrum systems," *Proc. IEEE*, vol. 64, p. 756, May 1976.
[15] A. Hamptschein, "Practical high performance concatenated coded spread spectrum channel for JTIDS," *Conf. Rec.*, *NTC'77*, vol. 3, pp. 35:4 - 1-35:4-8, Dec. 1977.
[16] H. R. Mathwich, J. F. Balcewicz, and M. Hecht, "The effect of tandem band and amplitude limiting on the E_b/N_o performance of minimum (frequency) shift keying (MSK)," *IEEE Trans. Commun.*, vol. COM-22, pp. 1525-1540, Oct. 1974.
[17] W. R. Smith, "SAW filters for CPSM spread spectrum communication," in *Proc. 1977 Ultrasonics Symp.*, p. 524.
[18] J. Cafarella, J. Allison, W. Brown, and E. Stern, "Programmable matched filtering with acoustic-electric convolvers in spread spectrum systems," in *Proc. 1975 Ultrasonics Symp.*, pp. 205-208.
[19] R. E. Kahn, S. A. Gronemeyer, J. Burchfiel, and R. C. Kunzelman, "Advances in packet radio technology," *Proc. IEEE*, vol. 66, pp. 1468-1496, Nov. 1978.
[20] C. Atzeni, G. Manes, and L. Masatti, "Programmable serial processing by analog chirp-transformation using SAW devices," in *Proc. 1975 Ultrasonics Symp.*, pp. 371-376.
[21] C. R. Cahn, "Spread spectrum applications and state-of-the-art equipment," AGARD Lecture Series 58, NTIS #AD766-914, July 1973.
[22] C. H. Sequin and M. F. Tompsett, *Charge Transfer Devices.* New York, Academic, 1975, ch. 4.
[23] A. Gersho, "Charge transfer filtering," *Proc. IEEE*, vol. 67, pp. 196-218, Feb. 1979.
[24] J. B. G. Roberts, R. Eames, D. V. McCaughan, and R. F. Simons, "A processor for pulse-Doppler radar," *IEEE J. Solid-State Circuits*, vol. SC-11, pp. 100-104, Feb. 1976.
[25] D. L. Smythe, R. W. Ralston, B. E. Burke, and E. Stern, "An acoustoelectric SAW/CCD device," in *Proc. 1978 Ultrasonics Symp.*, Sept. 1978, pp. 16-19.
[26] F. G. Marshall, C. O. Newton, and E. G. S. Paige, "Surface acoustic wave multistrip components and their applications," *IEEE Trans. Microwave Theory Tech.*, vol. MTT-21, pp. 216-224, Apr. 1973.
[27] J. B. G. Roberts, "The roles for CCD and SAW in signal processing," in *Proc. AGARD Conf. on Impact of CCD and SAW on Signal Processing and Imagery in Advanced Systems*, no. 230, Oct. 1977.

Laurence B. Milstein was born in Brooklyn, NY, on October 28, 1942. He received the B.E.E. degree from the City College of New York in 1964 and the M.S. and Ph.D. degrees, both in electrical engineering, from the Polytechnic Institute of Brooklyn in 1966 and 1968, respectively.

From 1968 to 1974 he was employed by the Space and Communications Group of Hughes Aircraft Company and from 1974 to 1976 he was with the Department of Electrical and Systems Engineering at Rensselaer Polytechnic Institute. Since 1976 he has been with the Department of Electrical Engineering and Computer Sciences at the University of California, San Diego, working in the area of digital communication theory with special emphasis on spread spectrum communication systems, and he has been a consultant to industry in the areas of radar and communications.

Dr. Milstein is an Associate Editor for Communication Theory for the IEEE TRANSACTIONS ON COMMUNICATIONS, a member of the Communication Theory Technical Committee of ComSoc, a member of Eta Kappa Nu, Tau Beta Pi, and Sigma Xi, and a Senior Member of the IEEE.

Pankaj K. Das was born in Calcutta, India, on June 15, 1937. He received the B.Sc. (Hons.) degree in physics and the M.Sc. (Tech.) and Ph.D. degrees in radiophysics and electronics from Calcutta University in 1957, 1960, and 1964, respectively.

From 1964 to 1965 he was an Instructor and from 1965 to 1968 an Assistant Professor in the Electrical Engineering Department, Brooklyn Polytechnic Institute. He was an Associate Professor of Electrical Engineering at the University of Rochester from 1968 to 1974. In 1973, he was an Organization of American States Visiting Professor in Mexico City on leave of absence from Rochester. Returning from Mexico, he joined Rensselaer Polytechnic Institute, Troy, NY, as an Associate Professor in the Electrical and Systems Engineering Department. He is now a Full Professor and also an active faculty member of the Center for Biomedical Engineering. He has a wide interest in the general area of electron devices. His research includes surface acoustic wave devices, integrated optics, hot electron microwave devices, and the application of ultrasound in bioengineering and nondestructive testing.

Dr. Das is a Member of the IEEE.

Section 6

Secure Communications

EDITORS' COMMENTS

In addition to all the "natural" causes of degradation in the performance of a communication system, such as intersymbol interference, multipath, imperfect system components, etc., many communication systems are subject to intentional interference and/or tampering in the form of jamming, message falsification, eavesdropping, and so on. It is these latter concerns that have spawned the great interest in the two topics to be addressed in this section, namely cryptography and spread spectrum communications.

The uses of cryptography are typically the authentication of the source of a given message, the authentication of data itself in the message, and the guaranteeing of the secrecy of the message. While the general ideas involved in the use of cryptography have been known and applied by various government agencies for some time, there is a growing general interest in cryptography by a much broader cross-section of users as a result of the prevalence of computer-to-computer communication for such vital things as banking and credit verification.

The National Bureau of Standards has come out with a data encryption standard (DES) which attempts to ensure a certain degree of security in any crypto-system that meets the standard. The DES, as well as several other very important concepts in cryptography such as public key cryptography, are discussed in the five papers devoted to cryptography found in this section.

The first of these five papers is by C. M. Campbell and is entitled "Design and Specification of Cryptographic Capabilities." It provides a general overview of the uses and techniques of cryptographic systems, as well as a brief introduction to the DES. The next two papers, "The Use of Public Key Cryptography in Communication System Design," by L. M. Adleman and R. L. Rivest, and "An Overview of Public Key Cryptography," by M. E. Hellman, concentrate on the philosophy behind, and some of the details of, public key cryptography. These papers not only describe public key cryptosystems, but also provide examples of their use in communication systems. Finally, the last two papers are more specialized in their use of cryptographic systems. The first paper, "Security of Computer Communication," by D. K. Branstad, in addition to providing a short overview of cryptography, emphasizes its use in a computer communication network. The second paper, "An Approach to

Secure Voice Communication Based on the Data Encryption Standard," by M. J. Orceyre and R. M. Heller, provides yet another summary of the DES as well as a detailed description as to how a crypto-system might be combined with a digital voice link to provide the desired security in the voice transmission.

The three remaining papers in this section are devoted to spread spectrum communications. Spread spectrum systems have long been used by the military to provide an antijam capability to their communication links. Other uses of spread spectrum systems include accurate navigation and ranging, low-detectability of signal transmission, and multipath rejection. The use of code division multiple-access is another area of increased interest in recent years.

All of these applications make use of the basic nature of the spread spectrum system, which is spreading the bandwidth of the data over a frequency range much larger than is necessary to transmit the information itself, and then despreading the received waveform at the receiver. If any uncorrelated interference is present in the received signal, the despreading operation, while collapsing the bandwidth of the desired signal, actually spreads the bandwidth of the interference, and hence allows final low-pass filtering to remove most of the power in the interfering signal. There are a variety of different techniques that are used to accomplish the spreading operation, with the two most common ones being direct sequence (DS) modulation and frequency hopping (FH). In DS systems, the underlying data has superimposed upon it a coded signal with a much larger bit rate (hence a much larger bandwidth). This coded signal is often a so-called pseudonoise or PN sequence. Its descriptive name "pseudonoise" arises from the fact that it has many properties which make it appear to be a truly random binary sequence.

The technique of frequency hopping is accomplished by varying the carrier frequency of the information bearing waveform in a pseudo-random manner. The hopping can be done at a rate of once per information bit or many times per information bit. In the latter case, it is typically referred to as fast frequency hopping. It can also be done a rate smaller than the information rate, in which case it is referred to as slow frequency hopping.

In addition to DS and FH, there are other techniques such

as time hopping. Also, hybrid techniques combining two or more of the above spreading schemes are often used.

Of the three spread spectrum papers presented in this section, the first two are overview papers. The tutorial by R. C. Dixon, entitled "Why Spread Spectrum?" introduces the reader to both direct sequence systems and frequency-hopped systems, and explains what is meant by such ideas as processing gain and jamming margin. The paper by W. F. Utlaut, entitled "Spread Spectrum Principles and Possible Application to Spectrum Utilization and Allocation," introduces the reader to some of the basic properties of pseudonoise sequences, which are the most well-known of the various waveforms used to spread the spectrum. It also provides an overview of direct sequence and frequency hopped systems, and concludes by describing several applications of spread spectrum communications.

The final paper in the section is by A. J. Viterbi and entitled "Spread Spectrum Communications—Myths and Realities." This paper is somewhat more advanced than the other two articles on spread spectrum. It illustrates that when spreading techniques are combined with error correction techniques, the performance of the system under consideration is invariably improved.

DESIGN AND SPECIFICATION OF CRYPTOGRAPHIC CAPABILITIES *

Carl M. Campbell

A variety of techniques for communications security are based on the use of a block encryption algorithm such as the DES.

Abstract Cryptography can be used to provide data secrecy, data authentication, and originator authentication. Non-reversible transformation techniques provide only the last. Cryptographic check digits provide both data and originator authentication, but no secrecy. Data secrecy, with or without data authentication, is provided by block encryption or data stream encryption techniques. Total systems security may be provided on a link-by-link, node-by-node, or end-to-end basis, depending upon the nature of the application.

I. INTRODUCTION

Up to the present, cryptography has been a relatively unknown science, used primarily to secure sensitive governmental communications. However, with the introduction of the Data Encryption Standard (DES) we expect to see cryptography widely applied in data processing systems, especially in digital communications, to provide data security. It is thus essential that the designers of these systems gain an understanding of this new technology.

This paper was originally published in *Computer Science & Technology: Computer Security and the Data Encryption Standard, Proc. Conf. on Computer Security and the Data Encryption Standard*, National Bureau of Standards, Gaithersburg, MD, Feb. 15, 1977.

The author is a Consultant to the Interbank Card Association. He is at 809 Malin Road, Newton Square, PA 19073.

*Reprinted from *IEEE Communications Society Magazine*, November 1978, Vol. 16, No. 6, pp. 15–19.

II. USES OF CRYPTOGRAPHY

Cryptography can be used to provide three aspects of data security:

1) Data secrecy.
2) Data authentication.
3) Originator authentication.

The first use of cryptography, data secrecy, is relatively well understood, and will be an important use in an EDP environment.

Data authentication and originator authentication are less understood, but will be very important uses of cryptography in the future. To understand data authentication, assume that A is transmitting data to B. B wants assurance that the data it is receiving is precisely the data which A transmitted. Though conventional error control techniques can protect against communications errors, B is concerned that someone with a sophisticated "active wiretapping capability may have deliberately modified the data from A, and made the appropriate modifications in any associated error control fields. Cryptographically implemented data authentication provides assurance that the data was received as originated.

Originator authentication is similar to data authentication. This time *B* requires assurance that it is receiving data from the real *A* and not from an impostor who may have assumed *A*'s identity. Again, cryptography can provide the solution.

There are an almost unlimited number of ways in which cryptography can be applied. Some applications meet only one or two of the above objectives, and some meet them all.

III. ORIGINATOR AUTHENTICATION

A simple use of cryptography meets only the third objective, originator authentication. In this approach, Fig. 1, each authorized user of a system is given a secret **authorization code**. Each terminal incorporates a cryptographic capability into which he enters this code. The code is **nonreversibly transformed** into another code. This means that, given the transformed code, there is no way to determine the actual code except for an exhaustive trial and error procedure, which is presumed to be nonfeasible if the original code is quite long (approximately 56 bits) and reasonably random. The system's central processor stores, in a manner which may be nonsecure, each user's transformed code. A simple comparison is thus sufficient to authenticate the user.

Note that this approach does not require a unique terminal key, so imposes no "key management" requirements. Note also that it does not require any on-line cryptographic capability at the central facility.

IV. DATA AUTHENTICATION

A very useful cryptographic technique, **cryptographic check digits**, provides data authentication and can provide originator authentication, but provides no data secrecy. Cryptographic check digits may be likened to parity check digits or to a cyclic redundancy check in that a check field is added to the message by the originator and verified by the recipient. However, unlike a conventional error-control check field, the cryptographic check digit field is generated by a cryptographic algorithm and utilizes a secret key known (desirably) by originator and recipient alone. Thus the field protects not only against accidental garbles, but also against deliberate attempts to

modify the transmitted data. Without knowing the secret key, the one attempting such data modification would be unable to make the appropriate changes in the cryptographic check digits field which would be required for his modification to escape detection.

Note that originator authentication is provided if the recipient is certain that only the authorized originator possesses the secret key.

The DES may be used to generate cryptographic check digits, as, for example, is illustrated in Fig. 2. Each group of 64 message bits is passed through the algorithm after being combined with the output of the previous pass. The final DES output is thus a residue which is a cryptographic function of the entire message. All or part of this residue may be used as the cryptographic check digits.

Cryptographic check digits alone cannot detect the fraudulent replay of a previously valid message, nor the deletion of a message. To protect against these threats, each transmission of a message must be made unique. One technique is to insert a cryptographically-protected sequence number into the message. Another is to use a different key for each message.

V. DATA SECRECY

Secrecy of transmitted data may be provided by a number of techniques, some providing data authentication and some not. All of the suggested techniques utilize a secret key, and so provide originator authentication if this key is properly controlled.

A. Block Encryption

The Data Encryption Standard is inherently a block encryption algorithm, requiring blocks of precisely 64 bits. Given a plain-text block of 64 bits, a secret key, and the "encrypt" command, the DES algorithm produces 64 cipher bits. Given these 64 cipher bits, the same key, and the "decrypt" command, the algorithm produces the original 64 plain-text bits. Thus, as long as the block size is exactly 64 bits, block encryption with DES is extremely simple.

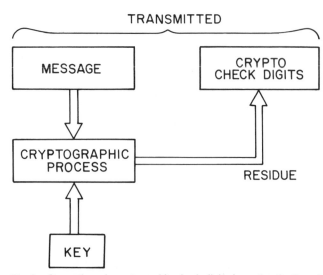

Fig. 2.   Generation of cryptographic check digits for authentication of originator data.

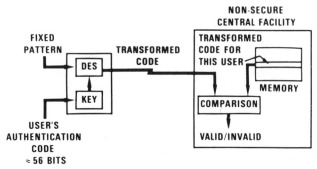

Fig. 1.   Non-reversible transformation for user authentication.

Short blocks: If the block size is less than 64 bits, these bits must be "padded" (with any fixed or, preferably, variable pattern) to make 64 bits if the algorithm is to be used in its normal block-encryption manner. All 64 of the resulting cipher bits must be transmitted to the recipient even though, for example, only 20 bits of underlying information are present. The recipient block-decrypts these 64 bits, resulting in 64 plain-text bits. All but 20 of these must be discarded, leaving the 20 original information bits.

The use of DES for a block size of less than 64 bits is thus somewhat inefficient, in that the full 64 bits must still be transmitted. Different techniques for using DES are possible, which overcome this disadvantage, but they introduce other disadvantages.

Multi-blocks: Where the block to be encrypted is long, it can be broken up into groups of 64 bit blocks, and each such block encrypted independently. This simple

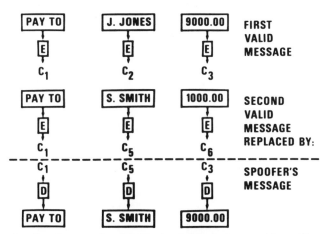

Spoofing. Known pairs of plaintext and ciphertext are used to modify enciphered message.

approach provides secrecy, but it does not provide a high degree of data authentication. For example, assume two block-encrypted messages, one reading: "PAY TO J. JONES $9,000.00" and the second: "PAY TO S. SMITH $1,000.00." If the "$9,000.00" and the "$1,000.00" should each fall precisely within a block, it would be possible to replace the cipher block for "$1,000.00" with that for "$9,000.00" so that when the recipient decrypts the second message it reads: "PAY TO S. SMITH $9,000.00."

This process, by which cipher is manipulated, is called "*spoofing.*" Note that the "spoofer" knows corresponding cipher and plain text, but does not know the secret key. His objective is to intercept, modify, and then retransmit the cipher, all in such a manner that his deception is not detected.

Encryption techniques can be devised which prevent "spoofing," but in order to do so it is necessary to introduce something called **garble extension**. This means that if any portion of the cipher becomes garbled (i.e., changed) the decryption by the recipient of a certain amount of subsequent cipher is also garbled.

Fig. 3 illustrates one method by which garble extension, and hence spoofing prevention, can be incorporated into a

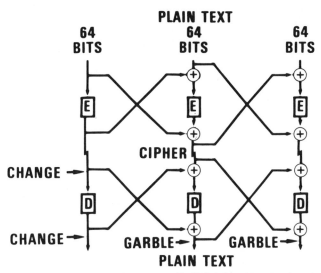

Fig. 3. Block interconnections to provide "infinite" garble extension. *E* and *D* denote respectively the block encryption and decryption operations.

block encryption system. The "*E*" boxes perform block encryption, and the "*D*" boxes block decryption. The "+" function indicates exclusive-or. The approach of Fig. 3 provides "infinite" garble extension. That is, any change to the cipher garbles the decryption of all subsequent cipher. Infinite garble extension has the features that the originator can place in the final block a pattern expected by the recipient. If the recipient finds the expected pattern at the end of the message, he is assured that the entire message, regardless of length, was received precisely as originated.

B. Data-Stream Encryption

The term **data-stream** refers to the serial flow (serial by bit, by character, or any other increment) of data, as over a communications line. **Data-stream encryption** refers to the encryption of such data in real-time, for subsequent "data-stream decryption," also in real-time. It is possible to use block encryption for data-stream encryption, but this is not desirable. In DES block encryption, the first bit cannot be encrypted until the 64th bit has been received, so that a block-encryption technique in a data-stream environment inherently imposes a delay of 64 bit times. Block decryption imposes an equal delay. Thus, communications delays would be unacceptably increased were block techniques to be used.

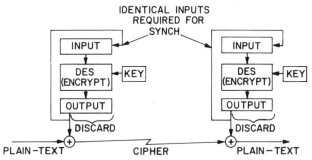

Fig. 4. Internal feedback configuration—The DES generates a stream of pseudorandom "encrypting" bits.

Fortunately, DES can be applied to a data-stream environment so as to minimally impact communications delays. Two such techniques are **internal feedback** and **cipher feedback**.

Internal Feedback: The internal-feedback approach to data-stream encryption uses DES to generate a stream of pseudorandom **encrypting bits**. These bits are exclusive-ored with the plain-text bits to form the cipher bits, as illustrated in Fig. 4. The decryption process operates the same way, with the exact same pseudorandom stream of encrypting bits being generated. Exclusive-oring these bits with the cipher bits then produces the original plain-text bits.

To use DES in this manner, any number of the 64 output (i.e., cipher) bits may be used. For simplicity of explanation, it is assumed that only 1 bit is used, and the other 63 discarded. The selected bit is not only used to encrypt the plain-text data, but is also fed back as the input to DES, and another algorithm cycle initiated. Thus, one algorithm cycle is required per encrypting bit.

To ensure that the decryption process generates the same pseudorandom encrypting bits as does the encryption process, the DES input registers of the two devices must commence operation with the same **initial fill**. The process by which this is accomplished is called **crypto synchronization.**

Cipher Feedback: This approach to data-stream encryption is very similar to the internal feedback approach, the difference being that cipher bits, rather than "encrypting bits," are used as the DES input. Note that this approach, Fig. 5, if used in a 1-bit feedback mode, is **self synchronizing** because after 64 bit times, the DES input register of the decryption device will contain the same data as does the input register of the encryption device. Note also that the approach provides garble extension, thus providing anti-spoofing protection.

VI. SYSTEM PHILOSOPHIES

There are three basic approaches to incorporating encryption into a communications system: link-by-link, node-by-node, and end-to-end encryption.

Link-by-link encryption: Fig. 6, is the technique most commonly used today. It may be implemented in a transparent manner with currently available devices, which are placed in series with the circuit between data terminal equipment and data communications equipment. This approach has the disadvantage that it allows all traffic to pass through the CPU of any node in plain-text.

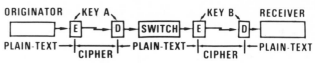
Fig. 6.   Link-by-link encryption.

Node-by-node encryption: Fig. 7, is a modified version of link-by-link encryption to overcome this disadvantage. Each link uses a unique key, but the "translation" from one key to the next occurs within a single "security module" which might serve as a peripheral device to the node's CPU. In this way plain-text data does not traverse the node, but exists only within this physically secure module. Note that enough message data must remain encrypted so that the node's CPU can properly route the message.

End-to-end encryption: Fig. 8, requires a **Key Control Center,** located somewhere within the communication system. Each end-point in the system holds a unique **long-term** key, and this center alone holds a copy of each such key. When one end-point wishes to communicate to another, a request to this effect is sent to the Key Control Center. This center then generates a temporary **per conversation** key, encrypts this in the long-term key of the originator, and also in the long-term key of the recipient, and sends the appropriate version to each. The originator decrypts this just-received encrypted temporary key using its long-term key, the recipient does likewise with its long-term key, and the two parties then converse with end-to-end encryption using this temporary key.

VII. PROCUREMENT CONSIDERATIONS

For retrofitting an existing system, link-by-link encryption utilizing transparent link encryption devices is a reasonable approach. DES feedback is a desirable choice for these devices.

For a new system, in which cryptography can be "designed in" rather than "added on," block-encryption techniques should be considered because of their more efficient use of the algorithm, and their absence of initial synchronization requirements. For a transaction-

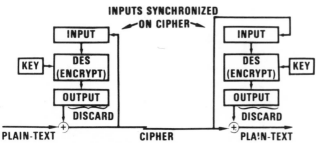
Fig. 5.   Cipher feedback configuration. Cipher bits are used to feed the DES input register.

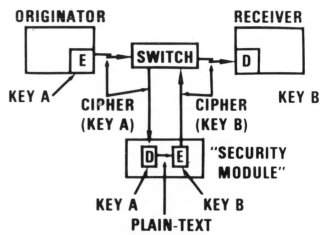
Fig. 7.   Node-by-node encryption.

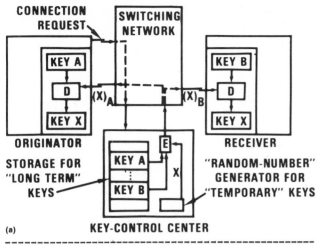

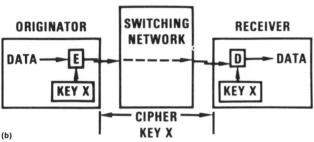

Fig. 8. End-to-end encryption. (a) Connection set-up. (b) Data transfer.

oriented system, in which messages are very short and routed to varying destinations, the node-by-node approach appears preferable because it does not impose any per-conversation overhead for key distribution. However, for a "session"-oriented environment in which conversations may be relatively long, end-to-end encryption appears to be the obvious choice.

REFERENCES

[1] D. K. Branstad, "Encryption protection in computer data communications systems," presented at the *4th Data Communications Symp.*, Quebec, Canada, Oct. 7–9, 1975.
[2] S. T. Kent, "Encryption-based protection protocols for interactive user-computer communications," Lab. Comput. Sci. Mass. Inst. Technol., Cambridge, May, 1976 Tech. Rep. 162.
[3] D. J. Sykes, "Protecting data by encryption," *Datamation Magazine*, Aug. 1976.

Carl M. Campbell, Jr. received the B.S. and M.S. degrees from Princeton University, Princeton, NJ.

He has been a Consultant since 1976 in the applications of cryptography and is assisting Interbank Card Association in the development of security techniques for electronic funds transfer systems. For the preceding 20 years he was with the Burroughs Corporation, specializing in cryptography and communications security, as well as mini- and micro-computer software development.

THE USE OF PUBLIC KEY CRYPTOGRAPHY IN COMMUNICATION SYSTEM DESIGN *

Leonard M. Adleman
and Ronald L. Rivest

A public key cryptosystem can be synergistically combined with a traditional system to obtain the best features of both approaches.

Abstract   Since the time of Caesar, cryptography has been used in the design of secure communications systems. Recently, Diffie and Hellman [2] have introduced a new type of cryptographic method, based on "trapdoor" functions, which promises to be of great value in the design of such systems. We present a review of public key cryptosystems, followed by examples of communications systems which make particularly elegant use of their properties.

I. A REVIEW OF PUBLIC KEY CRYPTOSYSTEMS

Public key cryptosystems were introduced by Diffie and Hellman in [2] where the interested reader will find a complete, easily readable exposition. The reader already familiar with such systems may prefer to skip to the examples in Section III.

In a traditional (nonpublic key) cryptosystem there is a general encryption procedure E into which a key K and a message M may be put in order to produce a cipher text C: formally $E(K,M) = C$. There is also a general decryption procedure D into which a key K and a cipher text C may be put in order to produce a message M: $D(K,C) = M$. Any such system has the following properties.

1) Decryption reverses encryption:

$$D(K,E(K,M)) = M.$$

2) It is "impractical" to decrypt without the appropriate key.

In a traditional system, if party A wishes to communicate with party B over a tapped line, the following steps are taken:

The research of L. M. Adleman was sponsored in part by NSF Grant MCS78-04343 and in part by the Office of Naval Research Grant N00014-67-0204-0063. The research of R. L. Rivest was sponsored in part by NSF Grant MCS76-14294. This paper was prepared for presentation at the National Telecommunications Conference (NTC'78), Birmingham, AL, December 4–6, 1978, and will appear in the *NTC'78 Conference Record*.

L. M. Adleman is with the Department of Mathematics, Massachusetts Institute of Technology, Cambridge, MA 02139.

R. L. Rivest is with the Department of Electrical Engineering and Computer Science, Massachusetts Institute of Technology, Cambridge, MA 02139.

*Reprinted from *IEEE Communications Society Magazine*, November 1978, Vol. 16, No. 6, pp. 20–23.

1) A and B communicate a key K which is unknown to all others (this may be accomplished via secure courier for example).

2) A encrypts his message M using E and K and sends the resulting ciphertext $C = E(M,K)$ to B.

3) B uses D and K to regain the message $M = D(K,C)$.

Clearly, to maintain the security of such a system, the key K must be kept secret.

A public key cryptosystem differs only slightly in overall plan from a traditional system. There is a general encryption procedure E into which a key K and a message M may be put to produce a cipher text $C = E(K,M)$. There is also a general decryption procedure D which takes cipher text and keys and produces messages. However, unlike the traditional system in which the same key is used to encrypt and decrypt, in a public key system each key K used to encrypt has a mate $K' \neq K$ which is used to decrypt. From now on we will denote the key used to encrypt by K_E and the one used to decrypt by K_D. As with a traditional system we want the following properties:

a) Decryption reverses encryption:

$$D(K_D, E(K_E, M)) = M.$$

b) It is "impractical" to decrypt without the appropriate key. (In particular K_E cannot be used to decrypt:

$$D(K_E, E(K_E, M)) \neq M.)$$

In addition, we will require the following properties:

c) It is practical (easy on a computer) to generate mated pairs $<K_E, K_D>$.

d) It is "impractical" to obtain K_D from K_E.

It is property d) which is the source of a public key cryptosystem's somewhat paradoxical properties.

If A wishes to communicate with B over a tapped line using a public key cryptosystem, the following steps are taken:

1) B generates an encryption key K_E and its decryption key mate K_D. K_D is kept by B and remains unknown to all others including A.

2) B sends K_E to A (the tapped line will suffice here since K_E need not be kept secret).

3) A encrypts his message M using E and K_E and sends the resulting ciphertext $C = E(K_E, M)$ to B.

4) B uses K_D and D to decrypt C and obtain $M = D(K_D, C)$.

Notice that the security of this system does not depend on keeping K_E secret. Even if K_E is publicly revealed, the security of the system is not endangered, since we have demanded that decrypting with K_E will not work $(D(K_E, E(K_E, M) = M)$, and it is "impractical" to obtain K_D from K_E. It is for this reason that the term "public key" is used.

For subtler applications (e.g., signatures [2],[6]) public key cryptosystems with additional properties are needed. In particular

e) Encryption reverses decryption

$$E(K_E, D(K_D, M)) = M.$$

f) It is "impractical" to encrypt without the appropriate key. (In particular K_D cannot be used to encrypt: $D(K_D, E(K_D, M)) \neq M.$)

g) It is "impractical" to obtain K_E from K_D.

For obvious reasons we will call a system with properties a)–g) a "double public key cryptosystem".

II. AN EXAMPLE OF A DOUBLE PUBLIC KEY CRYPTOSYSTEM

Several public key cryptosystems have been proposed in response to the Diffie–Hellman paper [5],[6]. Below we present an outline of the double system due to Rivest et al. [6]. The interested reader is encouraged to see [6] since important details are omitted here in order to facilitate the exposition.

1) To establish a mated pair K_E and K_D, the user first produces three large prime numbers p, q, and e. He next computes a number d such that $d \cdot e$ has a remainder of 1 when divided by $(p-1) \cdot (q-1)$ (i.e., $de = 1$ MOD $((p-1)(q-1))$). K_E is the pair of numbers $<e,n>$ and K_D is the pair of numbers $<d,n>$ where $n = p \cdot q$. (It is important that n be the result of multiplying p times q and not p and q themselves.)

2) The encryption procedure E takes a message M (thought of as a binary number, say in ASCII) and an encryption key $K_E = <e,n>$ and produces the cipher text C by raising M to the eth power and taking the remainder when divided by n ($C = M^e$ MOD (n)).

3) The decryption procedure D takes a cipher text C and a decoding key $K_D = <d,n>$ and decrypts by raising C to the dth power and taking the remainder when divided by n ($M = C^d$ MOD (n)).

Fast methods of finding large primes p, q, and e, of computing the appropriate d from them, and of encrypting and decrypting are given in [6]. Also see [6] for examples and arguments concerning the security of this double public key system.

III. EXAMPLES OF THE USE OF PUBLIC KEY CRYPTOSYSTEMS

A. Read Only Communications

This application comes from an article by Gina Bara Kolata which appeared recently in Science[1] [4].

As part of the Nuclear Test Band Treaty it has been suggested that seismographic devices be buried in Russian soil to monitor earth tremors and thereby detect nuclear activity (no doubt Russian devices would be placed in the U.S. as well). Apparently, the technology exists for implanting the devices and making them tamperproof; however, there is a concern over the security of the transmissions from them. Some method of protecting the transmissions from unauthorized insertion and deletion is necessary lest false transmissions indicating a halcyon state be sent while in fact testing is taking place. The

---

[1]The scheme described may be due to Gus Simmons of Sandia.

obvious answer is for the U.S. to encrypt the transmissions thereby inhibiting tampering. Unfortunately, this creates a new problem since Russia has no assurance that only seismic information and not "spy" data is being transmitted. Simple monitoring will not help since the Russians cannot read the encrypted transmissions. In one proposed solution, based on traditional cryptography, Russia would record the encrypted transmissions, then at the end of each month the United States would surrender the encoding key used that month, enabling Russia to confirm in retrospect that only legitimate information was sent. Apparently, in these circles a month lag time is unsatisfactory, and any attempt to make the key exchange more frequent creates unacceptable key management risks.

The solution to the problem makes use of the special properties of a double public key cryptosystem. In this solution, the United States generates a mated pair of keys K_E and K_D. K_D is revealed to Russia (so in this system K_D is the "public key") while K_E is secured inside the seismographic device. All transmissions are encoded using K_E. Since both the U.S. and Russia have the decoding key K_D, each can monitor and decode the resulting outputs. However, Russia has not been given the private encoding key

---

Extremely high encryption rates without the problems of key distribution can be achieved by combining traditional systems with public key cryptosystems.

K_E, cannot obtain it from K_D, and therefore is unable to tamper with the transmissions. Thus, Russia has the facility to read the encrypted language, but not to write it, and apparently all design constraints have been satisfied.

B. Securing Automatic Teller Machines

This system was designed by researchers at Interbank.

Automatic teller machines are in widespread use in this country. Twenty-four hours a day a user can approach a machine (usually located on the external wall of a bank) and using a protocol typically involving passwords and magnetic cards cause the device to deliver cash. Usually the ATM is connected via telephone lines to a central computer which does bookkeeping, and, when appropriate conditions are met, transmits commands for the release of money from the ATM. While there are issues of security concerning the use of passwords and magnetic cards, these will not concern us. We are interested in securing the line between the ATM and the central computer against insertion of messages which will cause illegitimate release of cash from the ATM. A traditional system of encryption along the lines works well here. Each ATM shares a key with the central computer, and this is used to encrypt along the line, thus inhibiting insertions. Unfortunately, this solution has associated key distribution problems. How is the key brought to the

ATM? Transmission in the clear over telephone lines is obviously unacceptable. Delivery by couriers invites bribery and theft. Hard wiring of the key at the time the ATM is built creates security problems in the manufacturing environment, and does not allow for key changes.

The solution proposed by the researchers at Interbank makes elegant use of public key techniques. The central computer generates a mated pair of keys K_E and K_D. K_D is kept by the central computer and security measures must be taken to keep it secret. Each ATM is provided with the corresponding public key K_E. Since the security of the system will not depend on keeping K_E secret, there is no serious problem in distributing it to the ATM's. At the onset of a commercial transaction, the ATM generates a random number R to be used for this transaction only.[2] The ATM stores R, encrypts the message "This is ATM x the current transaction number is R" using K_E, and sends the resulting ciphertext to the central computer. The computer decrypts the ciphertext using K_D and stores R. R is then used as a key in a traditional (or public key) cryptosystem for encrypting and decrypting all communications between the computer and the ATM until a new transaction begins, at which time the ATM independently generates a new R and the process begins anew.

Two rules govern the use of the keys

a) The central computer ignores all messages it receives which are encrypted using K_E except those of the form "This is ATM＿＿ the current transaction number is ＿＿" (we are not assuming the computer can distinguish messages (even in the correct form) which come from real ATM's and those which come from intruders).

b) The ATM ignores all messages it receives except those encrypted under the current R.

Surprisingly the key distribution problems have been solved. Even if the encryption key K_E is publicly revealed, it is of no value in defeating the system. If a prospective thief knows K_E, how could he cause money to be issued by the ATM? He does not know the current R since it has only appeared on the line encrypted using K_E and he does not know the decryption key K_D. By rule a) he can use K_E in just one way, to send the computer an encrypted message "This is ATM x the current transaction number is R'" for some R' of his choice. This will cause the computer to begin communicating with the ATM using R' instead of R, but by rule b), the ATM will steadfastly ignore all messages encrypted under R' and will therefore not release money.

This example illustrates how public key and traditional systems can be synergistically combined. A traditional system may possess valuable properties (for example, extremely high encryption rates) unavailable in public key systems, but the traditional system may also have associated problems (for example, key distribution) which can be solved using public key methods.

---

[2]There are technical problems involved in generating random numbers which must be considered in the implementation of such a system.

C. Distribution of Session Keys

A recent report by the MITRE Corporation [7] has dramatized the insecurity of telephone communications by revealing the ease with which microwave transmissions can be monitored. Apparently, for approximately $55 000 ($35 per line) an entire microwave link can be tapped (the ability to insert or delete messages in an undetectable manner is probably vastly more expensive). In response to this threat, several systems using traditional encryption have been developed [1],[3]. Typically, these systems involve the use of a hierarchy of keys: A "master key" which remains fixed for long periods (months or years) and "session keys" which are used for shorter periods (hours or days). A typical system for link encryption would involve the following steps:

1) A single master key is securely distributed to each node in the system (alternatively a different master key can be used for each pair of nodes).

2) When a session begins, say once a day, session keys are randomly generated and distributed to the nodes under encryption with the master key. This process is complete when each pair of communicating nodes shares a unique key.

3) Messages are sent node to node, encrypted under the current session key shared by those nodes.

4) At the end of the session all session keys are destroyed.

The main advantages of these systems are

1) The master key is rarely used and when used only random numbers are encrypted. This reduces the key's vulnerability to cryptanalytic attack.

2) The session keys are used only for one day and if lost do not compromise communications on other days.

The main disadvantages of these systems are

1) The master key must be distributed securely.

2) The master key requires long-term protection and its loss compromises all session keys and in turn all communications.

With a public key approach these disadvantages can be minimized. For example,

1) At the beginning of each sesson, selected nodes generate mated pairs of keys $<K_E, K_D>$.

2) Each selected node sends (over insecure line) K_E to its neighbors.

3) Each neighbor responds by randomly generating a "session key" and sending it to the selected node encrypted under K_E. The selected node decrypts it using

K_D. Again the process is complete when each pair of communicating nodes shares a unique key.

4) All K_D and K_E are destroyed.

5) Messages are sent node to node encrypted under the current session key.

6) At the end of the session all session keys are destroyed.

In this system there are no long-term keys. No key, public or traditional, is kept for longer than the length of a session. Thus we have the advantages of session keys without the disadvantages of long-term master keys.

REFERENCES

[1] F. Heinrich, "The network security center: A system network approach to computer network security," NBS Special Pub. 500-21-Vol 2.

[2] W. Diffie and M. Hellman, "New directions in cryptography," IEEE Trans. Inform. Theory, vol. IT-22, Nov. 1976.

[3] W. F. Ehrsam et al., "A cryptographic key management scheme for implementing the data encryption standard [DES]," IBM Sys. J. vol. 17, no. 2, 1978.

[4] G. B. Kolata, "Cryptology: A secret meeting at IDA," Science, Apr. 14, 1978.

[5] R. Merkle and M. Hellman, "Hiding information and signatures in trap door knapsacks," IEEE Trans. Inform. Theory, vol. IT-24, pp. 525–530, Sept. 1978.

[6] R. Rivest et al., "A method for obtaining digital signatures and public key cryptosystems," Commun. ACM, Feb. 1978.

[7] C. W. Sanders et al., "Selected examples of possible approaches to electronic communication interception operations," MITRE Tech. Rep. MTR-7461, Jan. 1977.

Leonard Adleman received his Ph.D. degree at the University of California, Berkeley, in the Department of Electrical Engineering and Computer Science in 1976. He is currently assistant Professor of Mathematics at MIT, Cambridge, where he is also a member of the Laboratory for Computer Science. His area of specialization has been computational complexity with particular emphasis on number theoretic problems. More recently his interests have included public key cryptosystems and both theoretical and applied aspects of cryptography.

Ronald L. Rivest is currently Associate Professor of Computer Science at MIT, Cambridge, where he has been for five years. He obtained his Ph.D. in 1974 from Stanford University, California. His work has been primarily in the area of computational complexity and cryptography.

AN OVERVIEW OF PUBLIC KEY CRYPTOGRAPHY *

Martin E. Hellman

With a public key cryptosystem, the key used to encipher a message can be made public without compromising the secrecy of a different key needed to decipher that message.

I. COMMERCIAL NEED FOR ENCRYPTION

Cryptography has been of great importance to the military and diplomatic communities since antiquity but failed, until recently, to attract much commercial attention. Recent commercial interest, by contrast, has been almost explosive due to the rapid computerization of information storage, transmission, and spying.

Telephone lines are vulnerable to wiretapping, and if carried by microwave radio, this need not entail the physical tapping of any wires. The act becomes passive and almost undetectable. It recently came to light that the Russians were using the antenna farms on the roofs of their embassy and consulates to listen in on domestic telephone conversations, and that they had been successful in sorting out some conversations to Congressmen.

Human sorting could be used, but is too expensive because only a small percentage of the traffic is interesting. Instead, the Russians automatically sorted the traffic on the basis of the dialing tones which precede each conversation and specify the number being called. These tones can be demodulated and a microprocessor used to activate a tape recorder whenever an "interesting" telephone number (one stored in memory) is detected. The low cost of such a device makes it possible to economically sort thousands of conversations for even one interesting one.

This work was supported in part under NSF Grant ENG 10173.

The author is with the Department of Electrical Engineering, Stanford University, Stanford, CA 94305.

*Reprinted from *IEEE Communications Society Magazine*, November 1978, Vol. 16, No. 6, pp. 24–32.

This problem is compounded in remote computing because the entire "conversation" is in computer readable form. An eavesdropper can then cheaply sort messages not only on the basis of the called number, but also on the content of the message, and record all messages which contain one or more keywords. By including a name or product on this list, an eavesdropper will obtain all messages from, to, or about the "targeted" person or product. While each fact by itself may not be considered sensitive, the compilation of so many facts will often be considered highly confidential.

It is now seen why electronic mail must be cryptographically protected, even though almost no physical mail is given this protection. Confidential physical messages are not written on postcards and, even if they were, could not be scanned at a cost of only $1 for several million words.

II. THE COST OF ENCRYPTION

Books about World War II intelligence operations make it clear that the allies were routinely reading enciphered German messages. The weakness of the Japanese codes was established by the Congressional hearings into the Pearl Harbor disaster, and while it is less well publicized, the Germans had broken the primary American field cipher.

If the major military powers of World War II could not afford secure cryptographic equipment, how is industry to do so in its much more cost-conscious environment?

Encryption is a special form of computation and, just as it was impossible to build good, inexpensive, reliable, portable computers in the 1940's, it was impossible to build good (secure), inexpensive, reliable, portable encryption units. The scientific calculator which sells for under $100 today would have cost on the order of a million dollars and required an entire room to house it in 1945.

While embryonic computers were developed during the war (often for codebreaking), they were too expensive, unreliable, and bulky for field use. Most computational aids were mechanical in nature and based on gears. Similarly, all of the major field ciphers employed gear-based devices and, just as Babbage's failure indicates the difficulty of building a good computer out of gears, it is also difficult to build a good cryptosystem from gears. The development of general-purpose digital hardware has freed the designers of cryptographic equipment to use the best operations from a cryptographic point of view, without having to worry about extraneous mechanical constraints.

As an illustration of the current low cost of encryption, the recently promulgated national Data Encryption Standard (DES) can be implemented on a single integrated circuit chip, and will sell in the $10 range before long. While some have criticized the standard as not being adequately secure [1], this inadequacy is due to political considerations and is not the fault of insufficient technology.

III. KEY DISTRIBUTION AND PUBLIC KEY SYSTEMS

While digital technology has reduced the cost of encryption to an almost negligible level, there are other major problems involved in securing a communication network. One of the most pressing is key distribution, the problem of securely transmitting keys to the users who need them.

The classical solution to the key distribution problem is indicated in Fig. 1. The key is distributed over a secure channel as indicated by the shielded cable. The secure channel is not used for direct transmission of the plaintext message P because it is too slow or expensive.

The military has traditionally used courier service for distributing keys to the sender and receiver. In commercial systems registered mail might be used. Either way, key distribution is slow, expensive, and a major impediment to secure communication.

Keys could be generated for each possible conversation and distributed to the appropriate users, but the cost

would be prohibitive. A system with even a million subscribers would have almost 500 billion possible keys to distribute. In the military, the chain of command limits the number of connections, but even there, key distribution has been a major problem. It will be even more acute in commercial systems.

It is possible for each user to have only one key which he shares with the network rather than with any other user, and for the network to use this as a master key for distributing conversation specific keys [2], [3]. This method requires that the portion of the network which distributes the keys (known as the key distribution center or node) be trustworthy and secure.

Diffie and Hellman [4] and independently Merkle [5] have proposed a radically different approach to the key distribution problem. As indicated in Fig. 2, secure communication takes place without any prearrangement between the conversants and without access to a secure key distribution channel. As indicated in the figure, two way communication is allowed and there are independent random number generators at both the transmitter and the receiver. Two way communication is essential to distinguish the receiver from the eavesdropper. Having random number generators at both ends is not as basic a requirement, and is only needed in some implementations.

The situation is analogous to having a room full of people who have never met before and who are of equal mathematical ability. I choose one other person in the room and, with everyone else listening, give him instructions which allow the two of us to carry on a conversation that no one else can understand. I then choose another person and do the same with him.

This sounds somewhat impossible and, from one point of view, it is. If the cryptanalyst had unlimited computer time he could understand everything we said. But that is also true of most conventional cryptographic systems— the cryptanalyst can try all keys until he finds the one that yields a meaningful decipherment of the intercepted message. The real question is whether we can, with very limited computations, exchange a message which would take the cryptanalyst eons to understand using the most powerful computers envisionable.

A public key cryptosystem [4] has two keys, one for enciphering and one for deciphering. While the two keys effect inverse operations and are therefore related, there must be no easily computed method of deriving the deciphering key from the enciphering key. The enciphering key can then be made public without compromising the deci-

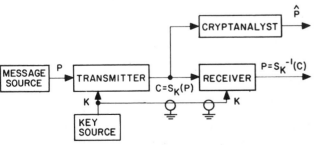

Fig. 1. Conventional Cryptographic System.

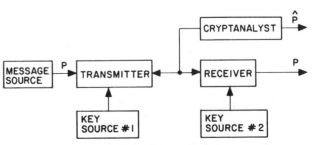

Fig. 2. Public Key Cryptographic System.

phering key so that anyone can encipher messages, but only the intended recipient can decipher messages.

The conventional cryptosystem of Fig. 1 can be likened to a mathematical strongbox with a resettable combination lock. The sender and receiver use a secure channel to agree on a combination (key) and can then easily lock and unlock (encipher and decipher) messages, but no one else can.

A public key cryptosystem can be likened to a mathematical strongbox with a new kind of resettable combination lock that has two combinations, one for locking and one for unlocking the lock. (The lock does not lock if merely closed.) By making the locking combination (enciphering key) public anyone can lock up information, but only the intended recipient who knows the unlocking combination (deciphering key) can unlock the box to recover the information.

Public key and related cryptosystems have been proposed by Merkle [5], Diffie and Hellman [4], Rivest *et al.* [6], Merkle and Hellman [7], and McEliece [8]. We will only outline the approaches, and the reader is referred to the original papers for details.

Electronic mail unlike ordinary mail is machine readable and can be automatically scanned for sensitive messages.

The RSA (Rivest *et al.*) scheme [6] is based on the fact that it is easy to generate two large primes and multiply them together, but it is much more difficult to factor the result. (Try factoring 518940557 by hand. Then try multiplying 15107 by 34351.) The product can therefore be made public as part of the enciphering key without compromising the factors which effectively constitute the deciphering key. By making each of the factors 100 digits long, the multiplication can be done in a fraction of a second, but factoring would require billions of years using the best known algorithm.

As with all public key cryptosystems there must be an easily implemented algorithm for choosing an enciphering-deciphering key pair, so that any user can generate a pair, regardless of his mathematical abilities. In the RSA scheme the key generation algorithm first selects two large prime numbers p and q and multiplies them to produce $n = pq$. Then Euler's function is computed as $\phi(n) = (p - 1)(q - 1)$. ($\phi(n)$ is the number of integers between 1 and n which have no common factor with n. Every p^{th} number has p as a common factor with n and every q^{th} number has q as a common factor with n.) Note that it is easy to compute $\phi(n)$ if the factorization of n is known, but computing $\phi(n)$ directly from n is equivalent in difficulty to factoring n [6].

$\phi(n)$ as given above has the interesting property that for any integer a between 0 and $n - 1$ (the integers modulo n) and any integer k

$$a^{k\phi(n) + 1} = a \mod n. \tag{1}$$

Therefore, while all other arithmetic is done modulo n, arithmetic in the exponent is done modulo $\phi(n)$.

A random number E is then chosen between 3 and $\phi(n) - 1$ and which has no common factors with $\phi(n)$. This then allows

$$D = E^{-1} \mod \phi(n) \tag{2}$$

to be calculated easily using an extended version of Euclid's algorithm for computing the greatest common di-

---

THE RIVEST–SHAMIR–ADLEMAN PUBLIC KEY SCHEME

Design

Find two large prime numbers p and q, each about 100 decimal digits long. Let $n = pq$ and $\psi = (p-1)(q-1)$.

Choose a random integer E between 3 and ψ which has no common factors with ψ. Then it is easy to find an integer D which is the "inverse" of E modulo ψ, that is, $D \cdot E$ differs from 1 by a multiple of ψ.

The public information consists of E and n. All other quantities here are kept secret.

Encryption

Given a plaintext message P which is an integer between 0 and $n-1$ and the public encryption number E, form the ciphertext integer

$$C = P^E \mod n.$$

In other words, raise P to the power E, divide the result by n, and let C be the remainder. (A practical way to do this computation is given in the text of Hellman's paper.)

Decryption

Using the secret decryption number D, find the plaintext P by

$$P = C^D \mod n.$$

Cryptanalysis

In order to determine the secret decryption key D, the cryptanalyst must factor the 200 or so digit number n. This task would take a million years with the best algorithm known today, assuming a 1 μs instruction time.

visor of two numbers [9, p. 315, problem 15; p. 523, solution to problem 15].

The information (E,n) is made public as the enciphering key and is used to transform unenciphered, plaintext messages into ciphertext messages as follows: a message is first represented as a sequence of integers each between 0 and $n - 1$. Let P denote such an integer. Then the corresponding ciphertext integer is given by the relation

$$C = P^E \bmod n. \qquad (3)$$

The information (D,n) is used as the deciphering key to recover the plaintext from the ciphertext via

$$P = C^D \bmod n. \qquad (4)$$

These are inverse transformations because from (3), (2), and (1)

$$C^D = P^{ED} = P^{k\phi(n) + 1} = P. \qquad (5)$$

As shown by Rivest *et al.*, computing the secret deciphering key from the public enciphering key is equivalent in difficulty to factoring n.

As a small example suppose $p = 5$ and $q = 11$. Then $n = 55$ and $\phi(n) = 40$. If $E = 7$ then $D = 23$ ($7 \times 23 = 161 = 1 \bmod 40$). If $P = 2$, then

$$C = 2^7 \bmod 55 = 18 \qquad (6)$$

and

$$C^D = 18^{23} \bmod 55 \qquad (7)$$

$$= 18^1 18^2 18^4 18^{16} \qquad (8)$$

$$= 18 \ 49 \ 36 \ 26 \bmod 55 \qquad (9)$$

$$= 2 \qquad (10)$$

which is the original plaintext.

Note that enciphering and deciphering each involve an exponentiation in modular arithmetic and that this can be accomplished with at most $2(\log_2 n)$ multiplications mod n. As indicated in (8), to evaluate $Y = a^X$, the exponent X is represented in binary form, the base a is raised to the 1st, 2nd, 4th, 8th, etc. powers (each step involving only one squaring or multiplication), and the appropriate set of these are multiplied together to form Y.

Merkle and Hellman's method [7] makes use of trap-door knapsack problems. The knapsack problem is a combinatorial problem in which one is given a vector of n integers, a, and an integer S which is a sum of a subset of the $\{a_i\}$. The problem is to solve for the subset, or equivalently, for the binary vector x which is the solution to the equation

$$S = a * x. \qquad (11)$$

While the knapsack problem is very difficult to solve in general, there are specific cases which are easy to solve. For example, if the knapsack vector is

$$a' = (171, 197, 459, 1191, 2410) \qquad (12)$$

then, given any S', x is easily found because each component of a' is larger than the sum of the preceding components. If $S' = 3798$, then it is seen that x_5 must be 1 because, if it were 0, $a_5' = 2410$ would not be in the sum and the remaining elements sum to less than S'. After subtracting the effect of a_5' from S', the solution continues recursively and establishes that $x_4 = 1$, $x_3 = 0$, $x_2 = 1$, and $x_1 = 0$.

The knapsack vector

$$a = (5457, 4213, 5316, 6013, 7439) \qquad (13)$$

does not possess the property that each element is larger than the sum of the preceding components, and the simple method of solution is not possible. Given $S = 17665$, there is no obvious method for finding that $x = (0,1,0,1,1)$ other than trying almost all 2^5 subsets.

But it "just so happens" that if each component of a is multiplied by 3950 modulo 8443 the vector a' of (12) is obtained. By performing the same transformation on S, the quantity $S' = 3798$ is obtained. It is now seen that there is a simple method for solving for x in the equation

$$S = a * x \qquad (14)$$

by transforming to the easily solved knapsack problem

$$S' = a' * x. \qquad (15)$$

The two solutions x are the same provided the modulus is greater than the sum of the $\{a_i\}$.

The variables of the transformation (the multiplier 3950 and the modulus 8443) are secret, trap-door information used in the construction of the trap-door knapsack vector a. There is no apparent, easy way to solve knapsack problems involving a unless one knows the trap-door information.

When a is made public anyone can represent a message as a sequence of binary x vectors and transmit the information securely in the corresponding sums, $S = a * x$. The intended recipient uses his trap-door information (secret deciphering key) to easily solve for x, but no one else can do this. Of course the a vector must be significantly longer than that used in this small, illustrative example.

McEliece's public key cryptosystem [8] is based on algebraic coding theory. Goppa codes are highly efficient error correcting codes [10], but their ease of error correction is destroyed if the bits which make up a codeword are scrambled prior to transmission. To generate a public enciphering key, a user first selects a Goppa code chosen at random from a large set of possible codes. He then selects a permutation of the codeword bits, computes the generator matrix associated with the scrambled Goppa code and makes it public as his enciphering key. His secret deciphering key is the permutation and choice of Goppa code.

Key distribution, the secure transmission of keys to the users who need them, is a major problem in securing a communication network.

Anyone can easily encode information (scrambling does not greatly increase the difficulty of encoding since the scrambled code is still linear), add a randomly generated error vector, and transmit this. But only the intended recipient knows the inverse permutation which allows the errors to be corrected easily.

McEliece estimates that a block length of 1000 bits with

500 information bits should foil cryptanalysis using the best currently known attacks.

The other two known methods for communicating securely over an insecure channel without securely transmitting a key are not true public key cryptosystems. Rather, they are public key distribution systems which are used to securely exchange a key over an insecure channel without any prearrangement, and that key is then used in a conventional cryptosystem.

Merkle's technique [5] involves an exchange of "puzzles." The first user generates n potential keys and hides them as the solutions to n different puzzles, each of which costs n units to solve. The second user chooses one of the n puzzles at random, solves it, and sends a test message encrypted in the associated key. The first user determines which key was chosen by trying all n of them on the test message.

The cost to the first user is proportional to n. He must generate and store n keys, generate and transmit n puzzles, and try n keys on the test message. The cost to the second user is also proportional to n because he must solve one puzzle which was designed to have solution cost equal to n.

The cost to an eavesdropper appears to grow as n^2. He can try solving puzzles at random and see if the associated key (solution) agrees with the test message. On the average, he must solve n/2 puzzles, each at a cost of n.

Diffie and Hellman [4] describe a public key distribution system based on the discrete exponential and logarithm functions. If q is a prime number and a is a primitive element, then X and Y are in a 1:1 correspondence for $1 \leqslant X, Y \leqslant (q - 1)$ where

$$Y = a^X \mod q \qquad (16)$$

and

$$X = \log_a Y \quad \text{over } GF(q). \qquad (17)$$

While the discrete exponential function (16) is easily evaluated, as in (7) and (8), no general, fast algorithms are known for evaluating the discrete logarithm function (17). Each user chooses a random element X and makes the associated Y public. When users i and j wish to establish a key for communicating privately they use

$$K_{ij} = a^{X_i X_j} \qquad (18)$$

$$= (Y_i)^{X_j} = (Y_j)^{X_i}. \qquad (19)$$

Equation (19) demonstrates how both users i and j use the easily computed discrete exponential function to calculate K_{ij} from their private and the other user's public information. An opponent who knows neither user's secret information can compute K_{ij} if he is willing to compute a discrete logarithm, but that can be made computationally infeasible using the best currently known algorithms [11].

The various public key systems are compared in Section V.

IV. DIGITAL SIGNATURES

Business runs on signatures and, until electronic communications can provide an equivalent of the written sig-

nature, it cannot fully replace the physical transportation of documents, letters, contracts, etc.

Current digital authenticators are letter or number sequences which are appended to the end of a message as a crude form of signature. By encrypting the message and authenticator with a conventional cryptographic system, the authenticator can be hidden from prying eyes. It therefore prevents third party forgeries. But, because the authentication information is *shared* by the sender and receiver, it cannot settle disputes as to what message, if any, was sent. The receiver can give the authentication information to a friend and ask him to send a signed message of the receiver's choosing. The legitimate sender of messages will of course deny having sent this message, but there is no way to tell whether the sender or receiver is lying. The whole concept of a contract is embedded in the possibility of such disputes, so stronger protection is needed.

A true digital signature must be a number (so it can be sent in electronic form) which is easily recognized by the receiver as validating the particular message received, but which could only have been generated by the sender. It may seem impossible for the receiver to be able to recognize a number which he cannot generate, but such is not the case.

While there are other ways to obtain digital signatures, the easiest to understand makes use of the public key cryptosystems discussed in the last section. The ith user has a public key E_i and a secret key D_i. This notation was chosen because E_i was used to encipher and D_i was used to decipher. Suppose, as in the RSA scheme, the enciphering function is **onto**, that is, for every integer C less than n, there exists an integer m for which $E_i(m) = C$. Then, we can interchange the order of operations and use D_i first to sign the message and E_i second to validate the signature. When user i wants to sign and send a message M to user j, he operates on M with his secret key D_i to obtain

$$C = D_i(M) \qquad (20)$$

which he then sends to user j. User j obtains i's public key E_i from a public file and operates with it on C to obtain M

$$E_i(C) = E_i[D_i(M)] = M \qquad (21)$$

User j saves C as proof that message M was sent to him by user i. No one else could have generated C, because only i knows D_i. And if j tries to change even one bit in C, he changes its entire meaning (such error propagation is necessary in a good cryptosystem).

If i later disclaims having sent message M to user j, then j takes C to a "judge" who accesses the public file and checks whether $E_i(C)$ is a meaningful message with the appropriate date, time, address, name, etc. If it is, the judge rules in favor of j. If it is not, the ruling is in favor of i.

Digital signatures have an advantage over written signatures because written signatures look the same, independent of the message. My signature is supposed to look the same on a $100 check as on a $1000 check, so a dishonest recipient can try to alter the check. Similarly, if a photostat of a contract is acceptable as proof, a dishonest person can alter the contract and make a copy which hides the alteration. Such mischief is impossible with digital signatures, provided the signature system is truly secure.

The disadvantage of digital signatures is that the ability to sign is equivalent to possession of a secret key. This key will probably be stored on a magnetic card which, unlike the ability to sign one's name, can be stolen.

V. COMPARISON OF PUBLIC KEY SYSTEMS

This section compares the public key systems which have been proposed. Speed, ease of signature generation, and certain other characteristics can be compared more readily than the all important question of security level. We can compare the security level using the best known methods for breaking each system, but there is the danger that better methods will be found which will change the relative rankings.

If signatures are desired, attention should be directed primarily to the RSA [6] and trap-door knapsack systems [7]. The RSA scheme yields signatures directly. While the trap-door knapsack signature method described in [7] is not direct, Merkle and Reeds have developed a method for generating "high density" trapdoor knapsacks which simplify signature generation, and Shamir has recently suggested a direct method for obtaining signatures. Both of these approaches are not yet published.

The $a^{(X_1 X_2)}$ and Goppa code methods do not appear to lend themselves to signatures, but Merkle has developed a puzzle-like technique for generating signatures.

So far, as storage requirements for the public file, the $a^{(X_1 X_2)}$ and RSA schemes are most interesting. Each requires on the order of 500 bits of storage per user. The trap-door knapsack scheme requires on the order of 100 kbits of storage per user, and the Goppa code method requires on the order of a megabit per user. Merkle's puzzle scheme is not really suited to public file storage and rather depends on transmission of public information at the start of each new conversation. The transmitted information must be on the order of a gigabit before significant levels of security are afforded.

Instead of storing each user's public key in a public file (similar to a telephone book), Kohnfelder [12] has suggested having the system give each user a signed message, or certificate, stating that user's public key. The certificate could be stored by the user on a magnetic card, and transmitted at the start of a conversation. This method converts public file storage requirements into transmission requirements. The system's public key would be needed to check the certificate and could be published widely. Protecting the system's secret key might be easy because no one else ever has to use it and it could be destroyed after it was used to certify a group of users.

Computation time on the part of the legitimate users is smallest with the trap-door knapsack method. The $a^{(X_1 X_2)}$ and RSA schemes each require several hundred times as much computation, but are still within reason. Merkle's technique requires even more computation. The Goppa code technique is extremely fast for enciphering, requiring approximately 500 XOR's on 1000 bit vectors, but I have not yet estimated its deciphering requirements.

Turning to security level, Merkle's puzzle method [5] has the advantage of being the most solid method for communicating securely over an insecure channel. That is, it is extremely doubtful that a better method will be found for breaking it. Unfortunately, it is also the least secure using the best known algorithm. Its work factor (ratio of cryptanalytic effort to enciphering and deciphering effort, using the best known algorithms) is only $n^2 : n$. Since encryption should cost on the order of $0.01 and cryptanalysis should cost on the order of $10 million or more, this ratio needs to be 10^9 or more and corresponds to $n = 10^9$. If all of the enciphering and deciphering effort were in computation, this might be possible in the near future (a $10 microprocessor can execute on the order of 1 million instructions per second), but Merkle's method requires n transmissions as well as n operations on the part of the legitimate users. Current technology therefore limits Merkle's scheme to $n \leq 10\,000$ which corresponds to approximately 500 kbits of transmission. If fiber optic or other low cost, ultra-high bandwidth communication links become available, Merkle's technique would become of greater practical interest.

Diffie and Hellman's exponentiation method [4] requires the legitimate users to perform an exponentiation in modular arithmetic while the best known cryptanalytic method requires the computation of a logarithm in modular arithmetic. Exponentiation is easily accomplished in at most $2b$ multiplications, much as in (8), where b is the number of bits in the representation of the modulus. Each multiplication can be accomplished with at most $2b$ additions or subtractions, and each of these operations involves at most b gate delays for the propagation of carry signals. Overall, an exponentiation in modular arithmetic can be accomplished in at most $4b^3$ gate delays.

Computation of a logarithm in modular arithmetic is much more complex, and the best currently known algorithm [11] requires $2^{b/2}$ or more operations provided the modulus is properly chosen. Each operation involves a multiplication, or $2b^2$ gate delays. The work factor is therefore exponential in b.

If $b = 500$, then 500 million gate delays are required at the legitimate users' terminals. With current technology this can be accomplished in several seconds, a not unreasonable delay for establishing a key during initial connection. Using $b = 500$ results in the cryptanalyst having to do more than 10^{75} times as much work as the legitimate users, a very safe margin. The real question is whether better methods exist for computing logarithms in modular arithmetic, or if it is even necessary to compute such a logarithm to break this system.

The following table gives the number of operations and time required for cryptanalysis for various values of b assuming a 1 μs instruction time:

b (bits)	100	200	300	500	750	1000
Operations	1.1×10^{15}	1.3×10^{30}	1.4×10^{45}	1.8×10^{75}	7.7×10^{112}	3.3×10^{150}
Time (yrs.)	36	4×10^{16}	5×10^{31}	6×10^{61}	2×10^{99}	1×10^{137}

The storage requirements of this system are small. The public file stores a single b-bit number for each user and only several b-bit words of memory are required at the transmitter and receiver, so that single chip implementation is possible for b on the order of 500.

The RSA system [6] also requires that the legitimate users perform a modular exponentiation, but cryptanalysis is equivalent to factoring a b-bit number. Schroeppel has developed a new, as yet unpublished factoring algorithm which appears to require approximately $\exp\{[\ln(n) \ \ln(\ln n)]^{1/2}\}$ machine cycles where $n = 2^b$ is the number to be factored. The following table gives the number of operations and time to factor a b-bit number again assuming a 1 μs instruction time:

b (bits)	100	200	300	500	750	1000
Operations	2.8×10^7	2.3×10^{11}	2.9×10^{14}	3.6×10^{19}	5.8×10^{24}	1.8×10^{29}
Time	30 s	3 days	9 yr	1 Myr	2 Gyr	6×10^{15} yr

Public file storage for the RSA scheme is reasonable, being several hundred to a thousand bits per user. Memory requirements at the transmitter and receiver are also comparable to the $a^{(X_1 X_2)}$ scheme, so that a single chip device can be built for enciphering and deciphering.

The best known method of cryptanalyzing the trap-door knapsack system requires on the order of $2^{n/2}$ operations where n is the size of the knapsack vector. Enciphering requires at most n additions, so the work factor is exponential. If n is replaced by b, the first table above gives the cryptanalytic effort required for various values of n, so $n \geqslant 200$ provides relatively high security levels. Since each element of the a vector is approximately $2n$ bits long, if $n = 200$, the public storage is approximately 80 kbits/user. Memory requirements at the transmitter and receiver are on the same order.

Both enciphering and deciphering require less computation than either the $a^{(X_1 X_2)}$ or RSA scheme. Enciphering requires at most n additions and deciphering requires one multiplication in modular arithmetic, followed by at most n subtractions.

Until electronic communications can provide an equivalent of the written signature, it cannot fully replace the physical transportation of documents, letters, contracts, etc.

Care must be exercised in interpreting these tables. First, they assume that the cryptanalyst uses the best currently known method, and there may be much faster approaches. For example, prior to the development of

Schroeppel's algorithm, the best factoring algorithm appeared to require $\exp\{[2 \ln(n) \ \ln(\ln n)]^{1/2}\}$ operations. When $b = 200$, that would have predicted that 360 yr, not 3 days, would be required for cryptanalysis. There is the danger that even faster algorithms will be found, necessitating a safety margin in our estimates. As demonstrated by this example, the safety margin is needed in the exponent, not the mantissa.

A similar comment applies to the seemingly higher security level afforded by the $a_i^{(X_1 X_2)}$ and trap-door knapsack methods when compared to the RSA scheme. For a given value of b the two tables show that the RSA scheme requires much less computation to break, using the best currently known techniques. But it is not clear whether this is because factoring is inherently easier than computing discrete logarithms or solving knapsack problems, or whether it is due to the greater study which has been devoted to factoring.

As computers become faster and more parallel, the time for cryptanalysis also falls. A 1 ns computer with million-fold parallelism might reduce the time estimates given in the tables by a factor of 10^9.

VI. CONCLUSIONS

We are in the midst of a communications revolution which will impact many aspects of people's every day lives. Cryptography is an essential ingredient in this revolution, and is necessary to preserve privacy from computerized censors capable of scanning millions of pages of documents for even one sensitive datum. The public key and digital signature concepts are necessary in commercial systems because of the large number of interconnections which are possible, and because of the need to settle disputes.

A major problem which confronts cryptography is the certification of these systems. How can we decide which proposed systems really are secure, and which only appear to be secure? Proofs are not possible using the currently developed theory of computational complexity and, while such proofs may be possible in the future, something must be done immediately. The currently accepted technique for certifying a cryptographic system as secure is to subject it to a mock attack under circumstances which are extremely favorable to the cryptanalyst and unfavorable to the system. If the system resists such a concerted attack under unfavorable conditions, it is hoped that it will also resist attacks by one's opponents under more realistic conditions.

Governments have built up expertise in the certification area but, due to security constraints, this is not currently available for certification of commercially oriented systems. Rather, this expertise in the hands of a foreign government poses a distinct threat to a nation's businesses. It has even been suggested that poor or nonexistent encryption will lead to international economic warfare, a concern of importance to national security. (There is speculation that this occurred with the large Russian grain purchases of several years ago.)

There is a tradeoff between this and other national security considerations which needs to be resolved, but

THE KNAPSACK PROBLEM

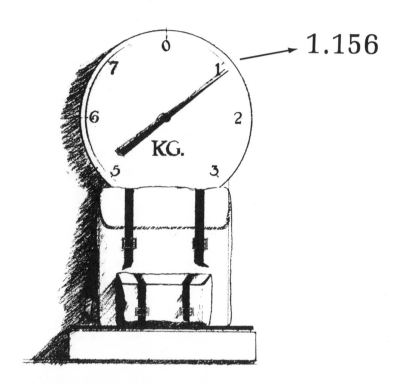

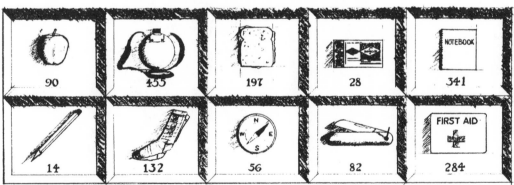

The knapsack is filled with a subset of the items shown, with weights indicated in grams. Given the weight of the filled knapsack, 1156 grams, can you determine which of the items are contained in the knapsack? (The scale is calibrated to deduct the weight of the empty knapsack.)

This simple version of the classic knapsack problem generally becomes computationally infeasible when there are 100 items rather than 10 as in this example. However, if the set of weights for the items happens to have some nice properties known only to someone with special "trap-door" information, then that person can quickly decipher the secret information, i.e., a 100 bit binary word that specifies which of the items are in the knapsack.

ART: Jeff Wyszkowski

the handling of the national data encryption standard indicates that public discussion and resolution of the tradeoff is unlikely unless individuals make their concern known at a technical and political level.

REFERENCES

[1] W. Diffie and M. E. Hellman, "Exhaustive cryptanalysis of the NBS data encryption standard," *Computer*, pp. 74–84, June 1977.

[2] D. Branstad, "Encryption protection in computer data communications," presented at the IEEE Fourth Data Communications Symposium, Oct. 7–9, 1975, Quebec, Canada.

[3] *IBM Syst. J.*, *(Special Issue on Cryptography)*, vol. 17, no. 2, 1978.

[4] W. Diffie and M. E. Hellman, "New directions in cryptography," *IEEE Trans. Inform. Theory*, vol. IT-22, pp. 644–654, Nov. 1976.

[5] R. C. Merkle, "Secure communication over an insecure channel," *Common. Ass. Comput. Mach.*, vol. 21, pp. 294–299, Apr. 1978.

[6] R. L. Rivest, A. Shamir, and L. Adleman, "On digital signatures and public key cryptosystems," *Commun. Ass. Comput. Mach.*, vol. 21, pp. 120–126, Feb. 1978.

[7] R. C. Merkle and M. E. Hellman, "Hiding information and signatures in trap-door knapsacks," *IEEE Trans. Inform. Theory*, vol. IT-24, pp. 525–530, Sept. 1978.

[8] R. J. McEliece, "A public key system based on algebraic coding theory," JPL DSN Progress Rep., 1978.

[9] D. E. Knuth, *The Art of Computer Programming, Vol. 2, Seminumerical Algorithms.* Reading, MA: Addison-Wesley, 1969.

[10] E. R. Berlekamp, "Goppa codes," *IEEE Trans. Inform. Theory*, vol. IT-19, pp. 590–592, Sept. 1973.

[11] S. C. Pohlig and M. E. Hellman, "An improved algorithm for computing logarithms over $GF(p)$ and its cryptographic significance, *IEEE Trans. Inform. Theory*, vol. IT-24, pp. 106–110, Jan. 1978.

[12] L. Kohnfelder, *Towards a Practical Public-Key Cryptosystem*, M.I.T. Lab. for Comput. Sci., June 1978.

Martin E. Hellman (S'63–M'69) was born in New York, NY, on October 2, 1945. He received the B.E. degree in electrical engineering from New York University, New York, NY, in 1966, and the M.S. and Ph.D. degrees in electrical engineering from Stanford University, Stanford, CA, in 1967 and 1969, respectively.

During 1968 to 1969 he was at the IBM Thomas J. Watson Research Center. From 1969 to 1971 he was an Assistant Professor of Electrical Engineering at M.I.T. He is currently an Associate Professor of Electrical Engineering at Stanford University, doing research in the areas of cryptography and information theory.

Dr. Hellman is a member of Tau Beta Pi, Eta Kappa Nu and Sigma Xi, and consults in the areas of communications and information theory. He is a past president of the San Francisco Section's Information Theory Chapter and is a member of the Information Theory Group's Board of Governors.

SECURITY OF COMPUTER COMMUNICATION *

Dennis K. Branstad

Cryptographic algorithms and systems can provide high levels of security when computers are connected in a distributed processing network.

Security of digital computer communications is rising on the list of computer security requirements as the interconnection of computers becomes more commonplace. The reasons for interconnecting computers include improved service, improved reliability, and improved information. However, there are risks in connecting computers. A recent newspaper article reports a story about a woman who was told that she could not make a deposit in her bank account on a certain day from a branch bank. When she asked why, the answer was "because its raining too hard." The reply was absolutely accurate. When the intensity of the rain exceeded some threshold, the communications between the terminal in the branch bank and the computer in the central bank became garbled because of wet telephone connections and all interbank operations had to be suspended.

The security provided by the techniques described in this paper will do nothing to alleviate this problem. In fact, the techniques cannot solve many of the practical security problems encountered in computer operations. They will provide, however, high levels of security against certain types of real threats, both existing and potential, that arise when terminals and computers are connected in a distributed processing network.

Cryptography is a method of secret writing. It is a way of hiding the content of a message through a position-scrambling process or through some other method of transformation. The original message is typically called **plaintext** and the transformed message is called **ciphertext**. The objective of cryptography is to scramble the message in such a way that an unauthorized and unintended recipient of the ciphertext cannot recover the underlying plaintext, while an authorized and intended recipient can easily do so. Cryptography cannot alleviate the communication problem caused by the rain because it is necessary that the ciphertext get from a transmitter to the receiver unperturbed. Cryptography cannot prevent intentional jamming of communications or accidental modification of communications. It will, however, help to detect them and prevent unwanted side effects of their occurrence. The prime benefits of cryptography are the prevention of theft and the intentional, unauthorized modification of messages in a communication system.

Transforming plaintext to ciphertext is called **enciphering** or **encrypting**. The method used for transformation is known as an encryption algorithm. It expresses the set of rules for performing the transformation. For communications purposes an inverse transformation must be used to convert the ciphertext to plaintext. The operation is called **deciphering** or **decryption**. The decryption transformation need not be identical to encryption and may appear quite different, but still will be its inverse. In addition, the enciphering operation need not be the inverse of the deciphering operation, i.e., the encryption operation must be done first. Classical cryptographic systems, however, use enciphering and deciphering transformations which are the inverses of each other and are similar, if not identical.

Security objectives which can be met through the use of cryptography include: 1) prevent disclosure of plaintext, 2) prevent selective modification of plaintext, 3) prevent insertion of false plaintext, 4) detect modification of ciphertext. When used in conjunction with a message numbering system, cryptography can: 5) detect deletion of messages, 6) detect replay of messages. In

The author is with the Institute for Computer Sciences and Technology, National Bureau of Standards, Washington, DC.

*Reprinted from *IEEE Communications Society Magazine*, November 1978, Vol. 16, No. 6, pp. 33-40.

Security in computer communications serves not only to prevent disclosure of the original message; other objectives include protection against alterations or introduction of false messages to the plaintext

all cases, it is assumed that an intelligent opponent or antagonist has physical access to the communication line being protected, but does not have access to the terminals on either end of the line. Physical security must be provided to prevent access to the terminals.

CRYPTOGRAPHIC ALGORITHMS

Cryptographic algorithms are easy to develop. There exist many mathematical transformations which are inverses of each other. Addition and subtraction are inverse transformations; the Exclusive-Or (Binary addition without carry) operation is its own inverse. The Exclusive-Or (**XOR**) operation is often used as a fundamental operation in a cryptographic algorithm. Random numbers, when XORed to data, make the results random. The same random number, if XORed to data twice, results in the original data. This fact is used in many cryptographic algorithms in which a pseudorandom number is generated and XORed to the data during encryption. The identical pseudorandom number is generated and XORed to the ciphertext during decryption. The security of the data depends entirely on the cryptographic characteristics of the random number generator.

Even though cryptographic algorithms are easy to develop, the level of security that they provide is difficult to measure. However, a good cryptographic algorithm has the following general characteristics.

1) The encryption and decryption operations should be simple and efficient for authorized users.

2) The decryption operation should be very difficult for unauthorized users.

3) The security of the data should not require that the algorithm itself be kept secret.

4) The efficiency and security of the algorithm should not be data dependent.

Cryptographic algorithms use a combination of functions operating on some data (plaintext) under the control of a parameter, known as a **key**, to transform the data into ciphertext. Binary algorithms operate on binary data of a fixed or maximum length. The result of the encryption operation is generally of the same length as the original data. If the result is shorter and data compression techniques have not been used, information has been lost and the original data cannot be recovered. Algorithms which discard information on purpose are called one-way algorithms and can be useful in certain applications, e.g., password encryption in which the original password can never be recovered from the ciphertext, but can be matched with the shorter ciphertext form for user authentication. If the result of the encryption operation is longer than the original data, more bits must be trans-

mitted or stored, resulting in reduced effective communication speed or in increased data storage costs.

Cryptographic algorithms can be categorized in various ways. One method relates to how the key is used in the algorithm. Almost all algorithms use one or more parameters which are selected by a group of users to be used for some time period. The parameters chosen are called the **cryptographic key** and the time period is called the **cryptographic period.** The key should be chosen from all the possible keys in a random fashion and the choice should be unrelated to previously used keys. The cryptographic period may be fixed (one day, one week, etc.) or may vary, depending on specified events (users resigning, possible compromises). When the same key is used for both the enciphering and deciphering operations, it is called a **secret key algorithm**. If different keys are used for the two operations and one is kept secret but the other is made public, then it is called a **public key algorithm** [7],[15]. (See article by Hellman, this issue.) The former case allows the same key to be used for both encryption and decryption by both users of a communication link. The latter case requires each user to have two keys; the public key of the other person and his own secret key. Each person must therefore know and protect a secret key even in a public key system.

CRYPTOGRAPHIC SYSTEMS

While a cryptographic algorithm specifies the mathematics or logic of the encryption and decryption transformations, the **cryptographic system** specifies how the

The security provided by a cryptographic system depends greatly on how the cryptographic algorithm is implemented and used in a particular application over and above the protection provided by the algorithm itself.

algorithm shall be implemented and used in a particular application. While there are numerous cryptographic algorithms which may be developed, there are even more numerous ways that they may be applied in any particular environment. The security provided by a cryptographic system depends greatly on the system itself over and above the cryptographic protection provided by the algorithm. The security can be either greater than or less than that fundamental to the algorithm.

A cryptographic system can either be added onto a communication system or integrated into it. Historically, the communication system already existed and the cryptographic system was added onto it in what was hoped to be a transparent manner. Stand-alone units containing a cryptographic device (a device implementing the cryptographic algorithm) would be placed between the communication device (modem) and the terminal (keyboard, printer, telephone). Such units provided electrical power, physical security, interface logic, and key entry facilities. They were designed to operate in pairs on each end of a communication circuit. Ciphertext and cryptographic control information was passed between them. Plaintext appeared outside the secure communication circuit. Cryptographic keys were manually generated and securely distributed to the units where they were manually entered at the beginning of a fixed cryptographic period. Great care was taken to assure that only ciphertext was ever transmitted down the circuit and that the key could not be obtained from the unit in any manner. Ciphertext was often transmitted at a constant rate, even though no data was entering the unit for transmission. Switching of the ciphertext from one circuit to another was not possible.

Cryptographic systems can be integrated into a communication system at various levels in the protocol hierarchy. Fig. 1 shows the typical diagram used when discussing communication protocols. An integrated cryptographic system can provide more security and services than an add-on system. The services depend on the level of integration. The cryptographic system can be controlled by the user or the application system, or both. A user-controlled cryptographic system provides sufficient tools for a user to generate, distribute, and store keys, enable and disable the cryptographic functions, and maintain data that can be accessed only by that user. A system-controlled cryptographic system automatically provides these services for a user with little or no knowledge of such activity by the user. Security can be provided at the user level (user-to-user), the terminal

level (terminal-to-terminal or computer), or the system level (system-maintained data).

THE FEDERAL DATA ENCRYPTION STANDARD

A Data Encryption Standard (**DES**) was published by the National Bureau of Standards [11] which specifies a cryptographic algorithm for use by the Federal Government in certain information processing applications. The standard is a mathematical description of an encryption algorithm and its inverse decryption algorithm. Both are based on the use of a secret key which must be the same for the encryption and decryption operations. References [2]–[4], [13], and [14] contain detailed descriptions of the standard, the fundamental level of security provided by the algorithm, and general descriptions of its applications.

The algorithm is designed to encipher and decipher blocks of data consisting of 64 bits under control of a 64-bit key. For any possible key (56 bits of the key are used by the algorithms and 8 bits are used for error detection during key management), all possible combinations of 64-bit inputs result in all possible combinations of outputs. The decryption operation is the inverse transformation. For a particular key, the same input results in the same output. Since the result of either operation depends on every bit of input and every bit of the 56-bit effective key, any change causes every bit of the result to change with equal probability. Thus, a mistake

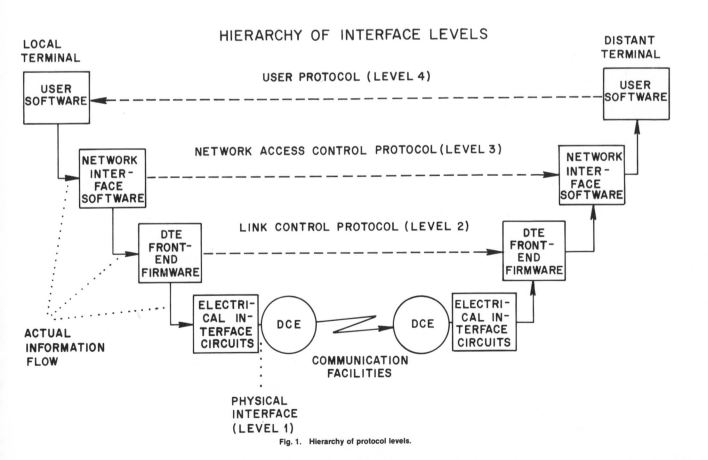

Fig. 1. Hierarchy of protocol levels.

in entering the key or the ciphertext or in the decipherment process results in a radically different output. This characteristic can be used to detect intentional or accidental errors in a communication system.

The DES algorithm has been implemented in integrated circuits which have been validated by NBS as complying with the standard [12]. To date, five devices have been validated. All produce identical results given the same data input and key. Each manufacturer, however, has provided various control functions for the devices, optimized various features of the devices (speed, cost, size, power consumption), and built them to fill an anticipated need in the marketplace. For communications purposes, additional standards are required to assure that these devices are able to communicate with other devices. Similar to the U. S. telephone system in which all (existing) telephones can communicate with one another, it is desirable that all computer terminals and peripheral devices be able to communicate with one another. However, similar to the early telephone system (At one time telephones in different areas, even though only across the street, could not communicate), many existing computer terminals and peripheral devices cannot interoperate. Communication standards and protocols have minimized the number of codes, speeds, physical interfaces, and protocols somewhat, but many devices still cannot communicate on a network without significant modifications or enhancements. Federal standards [11] and American National Standards Institute standards are being developed for the application of the Data Encryption Standard in data communications.

AN EXAMPLE OF NETWORK SECURITY

The National Bureau of Standards is developing a local data network intended to serve up to a thousand users of equipment having dissimilar device characteristics [5]. The objectives of the network include providing fully connected communication capability among terminals, microcomputers, minicomputers, large host computers, and laboratory instruments; providing flow control of data among communicating devices; providing service among some 20 buildings at NBS using a single coaxial cable; and providing selective cryptographic protection to the network. The remaining portion of this paper describes the security provisions planned for this network.

A user views the network as a long cable with a number of boxes connected into it. The boxes are microprocessor-based nodes called **Terminal Interface Equipments**, or **TIE's**. A TIE connects a network device (terminal, computer, or lab instrument) to the cable. Data travels through the cable at a signaling rate of one million bits per second. Data travels in logical entities called packets which appear (nearly) simultaneously at all TIE's. The function of the network is to transport data among TIE's. The purpose of network security is to assure that only the intended TIE can "read" the data, even though the normal protection features of the TIE have been circumvented and the data has been copied.

Every data packet on the network has the same form

(see Fig. 2). Each contains a destination address, a source address, a data field, an error detection field, and control information. The data field contains 0–128 bytes of information. Each address field contains 16 bits.

Each user of the network has a TIE connecting a device to the network. Each TIE can establish a single virtual connection to any other TIE to which data may be passed. A packet is assembled in the memory of a TIE before it is transmitted. Every TIE monitors the cable for a NOT BUSY condition. A TIE that has a packet ready for transmission begins sending when the NOT BUSY condition exists. Procedures are automatically invoked for retransmitting a message if more than one TIE attempts to initiate transmission concurrently. Every TIE monitors all transmissions, looking for packets having its destination address and originated by the TIE with which it has established a virtual connection.

The electronic equipment of a secure TIE consists of the network circuit board and the user circuit board of an ordinary TIE plus a security circuit board. The TIE is connected to the cable via the **network board**, which implements the protocols for network communication and collision avoidance. The **user board** contains a microprocessor and sufficient memory to hold the program which controls the TIE. The **security board** contains a DES device and a read-only memory chip containing the cryptographic key of the TIE as well as read-write memory for storing user keys and session keys. Data that is to be encrypted is sent to the security board from the user board. The presence, or absence, of the security board can be sensed by the user board. All TIEs have physical space for a security board, but only contain one if secure communication is desired for the associated device. Thus users not requiring secure communication pay no penalty for security.

CRYPTOGRAPHIC PROTOCOL

Cryptography is being integrated into the NBS network as a user/system controlled feature. Only the data field of any packet will be encrypted. A connection between any two devices can be established in either a regular or a secure mode. A secure connection requires that both TIE's in the connection contain security boards. A regular connection request has the following sample format:

CON:1723.

Assuming that the user is issuing the request from TIE 1711, a connection is established between TIE 1711 and TIE 1723.

The SYNC characters are used for synchronization. The destination address (two bytes) is followed by the source address (two bytes). CTRL is a control byte for the network. The DATA FIELD may contain 0–128 data bytes. The CRC is a 16-bit error detection field for the packet.

Fig. 2 Data packet in the NBS network

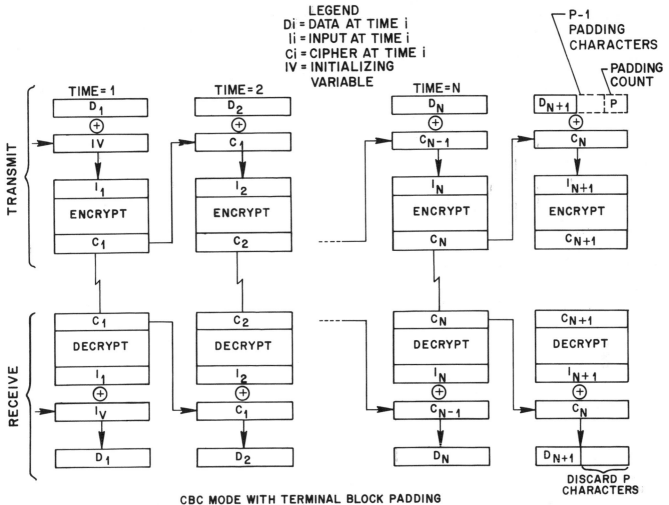

LEGEND
Di = DATA AT TIME i
Ii = INPUT AT TIME i
Ci = CIPHER AT TIME i
IV = INITIALIZING VARIABLE

CBC MODE WITH TERMINAL BLOCK PADDING

Fig. 3. Cipher Block Chaining

Assuming that a terminal is identified by its TIE and that a terminal also means a host computer, then a secure connection is similarly established in one of three ways: terminal–terminal, user–terminal, user–user. A mail system is also possible for secure messages for a particular person who is not present at the terminal. The message is stored in ciphertext form to be retrieved at a later time when the user is available.

Secure data will be encrypted with the DES using the **Cipher Block Chaining** [11],[16] mode of operation (Fig. 3). The data field of each packet will be encrypted as a separate entity. This formatted plaintext packet (Fig. 4) will be encrypted in its entirety before being

ID1 . . . ID4	DB1 DB2 . . . DBi	(PD1 . . . PDj)	CTL	EDC	ID1	EOM

ID1 through ID4 is an extendable packet identification number (0 - 268,435,455). DB1 through DBi are the data bytes to be transmitted. PD1 through PDj are 0-7 padding bytes used to assure that a multiple of 8 bytes are contained in the encrypted packet. CTL is an encryption control field. EDC is an encryption error detection code. ID1 is a repeat of the first byte of the packet identification number and EOM is an end-of-packet flag.

Fig. 4 Data field of an encrypted packet

transmitted. Anyone monitoring network traffic could receive the encrypted packet, but only the TIE with the correct key can decrypt the packet. The communication security problem is thus reduced to a key management problem.

Once a secure connection is established, it remains secure until the connection is broken. The same key is used for both directions in the connection. As soon as the connection is broken, the key is destroyed.

KEY MANAGEMENT

Key management is a critical aspect as a cryptographic system. Physical security and operational reliability are related aspects. Key management for the NBS secure network will be performed by two identical microprocessor systems dedicated to the task of key management. These two systems are called Network Security Centers or NSC's. The systems will be physically isolated and individually protected. They will appear to the network as two TIE's with a "rotary" address sequence (if the first TIE is busy, the second will automatically be tried). The hardware, software, and data bases of the two NSC's will be identical.

The key management scheme for the NBS network will

be based on a hierarchically structured set of cryptographic keys [10]. The hierarchy will include master keys, submaster keys, and session keys. Master keys will only be stored in physically secured memory within or close to the DES devices. Submaster keys will be enciphered under the master keys and will themselves be used to encipher other keys. Session keys will be generated by the NSC's and distributed to the security boards in the TIE's participating in a secure communication. Session keys will have a lifetime equal to the length of the session. Master keys will have a fixed, finite lifetime unless they have been compromised. Submaster keys will have a lifetime selected by a user.

The objective of the key management scheme is to establish a secure connection between TIE's in such a manner that the TIE's (hence the connected terminals) are authenticated, the users are authenticated (if requested by the initiating user) and a one-time session key is distributed to the TIE security boards for data protection. Several design approaches are currently under consideration. The operational implementation will depend on factors such as network security management, methods of manually protecting master keys, and overhead in establishing a secure connection.

SECURE CONNECTIONS

The key management system presently under consideration uses one key in each TIE for authentication of the TIE (and its attached terminal/computer) and one key for each user for authentication. All terminal and user keys are also stored in each NSC. A master key in each NSC is used to encrypt lists of all terminal keys and user keys. A user desiring a secure terminal to terminal connection issues the following exemplary command:

TTS:1723.

TTS stands for terminal–terminal secure and 1723 is the identifier of the terminal (TIE) that is being "called." The "calling" TIE contains a program to establish a secure connection to the "called" TIE in the following manner. Assume that the user is making the request from TIE 1711. A regular connection is made by TIE 1711 to the TIE of the NSC. The following message is transmitted to the NSC by TIE 1711:

TTS:1711:1723.

The NSC, upon receiving this message, acts as a service center in the establishment of a secure connection between terminal 1711 and terminal 1723. The NSC finds the terminal keys (encrypted under the NSC master key) for terminals 1711 and 1723 in its terminal key storage. Then using the DES it generates a random number which is defined to be the session key encrypted under the NSC master key. The NSC contains an integral DES device and does not require a security board in its TIE. The NSC decrypts TK1711 (the key for terminal 1711), TK1723, and the encrypted session key. After setting the parity bits of the session key, the NSC encrypts the session key (SK) using TK1711 and again using TK1723. It sends to terminal 1711 the following formatted message:

TTS:1711:1723:E-TK1711(SK):E-TK1723(SK).

E-TK1711(SK) means that the session key (SK) is encrypted (E) under the terminal key of 1711 (TK1711). TIE 1711 thus receives the session key encrypted under its own key as well as the key for TIE 1723.

TIE 1711 then generates a random number called the **initial vector (IV)**. The IV is used to randomize the first block of each packet transmitted during a secure connection. After a regular connection is made to TIE 1723, TIE 1711 transmits the following:

TTS:1711:1723:E-TK1723(SK):E-SK(IV).

TIE 1723 uses its stored terminal key to decrypt the SK. To verify that the SK received by TIE 1723 is the same as the SK received by TIE 1711, the IV is decrypted using SK1723 (i.e., the SK received by TIE 1723), the first 32 bits of the 64-bit IV are complemented, and the modified IV is encrypted under SK1723. TIE 1723 transmits the following formatted message to TIE 1711:

TTS:1723:1711:E-SK1723(modified IV).

TIE 1711 then decrypts the modified IV with SK1711 and compares it with the IV originally sent. The first 32 bits should be complemented and the last 32 bits should be identical. If correct, the following message is transmitted:

TTS:1711:1723:OK.

If incorrect, the following message is transmitted:

TTS:1711:1723:BAD.

If the connection is OK, all transmissions between the TIE's are encrypted in Cipher Block Chaining (CBC) mode using the SK and the IV. If the connection is BAD, the connection is aborted.

USER–USER CONNECTION

The previous section described a terminal–terminal connection in which the identity of the user on either end of the connection was unimportant. In many private communications, the identity of the calling or called party is important. The following describes the procedures planned for the secure NBS network in user-user connections. Terminal–user or user–terminal connections will not be described in this paper; they simply use the appropriate procedures established for each end of the secure connection.

A user (having identity U100) seeking to establish a private connection to another user (having identity U200) must supply the following formatted message to the TIE:

UUS:1723:U100:U200:UK100.

This requests that the "calling" TIE (assume TIE 1711) establishes a secure connection to U200 at terminal 1723 and supplies to TIE 1711 the user key for U100. It is assumed that U200 is present at terminal 1723.

TIE 1711 transmits the following formatted message to the NSC:

UUS:1711:1723:U100:U200.

The NSC finds the terminal keys for 1711 and 1723 in the encrypted terminal key storage and the user keys for U100 and U200 in the encrypted user key storage. The NSC generates a random session key encrypted under NSC master key. The five encrypted keys are used to produce the following formatted message which is sent to TIE 1711:

UUS:1711:1723:E-UK100(E-TK1711(SK)):
E-UK200(E-TK1723(SK)).

TIE 1711 then obtains the session key SK1711 by decrypting the received encrypted key using its TK1711 and UK100 contained in its key storage.

TIE 1711 produces the following message and, after establishing a connection to TIE 1723, transmits the following message:

UUS:1711:1723:U100:U200:E-UK200(E-TK1723(SK)):
E-SK1711(IV).

Upon receiving this message (in plaintext), TIE 1723 requests that U200 supply user key (U200) by printing the following message on the terminal associated with the TIE:

TIE: Connection Request from U100;
Please supply U200 Key:

Upon entry of the U200 key, TIE 1723 deciphers the session keys using TK1723 and UK200, deciphers the IV using the deciphered session key, modifies the IV as previously described, enciphers the modified IV, and transmits the following formatted message to TIE 1711:

UUS:1723:1711:U200:U100:E-SK1723(Modified IV).

TIE 1711 deciphers the modified IV with SK1711, compares it with the original IV, and the secure connection is considered established if the IV has been properly modified. A message is sent (in plaintext) to TIE 1723, informing it if the connection is OK or BAD. If OK, all further messages are enciphered in Cipher Block Chaining mode using the original IV.

Correctly receiving the modified IV at TIE 1711 verifies the identity of U100, U200, TIE 1711, TIE 1723 and verifies that the session key and the IV are correct and ready for use.

DESIGN RATIONALE

The initial design of the NBS secure network key management scheme is based on key management concepts [1],[2],[9],[10] that have been reduced to the minimum number of connections and messages necessary to establish a secure connection. It must be recognized that trade-offs are necessary between security and convenience in such a design. Convenience was emphasized in this example for easier understanding. Only five messages are required to establish and verify a secure connection. Because the only errors detected are key parity errors and improper key distribution errors, the only error reaction is to abort the attempted connection. As a result,

the cause of failure to establish a connection cannot be easily surmised in this simple design.

Implementation of these network security procedures (planned to be operational in 1980) will depend on additional reliability and security requirements identified during early network development. For example, additional anti-spoofing procedures will be provided to prevent a penetrator from recording an encrypted terminal–terminal connection procedure and subsequent communication and then replaying it back through the "calling" terminal after gaining physical access to the terminal (but not gaining access to the terminal key).

Requirements for physical security of the TIE's and the NSC's are expected to change during implementation of the network. Economy is presently a major design criteria. Physical locks on the TIE units which prevent unauthorized access to the TIE key will be provided. TIE keys will be generated at the NSC, written onto a Read Only memory (ROM) device, and physically transported to a TIE. User keys will be generated at the NSC and issued to users, either in printed form, on a magnetic striped card, or in a ROM device to be inserted in a port of the TIE or some combination thereof. Administrative procedures will be established so that user keys can be changed by both the user and the system.

Physical security of the NSC will include protection of the master key in the DES device and protection of the random access memory while the computer is operating. Keys stored on the disks will be encrypted with the master key, and hence will not require physical protection. However, if a master key is compromised, all key lists in both NSC's must be rebuilt and the master key in both NSC's must be changed. Dual NSC's are used for reliability purposes.

CONCLUSIONS

Security of computer communications can be provided through cryptography. Cryptographic algorithms and devices implementing them are available for this pur-

---

Through the use of small dedicated computers for storing and generating keys, the problems of key management can be reduced to manageable proportions.

---

pose. Communications protocols are being established which makes the transparent use of cryptography practical. Through the use of small, dedicated computers with sufficient data storage and processing capability for storing and generating keys, the problems of key management can be reduced to manageable proportions. If public-key cryptographic algorithms [7],[15] are developed which can be effectively used for key-management functions, the physical security requirement for NSC's will be eliminated or reduced to that of a public key distribution system. By enhancing the capability of the NSC and increasing the number of messages that it can handle, additional security features can be provided such as auditing, security surveillance, and host computer security interfacing. Comments and suggestions in these areas as well as in means of extending the protocol for establishing communication security will be welcomed by the author.

BIBLIOGRAPHY

[1] D. Branstad, "Security aspects of computer networks," in *AIAA Conf. Proc.*, Huntsville, AL, Apr. 1973.

[2] —, "Encryption protection in computer data communications," in *4th Data Commun. Symp. Proc.*, IEEE Computer Society, Oct. 1975.

[3] D. Branstad, J. Gait, and S. Katzke, "Report of the Workshop on Cryptography in Support of Computer Security," NBSIR 77-1291, National Bureau of Standards, Sept. 1977.

[4] D. Branstad, "Computer security and the Data Encryption Standard," NBS Special Publ. 500-27, National Bureau of Standards, Feb. 1978.

[5] R. Carpenter, J. Sokol, and R. Rosenthal, "A microprocessor-based local network node," in *COMPCON Proc.*, IEEE Computer Society, Sept. 1978.

[6] Data Encryption Standard, Federal Information Processing Standard (FIPS) Publ. 46, National Bureau of Standards, U.S. Department of Commerce, Jan. 1977.

[7] W. Diffie, and M. Hellman, "New directions in cryptography," *IEEE Trans. Inform. Theory*, Nov. 1976.

[8] ——, "Exhaustive cryptanalysis of the NBS Data Encryption Standard," *IEEE Computer*, June 1977.

[9] S. Ehrsam, S. Matyas, C. Meyer, and W. Tuchman, "A cryptographic key management scheme for implementing the Data Encryption Standard," *IBM Syst. J.* vol. 17, no. 2, 1978.

[10] J. Everton, "A hierarchical basis for encryption key management in a computer communications network," in *Trends and Applications: 1978 Proc.*, IEEE Computer Society, May 1978.

[11] Federal Standard 1026 (proposed), "Telecommunications: Compatibility requirements for use of the Data Encryption Standard," National Bureau of Standards, Oct. 1977.

[12] J. Gait, "Validating the correctness of hardware implementations of the NBS Data Encryption Standard," NBS Special Publ. 500-20, National Bureau of Standards, Nov. 1977.

[13] P. Meissner, "Report of the workshop on estimation of Significant Advances in Computer Technology," NBSIR 76-1189, National Bureau of Standards, Dec. 1976.

[14] C. Meyer, and S. Matyas, "Cryptography: A new dimension in computer data security," KETRON, Inc., 1978.

[15] R. Rivest, A. Shamir, and L. Adleman, "On digital signatures and public-key cryptosystems," Tech. Memo 82, MIT Lab. for Computer Science, Apr. 1977.

[16] L. Tuckerman, "Block-cipher cryptographic system with chaining," U.S. Patent 4 078 152 assigned to the IBM Corporation, Mar. 1978.

Dennis Branstad is the project leader for computer security at the Institute for Computer Sciences and Technology within the National Bureau of Standards. Since 1973, he has been responsible for the development of the Federal Data Encryption Standard. He has written several papers on the use of cryptography in computer network security applications. He has chaired or participated in several Federal computer security standards committees and in several ANSI activities developing cryptographic standards in financial transactions and data communications. He holds a Ph.D. degree in Computer Science from Iowa State University and is an Adjunct Professor of Computer Science at the University of Maryland.

AN APPROACH TO SECURE VOICE COMMUNICATION BASED ON THE DATA ENCRYPTION STANDARD*

M. J. Orceyre and R. M. Heller

I. INTRODUCTION

Telephone communications have been understood from their beginnings to be vulnerable to interception (unauthorized reception). In recent years, with increasing public and private sector reliance upon electronic media for communicating sensitive technical, financial, military, political, economic, and personal information, and with the rapidly increasing use of microwave and satellite telephone carrier media, concern about these vulnerabilities has mounted dramatically. Starting in mid-1977 there has been considerable attention given in the news media to the matter of wholesale interception by foreign governments of American private and commercial voice and data communications. Publicly available documents note the ease with which such common carrier transmissions can be "captured" for subsequent analysis and use by unauthorized listeners. Fig. 1 illustrates the many vulnerabilities of a typical public switched telephone network.

Within this broad framework, the matter of secure voice communication—enabling speakers to converse naturally over telephone media without fear that their conversation can be usefully intercepted—poses special problems and is receiving close attention within both the commercial and the Government sectors.

A. Encrypted Digital Speech

Among the many approaches to providing secure voice communications capabilities is the use of digital speech technology in combination with digital encryption techniques. This paper describes such an approach to secure voice—the application of the Data Encryption

The authors are with the IBM Corporation, Federal Systems Division, Gaithersburg, MD.

*Reprinted from *IEEE Communications Society Magazine*, November 1978, Vol. 16, No. 6, pp. 41–50.

Standard (DES) [1] to an efficient digital voice communication system such that:

1) interception of signals on any medium is extremely unlikely to be worthwhile since transformation of the intercepted signal into clear speech will be virtually impossible even with a very large cryptanalytic investment on the part of the interceptor, and

2) no great cost, performance, or human factors burdens are perceived by the communicating parties.

B. Basic Design Influences

In the authors' view, there are four considerations which should strongly affect secure voice system design:

1) *Network compatibility.* It is best if the secure voice system works with and has no effect upon the existing structure of the common carrier network. There should be no network changes and few if any service constraints imposed by the presence of secure voice devices.

2) *Human factors.* It is best if the secure voice system maintain the existing carrier speech quality, impose no awkwardness upon its users (such as button-pushing), and introduce no perceptible degradation of other service qualities (such as increased connection time, frequency of incidents requiring redialing, noticeable intermittent garbling, or loss of signal requiring speaker repetition, and so on.

3) *Economics.* Use of current public switched networks is desirable. Cost of the secure voice equipment is a strong acceptability factor and should be minimized. Placement of the secure voice equipment (at terminals, handsets, or at switches, or at switch-trunk interfaces) may well be determined not only by security requirements, but also by cost considerations, especially if, by placing devices at the switches, they may be shared among a large number

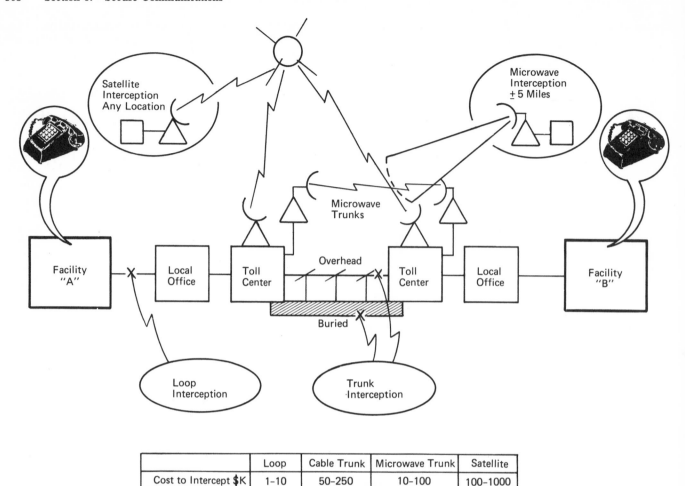

	Loop	Cable Trunk	Microwave Trunk	Satellite
Cost to Intercept $K	1–10	50–250	10–100	100–1000
Risk of Detection	High	High	Low	Low

Fig. 1. Switched Network Interception Vulnerabilities.

of telephones, thus reducing the number of devices required.

4) *Security*. The encryption algorithm and the manner of its use must be such that the cost to derive useful information from the intercept is prohibitive. The theoretical strength of the encryption algorithm must not in any way be diminished by the particular method of use, packaging, or required support procedures.

C. Assumptions

This paper assumes the existence of (and does not further discuss) an acceptable quality, low-bandwidth, realtime digital voice communication capability using low-cost microprocessor-based technology. The intent of the paper is to illustrate how a powerful encryption algorithm, the DES, can be integrated into such a digital voice system in order to satisfy the four fundamental design requirements outlined above.

It cannot be emphasized too strongly that the descriptions contained herein are illustrative of only some number of problems and solution approaches encountered in the design of a secure voice system. They are not intended, and should not be taken, as formal functional specifications representing a complete solu-

tion for specific system and cryptologic problems. The designer of a secure voice system must:

1) Assess all significant threats against his network.
2) Provide protective features to counter these threats.
3) Carefully evaluate both the short-term and long-term effectiveness of his system design.

The paper also assumes a context of a *secure voice device* (SVD) embodying digitization, modem, crypto, and control components whose intelligence and sensing capabilities enable it to be fully transparent to the public switched voice network (Fig. 2). The approach taken is simple; the device is able to sense that a completed connection has occurred (the dialed telephone has been reached and has gone off-hook), and at that moment it engages in transmission with its partner device at the other end. There is no opportunity for clear analog speech. In a fraction of a second (apart from existing network propagation delays) the devices establish full digital synchronization and crypto initialization, and secure digital speech is enabled.

D. Crypto-Related Concerns

The approach taken in this paper introduces no complications in the voice digitization process. The

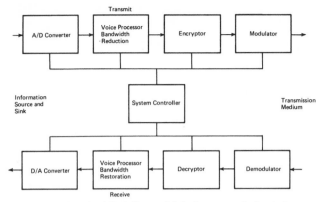

Fig. 2. Functional block diagram, full-duplex secure voice terminal.

digitization function need not be "aware" that encryption and decryption are taking place. At the control level, there is a design impact; in the design, by choice, a single controller component governs the flow of all digital streams within the secure voice device. The controller "sees" both the digitizing and the encrypting components as slave elements to itself.

Security considerations led to a design that can provide, with nearly identical devices, either true end-to-end (telephone to telephone) or approximate end-to-end (local switch to local switch) protection; in fact, both could be employed in different locations and communicate perfectly well with each other. The former is more secure in that clear speech exists only in the telephone device itself; the latter is less expensive in that it permits sharing the devices across the telephone population attached to that local switch. Crypto compatibility was the main concern, and it is achievable.

Of special concern in a digital speech switched network system are complications that can be introduced by the addition of a crypto function. Digitized encrypted voice can be extremely sensitive to line noise, bit errors, and losses of bit synchrony. Error conditions from which the other components of the secure voice device might easily recover can be propagated partially or even indefinitely by a crypto component, with resulting permanent loss of a useful clear signal. By the same token, the crypto can be integrated in such a way as never to introduce catastrophic error propagation; the proposed method accomplishes this while maintaining the full theoretical strength of the DES algorithm.

A crypto function must operate in absolute bit synchrony not only with its local speech digitizing component, but also with its remote crypto partner. This requires initialization exchanges between the crypto components at the start of transmission. It is during this necessary startup exchange, described later, that the crypto units are placed (by each other) in "logical" synchrony; that is, they are seeded with identical crypto keys and with identical initializing vectors (data input patterns), so that they can subsequently decipher each other's transmission.

Given the need to frequently and randomly change crypto keys and initializing vectors—since in the secrecy of these quantities lies the only security in the system—the following presentation is fairly detailed in discussing

the matter of secure crypto key distribution, storage and use, and the generation and updating of initializing vectors.

II. FUNDAMENTALS OF DES AND ITS EMPLOYMENT

A. Basic Crypto Concepts

The DES is a mathematical function for transforming a given pattern of 64 bits (input) into a different pattern of 64 bits (output). Certain parameters of the transform are controlled by a 64-bit key, of which 56 bits are used in the algorithmic transform and 8 bits are parity, used only for detecting errors within the key itself. The key is selected by an agency external to the algorithm, and it may be changed from one transform to the next if desired. The DES, of course, may be embodied in software, microcode, or hardware. The DES concerns only the transform of 64-bit patterns; it does not recommend or even address how the algorithm should be employed. There are, in fact, many different ways to configure the DES in a system to protect digital data, including digitized voice.

In this and the following section a brief review of certain terminology and concepts of cryptography is given in order to make the paper self-contained. Much greater detail on DES applications, including key management techniques, is provided in [2,3].

The simplest configuration involving the DES is given in Fig. 3. It is an example of data-dependent encipherment in a block-cipher form. A cipherstream output is considered to be *data dependent* (independent) if each bit in the stream is (is not) a function of other data bits.

The input data stream is divided into 64-bit blocks, each of which is successively transformed by the DES algorithm into an output ciphertext block of 64 bits. The decipherment operation is the reverse of encipherment. [1, p. 10].

The rationale for encipherment with chaining, the next higher level of DES application, is to prevent identical blocks of input data from appearing, after encipherment,

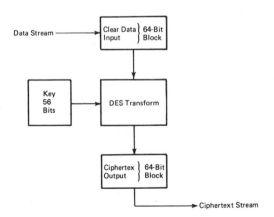

Notes:
(1) One or more bit errors in the received ciphertext stream result in loss of up to 64 clear bits (one whole block) in the deciphered cleartext stream.

(2) The insertion or deletion of bits in the received stream will permanently result in loss of crypto sync unless external protocols exist for resynchronizing.

(3) Synchronization between crypto units must be either exact, or to a multiple of 64 Bits.

Fig. 3. Data dependent encipherment: DES block cipher.

as identical blocks of output ciphertext. This result can be accomplished by *block-chaining feedback* (Fig. 4). The 64-bit output ciphertext is XORed (exclusive OR operation) with the next input data block. By convention, at the start of signaling, a 64-bit *initial vector* (*IV*) is chosen for use by both the enciphering and deciphering operations; it may be permanently fixed, or randomly generated each time as required. Deciphering is the reverse of enciphering—the same configuration is used, but with the DES in a decipher mode.

Another data-dependent chained encipherment variation is *stream enciphering with ciphertext feedback* (Fig. 5). *J* bits of the data stream ($1 \leqslant J \leqslant 64$) are XORed with *J* bits of the DES output. *J* bit feedback is from the ciphertext output into the DES input. In both this variation and block-chaining feedback, a single error in the received stream will cause error propagation over a 64-bit block. If bits are inserted in or deleted from the received bit stream, the $J = 1$ mode automatically returns to crypto sync after 64 correct bits have shifted into the receive DES input. Deciphering is the reverse of enciphering except that both the transmit and receive DES's are run in the same mode to produce the same bit stream as input to the XOR. The NBS Crypto Laboratory has implemented this approach with $J = 8$.

In order to prevent error propagation (which may be intolerable in a noisy narrowband digital voice environment) a data-independent stream cipher application is recommended.

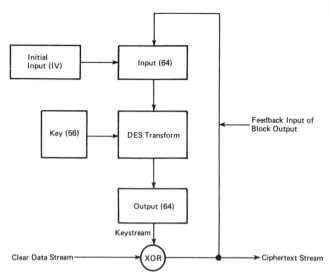

Fig. 5.   Data dependent encipherment: stream cipher with ciphertext feedback.

Fig. 6 depicts the DES algorithm used in a data-independent chaining mode. The DES algorithm in this employment is used solely as a random-bit stream generator. The successive outputs of the DES form a keystream which is XORed bit-for-bit against 64-bit blocks of the clear data stream. To decipher it is necessary only to XOR the identical DES output keystream against the received ciphertext stream. The keystream output is fed back upon itself, hence the name *key-autokey*.

This employment also requires that the deciphering DES function be seeded with the proper DES 56-bit key and initializing variable (IV); only then can it produce the required output random keystream. Synchronization is required between the two crypto functions to the exact bit.

Note that the ciphertext bit-error propagation effect does not occur. A one-bit error in the received ciphertext stream results in only a one-bit error in the cleartext

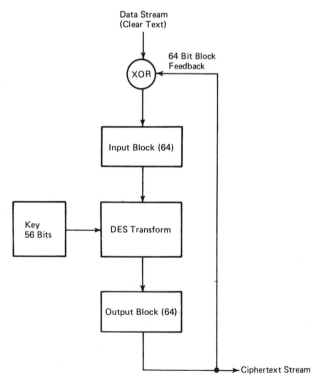

Notes:

(1) Each 64-Bit block of ciphertext output is fed back and XOR'ed against the next incoming 64-bit block of cleartext; the next input to DES is the result of this XOR.

(2) The error conditions are the same as for Figure 3.

(3) Synchronization must be exact (to the same bit) between crypto units.

Fig. 4.   Data dependent encipherment: block-chaining feedback.

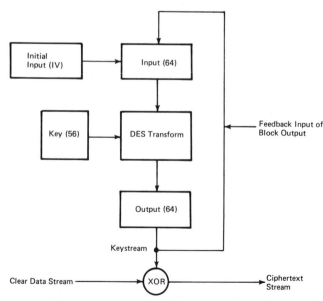

Fig. 6.   Data independent encipherment: key-autokey mode.

output of decipherment. However, external protocols are needed to protect against insertion or deletion of bits; such protocols normally exist in a communication system independently of any encryption mechanism.

There are many variations on the key-autokey theme. However, in all cases the important point to note is that, given system protocols that maintain synchrony and are present independently of any crypto function, the key-autokey crypto implementation cannot propagate bit errors—it adds no element of fragility to the system.

B. Crypto Key Management Considerations

Most cryptographic systems employ two different kinds of crypto keys:

1) *Master* keys, used only to encipher/decipher other keys, and

2) *Data* keys, used to encipher/decipher anything except other keys. A key in the process of being used to encipher/decipher data is sometimes called the working key.

An enciphered key is considered as secure (well-protected) as enciphered data. It may therefore be handled in the same manner as enciphered data. It may be transmitted throughout the network, kept in electronic storage (normally in tables of keys), and so on, as long as it remains enciphered.

A cardinal principle in a well-designed DES-based hardware implementation is that *no working key shall ever appear in the clear except within the DES engine itself* in any operational system. (The DES *engine* is comprised of the input, transform, output, and *Working Key Register* (WKR) elements.)

An illustration of a DES configuration with master key and data key is given in Fig. 7. This configuration can perform two types of crypto functions:

1) Enciphering/deciphering a working key (data key) using the master key, and

2) Using the working key to encipher/decipher data in normal operating mode.

In the first instance, the master key which has been stored in the *Master Key Register* (MKR), is loaded into the WKR. To decipher a key, the enciphered data key is loaded into the input register and a decipher-under-master key operation is performed; the DES output is gated directly into the WKR and only there; it displaces the master key value, and the device is ready for normal operation, enciphering or deciphering data. To encipher a key (performed only during a short, well-protected key generation operation), the clear key is loaded into the input register, the encipher-under-master-key operation performed, and the enciphered key is fed into the output register.

A crypto device should contain a relatively permanent master key that is used in a very limited way and is very carefully protected. In fact, once the assumption has been made that data keys, when enciphered, are themselves as safe as the data they are to protect, the protection

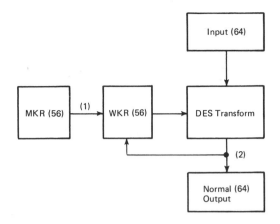

Notes:
(1) The master key can be moved only into the working key register, from which there is no exit other than into the DES implementation.

(2) The transform result (output) can be moved either into the normal output register or into the working key register but never into both.

Fig. 7. DES master key/data key implementation.

afforded by the entire crypto system can be seen to rest ultimately upon the secrecy of the master keys (refer to item 5 below).

DES crypto key management for both master keys and data keys includes the elements of:

1) *Key Generation*: how the secret crypto keys are created;

2) *Key Distribution*: how the keys, once created, are delivered (by manual or electronic means) to the geographically dispersed DES engines that will use them;

3) *Key Insertion*: how the keys, once delivered to their respective engines, are kept available to and actually loaded into the appropriate registers within the engines for use;

4) *Key Life or Persistence*: how often master keys and data keys should be changed to maintain the desired level of security; and

5) *Key Protection*: the protection of keys and related programs resident in hardware and software and the security aspects of personnel and procedures concerned with system operation and security. This can involve concerns ranging from component packaging to human factors engineering.

III. DES-BASED SECURE VOICE DEVICE (SVD) SYSTEM

A. SVD System Overview and Assumptions

The recommended approach to a secure voice capability via digital encryption assumes a basic functional configuration as illustrated in Fig. 8.

No assumption is made as to the position of circuit switches in the approach. As drawn, for example, there could be a switch at (1) or at (2) or at both or at neither; the same is true for the receiving side.

The assumption is made that the functional capability of the speech processor and the modem is such that signaling and synchronization requirements (modem-to-modem and speech processor-to-speech processor) are

Basic Configuration

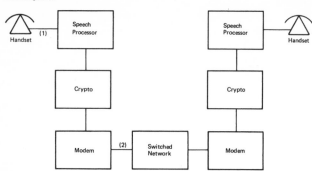

Fig. 8.   Basic SVD Crypto Approach

met independently of the crypto function. If this is true, then the recommended DES crypto function introduces no great additional synchronization problems. What is required is that the encryption function on the sending side and the decryption function on the receiving side begin to operate at precisely the same point (bit) in the speech-driven digital bit-stream. Once the modems and the speech processors are in sync, the crypto must start on the first bit of a suitably defined speech frame. This can be done simultaneously with establishing frame synchronization by a convention that starts the crypto a fixed number of bits after reception of an initial frame synchronization sequence.

The assumption is also made that each speech processor/modem set has its own crypto engine. This assumption is based on 1) the probability that the cost of the speech processor/modem set is such that the addition of a crypto engine and its controls to each set is not a relatively significant increase, and 2) the probability that there will be cases in which an isolated speech processor/modem set (at a remote telephone extension, for example) must have its own crypto engine and that uniform design of all secure voice devices is desirable.

In order to provide a network setting for the illustrated SVD approach the following situation is assumed:

1) Each physical location with one or a cluster of secure voice devices is called a *site*. A site may include a cluster of SVD's physically located at a circuit switch and it may also include one or more SVD's located with off-premise telephone extensions from that circuit switch.

2) All SVD's at a given site use the same DES master key. This key is changed infrequently (say quarterly) and can only be changed by a local physical action.

3) All SVD's at a given site use identical "*current site-key tables.*" This is a table containing one entry for each site; this entry is a 64-bit (8 byte) DES key enciphered under the given site's master key. This table can be sizable. In large networks there may be several thousand bytes of key table. The table is replaced at a more frequent rate than the site master key (say monthly).

4) All SVD's at a given site have identical "*future site-key tables.*" This table is the table that replaces the current site-key table at the next specified key change-over; it then becomes the next current site-key table. Its storage requirements (unless the number of sites has changed) are the same as the current site-key table.

5) The future site-key tables are filled with the next set of site keys at a key distribution time which occurs prior to key change-over, and is discussed fully in section C below.

B. SVD Key Management Procedures

This approach assumes that site-key tables are created at a single location for the entire network, and are distributed electronically from that location over the network to each other site. Key generation procedure is well described in [4]. Of special importance here is the method of site-key table distribution and verification.

Sites all have different master keys. For the site-key table to be usable at a given site, it must exist at that site as a table of keys enciphered under that site's unique master key. Therefore, prior to its distribution, the generated common table of new site keys must be enciphered, for each site, under that site's master key. When this step has been performed, there are N distinct site-key tables where N is the number of sites. Each table will be transmitted to its intended destination site.

A final step in the site-key table creation process is aimed at ensuring that errors in the subsequent distribution process will be detected; this is a verification procedure. A random 8-byte vector, R, will be generated at the distribution center and appended to each of the enciphered site-key tables. For each of the site-key tables, a distinct *verification vector*, VV_i, $i = 1,2, \cdots, N$, is computed. Each VV_i will be as long as the site-key table. This computation is accomplished by superencipherment of the already enciphered site-key tables using the block chaining feedback DES employment of Fig. 4 with R deciphered under the destination site's master key as the working key. One complete DES cycle is run with R as data input to provide the first 64-bit random feedback vector and then the entire site-key table is enciphered. The resulting VV_i (superenciphered site-key table) is kept with the corresponding enciphered site-key table at the distribution center. The actual steps performed in the key distribution and verification process are described below in Section C.

Since at key generation time, keys may briefly appear in the clear, site key and site-key table generation is conducted independently of the network and under the strictest physical security conditions. The process can be carried out using either a software or hardware DES. The enciphered site-key tables and their verification vectors can be produced well in advance of the time they must be distributed. When stored, the tables are in their enciphered form.

C. Key Distribution

Sites with secure voice equipment will require periodic updating of master keys and of site-key tables. These processes are different and are described below.

Master keys cannot be distributed electronically. By definition their security cannot depend upon their encipherment under other keys. No other key is considered to require more protection than the master key. In this system, as illustrated, anyone who possesses any site master key can, in theory, obtain all site keys for the system.

A master key is held on a separate pluggable *field-replaceable unit (FRU)* within each processor. All such devices at a site contain the same 64-bit master key. A site may have several spare master-key FRU's available (held under strict physical security) as replacement units.

The master keys at a site are updated by physical replacement of the FRU's with a new set. All must be replaced at very nearly the same time, since the site will have available a common new site-key table for use with the new master key.

The master key must always be available. It can be held only in its FRU, so it must be protected against such conditions as power loss. For simplicity in this system, the master-key FRU should be a nonvolatile ROM.

The master-key FRU should be secured within the processor by a physical lock and key arrangement.

Site-key tables are distributed electronically. The source may be the same site (and machine) at which the tables were created, or it may be a different site or machine.

The destination of the transmitted site-key table is each SVD at the site; at any single site, all SVD's use identical key tables, identically enciphered.

Each SVD with site-key table storage (space for the current table and the future table) must be capable of:

1) sensing a unique signaling stream indicating that a site-key table is about to be transmitted to it,

2) accepting the transmitted site-key table,

3) performing the specified verification of correct receipt, and

4) storing the received site-key table in the space reserved.

There are many variations to key distribution in this manner. A hierarchical scheme may be envisioned wherein each SVD can engage in this distribution process or each destination site can have a single table update function that receives the new table from the distribution site and then updates all devices at the site. However, all such approaches must include the following features and sequences of steps:

a) The distribution location possesses a copy of the new site-key table enciphered under the destination-site's master key.

b) The distribution location also possesses a copy of the verification vector for that site.

c) The distribution location establishes transmission linkage with the destination site SVD.

d) The distribution location then transmits the entire enciphered site-key table, including the random entry R.

e) Any error condition detected by the receiving site during this entire operation causes the reinitialization of the distribution process for that site.

f) The receiving site stores the transmitted table in the reserved space, except for R which is placed in the DES input register.

g) The receiving site performs a decipher-key operation, which leaves R undisturbed in the DES input register, but places a deciphered version of it in the DES Working-Key

Register. One DES cycle is then run to prepare the feedback loop for the subsequent operation.

h) The receiving SVD then proceeds with the steady-state crypto function generation of its VV. New site-key table entries are fetched, one at a time, from the stored table. XORed against the previous output of the DES cycle, run through another DES cycle, and transmitted back to the distribution location, until the last super-enciphered site-key has been so transmitted.

i) Any error condition detected by the distribution location, including failure of the transmitted verification vector to correspond exactly with the stored version causes reinitialization of the entire distribution process for that site.

D. SVD Crypto Configuration and Principles of Operation

The functional configuration of the SVD is presented in Figs. 9 and 10. Fig. 9 illustrates the crypto initialization and sync procedure flow and Fig. 10 shows the normal full-duplex, steady-state operation. This section discusses the SVD system blocks and basic principles of operation. Section E below gives details of the exact calling sequence.

For simplicity of illustration, the figures show one DES engine being shared for both transmit and receive in a full-duplex device. A nonerror-propagating, stream-cipher, key-autokey feedback encryption mode is used. By convention, both directions of transmission will use the called site's site key for encryption. In order to preclude the possibility of obtaining the information contained in a call by recording the enciphered digital voice stream and playing it back later through an SVD at the receiving site, the starting IV is formed in a special manner to be described below.

The Site-Key Store contains a table of all site keys in use during the current period. Each of the keys has been enciphered under the local-site master key. At both ends of the connection, the SVD Controllers, which are not illustrated, but which contain the coordinative intelligence for the SVD, cause the appropriate called site's key to be deciphered under the local site's master key and placed in the WKR's at the start of each conversation.

The *Transmit IV (TIV)* and *Receive IV (RIV)* Stores are 64-bit registers which feed the DES at both sites during transmit and receive operation respectively. TIV Store is a non-volatile register in order to permit creation of the starting IV (Fig. 9).

At initiation of a telephone conversation, if TIV_ϕ is the current contents of the originating site's TIV Store and TIV_A the current contents of the accepting site's TIV Store, then IV*, the starting IV for both sites is taken as IV* = $TIV_\phi \oplus TIV_A$. The TIV and RIV stores at both sites are reset to this value. In order to accomplish this computation, both sites must exchange the current contents of their TIV Stores. This exchange must be enciphered and is done using the block cipher mode with the called site's key as the working key. Of course, prior to the first use of the SVD's some random value must be placed in their TIV Stores. The contents of both the TIV and RIV Stores are continuously updated during normal steady-state operation (Fig. 10). The update occurs by

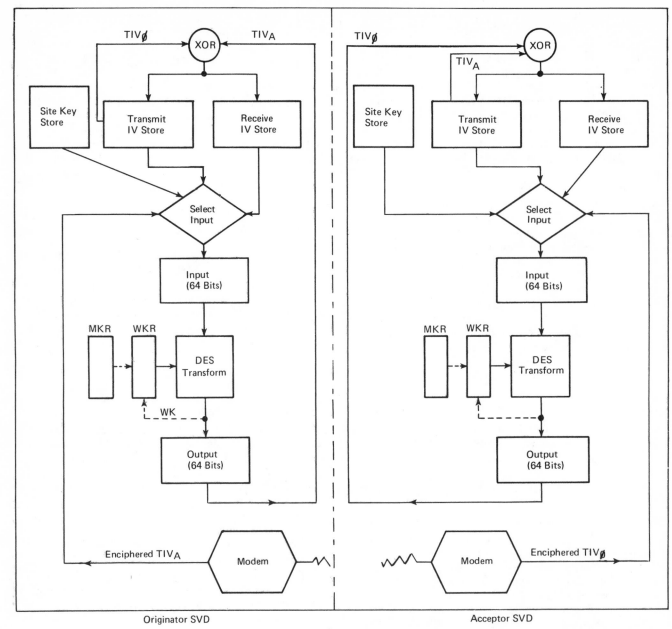

Fig. 9. SVD Crypto Initialization and Sync Configuration.

replacement of the current register contents with the result of an XOR of the current contents and the enciphered version of the current contents.

E. SVD System Calling Procedure

The actual series of steps performed by the communicating secure voice devices is described below. All signaling required to establish the connection is in the clear. The SVD's each contain four distinct logical functions: 1) the speech processing unit, 2) the DES digital crypto unit, 3) the modem unit, and 4) the controller unit that coordinates the operation of the SVD.

a) Initial dialing and connection signaling is in the clear. The controller routes this through the SVD directly or, depending on the architecture implemented, the controller does not see this signaling (because the SVD is not switched into the circuit until the circuit has been completed).

b) The signal for the Originator SVD to begin operation is architecture dependent. It may be the receipt of the "off hook" signal or the appropriate SVD modem tone from the Acceptor SVD. In no case can there be clear speech between parties connected by the voice circuit; the SVD's begin operation instantly upon circuit completion.

c) The modem synchronization sequence begins immediately. If synchronism cannot be established, the connection is terminated.

d) The Originator SVD Controller possesses the dialed number. Based on this quantity, it causes the appropriate associated site key to be loaded from the Site-Key Store into the crypto unit, deciphered under the local master key, and placed in the Working-Key Register. The Acceptor SVD Controller, knowing it is the acceptor of

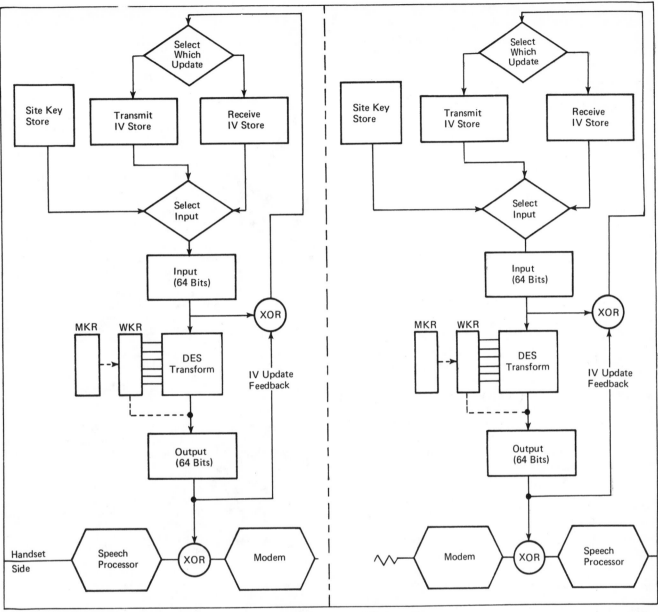

Fig. 10.   SVD key-autokey stream-cipher integration.

the call, causes its own site key to be loaded, deciphered, and placed in its own Working-Key Register. At this point (this may occur during the modem synchronization sequence) both the Originator and the Acceptor SVD's have crypto units ready for operation with the same key.

e) The Originator SVD Controller now causes the current contents of its Transmit IV Store, TIV_ϕ, to be loaded into its crypto unit and enciphered in a block cipher mode. The Originator SVD then transmits a "controller synchronization" pattern to inform the Acceptor SVD that the first bits of meaningful data are about to be transmitted, and appends to this pattern the enciphered version of TIV_ϕ.

f) The Acceptor SVD receives the controller synchronization pattern from the Originator and places the immediately following 64 bits in the input register of the crypto unit (Fig. 9). It then deciphers this value (which returns it to its original "clear" value) and xors it with

TIV_A, the current contents of its Transmit IV Store to get IV*. The Acceptor SVD then enciphers the current contents of its Transmit IV Store, transmits a controller synchronization pattern followed immediately by the enciphered TIV_A, loads IV* into both its Transmit IV Store and its Receive IV Store, and begins transmission of enciphered digital voice signals a fixed number of bits after the transmitted TIV_A.

g) The Originator SVD receives the controller synchronization pattern from the Acceptor, deciphers the immediately following 64-bit value. xor's it with the current contents of its Transmit IV Store, TIV_ϕ (thus replicating the New IV* that already exists in the Acceptor SVD), and loads IV* into both its IV Stores. The Originator SVD is now ready to commence steady-state operation, and begins transmission of enciphered digital voice signals following the transmission of another framing sequence.

h) Each SVD, upon receipt of the first 64 bits of enciphered speech data from the other SVD, enters the steady-state, stream-cipher, full duplex mode of operation. The TIV and RIV Stores are updated every 64 bits as described previously. The process is continued for each 64-bit block of enciphered speech data received by each SVD until the telephone conversation is concluded. In each SVD, of course, the deciphered speech data is served to the D-to-A side of the speech processor unit, which outputs clear analog speech for the telephone handset.

i) Upon completion of the conversation, the act of hanging up by either party causes the local SVD to be turned off or, depending on the architecture, switched off the voice circuit. In either case, the local SVD Controller must be sensitive either to the local "dead circuit" condition or to some "on-hook" signal generated by that condition.

IV. CONCLUSION

The SVD system approach developed in this paper attests to the feasibility of relatively low-cost, encrypted digital voice communications over the public switched network. The function can be added to the network with no effect upon the net—much as telephone answering machines are added today—with the exception that 4-wire telephone service is required. The human factor aspect can be very good—the existence of the system can be virtually imperceptible to its users—if in the design process this is a primary goal. Costs of digital secure voice devices are likely to be such, that for the next few years, casual purchase will be precluded; the devices are not toys, and will cost several thousands of dollars apiece. However, the protection gained is so great in relation to the cost, and the public awareness of telephone communication vulnerabilities is rising so fast that many unexpected public and private applications should emerge where the cost is clearly warranted. And finally, the security afforded by proper application of the DES algorithm should not be underestimated; the strength of the DES is vast relative to the cost of using it in microcircuitry.

REFERENCES

[1] Federal Information Processing Standard (FIPS) Publication 46, National Bureau of Standards, U.S. Dept. of Commerce, Jan. 1977.

[2] D. Branstad, "Security of computer communications," this issue, pp. 33–40.

[3] *IBM Syst. J.*, vol. 17, no. 2, 1978.

[4] S. M. Matyas and C. H. Meyer, "Generation, distribution and installation of cryptographic keys," *IBM Syst. J.*, vol. 17, no. 2, 1978.

Ralph Heller (M'60–SM'69) was born in New York City on April 4, 1934. He received the B.S. degree in mathematics from Long Island University and the M.S. degree in mathematics in 1957 from Yeshiva University, New York City. From 1957 to 1967, Mr. Heller was with the Advanced Development Subdivision of the Westinghouse Defense and Space Center, Baltimore, MD. His responsibilities included performance on and formulation of plans for research and development in the areas of error-correcting coding, digital communication technology and devices, channel characterization, signal processing, and computer reliability.

Since 1967 Mr. Heller has been with the IBM Federal Systems Division, Gaithersburg, MD. Currently he is an Advisory Engineer in the Telecommunications Department where he is responsible for system design and analysis in the areas of secure voice and data transmission, error-correcting coding, signal processing, and packet switching networks.

Mr. Heller holds five patents and has written numerous papers in the above areas. He has taught courses in coding and information theory at The Johns Hopkins University.

Michel J. Orceyre joined IBM in 1964 in the large systems development area in Poughkeepsie, NY, where he managed several systems software departments. In 1970 he became a member of an organization established by the company to develop requirements and basic architecture for security features and functions across the entire product line. This group instigated many research and development activities (including cryptographic work) throughout the corporation, and interacted widely, publicly and in private, with IBM customers and others who sought counsel in matters of data processing-related fraud and other losses, risk assessment techniques, and physical, procedural, personnel, operational, and technological countermeasures. Mr. Orceyre joined the Federal Systems Division (FSD) in Maryland in 1976, where he was involved in the design of the SATIN IV multilevel secure message switching system and in the creation of security-related portions of the Long-Range World Wide Military Command and Control System Architecture. He is now responsible for internal data processing security within FSD. Mr. Orceyre received the B.A. degree in 1959 from Holy Cross College, Worcester, MA.

WHY SPREAD SPECTRUM?*

Robert C. Dixon

INTRODUCTION

Through the properties of their coded modulation, spread-spectrum systems can provide multiple access, low interference to other systems, message privacy, interference rejection, and more. The up-to-date communicator should be aware of spread-spectrum techniques and how they can help solve difficult problems. These techniques are finding their way into many new programs. Here is an overview of the reasons why.

In recent years, a new class of communications systems has grown up around a modulation technique (or group of techniques) that comes under the general classification of "spread spectrum." This group of modulation techniques is characterized by having modulated signal spectra that do not resemble anything used before, in that they deliberately employ large bandwidths to send small amounts of information. In general, to be classified as spread-spectrum variety, the system must meet two criteria.

1) The transmitted bandwidth is much greater than the bandwidth of the information being sent.

2) Some function other than the information being transmitted is employed to determine the resultant transmitted signal bandwidth.

It is common in spread-spectrum systems to find transmitted RF signal bandwidths that are as much as 10^4 times the bandwidth of the information being sent. Some spread-spectrum systems have employed RF bandwidths 10^5 or 10^6 times their information bandwidth.

Why would a communications engineer in his right mind even consider employing such signals in a practical system? Because the process of spreading the signal bandwidth, and then collapsing it through coherent correlation with a local reference contained in the receiver offers a combination of advantages not available in any other way of communicating. These advantages are: selective addressing capability; code-division multiplexing; low-density output signals; inherent message privacy/security; high-resolution range measurement; and interference rejection.

To be sure, there are disadvantages, but these are often outweighed by the advantages. Two prime disadvantages may be listed for spread-spectrum systems: 1) they employ more bandwidth than a more "conventionally modulated" system using AM or FM, etc.; and 2) they are more complex in that they must include code sequence generators, correlators, code tracking loops, chirp or phase-coded matched filters, or other subsystems not necessarily needed in the more conventional systems.

The spread-spectrum techniques fall into one of three general categories, the first of which is called "direct-sequence" modulation. A second technique is denoted "frequency-hopping," while the third is the more familiar "chirp" technique often used in radar applications. Hybrid combinations of these have also been used in various systems to take advantage of specific properties.

DIRECT-SEQUENCE SYSTEMS

Direct-sequence spread-spectrum systems are so called because they employ a high-speed code[1] sequence, in addition to the basic information being sent, to double-sideband suppressed-carrier modulate their RF carrier. That is, the high-speed binary sequence phase (PSK) modulates the carrier, thereby determining the transmitted RF bandwidth and giving rise to the name "direct sequence." Binary code sequences as short as a few hundred bits or as long as 2^{89} (and longer) have been employed for this purpose, at code rates from fractions of a bit per second to several gigabits per second.

The result of modulating an RF carrier with such a code sequence is to produce a signal, centered at the carrier frequency, with a $(\sin x/x)^2$ frequency spectrum. The main lobe of this $(\sin x/x)^2$ spectrum has a bandwidth twice the clock rate of the modulating code, from null to null, and sidelobes whose null to null bandwidth is equal to the code's clock rate. Fig. 1 illustrates this most common type of direct-sequence-modulated spread-spectrum signal. Direct-sequence spectra vary somewhat in shape and in their relationship to the modulating code, depending upon the specific way in which the code is used to modulate the carrier (i.e., by phase-shift keying, pulse modulation, frequency-shift keying, etc.). The signal illustrated is that for a biphase, phase-shift-keyed signal, modulated by a code whose rate is R_{clock}.

The author is with Hughes Aircraft Co., Fullerton, Calif.

*Reprinted from *IEEE Communications Society Magazine,* July 1975, Vol. 13, No. , pp. 21-25. (short)

---

[1]The high-speed code sequences spoken of here are just long binary sequences (of ones and zeros) at bit rates usually in the range from one to a few hundred megahertz.

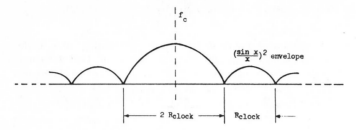

Fig. 1.   Direct-sequence signal spectrum.

FREQUENCY-HOPPING SYSTEMS

The wide-band frequency spectrum desired is generated in a different manner in a frequency-hopping system. It does just what its name implies. That is, it "hops" from frequency to frequency over a wide band. The specific order in which frequencies are occupied is a function of a code sequence, and the rate of hopping from one frequency to another is a function of the rate at which information is to be sent. The transmitted spectrum of a frequency hopping signal is quite different from that of a direct-sequence system. Instead of a $(\sin x/x)^2$-shaped envelope, the frequency hopper's output is flat over the band of frequencies used. Fig. 2 shows an ideal output spectrum of a frequency-hopping system. The bandwidth of a frequency-hopping signal is simply b times the number of frequency slots available, where b is the frequency separation between slots.

CHIRP SYSTEMS

Chirp systems are the only spread-spectrum systems that do not normally employ a code sequence to control their output signal spectra. Instead, a chirp signal is generated by sliding the carrier over a given range of frequencies in a linear or some other known manner during a fixed pulse period. This results in pulse-FM signal whose bandwidth is limited only by physical ability to shift a carrier frequency and by the ability to construct a receiver to demodulate it.

The idea behind chirp signals is that the receiver can employ a matched filter of a relatively simple design to reassemble the time-dispersed carrier power in such a way that it adds coherently and thus provides an improvement in signal to noise. Fig. 3 shows chirp signal waveforms.

HYBRID SYSTEMS

Spread-spectrum systems made up of combinations of the direct-sequence, frequency-hopping, and chirp systems are also practical and have been used in various configurations to

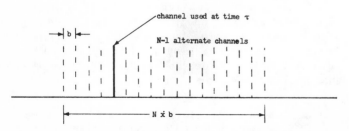

Fig. 2.   Frequency-hopping signal spectrum.

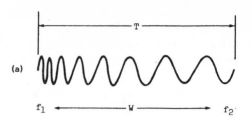

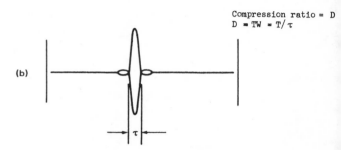

Compression ratio = D
D = TW = T/τ

Fig. 3.   Chirp signals. (a) Transmitted waveform. (b) Received waveform (dechirped).

exploit the properties available. Time-division multiplexing, or time hopping, is also applied in spread-spectrum systems to great advantage where a number of users must access a single link.

HOW SPREAD-SPECTRUM SIGNALING ADVANTAGES COME ABOUT

Selective addressing

Selective addressing is possible through use of the modulating code sequences to recognize a particular signal. Assignment of a particular code to a given receiver would allow it to be contacted only by a transmitter which is using that code to modulate its signal. With different codes assigned to all of the receivers in a network, a transmitter can select any one receiver for communication by simply transmitting that receiver's code; only that receiver will receive the message.

Code-division multiplexing

Code-division multiplexing is similar, in that a number of transmitters and receivers can operate on the same frequency at the same time by employing different codes. Either continuous transmission or time division is facilitated, since the synchronization inherent to transmission and reception of spread-spectrum signals provides an excellent time base for on and off timing.

Low density

Low-density transmitted signals are advantageous for prevention of interference to other systems as well as for providing a low probability of intercept. The low density of spread-spectrum signals is an inherent property which exists because of the bandwidth expansion. In a direct-sequence system, for instance, where the spectrum-spreading code is at a 20-Mbit/s rate, the transmitted output is at least 24 MHz wide (at the 3-dB points) and the transmitter's power is spread over this bandwidth. In that 24-MHz band, a 10-W transmitter would average a power density of approximately 4.16 μW/Hz. To a narrow-band receiver with a 50-kHz bandwidth, this 10-W signal would have less effect than a 200-mW transmitter of anything less than 50-kHz bandwidth. In addition, a spread-spectrum output signal appears to be incoherent and is therefore often less objectionable (from a subjective point of view) than a narrow-band signal.

Message privacy

Message privacy is inherent in spread-spectrum signals because of their coded transmission format. Of course, the degree of privacy, or security, is a function of the codes used. Spread-spectrum systems have been constructed to employ every kind of code from the relatively simple linear maximals[2] to the truly secure nonlinear encryption types. Proper design of the system can provide for substitution as required when higher or lower level message security is desired.

High-Resolution Range Measurements

Spread-spectrum signals of the direct-sequence type excel in their capability to provide high resolution range measurements. Again, this property is due to the high-speed codes used for modulation. Since synchronizing a spread-spectrum receiver depends on the receiver matching its code reference to the signal it receives to within one bit (typically, a spread-spectrum receiver's code will be matched to the incoming signal's code to within one tenth to one hundredth of a bit), then the inherent resolution capability of the signal is better than the range which corresponds to a bit period. Given that same system with a 20-Mbit/s code, the range between transmitter and receiver can easily be measured to within 50 ns, or 50 ft, and little difficulty is found in narrowing the resolution to 5 ft or less. An added advantage of spread-spectrum systems in the range area is that their range resolution is minimally affected by range. That is, a spread-spectrum ranging system that provides 50-ft basic resolution capability at 10 miles will also provide that same resolution capability at 100 miles or 500 miles. Direct sequence ranging techniques have been more than proven on deep space probes, where they provide accurate tracking for space probes millions of miles[3] away. In addition, spread-spectrum ranging has been employed in high-performance aircraft where accurate tracking has been demonstrated at 300-mi ranges with 2-W transmitter power.

Interference Rejection Capability

Spread-spectrum systems provide an interference rejection capability that cannot be matched in any other way. Both deliberate and unintentional interference are rejected by a spread-spectrum receiver, up to some maximum which is known as the "jamming margin" for that receiver. This jamming margin is also a function of the code sequence rate (in a direct sequence system) or the number of frequency channels available (in a frequency hopper). A chirp system's jamming margin is set by the frequency band it covers during its pulse time, or may be better expressed by its compression ratio. Chirp systems have received a great deal more attention in radar systems, to provide better transmitter power efficiency and range resolution, than in communications systems for interference rejection.

[2] Linear maximal codes, or *m*-sequences, are the longest sequences that can be generated by a given length shift register (2^n-1 bits for an n stage register).

[3] The code sequences used in ranging systems are chosen to be long enough so that no ambiguity exists at the maximum range for which they are to be employed.

PROCESS GAIN AND JAMMING MARGIN

Process Gain

Interference rejection, selective addressing, and code-division multiplexing occur as a result of the spectrum-spreading and consequent despreading necessary to the operation of a spread-spectrum receiver. In a particular system, the ratio of the spread or transmitted bandwidth to the rate of the information sent is called the "process gain" of that system. For a system in which the transmitted signal bandwidth is 20 MHz and the baseband is 10 kbits/s, process gain would be approximately $10 \log 2 \times 10^7/1 \times 10^4 = 33$ dB. This system would offer a 33-dB improvement in the signal-to-noise ratio between its receiver's RF input and its baseband output, less whatever might be lost in imperfect implementation. Table I compares the process gain (G_p) that can be expected from various types of spread-spectrum systems.

Jamming Margin

Jamming margin is determined by a system's process gain (jamming margin cannot exceed process gain), acceptable output signal-to-noise ratio, and implementation losses. This margin, sometimes called AJ margin, is the amount of interference that a receiver can withstand while operating and producing an acceptable output signal-to-noise ratio. For the above system, which has a 33-dB process gain, if the minimum acceptable output signal-to-noise ratio is 10 dB and implementation losses are 2 dB, then the jamming margin is 33-12 = 21 dB.

Figs. 4 and 5 show simplified direct-sequence and fre-

TABLE I
Spread-Spectrum System Process Gain

Technique	Approximate G_p
Direct Sequence	(RF bandwidth/information rate)
Frequency Hopping	Number of channels
Chirp	$TW = T/\tau$
Time Hopping	(1/duty factor)
Hybrids:	
DS/FH	$G_{pDS} + G_{pFH}$
DS/TH	(G_{pDS}/duty factor)
FH/TH	(G_{pFH}/duty factor)

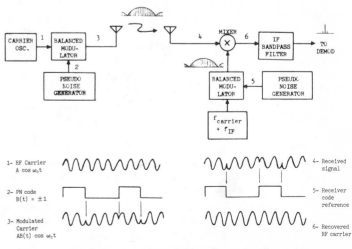

Fig. 4.   Simplified block diagram and waveforms for direct-sequence system.

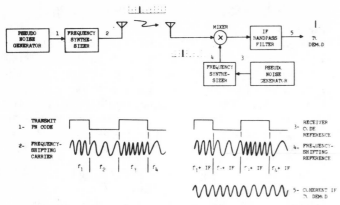

Fig. 5.   Simplified block diagram and waveforms for frequency hopper.

quency-hopping systems. Both operate in much the same way: In either the direct-sequence or the frequency-hopping transmitter, the spread-spectrum transmitted signal is generated as a function of the code sequence. (As it happens, the DS transmitter is directly modulated by the code, while the frequency hopper goes through a code-to-frequency translation.) In either case, the signal transmitted is a wide-band signal, with information imbedded in it, and that is the signal matched to the receiver's local reference.

Expressed as a formula,

$$M_j = G_p - (L_{\text{sys}} + S/N_{\text{out}})$$

where

M_j jamming margin
G_p process gain = (RF bandwidth/information rate)
L_{sys} system implementation losses
S/N_{out} acceptable receiver output S/N.

A 21-dB jamming margin would permit a receiver to operate in an environment in which its desired signal is 121 times smaller than the interference at its input. Expressed another way, an interfering transmitter can have 121 times more power output than the desired signal's transmitter (if their distances are equal) before it affects the receiver's operation.

Typical spread-spectrum transmitters are much simpler than their receiving counterparts. (Here we neglect any consideration of frequency translations and power amplifiers.) Information input to a spread-spectrum transmitter is usually digitized, if not already in a digital form, and imbedded within the code used for spectrum spreading. In the chirp systems, where there is no code, the chirp itself may be used, by sending a downchirp (decreasing frequency) to represent a one, and an upchirp to represent a zero. Once the information to be sent is imbedded in the code (by simple modulo-2 addition of the digitized information with a code), the code is used to balanced modulate a carrier or to control the frequency output of a frequency synthesizer.

Spread-spectrum signals act the same as other wide-band signals with respect to propagation and signal handling. Some cautions are in order, however: direct sequence signals must be carefully handled when it is intended that the modulated signal be multiplied in frequency, since it is possible to destroy the modulation in the multiplication process. Frequency hopping or chirp signals are broadened in their bandwidth by frequency multiplication, on the other hand.

Another consideration is that limiting may be detrimental to spread-spectrum signals, especially those of the direct-sequence type.

In general, one might say that direct-sequence transmitters exist only to generate the kind of signals that are needed by the receiver to allow it to discriminate against undesired inputs, and the transmitter must generate that signal and send it with minimum distortion. The signal then is at the mercy of the transmission medium and all of the would-be interferors within the receiver's field of view.

At the receiver, the local reference $f(t)$ (which is a replica of the transmitted signal, except that it does not contain the transmitter's imbedded information) is multiplied with the incoming signal, performing the operation $\int f(t) g(t-\tau) dt$. When the local code and received codes are matched, then $f(t) = g(t)$, and the choice $\tau = 0$ "despreads" the received signal, leaving only an information-modulated carrier, which is then demodulated to yield the desired information by the conventional methods. In the same process in which the desired signal is despread, or "correlated" by multiplication with the receiver's wide-band local reference, any nonsynchronous incoming signal is spread or decorrelated by being multiplied with that same wide-band local reference. Therefore, by passing the despread desired signal (now in a bandwidth commensurate with the information that was sent) through a bandpass filter that is just wide enough to pass the information-bearing carrier, the receiver rejects the undesired signal. This process allows rejection of most of the power contained in an undesired signal, since the undesired signal is forced to occupy a bandwidth that is equal to the covariance of the undesired signal and the local reference. (That is, any undesired signal has a bandwidth at least as wide as the receiver's local reference, once it is convolved with that local reference.) Therefore, the process gain that was previously discussed comes about through the remapping of a desired signal, to fit within a narrow-band filter that is able to reject almost all of the undesired signal input, which has been spread and (pseudo) randomized by convolution with the local reference.

The process is the same, whether in a direct sequence or frequency-hopping system, with a bandwidth trade being made to the advantage of the receiver through the convolution or correlation process. No bandwidth trade is made in chirp receivers, but an analogous process is carried out wherein the chirp filter compresses a desired input signal while dispersing those signals that are not matched to the filter. In each case, process gain is realized in the receiver through processing that takes advantage of the spread-spectrum waveform sent by the desired transmitter.

A FINAL WORD

This article does not really scratch the surface of either the details of spread-spectrum systems or the many applications for which they are useful. The techniques briefly described here will undoubtedly influence future communications system designs, however, since the tools offered are attractive and they offer alternatives not necessarily available otherwise. It behooves the communications engineer to become familiar with spread-spectrum techniques, for they are definitely a part of his future. As you read this article, new requirements are being formulated that cannot be met in any other way. It is not likely that standard broadcast or television signals will ever be spread spectrum, but they could be, were it not for the tremendous cost of changing over. We may all be certain, on the other hand, that all other systems we see in the future (especially those having military application) will be carefully considered for possible application of the spread-spectrum techniques that are briefly described here.

And just to prove that there really are such things as spread spectrum systems, illustrations of some representative equipment are shown in Figs. 6 through 9. Figs. 7 through 9 are courtesy of the Magnavox Research Laboratories.

Fig. 7. VHF spread-spectrum modem.

. Hughes PLRS user unit.

Fig. 8. UHF spread-spectrum set.

MODEM   R/T   INDICATOR

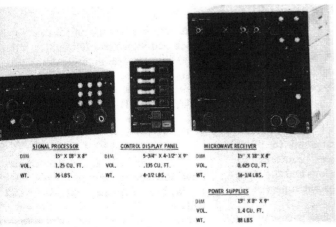

SIGNAL PROCESSOR		CONTROL DISPLAY PANEL		MICROWAVE RECEIVER	
DIM	15" X 18" X 8"	DIM	5-3/4" X 4-1/2" X 9"	DIM	15" X 18" X 4"
VOL.	1.25 CU. FT.	VOL.	.135 CU. FT.	VOL.	0.625 CU. FT.
WT.	36 LBS.	WT.	4-1/2 LBS.	WT.	16-1/4 LBS.

POWER SUPPLIES	
DIM	15" X 8" X 9"
VOL.	1.4 CU. FT.
WT.	88 LBS.

. Four-channel navigation receiver.

APPENDIX

SPREAD SPECTRUM RECEIVERS AS MATCHED FILTERS

All spread-spectrum systems employ "matched filter" reception, to discriminate against unwanted signals and provide for optimum detection of a desired signal in the face of noise or interference. A matched filter detector is one whose response is optimized for the signal desired, and whose output is a linear combination of its responses to desired and undesired signal inputs. Its transfer function is the complex conjugate of the signal to which it is matched, or put another way, a matched filter's impulse response is a time-reversed replica of its desired input signal.

The operation of matched filters has been described in [2] and [3] in detail. We will content ourselves here by pointing out that a matched filter's output signal is

$$\rho \leq \frac{2E}{N_0} \tag{1}$$

where E is the energy in the input signal, and N_0 is the single-ended noise power density, assuming white, stationary noise. This relation holds where $G(j\omega) = K\, S^*(j\omega)\exp\text{-}j\omega\delta$, that is, the transfer function of the filter is the complex conjugate of the input signal $S(j\omega)$, except for an amplitude function K and a delay function $\exp\text{—}j\omega\delta$.

It is obvious that chirp-type spread-spectrum systems employ a matched filter for signal recognition and reception, since the dispersive filters used in a chirp receiver can be used to generate the conjugate transmitted signal as well as detect the received signal. The other spread-spectrum systems are not quite so obviously matched filter users, but consider that they generate signals in their receivers that are the exact replica (codewise) of the expected input, and that these signals are convolved in the receiver with the incoming signal, forming the integral

$$\int_{-\infty}^{\infty} S(t)\, S(t-\tau)\, dt$$

as an output. Since this convolution process corresponds in the frequency domain to multiplying $S(j\omega)$ by $G(j\omega)$, as defined following (1), then it is seen that the direct-sequence and frequency-hopping spread-spectrum systems employ matched filter receivers just as the chirp systems do.

Some direct-sequence receivers employ devices such as surface acoustic wave delay lines, coded to match the desired received signal, that are indeed matched filters in the usual sense. Phase-coded matched filters of this type are limited in the number of code bits they can correlate simultaneously, however. This leads to implementation of systems that generate a local-reference coded signal to convolve with the incoming signal, using shift register sequence generators to generate the coded signals that the more readily recognized delay line filters cannot be implemented to do.

REFERENCES

[1] R. C. Dixon, *Spread-Spectrum Systems.* New York: Wiley, to be published.
[2] G. L. Turin, "An introduction to matched filters," *IRE Trans. Inform. Theory*, vol. IT-6, pp. 311-329, June 1960.
[3] H. M. Sierra, "The matched filter concept," *Electro-Technology*, Aug. 1964.

Robert C. Dixon has 23 years of overall experience, 16 of which have been devoted exclusively to the design and development of spread-spectrum communications and navigation systems. During these 16 years, he has participated in more than 50 separate programs in this field as program manager, system designer, subsystem design engineer, and system integrator. Mr. Dixon joined Hughes Aircraft Co., Ground Systems Group, where he works as a Senior Technical Staff Assistant, in 1975.

Prior to joining Hughes, Mr. Dixon was a Senior Research Engineer at Northrop, a Senior Staff Engineer at Magnavox Research Laboratories, Staff Engineer at TRW, and Senior Staff Engineer at Hoffman Electronics. He has made significant contributions to the development of a number of communications systems, primarily in the spread-spectrum field. Among these were the first producible spread-spectrum system, the first troposcatter spread-spectrum modem, the first spread-spectrum navigation system for high-performance aircraft, and the first satellite on-board spread-spectrum antijam demodulator.

Mr. Dixon has written numerous technical papers and reports, is the author of a book, *Spread-Spectrum Systems,* to be published by John Wiley & Sons, and is the originator of a short course having the same name as the book, which was presented at UCLA in March 1975. He holds the B.S.E.E. degree from Pacific States University, the M.S.S.E. degree from West Coast University, and Certificate in Business for Technical Personnel with the Professional Designation in Business, both from UCLA. In addition, Mr. Dixon has done graduate work in math and computer sciences at the University of Southern California.

SPREAD SPECTRUM: PRINCIPLES AND POSSIBLE APPLICATION TO SPECTRUM UTILIZATION AND ALLOCATION*

William F. Utlaut

Abstract Because of the continuing demand for more telecommunication capacity to serve the world's need for commerce and public safety, there is a continuing need for more efficient ways of sharing the radio spectrum. The conventional way of allocating the spectrum is by frequency division; however, for many kinds of services this is inefficient. Hence, it seems desirable to reexamine whether alternative procedures might not be necessary if the benefits of telecommunications are to be assured in the face of increased demand. Spread-spectrum techniques, which are based on principles antithetic to those currently used in spectrum allocation for reducing necessary bandwidth, seem to offer benefits for spectrum sharing, for some applications, superior to those of frequency division. This paper provides a tutorial summary of some of the principles upon which spread-spectrum systems have developed. It is hoped the reader might consider what role such techniques may play in future spectrum sharing and allocation opportunities.

I. INTRODUCTION

The use of radio telecommunications has grown enormously in the past few decades to provide a broad range of services required by modern civilization. One consequence of this growth is that portions of the radio spectrum are very crowded with users, and yet there is a requirement to meet legitimate needs of still more users. This presents major problems to members of the International Telecommunication Union (ITU) and national spectrum managers in developing procedures for sharing, allocating, and assigning the spectrum so as to satisfy the need to accommodate even more use of the spectrum.

One possibility of alleviating some of the demands for spectrum is, of course, to move to ever higher frequencies where there is less, or no, current usage. However, use of those higher frequencies may not permit the kind of system performance required, or equipment and techniques for operating at higher frequencies may not exist or are too costly for the service required. Thus, other solutions must be found.

The principal one employed today is that of reducing the bandwidth used to provide a service, either as a result of improved techniques for reducing necessary bandwidth or a forced reduction of assigned bandwidth by administrative decision.

Reducing assigned bandwidth also has obvious limitations, for the bandwidth assigned cannot continually be reduced without degrading system performance quality or requiring more costly equipment. Another solution which may permit an overlay of more users in a given band of the spectrum is to use a relatively new and evolving technique known as "spread spectrum." It is based upon a principle which is a direct antithesis of reducing bandwidth, for it utilizes modulated signal spectra that deliberately employ large emission bandwidths to send information requiring a bandwidth much less than is transmitted. The spectra have little resemblance to those of conventional modulation schemes. However, spread-spectrum techniques, through the properties of coded modulation, can provide systems which produce low

The author is with the Institute for Telecommunication Sciences, Office of Telecommunications, U.S. Department of Commerce, Boulder, CO 80302.

This paper was originally published in *ITU Telecommunication Journal*, vol. 45, pp. 20–32, Jan. 1978.

*Reprinted from *IEEE Communications Society Magazine*, September 1978, Vol. 16, No. 5, pp. 21–31.

interference to other systems, have high interference rejection capability, provide multiple access capability, and have other useful capabilities. The purpose of this tutorial paper is to describe some aspects of spread-spectrum techniques and to urge readers to consider how such techniques might be used to provide for greater use of the already crowded radio spectrum.

II. SPREAD-SPECTRUM CONSIDERATIONS

One of the important parameters in determining how well a communication system can perform is the signal-to-noise ratio. Because many telecommunication systems must operate with low carrier signal-to-noise ratio (for example, deep space probes or operations in high interference environments), effort has been given in past years to find techniques which would permit radar, ranging, and communication systems to operate under such conditions and with a high degree of resolution or accuracy of information transmission. In this effort, advances in statistical communication theory, coding theory, and the development of reliable and miniaturized digital components have all played a role.

Shannon's original work in the field of statistical communication theory [3] showed that the capacity of a channel to transfer error-free information is enhanced with increased bandwidth, even though the signal-to-noise ratio is decreased because of the increased bandwidth. This is the basis for spread-spectrum techniques.

Perhaps the most familiar example of the benefit of using a modulation carrier bandwidth significantly wider than the baseband bandwidth required to pass the information is found in conventional wideband FM. As is known, the bandwidth required by an FM signal is a function not only of the information bandwidth, but also of the amount of modulation. The noise and interference reduction advantage of FM over AM, however, becomes significant only when the frequency deviation from the unmodulated carrier is large in comparison with the modulating frequency. For such wideband FM, when the noise threshold is exceeded, a processing gain—an improvement in signal-to-noise ratio at the output of the processor over that at the input—can be obtained [4].

Since wideband FM produces a spectrum much wider than that required for the transmitted information, it might be considered a spread-spectrum technique. However, it differs from the spread-spectrum techniques to be described subsequently in that these latter techniques use some signal or operation other than the information being sent to broadband (spread) the transmitted signal, whereas in the FM case the information signal itself is used to broadband the signal transmitted. However, spread-spectrum techniques have a processing gain, analogous to that of wideband FM, permitting reduced carrier-to-noise, or low transmitted power levels, to produce satisfactory system performance.

Three general types of spread-spectrum techniques will be discussed to provide the reader with a more comprehensive view of spread-spectrum capabilities:

1) pseudorandom sequences, in which a carrier is modulated by a digital code sequence having a bit rate much higher than the information signal bandwidth;

2) frequency hopping, in which the carrier is frequency shifted in discrete increments in a pattern determined by a digital code sequence; and

3) frequency-modulation pulse compression, or "chirp," in which a carrier is swept linearly over a wide band of frequencies during a given pulse.

III. PSEUDORANDOM SEQUENCE SPREAD SPECTRUM [5]–[7]

Historically, the basic mathematical tool in radio communication theory has been the Fourier time and frequency analysis. The RF signal is generally regarded as a linear combination of sine waves, and the classical concept of modulation involves the variation of one of the three parameters (amplitude, frequency, or phase) associated with a pure sine wave so as to carry information. In recent years there has been an increasing emphasis on digital communications in which the digital signal may be regarded conceptually as a sequence of Ones and Zeros. In practice, these include a pulse train in which One is a pulse and Zero is a no-pulse, or using a high frequency CW signal for which a phase shift of +90° is a One and −90° a Zero, each lasting a unit of duration. The pulse sequence train can be viewed as a kind of amplitude-modulated square wave while the ±90° sequence may be regarded either as a phase-modulated sine wave or as a balanced, amplitude-modulated sine wave.

One of the important properties of sine waves is that all harmonics $\sin nx$ of the fundamental $\sin x$ are mutually orthogonal and the $\sin x$ is also orthogonal to two of its 90° phase shifts, $\cos x$ and $-\cos x$. (In statistical terms, orthogonal means uncorrelated.) It has been found that orthogonal properties of this sort are among the most desirable attributes of signals in a wide variety of communication situations. Whenever one has a set of possible messages to encode for a communication link, one desires that their encoded forms be as mutually distinct as possible. This is approximately achieved in the orthogonal, or uncorrelated, case, and it is upon this basis that major effort and advances in the development of error-correcting codes have their origins. Similarly, orthogonal codes play a major role in spread-spectrum techniques and permit a number of signals to be transmitted on the same nominal carrier frequency and occupy the same RF bandwidths. We shall discuss code generation and some properties of codes in the next section.

Fig. 1 shows a block diagram of a spread-spectrum system in a most general form. At the transmitter, a carrier, $A_0\cos\omega_0 t$, is modulated in a completely general way by any form of amplitude, angle modulation, or a combination of both to produce $S_1(t) = A_1(t)\cos[\omega_0 t + \varphi(t)]$. This modulated signal is then multiplied by a time function, $g_1(t)$, which spreads the energy of $S_1(t)$ over a bandwidth considerably wider than that of the modulated signal. The resulting signal, $g_1(t)S_1(t)$, then is transmitted over the radio channel where it combines linearly with other signals and noise. It then enters into the receiving system where it is processed to recover the wanted information signal. To do this, the composite signal entering the receiver is multiplied by an exact

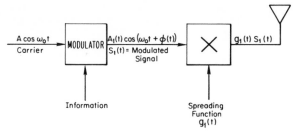

(a) Transmitter

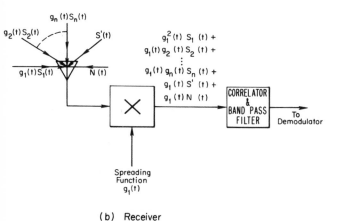

(b) Receiver

Fig. 1. Basic spread-spectrum concept.

replica of the spectrum-spreading function $g_1(t)$ so that at the output of the multiplier the signal terms are

wanted signal $= g_1^2(t)S_1(t)$
unwanted signals $= g_1(t)g_2(t)S_2(t) + \ldots + g_1(t)g_n(t)S_n(t)$
$\qquad\qquad\qquad + g_1(t)S'(t) + g_1(t)N$

where $g_1(t)g_n(t)S_n(t)$ represent other spread-spectrum signals, and $g_1(t)S'(t)$ and $g_1(t)N$ represent other signals and noise which have been spread over a broad spectrum at the receiver by the function $g_1(t)$.

The output of the multiplier is then passed through a correlator and a bandpass filter. Thus, if the function $g_1(t)$ is chosen so that $g_1^2(t) = 1$ and $g_1(t)g_n(t) = 0$, then the receiver would be able to extract only the wanted signal from all of the other spread-spectrum signals. The other signals, $S'(t)$ and noise, which are spread at the receiver by $g_1(t)$ over a wide bandwidth, will be significantly reduced in energy by the bandpass filter because it will have a bandwidth narrow enough to just pass the wanted signal. The wanted signal is "despread" by the multiplication with $g_1(t)$ in the receiver and, thus, has a bandwidth of the original information signal. Inasmuch as the process of spreading and despreading the radio frequency bandwidth is transparent to the wanted information signal, any type of coding to increase performance reliability, as error correction or detection coding, or to provide privacy may be used as desired in the usual manner.

The process just described, in which the information is used to modulate a carrier before it is spread by a code form, is only one way of embedding information in a spread-spectrum signal. A more common method is to

add the information to the spectrum-spreading code before it is used for spreading modulation. The information to be sent must be in some digital form in this process because addition to a code sequence involves modulo-2 addition to a binary code, as is shown schematically in Fig. 2. While phase-shift keying is used in this illustration, other forms of modulation could be used. (Phase-shift keying has been frequently used in spread-spectrum systems for satellite communications because the satellite channel in general has a non-linear amplitude transfer characteristic, and modulation techniques employed have tended to be of a constant envelope type, such as phase or frequency modulation.) The spread-spectrum signal is despread at the receiver, as before, by correlating the received signals with a local reference signal identical to that code used for signal spreading at the transmitter, and the wanted spread signal collapses to its original bandwidth. Signals which are not correlated with the spreading code are spread by the local reference signal to its bandwidth, or more, and a narrowband filter then suppresses the effects of all but the wanted signal. Because correlation plays a central role in this process, it is desirable to consider codes and their correlation functions further.

A. Spread-spectrum pseudorandom sequence codes

To classify a finite set of events of any sort as random refers to the *a priori* conditions under which the sequence was produced rather than the *a posteriori* consideration of what the sequence looks like or what properties it exhibits. Thus, if a sequence of Ones and Zeros is produced by flipping a coin (heads = 1, tails = 0), which selects from a sample space in an unpredictable fashion, it is a random process. However, if a similar sequence of Ones and Zeros is generated by a deterministic device, such as a shift register, it is not random, even if it looks so, but is designated as a pseudorandom sequence.

One of the simplest and most effective devices for generating deterministic sequences of pseudorandom Ones and Zeros is the shift register [8]. A shift register of degree n is a device consisting of n consecutive binary storage positions, which shifts the contents of each position to the next position down the line at the rate set by a timing device, or clock. A shift register can be converted

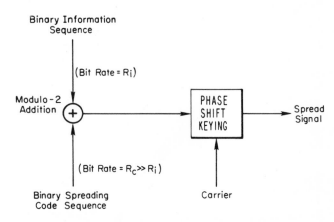

Fig. 2. Direct code modulation spread-spectrum schematic.

into a pseudorandom sequence generator by including a feedback loop, which computes a new term for the first stage based on the previous n terms. An example is shown in Fig. 3, where $n = 4$, and feedback from stages 3 and 4 is modulo-2 added and returned to stage 1. The rules of modulo-2 arithmetic are: $0 + 0 = 0$, $1 + 1 = 0$, $1 + 0 = 1$, and $0 + 1 = 1$. If the initial state of the shift register is 1 0 0 0 (reading from left to right), then the succession of states triggered by clock pulses would be: 1 0 0 0, 0 1 0 0, 0 0 1 0, 1 0 0 1, 1 1 0 0, 0 1 1 0, 1 0 1 1, 0 1 0 1, 1 0 1 0, 1 1 0 1, 1 1 1 0, 1 1 1 1, 0 1 1 1, 0 0 1 1, 0 0 0 1, 1 0 0 0, and the output (from position 4) at each

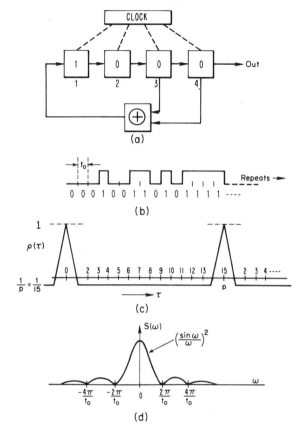

Fig. 3. Code sequence generator and spectrum (a) Four-stage shift register with feedback from stages 3 and 4, modulo-2 added. (b) Pseudorandom code sequence generated which repeats every 15 code elements. (c) Autocorrelation function for 15-element code. (d) Spectrum of pseudorandom code.

Maximum-Length Shift Register Sequences

The codes resulting from maximum-length linear shift register sequences have several important properties, as follows:

1) In each period of the sequence the number of Ones differ from the number of Zeros by at most 1 (Balance Property).

2) If a period of the sequence is compared term by term with any cycle shift of itself, the number of terms which are the same differ from those which are different by at most 1 (Correlation Property).

3) Among the runs of Ones and Zeros in each period, one-half of the runs of each kind are of length one, one-fourth of each kind are of length two, one-eighth of length three, and so on. Thus, the statistical distributions are well defined and always the same, but the relative position of the runs vary from code to code (Run Property).

3) Modulo-2 addition of a maximum-length linear code with a shifted replica of itself results in another replica with a phase shift different from either replica (Generic Property).

Using all of the possible linear combinations of feedback taps for an n-stage register, there are $[\varphi(2^n - 1)]/n$ maximal linear sequences that can be generated. Here $\varphi(2^n - 1)$ is an Euler number, the number of positive integers that are relatively prime to and less than $2^n - 1$. Table I shows some representative numbers of possible sequences that can be generated from shift registers with n stages. As can be seen, large numbers of sequences are available once shift register lengths greater than about 10 are used.

TABLE I

Shift Register Stages, n	Total Number of Sequences	Sequence Length
4	2	15
8	16	255
10	60	1023
12	144	4095
14	756	16383
15	1800	32767
16	2048	65535

---

state would be 0 0 0 1 0 0 1 1 0 1 0 1 1 1 1. This would continue repeating itself with a period of 15 states.

It may be observed that, while this looks like a random sequence, it is deterministic and has a finite periodicity. Given any linear shift register of degree n, the output sequence is always ultimately periodic, with a period $p = 2^n - 1$ at most. Any output sequence achieving a period of $p = 2^n - 1$ is called a maximum-length linear shift register sequence [8].

Maximum-length linear codes have been studied and used extensively for error-correcting codes, multiple-address coding, privacy encoding, and other similar purposes [9]–[11]. The code sequences used for spread-spectrum systems, while similar in nature, are generally of much greater length than those just mentioned, since they are intended for bandwidth spreading and not for direct transfer of information. The length and bit rate of the code sequence is important, for these parameters set bounds on the capacity of a system that can only be changed by changing the code. With this in mind, let us turn to a review of the correlation functions and power spectra of code sequences.

B. Correlation functions and power spectra, pseudorandom sequence codes

In order to calculate the performance of a spread-spectrum system, it is necessary to know the power spectrum of the transmitted signal and the cross correlation between the different pseudorandom codes used in spreading the spectrum. The spectrum may best be arrived at by making use of the well known Fourier transform relationship between the autocorrelation function and the power spectrum.

A normalized correlation function for functions of time can be defined as

$$\rho(\tau) = \lim_{T \to \infty} (1/T) \int_0^T F_1(t) F_2(t + \tau) \, dt. \qquad (1)$$

When F_1 and F_2 represent the same function, (1) gives the autocorrelation, whereas when they represent separate functions, the cross correlation is determined. For binary sequences, it turns out that ρ can be determined relatively simply by noting that the variable τ in (1) is the number of bits or digits by which the second sequence is shifted with respect to the first. Then by comparing the two sequences bit by bit, the normalized correlation function is determined from

$$\rho = \frac{\text{number of agreements} - \text{number of disagreements}}{\text{number of digits in the period of sequence}}. \qquad (2)$$

If, for example, (2) is applied to the code sequence of Fig. 3, it can be noted that for all positions of code shift, from one digit to 14 digits of shift, there are 7 agreements and 8 disagreements. Thus, $\rho = -1/15$ everywhere except for the position of no shift, where, naturally, all digits are in agreement and $\rho = 1$. Thus, the autocorrelation function repeats regularly with the same periodicity as the pseudorandom code. The same is true for other binary waveforms corresponding to a pseudorandom code, and they have an autocorrelation of unity at zero shift and 1/period at all other values of shift.

The power spectrum of a pseudorandom binary waveform with an autocorrelation function as just described is

$$S(\omega) = \left(\frac{p+1}{p^2}\right) \left[\frac{\sin \omega t_0/2}{\omega t_0/2}\right]^2 \sum_{\substack{n=-\infty \\ n=0}}^{\infty} \delta(\omega - 2\pi n/p t_0) + \frac{1}{p^2} \delta(\omega)$$

$$(3)$$

where p is the period of the sequence, t_0 is the period of one digit of the binary waveform, and $\delta(\cdot)$ is the Dirac delta function, or impulse function [12]. The autocorrelation function and power spectrum envelope for binary waveforms are illustrated in Fig. 3.

There are several points to be noted about the spectrum as given in (3). First, it is a line spectrum with frequencies at multiples of the fundamental frequency. Second, there is a scale factor inversely proportional to the period of the sequence. Thus, if the period of the sequence is doubled, the lines in the spectrum become twice as dense, but the power in each is reduced by a factor of 2. This is because the binary waveform is a constant amplitude square wave and hence has constant power. Third, the envelope of the spectrum, a $(\sin x/x)^2$ function, is determined by the digit period of the waveform. Therefore, the bandwidth spreading function is independent of the

length of the waveform code, but is determined solely by the digit period, i.e., how often the waveform switches. Finally, the dc term shows a power of $1/p^2$. Since this term reduces rapidly with increasing p, a carrier phase-reversed by such a pseudorandom sequence results in a spectrum about the carrier given by (3), but with a very small carrier term. This results because of the odd number of terms in the sequence and the Balance Property in which there is always a difference of 1 in the number of Ones and Zeros (see inset).

C. Direct code modulation spread spectrum, performance with interference

Following the discussion of the generalized view of the spread-spectrum process and after examining the process of code generation and its spectrum, we can now ascertain the interference rejection capability of the spread-spectrum receiver. In summary, the spread-spectrum operation is one in which the message modulation is multiplied by a frequency spreading pseudorandom code so as to be spread to a bandwidth between first nulls in the envelope equal to 2 times the code rate, i.e., $B_c = 2R_c = 2/t_0$. Spread-spectrum systems typically use an RF bandwidth which is 100 to 1000 times or more that required for the information rate. This signal is received along with additive uncorrelated interference and noise, as is suggested in Fig. 4. The mixture of received signals is correlated, i.e., multiplied with a replica of the same pseudorandom code used to spread the wanted message signal. This correlation process removes the coding from the wanted signal, leaving only the narrowband message signal, but since the unwanted signals are uncorrelated with the local code, at the output of the correlator, they are still spread over a wide band by the code. When the signal at the output of the correlator is passed through a narrowband filter, only that portion of the unwanted signal spectrum falling within the bandwidth of the filter will cause interference to the wanted signal. Clearly, the larger the ratio of the bandwidth of the spread signal to that of the information signal is, the smaller the effect of unwanted signal interference. We can note that at the correlator input, the signal-to-interference ratio is approximately

$$(S/I)_{\text{in}} = S/IB_c$$

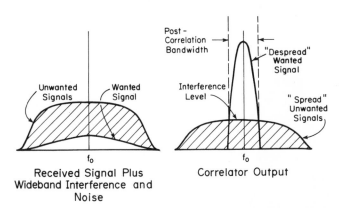

Fig. 4. Spread-spectrum correlator process on wanted and unwanted signals.

while at the output of the correlator and narrowband filter it is

$$(S/I)_{out} = S/IB_m$$

where I represents the spectral power density of the unwanted signal and B_c and B_m are, respectively, the spread-spectrum and information bandwidths. To the extent that the unwanted signal spectral power can be considered sufficiently noise-like for the contribution from each unwanted signal to be combined by linear addition of their powers, a processing gain for this spread-spectrum operation can be seen to be approximately

$$G_p = [(S/I)_{out}/(S/I)_{in}] = B_c/B_m = R_c/R_1. \quad (4)$$

While it might appear from (4) that the spread-spectrum process could permit operation when the power level of unwanted signal exceeds that of the wanted signal by the processing gain and that the processing gain can be increased to any desired level by increasing the spreading code rate, neither are possible. For even in an idealized situation, as the code rate is increased, the interference level produced by unwanted signals will decrease until it is small compared with the receiver thermal noise. Any further increase in code rate would not improve the overall wanted signal-to-interference ratio. Further, in the practical case, one must take into account the requirement for a useful signal-to-noise system output as well as internal losses of the processor. An interference margin [7] which might be attained then is

$$M_i = G_p - L - (S/N)_{out} (dB) \quad (5)$$

where

G_p = the spread-spectrum processing gain
L = the system implementation loss, and
$(S/N)_{out}$ = the operationally required ratio at the information output.

For example, if the spread-spectrum bandwidth is 1000 times that of the information bandwidth, so that $G_p = 30$ dB, and if $(S/N)_{out}$ is required to be 10 dB and L is taken as 2 dB, then the interference margin

$$M_i = 30 - 2 - 10 = 18 \text{ dB}.$$

Under the assumptions in this example, interference power could not exceed the wanted signal power by more than 18 dB and still maintain desired performance.

It was the intention of the preceding discussion to give the reader a general view of the direct code modulation method of spread spectrum, and it should be realized that a number of problems such as code length [13], code synchronization [14], less than ideal cross correlation of codes, and others necessary to adequately design a system have not been discussed. It is hoped that the reader has gained some insight into the process, however, and appreciates that the technique generates a signal which appears to other receivers (conventional as well as other spread spectrum) as a low-level noise-like signal. Given the proper receiver and code, a wanted signal can be obtained even when it is embedded in an interference power greatly exceeding that of the wanted signal. Further, be-

cause of the orthogonal nature of the pseudorandom codes, it is possible for a number of signals to be sent unambiguously, at the same frequency and at the same time. Code division multiplexing and selective addressing can be implemented by the code format when desired. Although direct code modulation has probably been the most widely used in practical systems, other than radars, another technique, usually called frequency hopping, is also important and can, in fact, show an improvement in communication capacity over that obtained with direct code modulation. At this stage it is more complicated and expensive to implement.

IV. FREQUENCY-HOPPING SPREAD SPECTRUM

The frequency-hopping (FH) spread-spectrum technique [7],[15] is similar to that of the pseudorandom direct code modulation technique. The main difference is in the way the transmitted spectrum is generated and in the way interference is rejected. Fig. 5 illustrates a generalized FH spread-spectrum system. As with the direct code method, a central feature is pseudorandom code generators at both the transmitter and receiver, capable of producing identical codes with proper synchronization. As before, there is no restriction on the choice of information modulation. In the FH method, the pseudorandom code sequence is used to switch the carrier frequency instead of directly modulating the carrier. When the local oscillator in the receiver is switched with a synchronized replica of the transmitted code, the frequency

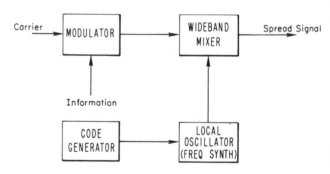

(a) Transmitter

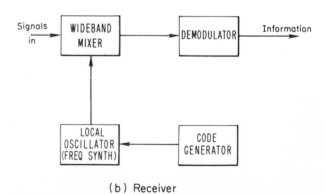

(b) Receiver

Fig. 5.   Frequency-hopping spread-spectrum schematic.

hops on the received signal will be removed, leaving the original modulated signal, which is demodulated in a conventional manner.

The bandwidth over which the energy is spread is essentially independent of the code clock rate and can be chosen by a combination of the number and size of frequency hops. An idealized power spectrum for an FH spread-spectrum method is shown in Fig. 6. It has a rectangular envelope and extends over a bandwidth $B_{rf} = (2^n - 1)\Delta f$ where, again, n is the number of stages used in the shift register to generate the frequency-hopping code and Δf is the frequency separation between discrete frequencies, which must be at least as wide as the information bandwidth, B_m. As in direct code modulation spread spectrum, a meaningful parameter, with respect to interference spread uniformly across the RF band, is the processing gain, which for an FH system is

$$G_p = B_{rf}/B_m = 2^n - 1, \text{ the number of channels used.}$$

While the direct code modulation spread-spectrum technique must have a code rate much higher than the message information rate to produce the degree of spectrum spreading necessary, such a high code rate is not necessarily required by the FH technique. In fact, the code rate can be lower than that of the information and, in part, depends upon the type of interference the designer expects the system to encounter.

Interference from unwanted spread-spectrum signals in FH systems occurs in either of two conditions which causes a frequency component of the "de-hopped" unwanted signal to fall in the passband of the receiver. One condition is that the receiver code shifts the local oscillator by the same amount that an unwanted signal is shifted. The degree to which this happens depends upon the degree of cross correlation between codes and is minimized by use of low cross correlation codes. A second condition of interference can occur if the receiving channel is nonlinear and intermodulation between interfering signals produces frequencies which, when mixed with the current value of local oscillator frequency, result in frequencies commensurate with the receiver pass band. A fixed frequency signal, of course, causes interference only when the code produces a local oscillator frequency which combines with it to produce a frequency capable of passing through the receiver pass band. At all other code positions the fixed frequency signal is translated to a different frequency which would not coincide with that of the receiver passband.

If frequency-hopping rates that are high in comparison

with the information rate are used, the interference produced by unwanted signals will be noise-like since each spectral line will be spread approximately as a $(\sin x/x)^2$ function in an amount proportional to the hopping rate. Thus, the spread will exceed the receiver information bandwidth in a way similar to the direct code modulation technique. In general, it is more difficult to use high hopping rates in FH systems because of limitations in the frequency synthesizer rate of frequency change. In part, this is caused by the desirability of preserving phase coherence from hop to hop in order to avoid phase modulation of the de-hopped wanted signal. If the hopping rate exceeds the information rate, any residual phase modulation may seriously degrade the subsequent demodulation of the information.

Frequency-hopping rates less than the information rate avoid this problem, but at low hopping rates, the spread on each line of the transmitted spectrum will be small compared with the information bandwidth and the interference from unwanted signals will tend to be coherent. However, the occurrence of interference will be intermittent, with periods of one hop suffering heavy interference and long periods of many hops (depending on code length) being free of interference. Operation on satellite links using a frequency-hopping spread-spectrum technique, with coded MFSK message modulation to generate a bandspread signal structure providing a high degree of multiple access has shown that the effect of other users was the same as that of Gaussian noise [16]. These operations included air-to-ground communications of voice and teletype messages via the satellite while using simple low gain antennas, and provided satisfactory performance while operating in the presence of other radio interference and multipath.

V. LINEAR FREQUENCY-MODULATION SPREAD SPECTRUM—"CHIRP"

The linear frequency-modulation technique of spectrum spreading [17] was developed a number of years ago to improve radar operation by obtaining the resolution of a short pulse, but with the detection capability of a long pulse. A long transmitted pulse is suitably modulated and the receiver is designed to act on the modulation to compress the pulse into a much shorter one. The transmitted waveform consists of a rectangular pulse of constant amplitude A and duration τ, as in Fig. 7. The frequency of the transmitted pulse is linearly increased from a frequency f_1 to f_2 during the pulse. The frequency-modulated signal passes through a filter in the receiver at a velocity of propagation proportional to the frequency. For this case, the filter speeds up the higher frequencies at the trailing edge of the pulse relative to those at the leading edge. The result is that the energy contained in the original long pulse is compressed into a shorter pulse of duration approximately $1/B$ where $B = f_2 - f_1$. This resulting pulse has a shape proportional to $\sin \pi Bt/\pi Bt$, and the instantaneous peak power of the compressed pulse is increased by the factor B over that of the long transmitted pulse. For this frequency-modulated spread spectrum then, a process gain, $G_p \cong B\tau$, is attained. Linear frequency modulation is not abso-

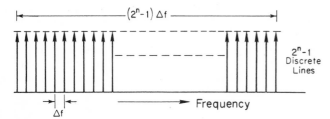

Fig. 6.   Ideal frequency-hopping spread spectrum.

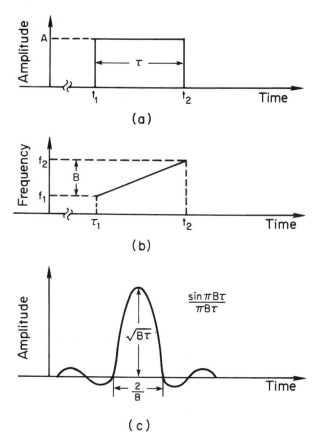

Fig. 7.   Linear FM spread spectrum, or pulse compression. (a) Transmitted pulse of amplitude *A* and duration τ. (b) Linear frequency variation over bandwidth *B* of transmitted pulse. (c) Output of receiver filter.

lutely necessary in a chirp system as the frequency modulation can be of almost any form, provided that the receiver filter is designed to match the transmitted waveform so as to reassemble, or despread, the time-dispersed carrier power in such a way that it adds coherently and thus provides an improvement in signal-to-noise ratio.

The chirp-type system has not generally been used in communication system operation, but in principal could be used, possibly, in a hybrid form, for such purposes. The present state of analysis suggests that a chirp-type communication system may face severe limitations because of the need for phase synchronization of a chirp signal set, and the fact that the optimum value of cross coherence is highly sensitive to synchronization channel signal-to-noise ratio and the spectral-to-Rayleigh signal power ratio of the radio channel in the transmission media [18]. Chirp systems as used in radar, do not normally employ a code sequence to control their output spectra, which the other techniques discussed do. However, it might be possible to do so, for example, by using an upward chirp to represent one state of a binary code and a downward chirp to represent the other state in communication systems. As further development of spread-spectrum techniques occurs, it is possible that a hybrid chirp technique would be found to be useful in communication operations [19].

VI. POSSIBLE APPLICATION OF SPREAD-SPECTRUM PRINCIPLES IN SPECTRUM ALLOCATION

Spread-spectrum techniques have been developed in the past to permit communication of message information under difficult conditions of very low signal-to-noise ratio (such as may be encountered due to high co-channel interference), low signal levels (such as may occur in systems used over very long paths, as, for example, deep space probes), or transmissions having low detectability. They were not developed with a primary objective of spectrum use improvement, and yet the objectives for which they were developed would appear to be desirable for spectrum allocation considerations. Two strategies for employing spread-spectrum techniques would seem to warrant further consideration, and experimental verification as to their usefulness, for increasing spectrum use efficiency and providing more users satisfactory system performance.

One strategy would be to overlay wideband spread-spectrum users in selected frequency bands now assigned for a number of narrowband users, and to use the frequencies simultaneously by both kinds of users. Under the right conditions, such simultaneous use of frequencies should be quite practical. The burden of achieving and demonstrating practicality would probably fall on spread-spectrum users because most of the spectrum in use today is already filled with narrowband users, and, traditionally, new users of the spectrum carry a burden of establishing sharing capability.

An approach to this would be to let present users continue operating as they are today and begin building an overlay of spread-spectrum users on the present networks. The spread-spectrum users would be expected, and likely required, to develop their systems so as to create minimum interference to the conventional systems as well as being able to provide satisfactory operational performance of the spread-spectrum systems in the presence of many interfering signals. That this strategy should be practical seems implicit, since, as summarized in preceding sections, the spread-spectrum signal is a low density signal (inherent because the signal power is spread over a wide bandwidth) which appears to be incoherent and to the conventional system is viewed as only a small noise increase.

As an illustration, consider a direct code modulation system using a spectrum-spreading code with a rate of, say, 1.5 Mbits/s. It would have a bandwidth between nulls of 3 MHz, and the power of the transmitter is spread over this bandwidth [see Fig. 3(d)]. In that 3 MHz, a 10 W transmitter, say, would average a power density of approximately 3.3 μW/Hz. Thus, this 10 W signal would have about the same effect in a receiver of 3 kHz bandwidth as would a 10 mW transmitter concentrating its power in a 3 kHz, or less, bandwidth. The reader can consider other examples with different combinations of code rate, transmitter power, and receiver bandwidths to see that properly designed spread-spectrum systems should result in minimal interference to existing narrowband systems.

Additionally, the second requirement that the spread-spectrum system be able to operate in the existing interference environment also seems implicit because of its inherent capability to provide interference rejection that is probably not matched in any other way. For, as discussed earlier, interference is rejected by the spread-spectrum receiver up to some margin level (5) which is a function of the code sequence rate, in a direct sequence system; the number of frequency channels used, in a frequency-hopping system; or in the compression ratio of a "chirp" system.

The main obstacle to the development of this strategy of overlaying spread-spectrum wide-bandwidth signals on existing narrowband systems will be the reticence of those currently authorized frequencies to want to share them when they appear to have nothing to gain. The best argument against this attitude is that the world society as a whole needs and will benefit from increasing communication capability. And to achieve that capability, careful planning, cooperation, and spectrum sharing on the part of all will be needed.

The second strategy, which is suggested for possible application of spread-spectrum principles to spectrum sharing, is to establish certain bands for spread-spectrum systems and, conceptionally, assign orthogonal codes rather than frequency channels. The number of maximal-linear code sequences that can be generated using a shift register with a different number of stages was illustrated in Table I.

This strategy could be implemented in an evolutionary manner, as an outgrowth of the first strategy, if, because of their advantages, the number of spread spectra overlaid in a band grew as narrowband systems diminished. Alternatively, when advantages of spread-spectrum techniques become more widely recognized, decisions could be made to implement spread-spectrum systems in certain bands with code assignments, as suggested.

One concern which legitimately arises is whether a multiple number of spread-spectrum users transmitting on the same nominal carrier frequency and occupying the same RF bandwidth permits as many total users of that band as conventional frequency division channelling. A universal response to this question does not seem possible because the relative number of users that can be contained by either approach depends upon the model assumed for user and signal power distribution message length distribution, and characteristics of equipment used.

The approach to determining how many spread-spectrum users can simultaneously occupy a given band is based upon the interference margin of the system. This interference margin determines the interference-to-signal ratio which can exist at the input to the receiver while maintaining adequate signal-to-interference at the output. The expression for interference margin was given in (5) as

$$M_i = G_p - L - (S/I)_{out}.$$

Recalling that $G_p = (S/I)_{out}/(S/I)_{in}$ and after some arithmetic manipulation, it can be shown that the interference tolerable at the receiver input is

$$I_{in_{max}} = M + I^L + S \text{ (dB)} \qquad (6)$$

where each quantity in capital letters is in decibels referenced to a common power level base. Unfortunately, (6) provides no information about the number of users possible, for the interference shown there could come from a 1000 signals of a given level, 100 signals at ten times that level, or 10 signals at 100 times that power level. In general, we can observe that the process is such that

$$10 \log \sum_{n=1}^{n} i_n = I_{in_{max}} \text{ (dB)}$$

where i_n is the interference contribution in watts from each of the n users. For a given situation, there will be some distribution for i_n, analytic or otherwise, and the problem is to find n such that the interference contributions from the n users sum up to the $I_{in_{max}}$ value, which gives the number of users allowed. For analytical studies, the distribution of i_n could be assumed as, say, a Poisson, chi-squared, Gaussian, or any other statistical distribution which approximates the situation being analyzed and, after proper evaluation of the cumulative distribution values, n could be found for the assumed interference distribution. As an example, if we use a mathematically convenient, but unlikely, distribution, that the same amount of interference is produced at each receiver by each user transmitter, so that $i_n = c$, then,

$$10 \log \sum_{c}^{n} i_n = 10 \log nc = I_{in_{max}}.$$

Because for the assumed interference distribution the value of c must be the same as the value of the wanted signal:

$$10 \log n = M + L = G_n - (S/I)_{out}.$$

Thus, if we were to consider a system using a 3 kHz information bandwidth, a 1.5 Mbit/s spreading code rate (RF bandwidth = 3 MHz), giving a $G_p = 30$ dB, and a required $(S/I)_{out} = 10$ dB, then

$$10 \log n = 30 - 10 = 20$$

and

$$n = 100 \text{ users.}$$

Because there are 1000 3 kHz channels possible in the 3 MHz bandwidth, this would represent poor spectrum use efficiency. In conventional allocation, of a single channel to a single user, a much larger percentage of the 1000 channels would normally be made available to users, the exact number being dependent upon how much spectrum is used to provide protection from equipment characteristics which cause interference over a band of frequencies wider than the required information bandwidth. The reason for the low efficiency of spectrum use in this illustrative case is the tacit assumption that users were on all the time, or at least with a high duty cycle. And, for such cases, bandwidth expansion systems are usually not spectrum use efficient. However, it was shown by Costas [20] almost two decades ago that for operations in which a number of stations must be permitted to transmit at any given time, but where each station is only transmitting a fraction of the total time, wide-band systems provide greater communication capacity and spectrum utilization than do narrowband systems.

In a congested band operation, in which a service is assigned various bands of frequencies and users are permitted to operate at any frequency within the band, Costas shows that the communication capacity using broadband systems exceeds that of a narrowband system by

$$C_B = C_N \left[1 / \alpha \, (S/N)_{min} \right] \qquad (7)$$

where

α = average fraction of time each station is actually transmitting,

$(S/N)_{min}$ = least favorable signal-to-noise ratio anticipated in the narrowband system, and

C_B, C_N = channel capacity per circuit in broadband and narrowband operation respectively.

From equation (7) it can be observed that as the number of users increases in a congested band operation, so that the S/N decreases, or as the duty cycle of operation decreases, broadband systems show increasing superiority over narrowband ones.

Costas also showed that, in an environment in which all users operated only on an assigned frequency, a similar relation existed and that low duty cycle operation broadband systems give greater communication capacity for a given bandwidth allocated to the service. The reason for this is that at low duty cycle, the narrowband system wastes spectrum because most of the allocated channels in the band will be idle at any one time. This cannot be avoided since each station must have access to communications at any time. The narrowband allocation eliminates interference between users, while in the broadband case, each station appears as "noise" to the others. The broadband system takes advantage of the low duty cycle which keeps the "noise" level low and increases the per-circuit capacity.

At high duty cycles, the narrowband system results in superior spectrum utilization, and it obviously makes sense to allocate spectrum using the conventional frequency division method for high duty systems, such as broadcast. For low duty rate operations, such as those that occur in mobile systems, systems in which a large number of users rely upon a common relay point, or systems which permit many users to operate on any of a number of frequencies in a band (amateur, Citizen's Band radio, etc.), bandwidth-expanding systems can easily prove to be the more efficient users of the spectrum.

From this perspective, let us make one last observation and return to the illustration used above to consider the relative communication capacity which the spread-spectrum and conventionally allocated narrowband systems would permit for low duty cycle users. To do this, let us make use of concepts developed in telephone traffic analysis of traffic intensity which can be supported at a given grade of service by a given number of trunk circuits. In this context, grade of service represents a measure of the probability that a call offered to a group of trunks, or circuits, will fail to find an idle circuit on the first attempt. Traffic intensity is the quantity of traffic per unit of time over the circuits, generally expressed in Erlangs, which is the intensity in a circuit that is continuously occupied. Utilizing tables of the Erlang B formula presented in the International Telegraph and Telephone

Consultative Committee (CCITT) *Blue Book* [21], we can estimate the traffic intensity for spread-spectrum and narrowband systems.

Let us assume two grades of service, 1 and 7 percent, i.e., that 1 in 100 or 1 in 14 attempts of establishing communication over the circuits would fail on the first attempt because all channels are occupied. If the 3 MHz band in the example above were divided in the conventional manner into 1000 3 kHz channels and a number of users permitted to use each channel, then, from data in the tables for the assumed grades of service, the resulting traffic intensity would be 10 and 80 Erlangs, respectively. Using the spread-spectrum technique, in the same bandwidth results in traffic intensity of 84.1 and 98.99 Erlangs for 1 and 7 percent grade of service. Thus, the assumed bandwidth-expanding allocation results in a significant increase in communication capacity over the conventional frequency division narrowband allocation, since with the spread-spectrum technique all channels are available to all users and the system is analogous to a multitrunk telephone system.

VII. CONCLUSIONS

The philosophy of spectrum allocation which has existed for many years has been one of sharing the inherent capacity of the radio spectrum among users by frequency division. As the number of users increased, methods were found to reduce the necessary bandwidth so that new users could be accommodated in the spectrum. As user population continues to increase, it can be questioned as to whether frequency division allocation can continue to be the only way of allocating the spectrum, because this approach may not always be the most efficient way of using spectrum.

The historical philosophy of spectrum allocation appears to have been based upon the way in which the radio art developed (in particular, the frequency selective filter), rather than on any fundamental physical principle. There are other ways in which the communication capacity of the spectrum can be shared. Frequency division represents a very poor choice for many applications if improvement in the direction of maximization of communication capacity in a given segment of the spectrum is an objective. Application of spread-spectrum principles is one way. And that it is a possibility, also results from the way the radio art has developed. For if development of solid-state microcircuits, digital techniques, coding theory, and other hardware and software capabilities did not exist at their current level, any suggestion of using spread-spectrum principles for spectrum utilization improvement would be meaningless. But because these advances have been made and are having impact upon the trend of system development (e.g., the trend toward digital systems), and because of the advancing demand for increased communications, it behooves the telecommunications community to reexamine the methods of spectrum sharing and to determine in light of current capabilities whether modification of the present method is not both necessary and desirable, if the benefits of communications to the world's society are to be maintained.

REFERENCES

[1] *Radio Regulations (Edition of 1976)*, published by the ITU, Geneva, 1976.

[2] *CCIR—XIIIth Plenary Assembly (Geneva, 1974)*, published by the ITU, Geneva, 1975.

[3] C. E. Shannon, "A mathematical theory of communication," *Bell Syst. Tech. J.*, vol. 27, pp. 623–656, 1948.

[4] M. Schwartz, *Information Transmission, Modulation and Noise.* New York: McGraw-Hill, 1959.

[5] R. B. Ward, "Digital communications on a pseudo-noise tracking link using sequence inversion modulation," *IEEE Trans. Commun. Technol.*, vol. COM-15, pp. 69–78, 1967.

[6] H. Blasbalg, H. F. Najjar, R. A. d'Antonio, and R. A. Haddad, "Air-ground, ground-air communications using pseudo-noise through a satellite," *IEEE Trans. Aerosp. Electron. Syst.*, vol. AES-4, pp. 774–790, 1968.

[7] R. C. Dixon, *Spread Spectrum Systems.* New York: Wiley, 1975.

[8] S. W. Golomb, *Shift Register Sequences.* San Francisco: Holden-Day, 1967.

[9] W. W. Peterson, *Error-Correcting Codes.* New York: Wiley, 1961.

[10] J. M. Aein, "Multiple access to a hard-limiting communication-satellite repeater," *IEEE Trans. Space Electron. Telem.*, vol. SET-10, pp. 159–167, 1964.

[11] D. R. Anderson and P. A. Wintz, "Analysis of a spread-spectrum multiple-access system with a hard limiter," *IEEE Trans. Commun. Technol.*, vol. COM-17, pp. 285–290, 1969.

[12] S. W. Golomb, *Digital Communications.* Englewood Cliffs, NJ: Prentice-Hall, 1964.

[13] R. Gold, "Optimal binary sequences for spread spectrum multiplexing," *IEEE Trans. Inform. Theory*, vol. IT-13, pp. 619–621, 1967.

[14] J. R. Sergo, Jr. and J. F. Hayes, "Analysis and simulation of a PN synchronization system," *IEEE Trans. Commun. Technol.*, vol. COM-18, pp. 676–679, 1970.

[15] H. H. Schreiber, "Self-noise of frequency hopping signals," *IEEE Trans. Commun. Technol.*, vol. COM-17, pp. 588–590, 1969.

[16] P. R. Drouilhet, Jr. and S. L. Bernstein, "TATS—A band-spread demodulation system for multiple access tactical satellite communication," in *IEEE EASCON'69 Conv. Rec.*, pp. 126–132.

[17] J. R. Klauder, A. C. Price, S. Darlington, and W. J. Abersheim, "The theory and design of chirp radars," *Bell Syst. Tech. J.*, vol. 29, pp. 745–808, 1960.

[18] A. J. Berni and W. D. Gregg, "On the utility of chirp modulation for digital signaling," *IEEE Trans. Commun.*, Vol. COM-21, pp. 748–751, 1973.

[19] J. Burnsweig and J. Wooldridge, "Ranging and data transmission using digital encoded FM-'chirp' surface acoustic wave filters," *IEEE Trans. Microwave Theory Tech.*, vol. MTT-22, pp. 272–279, 1973.

[20] J. P. Costas, "Poisson, Shannon, and the radio amateur," *Proc. IRE*, vol. 47, pp. 2058–2068, 1959.

[21] *CCITT—IIIrd Plenary Assembly (Geneva, 1964)*, Blue Book, vol. II, p. 239, published by the ITU, Geneva, 1965.

William F. Utlaut (SM'55–F'70) is Deputy Director of the Institute for Telecommunication Sciences, U.S. Department of Commerce, Boulder, CO, and prior to various departmental reorganizations, held similar positions in the National Bureau of Standards, ESSA, and OT since 1954. At earlier times he served on the Electrical Engineering Faculty at the University of Colorado, and was employed by the General Electric Company, Schenectady, NY. He has carried out research and engineering studies relating to various telecommunication systems. He is U.S. Chairman of CCIR Study Group 1, U.S. Commission C of URSI, a member of the National Committees of URSI and CCIR, and a Registered Professional Engineer in the State of Colorado.

Dr. Utlaut was Chairman of the Denver/Boulder Chapter of the IEEE Antennas and Propagation Group in 1963, Chairman of the ICC Technical Program in 1965, Chairman of the IEEE Denver Section from 1966 to 1967, and Chairman of ICC in 1969. In addition, he served as a member of the Administrative Committee of the IEEE Communication Technology Group from 1970 to 1971, and was a member of the Board of Governors of the Communications Society from 1972 to 1975.

SPREAD SPECTRUM COMMUNICATIONS—MYTHS AND REALITIES *

Andrew J. Viterbi

Coding is always beneficial and sometimes crucial for the suppression of interference in spread spectrum communications.

INTRODUCTION

Spread spectrum communication techniques date back to the early fifties. Since the earliest applications, system improvements have been more evolutionary than revolutionary. Like most improvements in electronic systems, these are due primarily to the availability of ever higher speed integrated circuit components, which translate in this case to wider spread spectra. In three decades the achievable spreading factor has grown by about three orders of magnitude[1] to the point that we are now limited more by bandwidth allocations than by technology limitations. Before we examine the quantitative effects of spreading, let us catalog briefly the multiple purposes of spread spectrum communications.

First, we note that spreading here refers to expansion of the bandwidth well beyond what is required to transmit digital data. Thus, a system transmitting data at a rate (R) of 100 Mbits/s using approximately 100 MHz of bandwidth (W) is not spread at all, while a system transmitting at 100 bits/s spread over a spectrum of about 100 MHz has a factor $W/R = 10^6$, or 60 dB of so-called *processing gain*.

The author is with the Linkabit Corporation, San Diego, CA 92121.
[1] Which parallels the evolution of data rate capabilities of digital communications.

*Reprinted from *IEEE Communications Society Magazine,* May 1979, Vol. 17, No. 3, pp. 11–18.

PURPOSES

The purpose and applicability of spread spectrum techniques is threefold:
- Interference Suppression
- Energy Density Reduction
- Ranging or Time Delay Measurement

Foremost among these is the suppression of interference which may be characterized as any combination of the following:

1) Other Users: intentional (hostile or unintentional),

2) Multiple Access: spectrum sharing by "coordinated" users,

3) Multipath: self-jamming by delayed signal.

Protection against in-band interference is usually called anti-jamming (A/J). This is the single most extensive application of spread spectrum communication. A similar application is that of multiple access by numerous users who share the same spectrum in a coordinated manner, in that each employs signaling characteristics or parameters (often referred to as codes) which are distinguishable from those of all other users. One reason for using this shared spectrum, so-called code-division multiple access (CDMA), is that by distinguishing signals in this way, separation in the more common dimensions of frequency or time is not required, and hence the usual transmission tolerances need not be imposed on these parameters.

The third form of interference suppressed by spread spectrum techniques is the self interference caused by multipath in which delayed versions of the signal, arriving via alternate paths, interfere with the direct path transmission.

While the second and third forms of interference would appear more benign than that of a hostile emitter, the technique and effect are the same. What makes the intentional interference more challenging is the game aspect of the problem and the fact that the interfering source is generally granted much more power than the communicator, which is usually not the case for cooperating users and even less so for multipath interference.

The second class of applications centers about the reduction of the energy density of the transmitted signal. This, too, has a threefold purpose:

    1) to meet international allocations regulations,
    2) to minimize detectability,
    3) for privacy.

Downlink transmissions from satellites must meet international regulations on the spectral density of the signals received on earth. By spreading this energy over a wider bandwidth, total transmitted power can be increased, and hence performance improved. Spreading also decreases the detectability of a signal by a regulatory body which employs spectral analysis to monitor or regulate emissions. (It is not known whether bootleg radio amateurs are using spread spectrum modulation to evade FCC regulations.) Even more promising is the potential for achieving privacy in communication by spreading one's signal sufficiently to "hide" in the background noise.

The application of spread spectrum for ranging or position location is rapidly gaining in importance. In simplest terms, position location consists of measuring the delay of a pulse or pulses. Error in delay measurement is inversely proportional to the bandwidth of the signal pulse. This is most easily seen by the simple example of Fig. 1. The accuracy of the measurement Δt is obviously proportional to the rise time of the pulse, which is inversely proportional to the bandwidth of the pulse signal. Of course, a one-shot measurement on a single pulse is not

very reliable. Rather, the spread spectrum signal used for ranging is a long sequence of polarity changes (binary PSK-modulated signal). Upon reception, this is correlated against a local replica and "lined up" to perform an accurate range or delay measurement.

BASIC TECHNIQUES

Having outlined the multiple uses of spectrum spreading, we must examine at least a superficial description of the concept before we can proceed to dispel myths and uncover realities about this increasingly popular technique. Fig. 2 is an all-purpose diagram to describe spread spectrum modulation. Multiplication of two unrelated signals produces a signal whose spectrum is the *convolution* of the spectra of the two component signals. Thus, if the digital data (binary) signal is relatively narrow-band compared to the spreading signal, the product signal will have nearly the spectrum of the wider (spreading) signal. So much for the modulator. At the demodulator, the received signal is multiplied by exactly the same spreading signal. Now if the spreading signal, locally generated at the receiver, is lined up (synchronized) with the received spread signal, the result is the original signal plus, possibly, some spurious higher frequency components outside the band of the original signal, and hence easily filtered to reproduce the original data essentially undistorted. If there is any undesired signal at the receiver, on the other hand, the spreading signal will affect it just as it did the original signal at the transmitter. Thus, even if it is a narrow-band signal in the middle of the band of interest, it will be spread to the bandwidth of the spreading signal.

The result is that the undesired (jamming) signal will have a bandwidth of at least W. If its power is J watts, its average density, which is essentially uniform and can be treated as wide-band noise, will be

$$N_0 = J/W \text{ watts/Hz.}$$

Let the desired component of the received signal have power S watts. Thus, if the data rate is R bits/second, the received energy per bit is

$$E_b = S/R \text{ watts} \cdot \text{second.}$$

Now it is generally recognized that digital communication system bit error rate performance is a direct function of the dimensionless ratio E_b/N_0, which for spread spectrum signals may thus be expressed as

$$\frac{E_b}{N_0} = \frac{S}{J} \frac{W}{R}$$

and hence, the jamming power-to-signal power ratio is

$$\frac{J}{S} = \frac{W/R}{E_b/N_0}. \tag{1}$$

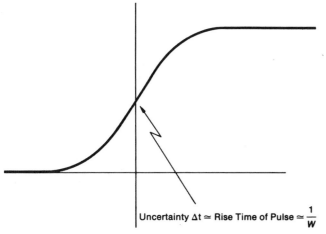

Uncertainty $\Delta t \simeq$ Rise Time of Pulse $\simeq \dfrac{1}{W}$

Fig. 1. Time delay measurement.

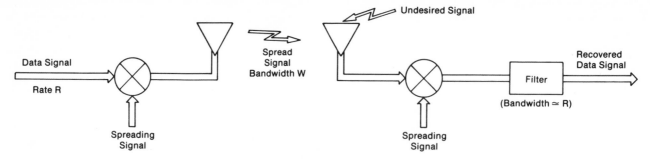

Multiplication ONCE Spreads Signal Bandwidth
Multiplication TWICE Followed by Filtering Recovers Original Signal
DESIRED Signal Multiplied TWICE, Undesired ONCE

Fig. 2. Basic spread-spectrum techniques.

This establishes that if E_b/N_0 is the *minimum bit energy-to-noise density ratio* needed to support a given bit error rate, *and if* W/R is the ratio of spread bandwidth to the original data bandwidth, also called the *processing gain*, then J/S is the *maximum tolerable jamming power-to-signal power ratio*, also known as the *jamming margin.*

We have come this far without even specifying the characteristics of the spreading signal. There are, in fact, two distinct classes of spreading techniques. The first is called *direct sequence* or *pseudonoise* (PN) spread spectrum. Here the spreading is achieved by multiplication by a binary pseudorandom sequence whose symbol (switching) rate is many times the binary data bit rate. The spreading sequence symbol rate is sometimes called the *chip rate.*

The second class utilizes a *frequency hopping* carrier. Here the spreading signal remains at a given frequency for each bit or even for several bits. Thus, locally it is no wider than the data signal, but when it hops to a new frequency, it may be anywhere within the "spreading" bandwidth W.

One fundamental difference between the two techniques is that direct sequence PN spread signals can be coherently demodulated. With frequency hopped signals, on the other hand, phase coherence is difficult to maintain when the signal frequency is hopped over a wide range; hence, this modulation is usually demodulated noncoherently.

We are now ready to explore several firmly entrenched items of common wisdom regarding the relative desirability of various features of spread spectrum systems. Often these attitudes are misguided, as we shall presently show. In all cases, the ideas hold for both classes of spread spectrum techniques, but for all but the last concept the arguments are somewhat simpler for direct sequence spreading, which we shall therefore consider.

We are ready now to reveal the first of four myths.

First Myth:

Error-correcting coding requires redundancy, which spreads bandwidth and thus reduces available processing gain for the available bandwidth.

Coding does not reduce the *effective* processing gain in a spread spectrum system.

Reality is, in fact, just the opposite. To see that coding does not reduce the *effective* processing gain, let us rewrite jamming margin (1) in terms of the *symbol[2] rate* R_s and the *symbol energy* E_s. These are related to the bit rate and the bit energy through the code rate r, defined as the number of data bits per transmitted symbol, or the inverse of the coding expansion factor. (For example, a rate 1/2 coded system transmits two code symbols for each data bit.) It follows that

$$R_s = R/r \qquad \text{and} \qquad E_s = E_b r.$$

Now if we repeat the previous dimensional argument replacing bits by symbols everywhere, we have

$$J/S = \frac{W/R_s}{E_s/N_0},$$

but substituting the preceding definitions for symbol rate and energy, we obtain for the maximum tolerable J/S ratio

$$J/S = \frac{W}{R/r} \frac{N_0}{E_b r} = \frac{W/R}{E_b/N_0}$$

which gets us back to (1). This may seem like sleight of hand, but it really is not. Moreover, although it will take some further reading to be convinced, we are really ahead of the game. For with coding, the required E_b/N_0 for a given level of performance (bit error rate) is actually *reduced.* Thus, for a given processing gain (W/R) the jamming margin is further *increased* by coding.

For those who are satisfied that spectrum spreading (especially direct sequence PN) techniques make the noise look "white" while the signal energy, without or with coding, can be fully recovered by the receiver's

[2]Symbol rate refers to the *code symbol* of the error-correcting code—not that of the PN spreading code, which is usually called chip rate.

"correlating" multiplier, the dispelling of the First Myth will come as no surprise. Yet it is often this sophisticated group who will fall prey to the

Second Myth:
  Error-correcting coding is effective only against uniform interference.

In particular, the myth continues; coding is not effective against *pulsed interference.* Yet, this is even more dramatically false than the First Myth. Let us consider what the effect of pulsed interference can be for an uncoded system. Suppose the jamming is present only a fraction $\rho < 1$ of the time, but that during this time, the noise density level is increased to a level N_0/ρ watts. This assumes spectrum spreading which turns the jamming signal into broad-band noise and an *average power* rather than a peak power limitation on the jammer. (While this may be slightly pessimistic for the communicator, any other assumption is a risky bet against technological progress.) Now it is well known that with coherent demodulation an *uncoded* BPSK modulated system produces a bit error rate P_b related to E_b/N_0 as

$$P_b = Q(\sqrt{2E_b/N_0}) \lesssim e^{-E_b/N_0} \qquad (2)$$

where

$$Q(x) = \frac{1}{\sqrt{2\pi}} \int_x^\infty e^{-u^2/2}\, du.$$

But if the noise is intermittent, and hence only with probability ρ corrupts a given transmitted bit[3] with the higher noise density N_0/ρ, the resulting bit error rate becomes

$$P_b = \rho Q \sqrt{2E_b\rho/N_0} \lesssim \rho e^{-\rho E_b/N_0}. \qquad (2')$$

Clearly, the jammer would choose the duty factor ρ which pessimizes performance—that is, maximizes bit error rate. In terms of the approximation, which is a strict upper bound, this occurs when

$$\rho = \frac{1}{E_b/N_0} \quad \text{provided } E_b/N_0 > 1$$

at which value

$$\max_{0<\rho<1} \ P_b \lesssim \frac{e^{-1}}{E_b/N_0}. \qquad (3)$$

(Note that although we worked with the approximation for its simplicity, had we used the exact (Q-error-function) expression, the worst case ρ would be nearly the same and the maximum bit error rate would not be significantly lower.)

---

[3]We assume for simplicity that a given interference pulse corrupts an integral number of bits. This is a reasonable assumption if the pulse width is many times the bit duration. Otherwise, the situation is actually less favorable to the jammer.

Coding Fundamentals

When a binary data stream must be transmitted over a noisy channel with a troublesome bit error rate, coding can be used to significantly reduce the error rate incurred by the message.

In **block coding** schemes, the message bit stream is partitioned into blocks of k bits, where k is the **block length.** Each such message block is replaced with an n bit code word (n is bigger than k) which is transmitted in its place. Thus, every n bit transmitted "contains" only k message bits so that the **rate** r of the code is k/n bits per code symbol.

A common noisy channel model is the so-called **Gaussian channel** where each bit, viewed as a square pulse of amplitude ± 1, is independently subjected to additive noise and an error occurs when the noise alters the pulse polarity.

As a result of errors, the received n-bit block can be any of 2^n possible words. Since there are only 2^k different code words that could have been transmitted (one for each k-bit message block) and 2^k is typically much less than 2^n, the number of possible received words 2^n is much greater than the number of code words 2^k. For each received code word, the decoder decides what was the most likely code word that was transmitted, and the receiver then identifies the corresponding k-bit message block. In this way error correction can be achieved.

A notably different approach is **convolutional coding**. Here the incoming message bit stream is applied to a K-stage shift register which is shifted b bits at a time. For each K message bits stored in the register, there are n linear logic circuits which operate on the register contents to produce n code bits of the encoded output stream. For each shift of the register, b new message bits are inserted and n code bits are delivered, so that the rate is b/n information bits per code symbol. In this case, a particular code bit depends on K message bits where K is called the **constraint length** of the code. Note also that a particular message bit remains in the register for K/b shifts, and thus influences the value of nK/b code bits.

Unlike block coding, the optimal decoding operation for convolutional codes requires a memory that stores, in effect, a function of the entire past history of the received bit stream. The performance (as measured by error rate) of a convolutional coding system improves as the complexity (i.e., memory) allowed for the decoder is increased. Several methods of decoding convolutional codes have been developed. The optimal (maximum likelihood) scheme is generally known as the Viterbi algorithm. Viterbi decoding for reasonably short constraint lengths is feasible to implement and high decoding speeds are achievable. For extremely low error probabilities, a large constraint length K is required. The computational complexity of Viterbi decoding for large K makes this approach impractical. Another approach, **sequential decoding,** then becomes more attractive. A third technique, **feedback decoding,** though inferior in performance against random errors, is particularly well suited to correcting systematic error bursts which may occur in fading channels.

Both fading and pulse jamming introduce memory in the channel and further modify the channel statistics. Yet the same coding techniques as used for the Gaussian channel are at least as effective here, provided interleaving is employed to reduce or eliminate this memory.

The result is quite dramatic. Pulse jamming—with spread spectrum but without coding—changes an exponential relation into an inverse-linear one. Numerically, if we desire bit error rate performance on the order of $P_b \simeq 10^{-5}$, stationary noise (or jamming) requires only $E_b/N_0 \simeq 10$ dB, while with pulse jamming we must have $E_b/N_0 \simeq 45$ dB, an increase in *required signal power of over three orders of magnitude!*

Amazingly, coding can almost fully restore this deplorable situation, but before we can explain why, we must briefly explore a summary of the general capabilities of coded systems.

ERROR CORRECTING CODING FUNDAMENTALS

All that we need to know about coding for the present purpose is that for practically any memoryless channel, there are many good binary codes of rate r bits/symbol for which the bit error rate is upper bounded[1] by either

$$P_b < 2^{-K(\alpha-1)}, \qquad \text{if it is a } block \text{ code of} \qquad (4)$$
$$\text{block length } K$$

or

$$P_b < \frac{2^{-K\alpha}}{[1 - 2^{-(\alpha-1)}]^2}, \qquad \text{if it is a } convolutional \text{ code} \qquad (5)$$
$$\text{of constraint length } K.$$

In either case

$$\alpha = \frac{r_0}{r} > 1 \qquad (6)$$

provided the code rate $r < r_0 < 1$.

Performance then depends strongly on the value of the parameter[4] r_0. This parameter, and consequently α, is increased if the decoder is furnished with everything the receiver "knows" about the channel; that is, for binary symbols, not only the receiver's "belief" that the transmitted symbol was a "zero" or a "one," but how strongly the receiver believes this. This confidence in the decision is called a "soft decision," "quality information," or simply a "metric."

Now for a uniform Gaussian channel, with soft decisions furnished to the decoder, the all-important parameter r_0 is a function of only the symbol energy-to-noise density. That is,

$$r_0 = 1 - \log_2 [1 + e^{-E_s/N_0}] \qquad (7)$$

where, as previously defined,

$$E_s = E_b r. \qquad (8)$$

---

[4] This parameter is also the so-called computational cutoff rate beyond which the sequential-decoding mean computational load becomes unbounded.

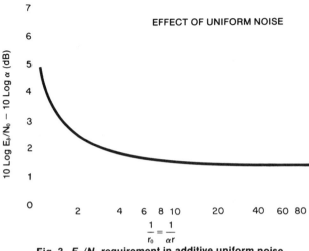

Fig. 3. E_b/N_0 requirement in additive uniform noise.

Combining (6), (7), and (8), we can relate E_b/N_0 to α and r_0 as

$$\frac{E_b}{N_0} = -\alpha \frac{\ln(2^{1-r_0} - 1)}{r_0}. \qquad (9)$$

This quantity expressed in decibels ($10 \log E_b/N_0 - 10 \log \alpha$) is plotted in Fig. 3 as a function of r_0. Of course, Fig. 3 or (9) is meaningful only when taken together with (4) or (5). The interpretation is that for a given acceptable complexity of implementation, which is roughly proportional to 2^K for either class of codes, and a given code rate r, we need to select a value of α to guarantee the required P_b, according to (4) or (5). This establishes r_0 according to (6), and finally we obtain E_b/N_0 by adding 10 log α to the ordinate of Fig. 3 for the given r_0. If the resulting r_0 is greater than 1, we must choose a smaller code rate r for which $r_0 < 1$ for the required α. Although (4) and (5) are only upper bounds, and hence pessimistic estimates of the value of coding, they are sufficient to establish its merit, and in fact are reasonably accurate. As a practical matter, commonly used and commercially available convolutional decoders require an E_b/N_0 which is about 0.5–2 dB above the curve of Fig. 3 to achieve $P_b \simeq 10^{-5}$.

For specific comparison, we have from (2) that to achieve $P_b = 10^{-5}$ on the uniform Gaussian channel without coding requires $E_b/N_0 = 9.6$ dB. With rate 1/2 convolutional codes, practical soft-decision decoders require between 3.5 and 4.5 dB for the same performance. Rate 3/4 codes require approximately 1 dB more. If the decoder is provided with only hard-decision inputs, it requires approximately 2 dB higher E_b/N_0.

Thus far we have considered only stationary wideband noise, whether of thermal origin, or so rendered by the direct sequence spread spectrum technique we have investigated. We now return to the pulse jammer and show how coding remedies the deplorable situation which we left before the present digression.

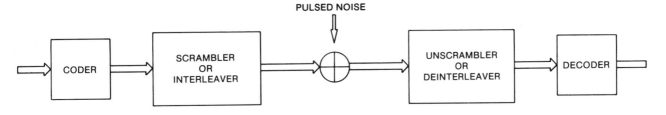

Fig. 4. Introduction of interleaving for disposing bursts.

SECOND MYTH REVISITED— APPROPRIATELY ARMED

Suppose now that we code as before, but for the nonuniform (pulse) jammer. Spreading causes this to appear at the receiver as wide-band noise of density level N_0/ρ, but for a reduced duty factor ρ. Suppose, as before, with little loss of reality, that an integral number of code symbols are affected by jamming. We cannot quite apply what we just learned about coding because the jamming pulses affect many contiguous symbols; so we can hardly call the channel memoryless as required. But this is easily remedied. Suppose we construct a device which randomly scrambles the order of the symbols prior to transmission, but after coding, and puts them back in the right order after reception, but before decoding (Fig. 4). (Scramblers and unscramblers are more commonly called interleavers and deinterleavers.) But the unscrambler which restores the transmitted symbols to their right place in order actually scrambles the regular jamming pulses into random patterns.[5] Scrambling or interleaving thus makes our system memoryless again and we can apply our new-found coding knowledge.

Without belaboring the exact details, arguing intuitively and believingly on the basis of (2) and (2'), let us replace e^{-E_s/N_0} by $\rho e^{-\rho E_s/N_0}$ in all the formulas of the previous (uniform noise) section. Thus, (7) is replaced by

$$r_0 = 1 - \log_2[1 + \rho e^{-\rho E_s/N_0}]. \qquad (7')$$

Combining (7') with (6) and (8), we get

$$\frac{E_b}{N_0} = -\frac{\alpha}{\rho r_0}\ln\left[\frac{2^{1-r_0}-1}{\rho}\right] \qquad (9')$$

which obviously reduces to (9) for uniform jamming ($\rho = 1$).

This is maximized by a jammer with duty cycle

$$\rho = (2^{1-r_0} - 1)e \qquad (10)$$

provided $r_0 > 1 - \log_2(1 + e^{-1}) = 0.548$ for which

[5] Note the parallel with spectrum spreading; the second multiplier unspread the signal and spread the interference; here the second device unscrambles the desired code sequence and scrambles the undesired pulsed interference.

$$\max_{0<\rho<1}\frac{E_b}{N_0} = \frac{\alpha e^{-1}}{r_0(2^{1-r_0}-1)} \qquad \text{for } r_0 > 0.548. \qquad (11)$$

Even the severe impact of pulsed jamming can be entirely contained by suitable coding.

If $r_0 \leq 0.548$, $\rho = 1$ maximizes E_b/N_0 and (9') reduces to (9). This says that if r_0 is small enough, *no penalty* is paid to a pulse jammer. Even for $r_0 > 0.548$, the penalty is small, as seen in Fig. 5, which shows the new pulsed noise case and reproduces the uniform noise case from Fig. 3.

Thus, *even more amazing than the original 35 dB loss to pulsed noise, coding recovers it all and then some.* In spite of this, there still are skeptics who believe in the

Third Myth:

Interleaving destroys memory. Memory can be exploited to correct errors. Hence, interleaving is bad.

The discussion leading to the system design of Fig. 3 should suffice to dispel such misgivings, particularly when we recall that soft decisions contain about all the information available about the channel (granted, of course, that memory may be exploited to extract such quality information). To definitely put to rest all discussion of the matter, let us consider a burst interference phenomenon which always affects B symbols, and let there be no overlaps of bursts (overlaps can only help the

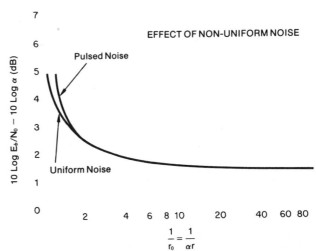

Fig. 5. E_b/N_0 requirement in pulsed and uniform noise.

communicator). In this case, the key parameter of (7′) becomes

$$r_0 = 1 - \frac{1}{B} \log_2 \{(1 - \rho) + \rho[1 + e^{-\rho E_s/N_0}]^B\} \qquad (7'')$$

which causes (9′) to be replaced by

$$\frac{E_b}{N_0} = -\frac{\alpha}{\rho r_0} \ln\left\{\left[\frac{2^{(1-r_0)B} - (1 - \rho)}{\rho}\right]^{1/B} - 1\right\} \qquad (9'')$$

which leads to the plots of Fig. 6. Clearly (7″) and (9″) reduce to (7′) and (9′) for $B = 1$, which is the case when interleaving is employed.

In fact, the situation is even worse than shown in Fig. 6. For as r decreases, the number of symbols per bit and, for a constant duration interference pulse, the number of symbols per pulse (B) increases. Thus, keeping B fixed, as is done for convenience in Fig. 6, gives misleadingly favorable results.

Our final "myth" happens fortuitously to coincide with reality. We call it, therefore, a "Folk Theorem." It concerns an interesting comparison of direct sequence (coherent) spreading with frequency hopping (noncoherent) spreading:

Fourth Folk Theorem: (Myth = Reality)

Performance of frequency-hopped spread spectrum is 3 dB worse than that of direct sequence (PN) spread spectrum.

The commonly invoked "mythical" argument is that noncoherent systems can utilize at best orthogonal signals (e.g., binary FSK modulation) instead of antipodal signals (binary PSK modulation) and this accounts for the 3 dB. The trouble with this argument is that it ignores the possiblity of higher signaling alphabets (such as MFSK) and, worse still, the real possibility that frequency-

hopped systems may be more vulnerable to nonuniform interference.

We note, in fact, that in frequency-hopped systems, the jammer need not pay the cost of a higher peak power signal, for if he jams just a fraction of the band,[6] $\rho < 1$, with power density N_0/ρ, he will appear to the receiver just as a partial time jammer. Note also that if the hopping rate is at least as great as the symbol rate, interleaving is unnecessary.

For alphabets of size q, (4) and (5) must be modified to become

$$P_b < q^{-K(\alpha-1)} \text{ block codes} \qquad (4')$$

and

$$P_b < \frac{(q - 1)q^{-K\alpha}}{[1 - q^{-(\alpha-1)}]^2} \text{ convolutional codes} \qquad (5')$$

which reduce to (4) and (5) when $q = 2$. There is nothing to gain by using multiple signal alphabets for coherent, direct sequence systems, but with frequency-hopped systems, we can show [2] that for the worst case partial-band jammer

$$\max_{0<\rho<1} \frac{E_b}{N_0} = \alpha \frac{(q - 1)4e^{-1}}{(\log_2 q)r_0(q^{1-r_0} - 1)}$$

provided

$$\frac{E_b}{N_0} \geqslant \frac{3}{r \log_2 q}.$$

For asymptotically large q, this approaches

$$\frac{E_b}{N_0} \simeq (4 \ln 2)\alpha$$

which is exactly a factor of 2 (3 dB) above the minimum of Fig. 3 which occurs as $r_0 \longrightarrow 0$. This comparison is shown in Fig. 7, which also shows the diminishing returns of using alphabet size $q > 8$. The asymptotic minimum is virtually reached for $q > 32$.

Notice that noncoherent frequency-hopped systems exhibit a minimum, while coherent direct sequence systems improve monotonically as $1/r_0$ increases. The explanation of this behavior is better understood by examining Fig. 8 which is a more detailed and more realistic examination of performance for an octal alphabet. Here the assumptions are more realistic. Specifically, channel quality information is limited to two bits (four levels) out of each of the q matched filters. This also allows for a practical automatic gain control (AGC) technique, which has not been mentioned up to this point.

The curve for $\rho = 1$ is, of course, for uniform noise.

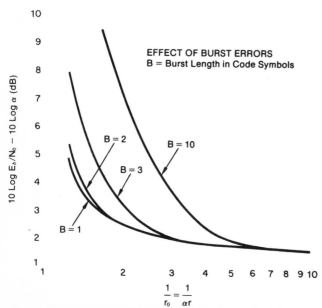

Fig. 6. E_b/N_0 requirement in pulsed noise without interleaving.

---

[6]Possibly varying this by hopping himself in order to defeat the obvious communicator strategy of determining the jammed region and staying out of it.

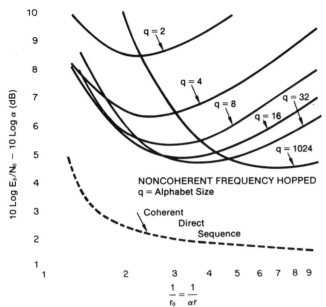

Fig. 7. E_b/N_0 requirement for noncoherent frequency hopped systems.

The increase at high rates (r and r_0 close to 1) is due to the lack of coding redundancy. The increase at low rates is due to the higher loss, characteristic of noncoherent combining of symbols in high diversity (here low rate) noncoherent communication systems. As ρ, the fraction of the interference bandwidth decreases, performance gets increasingly worse at high rates since the diversity is lacking to overcome the strong jammer. But as r_0 decreases, diversity becomes sufficient to fully defeat the low ρ jammer, as shown by the family of curves of Fig. 8.

SUMMARY

Beyond cataloging the many uses of spread spectrum communication, we have made no attempt to be uniform in our treatment of this extensive and many-faceted field. We have concentrated rather on its application for the suppression of interference, and have made three main points:

• Coding is always useful, and it may be critical to adequate performance of spread spectrum systems, particularly when the nature of the interference is partial-time, or in the case of frequency-hopped spreading, partial band. Proper interfacing of the decoder to the demodulator, in utilizing quality (soft decision) information, is important to ensure maximum benefit from coding.

• Interleaving or scrambling may be equally essential in the presence of burst interference.

• Direct sequence spread spectrum efficiency is about double that for frequency hopping. This is tantamount to doubling the processing gain W/R. However, frequency-hopping technology may have an edge in achievable band spreading of one or more orders of magnitude over direct sequence spreading technology which greatly overshadows the "system edge" of the latter.

Although the efficiency of direct sequence spread spectrum is about double that of frequency hopping, this advantage is overshadowed by the greater band spreading achievable with frequency hopping technology.

REFERENCES

[1] A. J. Viterbi and J. K. Omura, *Principles of Digital Communication and Coding.*   New York: McGraw-Hill, 1979.
[2] A. J. Viterbi and I. M. Jacobs, "Advances in coding and modulation for noncoherent channels affected by fading, partial-band and multiple-access interference," in *Advances in Communication Systems,* vol. 4. New York: Academic, 1975.

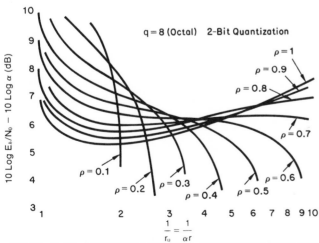

Fig. 8. E_b/N_0 requirement for octal noncoherent frequency hopped system in partial band interference with receiver quantization (ρ = interference fractional bandwidth).

Andrew J. Viterbi received the B.S. and M.S. degrees in electrical engineering in 1957 from M.I.T. In 1962 he received the Ph.D. degree in electrical engineering from the University of Southern California.

From 1957 to 1963, Dr. Viterbi was with the Communication Research Section of the Jet Propulsion Laboratory, Pasadena, CA, where he rose to the position of Research Group Supervisor. In 1963 he joined the faculty of the University of California, Los Angeles, where he became Professor of Engineering and Applied Science in 1969. In 1973 he took an industrial leave of absence to devote his full-time effors as Vice President of Linkabit, which he participated in founding in 1968. In 1975 he resigned from UCLA and accepted the title Adjunct Professor of Applied Physics and Information Science at University of California, San Diego. In 1974 he was elected to the newly created position of Executive Vice President of Linkabit.

Dr. Viterbi has been active in professional groups of the IEEE in various capacities on both the local and national levels. From 1964 through 1970, and again since 1976, he was a member of the Ad Com/Board of Governors of the Information Theory Group and served as Group Chairman in 1970. Since 1967 he has been a member of the Editorial Board and the journal *Information and Control;* from 1969 to 1977 a member of the Editorial Board of the PROCEEDINGS OF THE IEEE; and from 1971 to 1975 an Associate Editor of the IEEE TRANSACTIONS ON INFORMATION THEORY. He has served as a member of advisory committees and as a consultant for the U.S. Air Force, NASA, and the U.S. Army, including the Army Scientific Advisory Panel. In 1973 he was elected a Fellow of the IEEE.

Dr. Viterbi is the coauthor of two books on digital communication and author of the first text on coherent communication. He has received various awards for his publications, including the 1968 IEEE Information Theory Group Outstanding Paper Award and the 1975 Christopher Columbus International Communication Award.

INDEX

AUTHOR INDEX